Lecture Notes of the Institute
for Computer Sciences, Social Informatics
and Telecommunications Engineering   73

Editorial Board

Ozgur Akan
    Middle East Technical University, Ankara, Turkey
Paolo Bellavista
    University of Bologna, Italy
Jiannong Cao
    Hong Kong Polytechnic University, Hong Kong
Falko Dressler
    University of Erlangen, Germany
Domenico Ferrari
    Università Cattolica Piacenza, Italy
Mario Gerla
    UCLA, USA
Hisashi Kobayashi
    Princeton University, USA
Sergio Palazzo
    University of Catania, Italy
Sartaj Sahni
    University of Florida, USA
Xuemin (Sherman) Shen
    University of Waterloo, Canada
Mircea Stan
    University of Virginia, USA
Jia Xiaohua
    City University of Hong Kong, Hong Kong
Albert Zomaya
    University of Sydney, Australia
Geoffrey Coulson
    Lancaster University, UK

Patrick Sénac   Max Ott
Aruna Seneviratne (Eds.)

Mobile and Ubiquitous Systems: Computing, Networking, and Services

7th International ICST Conference, MobiQuitous 2010
Sydney, Australia, December 6-9, 2010
Revised Selected Papers

 Springer

Volume Editors

Patrick Sénac
LAARS, CNRS, Département de Mathématique
et Informatique, ENSICA
1 Place Emile Blouin
31056 Toulouse Cedex, France
E-mail: patrick.senac@isae.fr

Max Ott
NICTA, Australian Technology Park,
Eveleigh, NSW 1430, Australia
E-mail: max.ott@nicta.com.au

Aruna Seneviratne
UNSW, Sydney, NSW 2053
and
NICTA
13 Garden Street, Eveleigh, NSW 1430, Australia
E-mail: aruna.seneviratne@nicta.com.au

ISSN 1867-8211 e-ISSN 1867-822X
ISBN 978-3-642-29153-1 e-ISBN 978-3-642-29154-8
DOI 10.1007/978-3-642-29154-8

Springer Heidelberg Dordrecht London New York

Library of Congress Control Number: 2012934014

CR Subject Classification (1998): C.2, H.4, I.2, H.3, D.2, H.5

© ICST Institute for Computer Science, Social Informatics and Telecommunications Engineering 2012

This work is subject to copyright. All rights are reserved, whether the whole or part of the material is concerned, specifically the rights of translation, reprinting, re-use of illustrations, recitation, broadcasting, reproduction on microfilms or in any other way, and storage in data banks. Duplication of this publication or parts thereof is permitted only under the provisions of the German Copyright Law of September 9, 1965, in its current version, and permission for use must always be obtained from Springer. Violations are liable to prosecution under the German Copyright Law.
The use of general descriptive names, registered names, trademarks, etc. in this publication does not imply, even in the absence of a specific statement, that such names are exempt from the relevant protective laws and regulations and therefore free for general use.

Typesetting: Camera-ready by author, data conversion by Scientific Publishing Services, Chennai, India

Printed on acid-free paper

Springer is part of Springer Science+Business Media (www.springer.com)

Preface

Welcome to the proceedings of the 7th International ICST Conference on Mobile and Ubiquitous Systems (MobiQuitous) which was held in Sydney. The conference comprised four days of workshops, conference sessions, keynote presentations, poster discussions, networking and most importantly, enjoyment.

With every passing year, MobiQuitous has grown in reputation and stature and has become a very selective venue for research publications in the broad area of mobile computing algorithms, prototypes and applications. This year was no exception. After a careful and rigorous review process and a robust discussion among the Technical Program Committee (TPC), 24 high-quality technical papers were selected from 105 submissions from around the world. In addition, there were 12 work-in-progress papers and 27 posters. I am deeply indebted to Patrick Sénac and Max Ott, the TPC Chairs, for their diligent handling of the review process and for being so particular in preserving the quality of the selected papers. The conference program's ten sessions reflect this, covering a diverse range of very timely topics ranging from paper architectures to toolkits and mechanisms for privacy, energy efficiency and context awareness.

The main conference also featured two exciting and thought-provoking keynotes from speakers with very different backgrounds. The first was presented by Gordon Waddington from the University of Canberra, Australia, with a background in the movement sciences and rehabilitation fields as a sports physiotherapist and a clinical exercise physiologist. The talk titled "Towards a Personal Wellness Footprint," examined possibilities and research directions from the interface between current research in health and exercise science and new and emerging technology applications in human movement science. The second keynote was presented by Chris Winter, from the Australian Broadcasting Corporation's Innovation Division, who has a long background in new media, digital TV, technology marketing and radio. His talk titled "ABC Mobile—Connecting to Audiences in a World of Constant Change," examined how the traditional broadcasters are changing to adapt so as to provide content for mobile users.

The conference included excellent full-day workshops organized by Nikola Serbedzija (Fraunhofer FIRST), Martin Wirsing (LMU Munich), and Alois Ferscha (Universität Linz) on User-Centric Pervasive Adaptive Systems. It covered some of the most pressing issues in the design and development of pervasive systems, namely, context-aware systems, pervasive applications and ethical issues.

The organization and smooth running of Mobiquitous 2010 would not have been possible without the unselfish support we received from a number of people. It was a privilege to work with several other excellent and knowledgeable people and I am personally grateful to the TPC Chairs who enthusiastically handled all issues related to paper submissions, to Prashanthi Jayawardhene who single-handedly made Mobiquitous happen by making all the local arrangements and managing the day-to-day running of the conference, and to Christophe Dwertman, our Web chair, for maintaining a very high quality Web presence and for his unfailing promptness in responding to my many requests as well as providing network connectivity at the conference venue. Finally, I would like to acknowledge the support of the NSW Department of State and Regional Development, which gave us the opportunity to host the conference at very short notice on their premises with a magnificent 360-degree view of Sydney.

Aruna Seneviratne

Organization

Steering Committee Chair

Imrich Chlamtac   Create-Net, Italy

Steering Committee

Fausto Giunchiglia   University of Trento, Italy
Tom La Porta   Penn State, USA
Francesco De Pellegrini   Create-Net, Italy
Chiara Petrioli   Università di Roma "La Sapienza", Italy
Krishna Sivalingam   University of Maryland Baltimore, USA
Thanos Vasilakos   University of Western Macedonia, Greece

Workshop Chair

Emmanuel Lochin   ISAE, France

Publicity Chairs

Kanchana Kanchanasut   AIT, Thailand
Sebastien Ardon   NICTA, Australia

Conference Coordinator

Mona Hezso   ICST

Web Chair

Christoph Dwertmann   NICTA, Australia

Technical Program Committee

General Chair

Aruna Seneviratne   NICTA, Australia

TPC Chairs

Max Ott   NICTA, Australia
Patrick Senac   LAAS, CNRS, France

TPC Members

Sebastien Ardon	NICTA, Australia
Cristian Borcea	NJIT, USA
Jean-Yves Le Boudec	EPFL, Switzerland
Paul Castro	IBM Watson, USA
Jon Crowcroft	University of Cambridge, UK
Yong Cui	Tsinghua University, China
Walid Dabbous	INRIA, France
Marcelo Dias de Amorim	UPC, France
Michel Diaz	LAAS, CNRS, France
Otto Duarte	UFRJ, Brazil
Andrzej Duda	Grenoble Institute of Technology, France
Chris Gniady	University of Arizona, USA
Tao Gu	University of Southern Denmark
Marco Gruteser	WINLAB, Rutgers University, USA
Qi Han	Colorado School of Mines, USA
Jussi Kangasharju	University of Helsinki, Finland
Ahmed Karmouch	University Ottawa, Canada
Markku Kojo	University of Helsinki, Finland
Charles Krasic	University of British Columbia, Canada
Henrik Lundgren	Thomson, France
Cecilia Mascolo	University of Cambridge, UK
Rene Mayrhofer	University of Vienna, Austria
Iqbal Mohomed	Microsoft Research, USA
Tamer Nadeem	Siemens Research, Germany
June-Hwan Song	KAIST, Korea
Danny Soroker	IBM Watson, USA
Patrick Stuedi	Microsoft Research, USA
Alex Varshavsky	AT&T Labs, USA

Table of Contents

Main Conference

Safe Execution of Dynamically Loaded Code on Mobile Phones 1
  *Glen Pink, Simon Gerber, Michael Fry, Judy Kay,
  Bob Kummerfeld, and Rainer Wasinger*

Towards Enabling Next Generation Mobile Mashups 13
  *Vikas Agarwal, Sunil Goyal, Sumit Mittal, Sougata Mukherjea,
  John Ponzo, and Fenil Shah*

Mirroring Smartphones for Good: A Feasibility Study 26
  Bo Zhao, Zhi Xu, Caixia Chi, Sencun Zhu, and Guohong Cao

μC-SemPS: Energy-Efficient Semantic Publish/Subscribe for
Battery-Powered Systems 39
  Davy Preuveneers and Yolande Berbers

Collaborative Algorithm with a Green Touch 51
  *Luciana Oliveira, Djamel Hadj Sadok, Glauco Gonçalves,
  Renato Abreu, and Judith Kelner*

Evaluating Mobile Phones as Energy Consumption Feedback Devices ... 63
  *Markus Weiss, Claire-Michelle Loock, Thorsten Staake,
  Friedemann Mattern, and Elgar Fleisch*

Developing Pervasive Systems as Service-Oriented Multi-Agent
Systems 78
  Jorge Agüero, Miguel Rebollo, Carlos Carrascosa, and Vicente Julián

Adaptation Support for Agent Based Pervasive Systems 90
  *Kutila Gunasekera, Shonali Krishnaswamy, Seng Wai Loke, and
  Arkady Zaslavsky*

Mining Emerging Sequential Patterns for Activity Recognition in Body
Sensor Networks 102
  *Tao Gu, Liang Wang, Hanhua Chen, Guimei Liu,
  Xianping Tao, and Jian Lu*

Indoor Cooperative Positioning Based on Fingerprinting and Support
Vector Machines 114
  Abdellah Chehri, Hussein Mouftah, and Wisam Farjow

Crowd Sourcing Indoor Maps with Mobile Sensors 125
  Yiguang Xuan, Raja Sengupta, and Yaser Fallah

Real Time Six Degree of Freedom Pose Estimation Using Infrared Light
Sources and Wiimote IR Camera with 3D TV Demonstration 137
 Ali Boyali, Manolya Kavakli, and Jason Twamley

VLOCI: Using Distance Measurements to Improve the Accuracy of
Location Coordinates in GPS-Equipped VANETs 149
 Farhan Ahammed, Javid Taheri, Albert Y. Zomaya, and Max Ott

On Improving the Energy Efficiency and Robustness of Position
Tracking for Mobile Devices 162
 Mikkel Baun Kjærgaard

An ETX Based Positioning System for Wireless Ad-Hoc Networks 174
 A.K.M. Mahtab Hossain, Preechai Mekbungwan, and
 Kanchana Kanchanasut

A Dynamic Authentication Scheme for Hierarchical Wireless Sensor
Networks 186
 Junqi Zhang, Rajan Shankaran, Mehmet A. Orgun,
 Abdul Sattar, and Vijay Varadharajan

Anonymity-Aware Face-to-Face Mobile Payment 198
 Koichi Kamijo, Toru Aihara, and Masana Murase

LINK: Location Verification through Immediate Neighbors
Knowledge 210
 Manoop Talasila, Reza Curtmola, and Cristian Borcea

Passport/Visa: Authentication and Authorisation Tokens for
Ubiquitous Wireless Communications 224
 Abdullah Almuhaideb, Phu Dung Le, and Bala Srinivasan

Virtualization for Load Balancing on IEEE 802.11 Networks 237
 Tibério M. de Oliveira, Marcel W.R. da Silva,
 Kleber V. Cardoso, and José Ferreira de Rezende

A Packet Error Recovery Scheme for Vertical Handovers Mobility
Management Protocols 249
 Pierre-Ugo Tournoux, Emmanuel Lochin, Henrik Petander, and
 Jérôme Lacan

A Quantitative Comparison of Communication Paradigms for
MANETs 261
 Justin Collins and Rajive Bagrodia

Multi-modeling and Co-simulation-Based Mobile Ubiquitous Protocols
and Services Development and Assessment 273
 Tom Leclerc, Julien Siebert, Vincent Chevrier,
 Laurent Ciarletta, and Olivier Festor

TERMOS: A Formal Language for Scenarios in Mobile Computing
Systems 285
  Hélène Waeselynck, Zoltán Micskei, Nicolas Rivière,
  Áron Hamvas, and Irina Nitu

Work in Progress

Enforcing Security Policies in Mobile Devices Using Multiple
Personas 297
  Akhilesh Gupta, Anupam Joshi, and Gopal Pingali

Scalable and Efficient Pattern Recognition Classifier for WSN 303
  Nomica Imran and Asad I. Khan

Pervasive Integrity Checking with Coupled Objects................... 305
  Paul Couderc, Michel Banâtre, and Fabien Allard

Service Discovery for Service-Oriented Content Adaptation 308
  Mohd Farhan Md Fudzee, Jemal Abawajy, and Mustafa Mat Deris

A Hybrid Mutual Authentication Protocol for RFID.................. 310
  Harinda Fernando and Jemal Abawajy

E2E Mobility Management in Ubiquitous and Ambient Networks
Context 312
  Rachad Nassar and Noëmie Simoni

Energy-Aware Cooperative Download Method among Bluetooth-Ready
Mobile Phone Users 324
  Yu Takamatsu, Weihua Sun, Yukiko Yamauchi,
  Keiichi Yasumoto, and Minoru Ito

Measuring Quality of Experience in Pervasive Systems Using
Probabilistic Context-Aware Approach 330
  Karan Mitra, Arkady Zaslavsky, and Christer Åhlund

Task-Oriented Systems for Interaction with Ubiquitous Computing
Environments 332
  Chuong C. Vo, Torab Torabi, and Seng Wai Loke

Context Data Management for Mobile Spaces....................... 340
  Penghe Chen, Shubhabrata Sen, Hung Keng Pung,
  Wenwei Xue, and Wai Choong Wong

Efficient Intrusion Detection for Mobile Devices Using Spatio-temporal
Mobility Patterns 342
  Sausan Yazji, Robert P. Dick, Peter Scheuermann, and
  Goce Trajcevski

Posters

A Study on Security Management Architecture for Personal
Networks 344
   *Takashi Matsunaka, Takayuki Warabino, Yoji Kishi,
Takeshi Umezawa, Kiyohide Nakauchi, and Masugi Inoue*

Preliminary Results in Virtual Testing for Smart Buildings 347
   *Julien Bruneau, Charles Consel, Marcia O'Malley, Walid Taha, and
Wail Masry Hannourah*

A Model-Based Approach for Building Ubiquitous Applications Based
on Wireless Sensor Network 350
   *Taniro Rodrigues, Priscilla Dantas, Flávia C. Delicato,
Paulo F. Pires, Claudio Miceli, Luci Pirmez, Ge Huang, and
Albert Y. Zomaya*

Probabilistic Distance Estimation in Wireless Sensor Networks 353
   Ge Huang, Flávia C. Delicato, Paulo F. Pires, and Albert Y. Zomaya

Context Aware Framework 358
   Sridevi S., Sayantani Bhattacharya, and Pitchiah R.

MOHA: A Novel Target Recognition Scheme for WSNs 364
   Mohammed Al-Naeem and Asad I. Khan

Policy-Based Personalized Context Dissemination for Location-Aware
Services 366
   *Yousif Al Ridhawi, Ismaeel Al Ridhawi, Loubet Bruno, and
Ahmed Karmouch*

Automatic Generation of Radio Maps for Localization Systems 372
   Ahmed Eleryan, Mohamed Elsabagh, and Moustafa Youssef

Monitoring Interactions with RFID Tagged Objects Using RSSI 374
   Siddika Parlak and Ivan Marsic

AmICA – A Flexible, Compact, Easy-to-Program and Low-Power WSN
Platform 381
   *Sebastian Wille, Norbert Wehn, Ivan Martinovic, Simon Kunz, and
Peter Göhner*

The Use of GPS for Handling Lack of Indoor Constraints in Particle
Filter-Based Inertial Positioning 383
   Thomas Toftkjær and Mikkel Baun Kjærgaard

F4Plan: An Approach to Build Efficient Adaptation Plans 386
   *Francoise André, Erwan Daubert, Grégory Nain, Brice Morin, and
Olivier Barais*

Workshop

Context Acquisition and Acting in Pervasive Physiological
Applications.. 393
    Andreas Schroeder, Christian Kroiß, and Thomas Mair

Linking between Personal Smart Spaces........................... 401
    *Sarah Gallacher, Elizabeth Papadopoulou, Nick K. Taylor,
M. Howard Williams, and Fraser R. Blackmun*

A Smart-Phone-Based Health Management System Using a Wearable
Ring-Type Pulse Sensor 409
    *Yu-Chi Wu, Wei-Hong Hsu, Chao-Shu Chang, Wen-Ching Yu,
Wen-Liang Huang, and Meng-Jen Chen*

An Adaptive Driver Alert System Making Use of Implicit Sensing and
Notification Techniques 417
    *Gilbert Beyer, Gian Mario Bertolotti, Andrea Cristiani, and
Shadi Al Dehni*

Defining the Criteria for Supporting Pervasiveness in Complex Adaptive
Systems 425
    Shiva Mir

User Centric Systems: Ethical Consideration 431
    Nikola Serbedzija

Author Index.. 439

Safe Execution of Dynamically Loaded Code on Mobile Phones

Glen Pink, Simon Gerber, Michael Fry, Judy Kay,
Bob Kummerfeld, and Rainer Wasinger

University of Sydney, Australia
{gpin7031,sger6218,Michael.Fry,Judy.Kay,Bob.Kummerfeld,
Rainer.Wasinger}@sydney.edu.au

Abstract. Mobile phones are *personal* devices, and as such there is an increasing need for personalised, context-aware applications. This paper describes DCEF (Dynamic Code Execution Framework), a framework which allows applications to securely execute dynamically loaded code, providing new functionality such as client-side personalisation. DCEF ensures the user's personal information remains safe while executing code from potentially untrusted sources. Our contributions are: the abstract design of DCEF; an evaluation of the security of our design; the implementation of DCEF; a demonstration that runtime performance is acceptable and validation of DCEF by using it to create an application which provides personalised information delivery about cultural heritage and museum sites.

Keywords: Client-side user modelling, security frameworks, personalised mobile applications.

1 Introduction

Mobile phones have become important personal devices. There were over 4.1 billion mobiles worldwide in 2009 [13], with smart phones making up 5%. This is predicted to rise to anywhere from 23% by 2013 (Juniper)[1] to 37% by 2012 (Gartner)[2]. This suggests a future where many users could benefit from the ability to easily download third party applications.

However, this can be risky because the developers of an application may have malicious intent, or write 'misbehaving code' that unintentionally damages the phone. This creates a tension between exploiting the value of smart phone applications and the risk of exposing users to misbehaving or malicious code. One attempt to reducing this risk is the *app store model*; third-party providers can only release code vetted by the app store provider, who ensures that it can be

---

[1] Press Release: Smartphones to Account for 23% of All New Mobile Phones by 2013, as Application Stores Help Drive Demand, According to Juniper Research: http://juniperresearch.com/viewpressrelease.php?pr=131
[2] Gartner Says PC Vendors Eyeing Booming Smartphone Market: http://www.gartner.com/it/page.jsp?id=1215932

trusted. Our work aims to support a more dynamic model for application delivery, allowing phones to safely download and execute dynamic code with no additional risk of misbehaving code.

A significant driver for this requirement is personalisation. To date, personalised services on mobile phones have predominantly used server-side personalisation, storing the individual's user-model on a central server. This is in part due to the limited storage and power of older phones. The improved capabilities of smart phones can support *client-side* personalisation, making use of potentially private information, such as the user's preferences, recent activity and current context, without allowing the data to leave the phone. We illustrate our vision with the following scenario:

Alice installs the MuseumGuide application from an app store on her smart phone. It immediately asks if it may access her age, education level and interests. It assures her that it keeps these on her phone and does not release them. When she visits a new museum, she can load a specific guide for that site. So arriving at the Nicholson Museum, she confidently downloads their tour and receives a personalised experience, tailored to her particular interests.

This scenario begins with a trusted application, the *MuseumGuide*, being downloaded from an app store. Note that it gives the user control over what personal information they allow it to use. After installation, our framework enables Alice (via the MuseumGuide) to download arbitrary code at any new tourist site, such as the Nicholson Museum. The code is provided by a third party, in this case the museum, as opposed to being shipped with the MuseumGuide. We call these dynamic third-party applications *phonelets*. Alice's phone can dynamically load phonelets without her needing to take any action. Importantly, phonelets cannot access Alice's private data, corrupt or release that information.

This paper presents DCEF, our Dynamic Code Execution Framework. It enables a mobile phone to dynamically load code but it prevents that code from doing damage. The code is permitted to access, under controlled conditions, private data on the device so that it can personalise the user experience.

The next section describes DCEF and its implementation. We then present a qualitative evaluation of its security, an empirical evaluation of its performance and demonstrate its use with a low-fidelity prototype. We then review related work and conclude with a summary of our contributions.

2 The Framework and Its Implementation

DCEF is a generalised framework that can support a broad class of trusted applications and associated phonelets. This section discusses the design goals and implementation of the framework.

2.1 Android

DCEF was developed for the Android platform primarily due to the open and accessible nature of the framework. Android applications have two core components: Activities (user interface screens) and Services (background workers).

Activities and Services can be started via asynchronous messages called Intents. Activities can directly communicate with Services via Android's Inter-Process Communication [IPC] mechanism[3]. This isolation of applications and processes, along with Android permissions, effectively creates a sandbox within which each application operates. By default, no Android application can perform any operation that affects any other application, the OS, or the user[4]. Each application runs as a unique user and can only read and write to their own allocated memory. However, applications can explicitly mark their files as globally accessible and users can allow applications to access more sophisticated phone functionality. This access is provided through the use of Android permissions.

The permissions that an application requires, such as the ability to make phone calls, are defined by the developer when writing the application. When the application is installed the user is informed of which permissions it requires and must decide whether to proceed with the installation. Any dynamically loaded code executed by an application exists within its sandbox and will inherit all of its permissions. It will be able to read and write any file, and access any phone features accessible to the loading application. If the loading application has no Android permissions then there is little risk. However, useful applications will want some access to phone functionality. The trade-off is that they become vulnerable to misbehaving code. At worst, access to the "android.permission.INTERNET" permission would allow dynamic code full access to the Internet, creating major vulnerabilities. Hence we need to separate the permissions that dynamic code is given from the permissions of the application that is loading it. This was one of the primary design goals of our framework.

2.2 Architecture

The DCEF has four components:

1. Third-party applications which request code to be downloaded and executed.
2. The application which dynamically loads the code and executes it (Code Execution Service).
3. The application which downloads and stores code for dynamic execution (Download Service).
4. Third party web server or other location where the dynamic code to be loaded is hosted.

These components communicate using Android's IPC mechanism, except for 3 and 4 which communicate with each other via standard sockets. We now describe these components in the order they are invoked. This is shown visually in Figure 1.

The user must first install an application, obtained from a trusted third party. This application will provide a personalised service to the user via dynamically

---

[3] Android Developer, Designing a remote interface using aidl:
   http://developer.android.com/guide/developing/tools/aidl.html
[4] Android Developer, Security and Permissions:
   http://developer.android.com/guide/topics/security/security.html

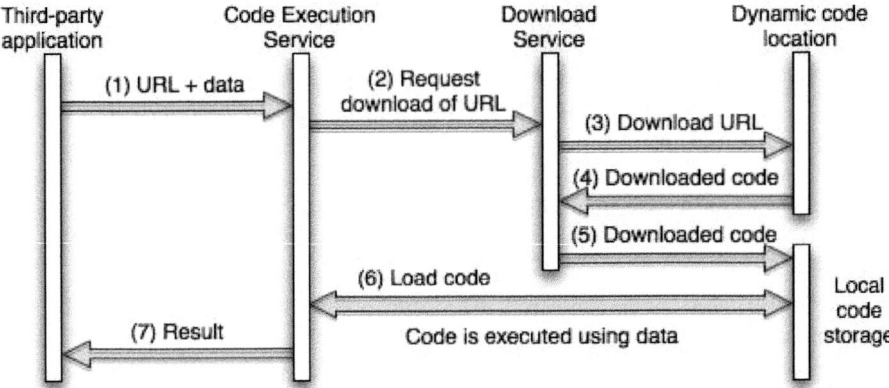

Fig. 1. Framework architecture

loaded code. The application will be obtained by an established method such as an app store. The application will require certain Android permissions, e.g. access to phone sensors, files, profile data, etc. The user is asked explicitly to grant these permissions when the application is installed.

We refer to this application as a 'stub' because, when executed, it acquires and executes dynamic code to realise its functionality. A URL defines the location of the dynamic code. The code resides as a JAR file on a web server and is downloaded using HTTP. Therefore, the application must acquire the download URL. This could be achieved, for example, by having the user scan a QR code. The application also obtains any on-phone data needed for the service. Step 1 in Figure 1 shows the URL and data being passed to the Code Execution Service [CES] via Android IPC.

The CES is the key component of the system. It manages the loading and execution of code and provides an interface for the third party applications. Importantly, the CES has *no permissions*, sandboxing potentially misbehaving code. However, in order to download code, we require the 'Internet' permission. To achieve this, the CES invokes the Download Service (Step 2 in Figure 1), passing it the URL of the code to download.

The Download Service exposes an interface via the standard Android IPC mechanism. This interface exposes two methods. The first downloads a Jar from a given URL, and the second returns a list of available Jars and dynamically loadable classes. Downloaded jars are stored in the Download Service's private data directory. This Jar is deliberately made world readable so it can be accessed by the CES, as shown by Steps 3-5 in Figure 1. The Download Service requires the Android 'Internet' permission to perform its task. Hence the Download Service exists as a separate application from the CES. We must also protect the dynamically loaded code from being modified. Since the Jars are only writable by the Download Service itself, this requirement is already met.

Finally, the CES loads the code dynamically in a separate thread and executes it using the data provided by the third party application (Step 6). The thread is monitored for attacks such as denial-of-service (described below). The result of the code execution is then returned to the third party application (Step 7).

The interaction between the Download Service and the CES provides a mechanism for allowing the stub to execute dynamic code on the device while preventing misbehaving code from damaging the phone or leaking data.

3 Evaluation

Securing the system against misbehaving code is our highest priority, so we first present a security analysis. We then report measures of system performance and describe a demonstrator application.

3.1 Security

Fundamental to securing an application is the Principle of Least Privilege [10]. Every dynamically loaded piece of code should have access to just the functionality it needs and that the user allows. Significant permissions for protecting user data are those which allow user data to be transferred off the device. On the Android platform, these are CALL_PHONE, CALL_PRIVILEGED, SEND_SMS and INTERNET, the last being critical. If it is granted, any number of sockets can be opened to send or receive data across the Internet. These are not the only permissions that can be used maliciously. Any permission that allows access to any phone component can be used maliciously, e.g. for a denial of service attack.

Accordingly, when designing DCEF, we gave dynamic code no Android permissions except for those required to communicate with calling applications. This minimises the effect the code can have on the device. We created a sandbox, the CES, in which dynamic code runs. However, one element of extended functionality can still be used by dynamic code, indirect communication with other processes. This is addressed in the third of the conditions set out below.

The following is a set of four conditions which, if met, allow us to reasonably conclude that DCEF is secure against misbehaving code. We provide our justification for choosing these requirements and explain how our framework meets the requirements.

Dynamically loaded code may only interact with the phone via DCEF: This is achieved by sandboxing the executing code. We prevent all access to the phone's data and restrict access to all functionality that would require Android permissions.

Since CES has no permissions, the dynamically loaded code has no permissions. This protects most of the phone from manipulation. The only application it can supply data to is the DownloadService, which has no other functionality. It cannot affect the data of any other application. Dynamically loaded code can write to globally writable locations, such as an SD card or any file an application chooses to create as 'world writable', but this is the case with any application.

This requirement prevents access to data on the phone or removable storage. It also prevents direct access to all phone functionality, but not indirect access via other processes. Nor does it prevent Intents from being fired. This is discussed below.

Dynamically loaded code must not be able to interact with other dynamic code: Since multiple pieces of dynamic code can be executed concurrently by the CES, they must not be allowed to interact with each other or they could potentially corrupt data or collect information they are not allowed to access. Processes are prevented from interacting with each other by Android. Dynamically loaded code cannot overwrite the private data used by other dynamically loaded code files as it is executed within CES. This is because code storage is controlled by the Download Service and dynamically loaded code does not have permission to store data there. As a result, when designing applications that use DCEF, any persistence of data should be provided by the 'stub' application as any dynamic code can overwrite data stored by the CES.

This requirement restricts direct and indirect access to code loaded dynamically. This still leaves direct and indirect access to processes in memory.

Dynamically loaded code must not be able to access the memory of other processes: We must prevent direct access to other processes, otherwise dynamic code could simply escalate its privileges by accessing another process with more permissions (and of course corrupt those processes themselves). This requirement is provided by Android as previously discussed.

Android prevents direct process access, but indirect process interaction is still possible by using Intents[5]. However, Intents may still be allowed within dynamic code. This does not allow dynamic code access to any data that it should not have access to, as any data provided by Intents is globally accessible. Hence, it does not compromise the system to allow dynamic code to also have this privilege.

The device, user and user experience should otherwise be protected from misbehaving dynamic code: By meeting the above requirements, many of the security and privacy risks of executing code downloaded from the internet are avoided. However, there are still risks of denial of service attacks. Code can be loaded which can, if not otherwise restricted, execute for any amount of time. The dynamically loaded code can also create Intents. Rapidly creating Intents acts as a denial of service attack against the whole device. In order to deal with denial of service attacks, the thread in which the dynamic code is executed is monitored and given a run-time limit.

We could also, as future work, limit the resources that the code can access to mitigate any damage that denial of service attacks may cause.

---

[5] Android Developer, Intents and Intent Filters:
http://developer.android.com/guide/topics/intents/intents-filters.html

In summary, DCEF separates the dynamic code from the device. To achieve this, DCEF relies on Android acting as documented without bugs or exploits. There may be unforeseen highly sophisticated attacks. For example, it is conceivable that a timing attack could be implemented by making dynamically loaded code run for certain periods of time. An application could be structured so that it could make download requests at particular intervals, potentially signalling an external server by timing requests appropriately. Our security analysis demonstrates that DCEF meets the defined requirements and that these deal with important security risks.

3.2 Performance

We evaluated the device footprint and execution time. Both the Download Service and CES applications have small footprints: their sizes on disk are 12kB and 13kB respectively, modest demands on a typical smart phone.

We then built and tested two demonstrator applications, DynamicLocator and MuseumGuide. We now report results for MuseumGuide, which implements the functionality described in the introductory scenario. We present measurements of its download and execution times. Results for the second application were excluded as they were similar.

Performance was measured internally using a code profiler, adding a small overhead. We determined and report below on the time for: downloading within the Download Service (Actual Download Time); the downloading method using IPC (Extra Download Process Time); loading the class (ClassLoad Time); instantiating the class and running it dynamically (Dynamic Process Execution); doing other tasks in the framework (Other Execution Time); and other processing in the third party application including displaying output (Other Process Time).

Measurement starts once a user initiates a download and stops when the tour is displayed. This includes time required to render text and images for ten items.

Table 1. Aggregated Download Times for 3G

Jar Size (KB)	Tests	Average download time (s)	Standard deviation
1.4	10	2.436	0.730
41	40	2.769	1.832
140	40	2.804	0.641

Table 2. Aggregated Download Times for Wifi

Jar Size (KB)	Tests	Average download time (s)	Standard deviation
41	40	0.784	0.480
140	40	0.617	0.659

Fig. 2. Execution Time, Museum Guide Using 3G

Tests were run on a G1 phone[6] running other processes in the background as would be normal for a phone in use. The size of the jar for the Museum Guide was 140KB. The significance of the download size is discussed below. Typically jars would contain only small amounts of code, and no media which is downloaded separately. Download was initially performed over a commercial 3G network. Tests were run 40 times and are displayed in Figure 2.

The total elapsed time is typically only a few seconds, usually less than 4.5s, with the download time dominating. The variances can be attributed to other processes running on the phone and latency variations in the 3G network. 'Other Process Time' takes the next largest amount of time, but this is part of the third party application so is not strictly a cost of DCEF. The processing time, which represents the actual dynamic loading and execution of code, is minor compared to the download time. Overall, the times are of a similar order to other applications and should be acceptable.

The jars used in the tests were larger than necessary, potentially hundreds of times larger in the case of MuseumGuide. This was for ease of testing and implementation of testing components. To compare the effect of download size a set of tests was run on the download component with Jars of 1.4KB and 41KB. The results are in Table 1. This indicates the size of the file has a modest effect on total time. Rather, download times for smaller files are dominated by latencies in the 3G network.

We then ran the same set of tests over a standard WPA2 encrypted Wifi network. This yielded the results in Figure 3 and in Table 2 (for two file sizes). Downloads were much faster on the Wifi network. In fact, the average download of the larger file was faster than the smaller one, suggesting that latency

---

[6] HTC G1 specifications: http://www.htc.com/www/product/g1/specification.html

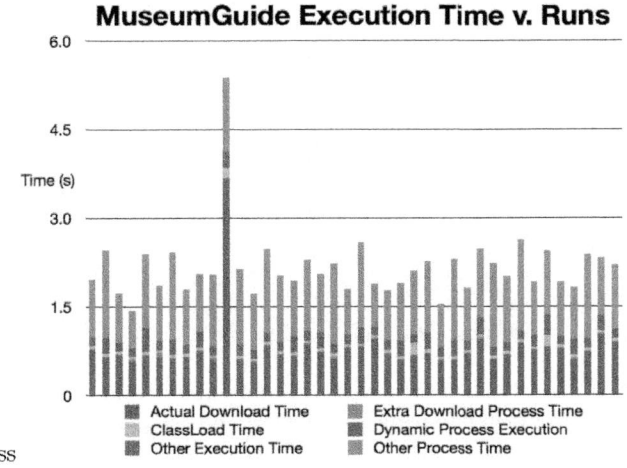

Fig. 3. Museum Guide Using WiFi

and other factors have more effect on the download speed than the size of the downloaded file. It is notable that there are still spikes in total time when using Wifi, indicating that variables on the phone and/or the server affect the response times. Overall, these results indicate DCEF provides adequate response times.

3.3 Demonstrator Application

While the last section referred to the MuseumGuide's security and performance, we now provide a brief description of it. It demonstrates the use of the DCEF to create a valuable class of applications which provide personalised information about a museum or cultural heritage site, as described in our introductory scenario. This makes use of PersonisJ, a client-side user-modelling framework [4]. PersonisJ holds and manages arbitrary information about the user, such as their name, age, interests and location. The user first downloads the MuseumGuide application from a trusted source. At this point, the user is informed that MuseumGuide would like to use the PersonisJ user-model (these are specific permissions defined within the PersonisJ service; see [4] for more details) and that the MuseumGuide needs to access the Internet.

Later, as the user travels to new museums, they can download phonelets implementing the personalised tour created by each museum. While MuseumGuide (i.e. the 'third party app' shown on the left in Figure 1) has the privileges needed to access PersonisJ and the internet, the tour app for a given museum (i.e. the 'dynamic code' shown on the right in Figure 1) does not have access to the user model or the internet. We implemented both the MuseumGuide and a demonstration phonelet for the Nicholson Museum at Sydney University. Figure 4 illustrates this in use, with the leftmost screen showing the alert that MuseumGuide issues when it detects a nearby museum that matches the user's preferences. Upon clicking the notification, the user is invited to download the phonelet with

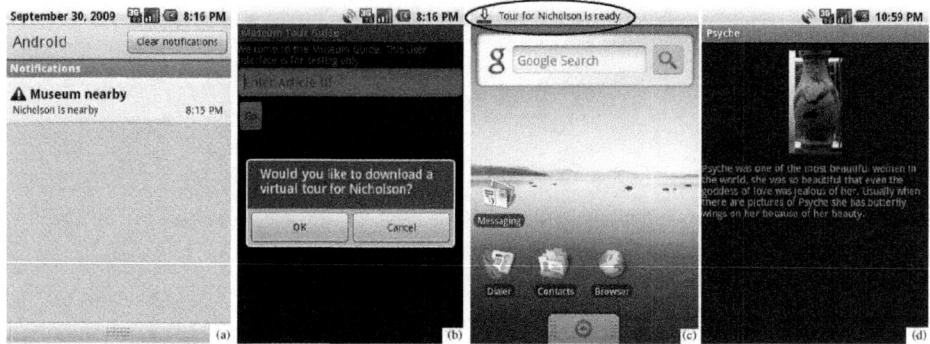

Fig. 4. MuseumGuide application showing: a) museum proximity notification, b) download prompt, c) 'download complete' notification, and d) personalised museum tour

the personalised tour (Figure 4b). At this point, the tour's content and code are downloaded and run in DCEF's safe execution environment and the user is informed when their tour is ready (Figure 4c). The phonelet is provided with the user model, via DCEF, which it uses to generate the personalised tour. Figure 4d shows an example of one screen of a personalised tour.

4 Related Work

There has been considerable previous work on the broader topic of personalisation, such as that reviewed in [5] for the case of e-commerce. To date, the focus has been on server side personalisation, as in [7], an m-commerce application which targets promotions to users based on their preferences. Our work breaks new ground in supporting client-side personalisation on mobile devices. This goes beyond the recognition of the notion of portable profiles on mobile devices described in a survey of issues [12] for mobile personalisation. The similar task of customising multimodal interactions with smart phones was developed in [6]. However, this did not provide customisation of applications. Similarly, several mobile application platforms allow users to download *data* from third parties (for example, Layar[7], Wikitude[8], and Urban Spoon[9]). While these allow third parties to link into their framework with 'content' layers, they do not allow third parties to provide supporting 'code', and thus lack the flexibility and ability to support personalised applications, such as those enabled by DCEF. Many researchers have noted the need to ensure the privacy of personal preferences, especially in so-called smart environments [1,3], with broad recommendations that point to the potential value of client-side personalisation on mobile phones.

Our framework aims to ensure security for client-side personalisation and other facilities that can be supported by dynamically loaded code on mobile

---

[7] Layar: http://www.layar.com/
[8] Wikitude: http://www.wikitude.org/
[9] Urban Spoon: http://www.urbanspoon.com/choose

phones. There are various attack vectors on mobile phones [2,11], such as Bluetooth, messaging and Internet access. These vectors allow the placement and execution of malicious code on a device. However, because such code must be platform specific, the diversity of current phone platforms has, so far, afforded some protection. The main platform targeted has been Symbian OS [11]. This is unsurprising, since for much of the last decade it has been the dominant platform in the mobile phone market. Trojans masquerading as applications have been detected for Symbian OS[10]. Java trojans have also been developed and further potential for malware exists for mobile Java platforms [8,9]. Even so, the actual number of attacks on mobile phones has been infinitesimal compared to the PC world. With the recent explosive growth in smart phone sales, this may be about to change.

Perhaps the greatest potential for the spread of malware on smart phones lies in the form of third party applications. We have already noted that one approach, adopted for the iPhone, relies on manual filtering mechanisms. The Apple App Store acts as a trusted third party to verify that applications are safe[11]. This does not guarantee security as errors can occur in the verification process. Defences against malware derived from the PC world are beginning to be adopted for mobiles, such as virus detection. However, to the best of our knowledge, there have not been any developments aimed at permitting the secure, automatic download and execution of code.

5 Conclusions

We have described the motivation for a secure dynamic code loading framework and described DCEF, which combines the use of an execution proxy with a code download mechanism. This framework enables a mobile phone to dynamically load third party 'phonelets' into trusted 'stub' applications. Through the use of an execution proxy, thread monitoring and other mechanisms, this dynamic code protects the phone against misbehaving code. Importantly, it ensures that both the 'stub' and 'phonelet' have only the access rights the user has explicitly granted. We have demonstrated the security of DCEF in terms of four key conditions that the framework satisfies: dynamically loaded code may only interact with the phone via DCEF; dynamically loaded code must not be able to interact with other dynamic code; dynamically loaded code must not be able to access the memory of other processes; and the device, user and user experience should otherwise be protected from misbehaving dynamic code. We have reported analysis of DCEF performance in terms of its modest memory and time demands. We have also provided details of one of the demonstrator applications that makes use of DCEF.

---

[10] Mobile malware evolution: An overview:
http://www.viruslist.com/en/analysis?pubid=200119916
[11] Apple iPhone SDK Agreement:
http://blog.wired.com/gadgets/files/iphone-sdk-agreement.pdf

Our work makes an important contribution, the provision of a framework for the secure and automatic execution of third-party code on mobile devices. Using DCEF, programmers can develop applications that acquire new functionality on-the-fly. This enables an important and large class of applications, namely those that provide dynamic on-device, client-side personalisation, to operate without compromising a user's privacy. DCEF also provides the ability to enhance functionality of a user's device without the user having to intervene.

References

1. Armac, I., Rose, D.: Privacy-friendly user modelling for smart environments. In: Mobiquitous 2008, pp. 1–6. ICST, Brussels (2008)
2. Chen, T., Peikari, C.: Malicious Software in Mobile Devices. In: Handbook of Research on Wireless Security, p. 1 (2008)
3. Cottrill, C., Thakuriah, P.: GPS use by households: early indicators of privacy preferences regarding ubiquitous mobility information access. In: Cahill, V. (ed.) MobiQuitous. ACM, New York (2008)
4. Gerber, S., Fry, M., Kay, J., Kummerfeld, B., Pink, G., Wasinger, R.: PersonisJ: Mobile, Client-Side User Modelling. In: User Modeling, Adaptation and Personalization (UMAP), pp. 111–122. Springer, Heidelberg (2010)
5. Goy, A., Ardissono, L., Petrone, G.: Personalization in e-commerce applications. In: Brusilovsky, P., Kobsa, A., Nejdl, W. (eds.) Adaptive Web 2007. LNCS, vol. 4321, pp. 485–520. Springer, Heidelberg (2007)
6. Korpipaa, P., Malm, E., Rantakokko, T., Kyllonen, V., Kela, J., Mantyjarvi, J., Hakkila, J., Kansala, I.: Customizing user interaction in smart phones. IEEE Pervasive Computing, 82–90 (2006)
7. Kurkovsky, S., Harihar, K.: Using ubiquitous computing in interactive mobile marketing. Personal and Ubiquitous Computing 10(4), 227–240 (2006)
8. Reynaud-Plantey, D.: New threats of Java viruses. Journal in Computer Virology 1(1-2), 32–43 (2005)
9. Reynaud-Plantey, D.: The Java Mobile Risk. Journal in Computer Virology 2(2), 101–107 (2006)
10. Saltzer, J.H., Schroeder, M.D.: The protection of information in computer systems. Proceedings of the IEEE 63(9), 1278–1308 (1975)
11. Töyssy, S., Helenius, M.: About malicious software in smartphones. Journal in Computer Virology 2(2), 109–119 (2006)
12. Uhlmann, S., Lugmayr, A.: Personalization algorithms for portable personality. In: MindTrek 2008, pp. 117–121. ACM, New York (2008)
13. Union, I.T.: Measuring the Information Society 2010. International Telecommunication Union (2010)

Towards Enabling Next Generation Mobile Mashups

Vikas Agarwal[1], Sunil Goyal[1], Sumit Mittal[1],
Sougata Mukherjea[1], John Ponzo[2], and Fenil Shah[2,*]

[1] IBM Research - India, New Delhi
{avikas,gsunil,sumittal,smukherj}@in.ibm.com
[2] IBM T.J. Watson Research Center, NY, USA
{jponzo,fenils}@us.ibm.com

Abstract. Evolution of Web browser functionality on mobile devices is the driving force for 'mobile mashups', where content rendered on a device is amalgamated from multiple Web sources. From richness perspective, such mashups can be enhanced to incorporate features that are unique to the mobile setting - (1) native Device features, such as location and calendar information, camera, Bluetooth, etc. available on a smart mobile platform, and (2) core Telecom network functionality, such as SMS and Third Party Call Control, exposed as services in a converged IP/Web network setup. Although various techniques exist for creating desktop-based mashups, these are insufficient to utilize a three-dimensional setting present in the mobile domain - comprising of the Web, native Device features and Telecom services. In this paper, we describe middleware support for this purpose, both on the server side dealing with processing and integration of content, as well as on the device side dealing with rendering, device integration, Web service invocation, and execution. Moreover, we characterize how various components in this middleware ensure portability and adaptation of mashups across different devices and Telecom protocols. Based on our approach, we provide an implementation of mashup framework on three popular mobile platforms - iPhone, Android and Nokia S60, and discuss it's utility.

1 Introduction

'Mashups' are applications created by integrating offerings from multiple Web sources, and have become very popular in recent years. Further, adoption of *mobile mashups* is being driven by the evolution of Web browsers on the mobile device. Most modern smart phones today have browsers that are HTML and JavaScript standards compliant, and provide a rich, powerful Web browsing experience to the mobile users. With rapid enhancements in processing power, memory, display and other features of mobile phones, and with continuous improvement in mobile network bandwidth, mobile mashups bear the potential of being as successful as the desktop ones.

---

* This author is currently working at Google Inc., USA.

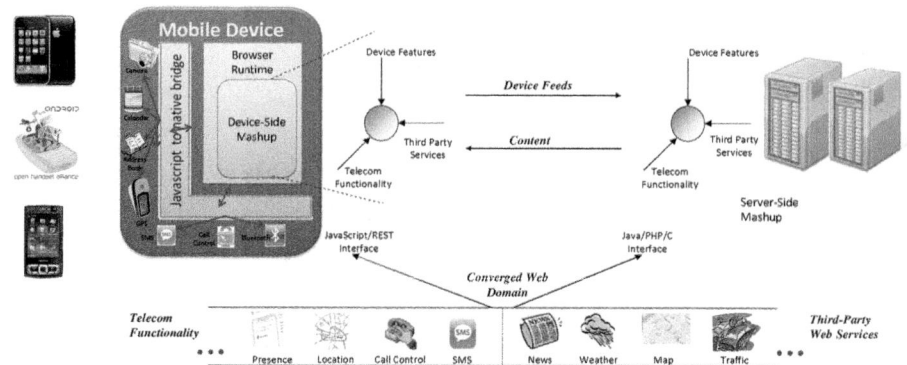

Fig. 1. Overview of a Mobile Mashup

To help developers create rich mobile applications, popular platform vendors like Nokia, Blackberry and Android expose interfaces for a variety of features available on the mobile handset, for both data (user's contacts, calendar, geographic location, etc.) and functionality (making calls, sending SMSes, using the camera, etc.) [2]. On similar lines, there is willingness by Telecom operators to move from a *walled-garden* model to an *open-garden* model [12], whereby they expose core functionalities of Location, Presence, Call Control, etc. as services to developers. From the perspective of mobile mashups, it is desirable that such Device and Telecom features are utilized to enhance the entire mashup experience of a mobile user. For example, location and presence information available on mobile phones/Telecom can be used to design interesting mashups related to directory services, workforce management, social networking, etc. Similarly, camera on the phone can help enrich a retail mashup with the facility to retrieve product information based on scanning of bar-codes.

Various tools and technologies exist [4,11,18] that help developers create mashup applications for the desktop. However, these fall short in the context of a three-dimensional setting present in the mobile domain - native Device features, Telecom services and Web-based offerings. In this paper, we start with our view of a next generation mobile mashup consisting of device side and server side mashup components, as depicted in Figure 1. The server side mashup executes on a Web server and deals with processing and integration of content received from each dimension. The device side portion, on the other hand, resides on the browser environment of different platforms, such as iPhone, Android or Nokia S60, and renders the content received from server. Through client-side scripts (JavaScript and AJAX), this portion executes Telecom functionality - SMS, Third Party Call Control, etc., Web offerings - Map information, News, etc., as well as various features of the device, such as Camera and Bluetooth. Moreover, this portion also participates actively in a mashup by feeding device information, such as Location and Calendar, to the server side component for inclusion in the processing logic.

We argue that in order to enable the above mashup scenario, the following novel characteristics need to be incorporated -

(1) **Two-way Mashup Model**. In current mashups, the client is responsible only for rendering the content it receives from the server. However, for mobile mashups, the device can act as an *active component*, augmenting a mashup with its own information and features. In essence, a client[1]-server architecture for mashups requires middleware enhancements to provision (a) a *two-way* flow of information - from device to server, apart from server to device, and (b) a *two-way* mashup support where the device and Telecom features can be used 'both' on the server (during processing and content integration) and the client (during mashup rendering and execution).

(2) **Device Integration**. On most platforms, APIs for device features are available in native language (J2ME, C++, Objective C, etc.), while for mashup development the interfaces are required along the Web programming model, i.e. JavaScript. Therefore, on a given mobile platform, we first need a *bridge* that takes the native APIs and exposes them in a Web mashable form. We emphasize that such a bridge needs to have three distinct attributes -

(i) support for a *bidirectional communication* between the mobile browser and the native APIs, so that a) input objects required for invoking a native feature can be passed from within the mashup, through the browser, to the underlying platform, and b) output objects (return values, exceptions) available as a result of native invocation can be marshaled back to the mashup through the browser.

(ii) provisioning a *channel for callbacks*, in case result of an API invocation is not immediately available or possible - for example, periodic proximity alerts. In essence, events generated in the native context on a mobile platform should be channeled through appropriate signaling and notification to the invoking browser context of a mashup.

(iii) handling immense *fragmentation* in syntax and semantics of various device features across different platforms - starting from diversity in the 'name' of the interface, to 'name', 'data type' and 'ordering' of attached parameters, to the differences in the set of 'exceptions' thrown by each platform.

(3) **Telecom Support**. A number of legacy protocols exist in the Telecom domain, while new standards continue to be drafted and absorbed by different operators. Therefore, to reduce burden on a mashup developer, we need an *abstraction* layer that lets various Telecom features be invoked without requiring the developer to know the underlying protocol specifics. Moreover, for a given feature, this layer should enable a *seamless switch* among different protocols; especially required in scenarios where the Telecom networks are gradually evolving to move from legacy interfaces to standards like Parlay-X [14] and SIP [8].

2 Mobile Mashup Framework

In this section, we describe a framework that enables realization of the mobile mashup view outlined above. Our framework consists of middleware components, both on the device side, as well as the server side. We describe these components next, while characterizing the role played by each component in the mobile mashup setting.

---

[1] in this case, the device

2.1 Mobile Device Middleware

As shown in Figure 2, the device side middleware runs on top of existing platform middleware, and provides an **Enhanced Browser Context** in which the client side mashups execute. In essence, the enhanced browser context uses the existing **Browser Runtime** available on a platform to render the HTML pages and to execute associated JavaScript code. It further uses a **Native-to-Mashup Bridge** to allow access to device features, such as Camera, Calendar, Contacts, etc., using JavaScript interfaces from within a mashup. As discussed in the previous section, this bridge provides a bi-directional communication capability between the browser context and the native APIs (allowing the browser to invoke native APIs, pass inputs and receive outputs), and also supports passing of events from native APIs to the browser through JavaScript callbacks. The bridge can be realized in multiple ways, - 1) through *modification of the source code* of an existing browser to allow access to native capabilities via JavaScript, 2) through *creation of a plugin* for the browser that adds device features without modifying the browser source code, and 3) by *embedding a browser runtime* (where it is available as a 'class' in the native programming language) within a native mobile application, and enabling access to native features for mashups rendered through the embedded runtime. Depending on the specifics of a given platform, one or more of these techniques could be used.

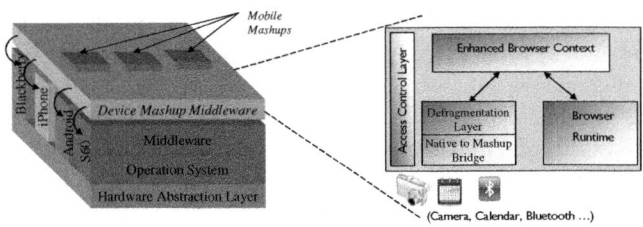

Fig. 2. Middleware Support on Device

Another component of device mashup middleware is the **De-fragmentation Layer**. This layer builds on the JavaScript interfaces exposed by Native-to-Mashup bridge and exposes a consistent set of interfaces across various platforms. Our generic model for common JavaScript interfaces is -

<*interface name*> (<*arguments 1 ... n*>, <*callback*>, <*errorCallback*>, <*options*>)

For instance, consider the following listing that defines a uniform API across Android, iPhone, and Nokia S60 for invoking periodic location updates -

startUpdatesOnTimeChange (timeFilter, callback, errorCallback, options)

Here, semantics of the above API as well as data structures of the parameters involved, such as `timeFilter`, are uniform across the platforms. Similarly, updates for location information, encapsulated in a `Location` object are provided through a commonly defined `callback` method, while errors are propagated with the help of a common `errorCallback` method. Note that a generic `options` parameter is also provided with the interface. Similar to what we argue in one of our earlier works [2], this parameter is optional and can be used to configure platform-specific attributes, such as *Criteria* in the case of S60, *Location Provider* in the case of Android, and *accuracy* for iPhone. The structure and values of this parameter are platform dependent, and therefore should be strictly used only when a developer wishes to fine-tune mashups on a particular platform. Normally, null value should be used for this parameter, in which case default values for various attributes would be used on the corresponding platform.

Accessing a device or Telecom feature might have monetary cost associated to it, such as sending an SMS, making a Call, etc. or mashup might deal with sensitive personal information stored on the device, for instance Location and Calendar entries. Therefore, a component that becomes intrinsic to the entire set-up is the **Access Control Layer** that performs the task of regulating access to these features based on user defined policies. More specifically, this component intercepts each request from within a mashup application for invoking a feature, and performs appropriate checks to determine whether the desired access is allowed. These policies take into consideration different factors such as frequency of access, time of day, user's current location, etc. to take automatic decisions or prompt the user for explicit approval. A user should be able to configure various policies when the mashup is first accessed and refine the same over time.

2.2 Server Side Middleware

As shown in Figure 3, the server side middleware consists of two blocks - 1) a **Telecom block** to enable access to Telecom network functionalities, and 2) a **Device block** to receive information feeds from mobile devices and to perform device specific adaptation of mashups.

At the heart of Telecom block is the **Protocol Binding** component that connects to various Telecom services using the underlying network protocols, such as SIP, Parlay-X and CORBA. Through these binding components, the framework removes the burden - from mashup developer - of knowing Telecom specifics, for instance session management for SIP, broker object for CORBA and SOAP headers for Parlay-X. Note that once the bindings are in place, the Telecom block provides mashable interfaces for different services in various programming

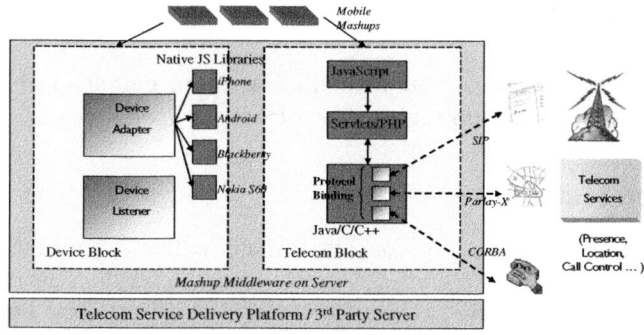

Fig. 3. Middleware Support on Server

languages - Java, C, C++, etc. While these interfaces can be directly used in a server side mashup, REST based interfaces are also exposed so that Telecom services can be invoked from client side mashups using JavaScript.

Binding stubs also enable seamless switching between different protocols. Consider a scenario where network location information can be fetched using both SIP and Parlay-X. In SIP, this information is obtained by subscribing to a Presence Server for the presence information, and parsing the returned document. Parlay-X, on the other hand, requires the request to be made using a SOAP envelope, and returns the location information encapsulated again under SOAP. In our framework, these disparate steps are wrapped under generic interfaces and corresponding stubs for various Telecom protocols are provided. Properties and attributes required for each protocol, such as service port information, tolerable delay, accuracy, etc. are configured using an *options* parameter similar to the device middleware model.

The device block consists of a **Device Listener** component that receives data feeds from a device side mashup and provides this information to the server side mashup for processing and integration. As mentioned earlier, these feeds enable the mobile device to participate actively in a mashup for server side content generation. Another component in this block is the **Device Adapter** that performs device specific adaptation of mashups. For example, appropriate JavaScript libraries (containing code to access device features) need to be included in a mashup page depending on the device where the mashup is being rendered. Also, the look and feel of a mashup is adjusted based on the device properties, such as screen size, resolution, etc. by including appropriate CSS files. Further, any other device specific adaptation, for example, altering the layout of a mashup to resemble native look-and-feel, is done using this component.

As shown in Figure 3, the server side middleware can be hosted either on the service delivery platform (SDP) of a Telecom operator or on a third party server. In the first scenario, Telecom features only from a single operator are available. In the second setup, however, multiple operators can make their offerings available, which makes it imperative for the Telecom block to provide uniform APIs across different providers for enabling easy portability of mashups.

3 Implementation

In this section, we describe an implementation of our mobile mashup framework consisting of device-side and server-side middleware components.

3.1 Device Middleware

On all the three platforms we consider in this paper - Android, iPhone, and Nokia S60 - it is possible to embed a browser engine inside a native application that allows rendering of Web content developed using artifacts like HTML and JavaScript. We extend this embedded browser model to develop mashup bridges for native device capabilities on each platform.

The **Android** platform provides a `WebView` class in Java for creating a browser instance inside a native Android application. Further, this class provides a generic API 'addJavaScriptInterface()' that allows addition of Java objects within a `WebView` browser instance, and lets them be treated as pure JavaScript entities. In essence, any such object now becomes a 'connection' between the browser context and the native Android platform; we use this facility as the basis to expose JavaScript functions for various device features. An issue that required resolution pertained to callbacks, since a JavaScript 'connection' object in the browser context lacks the ability to provide a callback function for the underlying Java object. To overcome this, we utilize the 'loadUrl()' method in the `WebView` class using which a JavaScript function, qualified as a URL, can be invoked from a Java object. Also, since only string arguments can be passed to the called JavaScript function using this technique, all parameters are marshaled as string serialized JSON objects.

The **iPhone** and **Nokia S60** platforms provide mechanism for attaching listeners to the embedded browser for receiving events corresponding to changes in *URL*, *title* and *status text* of the browser. We use this mechanism for accessing native APIs through JavaScript on these platforms. In essence, we first define listeners that listen for changes in URL, title or status text of the embedded browser. Specifically, on iPhone we attached listener for *URL* change event, whereas on Nokia S60 we intercepted the *title* change event. Now, whenever we need to call a native API from JavaScript we change the URL/title to *<mashupDomain>?Id=<id>&Name =<deviceFeature>&Values=<values>*. Here, `Id` is the request identifier, `Name` is the native feature being called and `Values` contains the parameters needed to invoke the feature. Execution of this code is trapped by above listeners, where the unique `mashupDomain` qualifier helps to deduce that a native service is being accessed. The listener extracts the values of the parameter and calls the specified native API. Moreover, these platforms provide a mechanism similar to that on Android for calling a JavaScript function from the native code - iPhone provides a method called *stringByEvaluatingJavaScriptFromString()* in the `UIWebView` class, whereas S60 provides a method called *setUrl()* in the `Browser` class. We use these methods to send the response back to JavaScript from the called native API.

Using the mashup bridges, we exposed various device features such as Location, Contacts, Camera, etc. through JavaScript interfaces within an embedded browser context on each platform. We then provided a de-fragmentation layer that absorbs differences in syntax and semantics of these interfaces across different platforms. Towards this, we build upon our previous work [2], and handle heterogeneity of mobile features using a three-phased process - 1) semantic phase, where we fix the structure of the interface, in terms of the method name, associated parameters (including their name, ordering and dimensions), as well as the return value, 2) syntactic phase, in which we remove differences in data structures of various objects, and 3) binding phase, which contains implementation of the common interface on top of the original platform offering, and also provides mechanisms to fine-tune an interface using platform specific attributes and properties. Due to lack of space, we omit further details here, and direct the interested reader to [2] for more information.

Finally, we designed an access control layer [1] that allows a user to configure access policies for various device and Telecom features. Currently, the implementation is available for Android platform, and allows policies defined around three basic tenets - domain (determined by URL) of the mashup in consideration, context of the user (determined through a combination of user's current location and the current time), and frequency of access (for example, how often can a feature be accessed). In essence, whenever a feature is invoked in a mashup, depending upon the policies configured for that mashup, the invocation is either allowed or denied. We are working towards making this access control layer available on other platforms as well.

3.2 Server Middleware

For Telecom support, we first defined common APIs for various services in Java and then created protocol bindings for different Telecom protocols. In particular, we took two real-life Telecom products - IBM Telecom Web Services Server (TWSS)[2] and IBM WebSphere Presence Server (WPS)[3], and built stubs for calling SMS, Location, Presence and Third-Party Call Control (3PCC) services. TWSS allows access to network services through standards-based Parlay-X Web Services, whereas WPS provides real-time presence information via the SIP protocol. Apart from this, we also created stubs for various network services exposed in CORBA by a Telecom network simulator - Open API Solutions (OAS)[4] version 2. As discussed earlier, these Protocol Binding stubs allow easy switching of Telecom protocols without requiring any changes in a mashup application.

For mashable client side APIs, we created JavaScript interfaces for Parlay-X based functionality by defining XML fragments containing the desired SOAP headers - sending the XML as AJAX requests to the server, and parsing the returned XML fragments that contained SOAP responses. However, for SIP based Presence service, this procedure does not work since SIP messages are exchanged over TCP/UDP. A JavaScript interface in this case was created by first implementing a servlet that talks to the Presence Server using SIP over UDP messages. The Presence JavaScript interface interacts with this servlet to fetch presence-related information. Interfaces for CORBA based OAS services were similarly developed using the servlet model.

For the device block implementation, we develop three JavaScript files containing code for invoking features of each device and three CSS files to tailor the mashup UI for the corresponding platform, one each for Android, iPhone and Nokia S60. The Device Adapter component is implemented as a servlet that detects the device platform using 'user-agent' field in the request header and adds appropriate JavaScript and CSS files. An option is also added to selectively disable certain device features that are not available - an instance in case is the SMS service which is not exposed in iPhone. The Device Listener component was also implemented as a servlet where information from the device can be submitted using the servlet *Post*

---

[2] http://www-306.ibm.com/software/pervasive/serviceserver/
[3] www.ibm.com/software/pervasive/presenceserver/
[4] www.openapisolutions.com

method. The device data is stored by the listener using a 3-tuple <*mobile phone#, device data, time-stamp*> and made available to various mashup applications.

3.3 Integration with Mashup Environments

We took two environments for creating mashup applications, Eclipse[5] - a popular open source meta application framework, and Lotus Mashups [11] - an integrated suite from IBM for mashing widgets and data feeds from multiple sources, and enhanced them to enable mashing of device and Telecom features in addition to web offerings. For Eclipse, these APIs were integrated via the 'Snippet Contributor Plugin' that provides drag-n-drop of APIs in the web editor. In Lotus Mashups, on the other hand, the APIs were made available in the required 'iWidget' format. We do not provide integration details here due to lack of space, and direct the interested readers to [3] for more information.

4 Discussion

We have developed several mashups that use Device and Telecom features enabled by our framework; Figure 4 presents snapshots of one such social networking mashup. As shown, the mashup brings together various offerings from the device (*Camera* to take pictures, *Contact List* to obtain a user's friends list, *Location Updates* to get GPS location periodically), Telecom network (*Call* and *SMS* to communicate with friends, *Presence Services* to get their presence data), and the Web (*Twitter* to publish and fetch tweets of friends, *Facebook* for photo and profiles, *Google Maps* for visual display of friends on a map).

Note that a single application was developed, which looks and behaves exactly the same across Android, iPhone and Nokia S60 platforms. The common interfaces provided by the framework hide semantic and syntactic heterogeneities of device features, thereby allowing a single code base and ease of programming. Moreover, when a new version of a platform introduces changes in device APIs, it is absorbed in our framework, alleviating the need of application maintenance 'as platforms evolve'. For example, APIs for accessing contact information were changed in Android platform moving from release 1.5 to 2.1 - these changes were accounted for in the mashup bridge implementation of the corresponding Android releases. Similar arguments apply to usage of various Telecom features.

The core of our device middleware is the bi-directional communication between the mobile browser and the native platform. Figure 5 shows the overhead associated with this communication for each platform. Here, we took three device features and measured the time for - 1) invoking a native device feature from JavaScript code, and 2) callback from native code to JavaScript. Each number reported is an average of ten execution traces. As the figure shows, the overhead of device middleware is very small, and indicates a fast transfer between the embedded browser context and the corresponding native feature. There are variations,

---

[5] www.eclipse.org

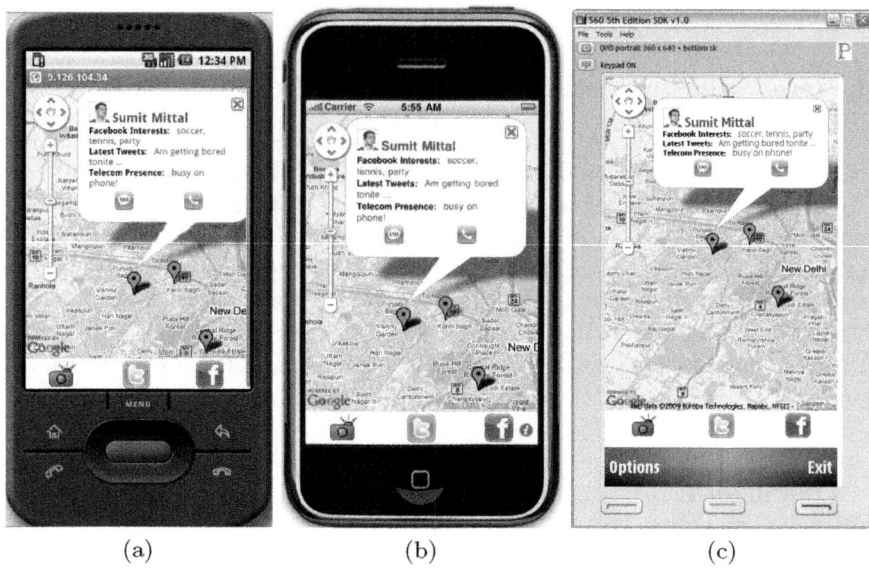

Fig. 4. Social Networking Mashup on (a) Android (b) iPhone (c) Nokia S60

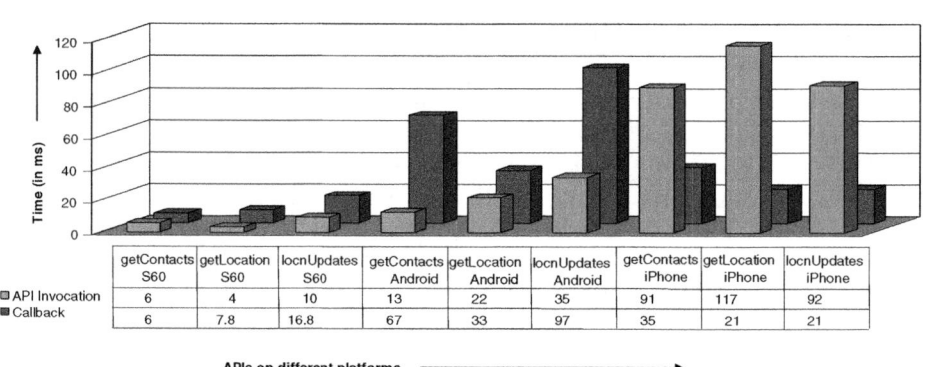

Fig. 5. Performance of Device Middleware

however, across the platforms and across the APIs on a given platform, due to differences in JavaScript processing engines, event passing and handling mechanisms in the embedded browser, de-fragmentation logic, etc.

On the server side, we did a similar performance evaluation of various components - device adapter, device listener, and Telecom block. The overhead of these components too was found to be small - of the order of a few milliseconds. From a broad perspective, considering the time a typical application spends in UI interaction, business logic, etc., we conclude that the cost of using our middleware components is negligible as compared to the total mashup runtime.

5 Related Work

There are several professional tools in the industry that facilitate the creation of Web based mashups. Examples are Yahoo Pipes [18], and IBM Lotus Mashups [11]. Academic research on mashup tools has also been undertaken. For instance, [10] presents an environment for developing spreadsheet-based Web mashups. None of these works, however, establish components for integrating device and Telecom features in a mobile mashup. [6] presents a mobile mashup platform that integrates data from Web-based services on the server side and utilizes users' context information from sensors to adapt the mashup at the client side. However, this framework is far from a comprehensive approach required for incorporating the three-dimensional mobile setting that we outline in this paper. Also, it currently runs on certain Nokia devices only.

Most smart-phone platforms today provide native APIs for several services, such as GPS location, address book, calendar and camera. We find that APIs for common services in even J2ME-based platforms have become fragmented on platforms that support new hybrid Java runtimes, such as Android. Fragmentation is further exacerbated on non-Java based platforms like iPhone, which uses the Objective-C language and the Cocoa user interface library. Various standardization efforts, such as OMTP Bondi [13], attempt to overcome this fragmentation. In [2], we presented a three-tiered model for absorbing heterogeneity in syntax, semantics and implementation of interfaces corresponding to device features across multiple platforms. Similarly, PhoneGap [15] is an effort towards enabling uniform JavaScript access to native APIs. From the perspective of our framework, we could use either of these approaches as building blocks in our device middleware stack.

Session Initiation Protocol (SIP) is a standard being widely adopted by Telecom operators to expose their core functionalities - voice service, SMS service, Call Control, etc. - using SIP. JSR-289 [9] has been proposed by Sun and Ericsson to enhance existing SIPServlet specification and support development of composed applications involving both HTTP and SIP servlets. Web21C [17] from British Telecom is a Web 2.0 based service aggregation environment that allows developers to integrate core Telecom functionality with other Web services into a single application. On the other hand, [5] gives a broad overview of existing approaches for enabling a unified Telecom and Web services composition. However, both fall short in describing a generic model for supporting mashable Telecom interfaces. In [12], we introduced SewNet which provides an abstraction model for encapsulating invocation, coordination and enrichment of Telecom functionalities; but did not deal with mobile mashups.

One of the major challenges in the area of mobile applications is the huge privacy and security implication around sensitive user information like location, contacts and calendar entries [7]. PeopleFinder [16] is a location sharing application that gives users flexibility to create rules with varying complexity for configuring privacy settings around sharing their location. In this paper, we have created a similar policy framework, but differ on two counts. Firstly, we move beyond location and cover other sensitive information as well. Secondly, we apply the policies to a generic mobile mashup setting, and not to a specific application.

6 Conclusion and Future Work

Evolution of mobile browsers is driving the adoption of mashups that are accessed through the mobile device. In this paper, we proposed a framework for creating next generation mobile mashups that amalgamate data and offerings from three dimensions: Device features, Telecom network, and Web accessible services. Towards this, we described middleware components, both on the server side as well as the device side, to provide support for mashing device and Telecom features. Our framework allows portability across different device platforms and different Telecom protocols. We demonstrated the utility of our framework using three popular platforms - Nokia S60, Android and iPhone.

In the future, we would like to extend our framework to cover more platforms, Device features and Telecom services, as well as enhance the existing security and privacy considerations. Moreover, we wish to integrate our framework with several mashup and mobile development environments. Finally, we intend to conduct user studies that will help us gain valuable feedback from the developer community with respect to further refinements and extensions.

References

1. Adappa, S., Agarwal, V., Goyal, S., Kumaraguru, P., Mittal, S.: User Controllable Security and Privacy for Mobile Mashups. Technical Report RI10011, IBM Research (October 2010)
2. Agarwal, V., Goyal, S., Mittal, S., Mukherjea, S.: MobiVine: A Framework to Handle Fragmentation of Platform Interfaces for Mobile Applications. In: Proceedings of 10th International Middleware Conference, Illinois, USA (November 2009)
3. Agarwal, V., Goyal, S., Mittal, S., Mukherjea, S., Ponzo, J., Shah, F.: A Middleware Framework for Mashing Device and Telecom Features with the Web. Technical Report RI10009, IBM Research (July 2010)
4. BEA AquaLogic Family of Tools, http://www.bea.com/framework.jsp?CNT=index.htm&FP=/content/products/aqualogic/
5. Bond, G., Cheung, E., Fikouras, I., Levenshteyn, R.: Unified Telecom and Web Services Composition: Problem Definition and Future Directions. In: Proceedings of the 3rd International Conference on Principles, Systems and Applications of IP Telecommunications, Georgia (2009)
6. Brodt, A., Nicklas, D.: The TELAR Mobile Mashup Platform for Nokia Internet Tablets. In: Proceedings of 11th International Conference on Extending Database Technology (EDBT), Nantes, France (March 2008)
7. Hypponen, M.: Malware Goes Mobile. Scientific American (November 2006)
8. Rosenberg, J., Schulzrinne, H., et al.: SIP: Session Initiation Protocol (2002), http://www.rfc-editor.org/rfc/rfc3261.txt
9. JSR 289, http://jcp.org/en/jsr/detail?id=289
10. Kongdenfha, W., Benatallah, B., Vayssiere, J., Saint-Paul, R., Casati, F.: Rapid Development of Spreadsheet-based Web Mashups. In: Proceedings of 18th International World Wide Conference (WWW), Madrid, Spain (April 2009)
11. Mashups, L.: http://www-01.ibm.com/software/lotus/products/mashups/

12. Mittal, S., Chakraborty, D., Goyal, S., Mukherjea, S.: SewNet - A Framework for Creating Services Utilizing Telecom Functionality. In: Proceedings of 17th International World Wide Conference, Beijing, China (April 2008)
13. OMTP Bondi, http://bondi.omtp.org/
14. Open Service Access (OSA); Parlay-X Web Services; Part 1: Common. 3GPP TS 29.199-01
15. PhoneGap, http://phonegap.com/
16. Sadeh, N., Hong, J., Cranor, L., Fette, I., Kelley, P., Prabaker, M., Rao, J.: Understanding and Capturing People's Privacy Policies in a Mobile Social Networking Application. Journal of Personal and Ubiquitous Computing 13(6) (August 2009)
17. Web 21C SDK, http://web21c.bt.com/
18. Yahoo Pipes, http://pipes.yahoo.com/pipes/

Mirroring Smartphones for Good:
A Feasibility Study[*]

Bo Zhao[1], Zhi Xu[1], Caixia Chi[2], Sencun Zhu[1], and Guohong Cao[1]

[1] Department of Computer Science and Engineering,
The Pennsylvania State University
[2] Bell Laboratories,
Alcatel-Lucent Technologies
{bzhao,zux103,szhu,gcao}@cse.psu.edu, chic@alcatel-lucent.com

Abstract. More and more applications and functionalities have been introduced to smartphones, but smartphones have limited resources on computation and battery. To enhance the capacity of smartphones, an interesting idea is to use Cloud Computing and virtualization techniques to shift the workload from smartphones to a computational infrastructure. In this paper, we propose a new framework which keeps a mirror for each smartphone on a computing infrastructure in the telecom network. With mirror, we can greatly reduce the workload and virtually expand the resources of smartphones. We show the feasibility of deploying such a framework in telecom networks by protocol design, synchronization study and scalability test. To show the benefit, we introduce two applications where both the computational workload on smartphones and network traffic in telecom networks can be significantly reduced by our techniques.

1 Introduction

Smartphones have become more and more intelligent and powerful. Many complex applications, which used to be only on PCs, have been developed and running on smartphones. These applications expand the functionalities of smartphones and make it more convenient for users to get connected. However, they also greatly increase the workload on smartphones and introduce a lot of data transmissions between smartphones and telecom networks. The heavy workload and traffic affect both smartphone users and Telecommunication Service Provider (TSP). For users, heavy workload and traffic drain smartphone battery quickly. As we tested on Android Dev Phone 1, scanning a folder with 200MB files would take about 40 minutes and cause 10% battery; and uploading a 20MB file would cost more than 10% battery.

Recently some research [1] [2] [3] [4] has been done leveraging cloud techniques to help smartphones. An augmented execution model is provided to off-load application execution from smartphone to its clones in cloud [2]. In the in-cloud antivirus system [3], smartphones can send suspicious files to the antivirus service in cloud for scanning so as to avoid performing resource consuming scanning applications locally at phone.In addition, Zhang *et al.* [4] suggested to build elastic applications which can

---

[*] This work was supported in part by NSF CAREER 0643906.

be partitioned into function independent components. Those components can be executed in cloud instead of running on smartphones. Cloudlets [1] are designed for applications that require real-time interactive response, such as augmented applications. In their framework, the smartphone delivers a virtual machine (VM) overlay to the cloudlet infrastructure. Via this overlay, smartphones deliver part of execution to launch VMs, which provide various services and return results to smartphones. Smartphones are connected with cloudlets via high speed wireless network connections.

Contributions: We propose a new framework which keeps a mirror for each smartphone on a computing infrastructure in the telecom network. With the mirror we can shift some computational workload from a smartphone to its mirror. Since a mirror is synchronized with its corresponding smartphone, some operations, such as file sharing and virus scanning, can be performed on the mirror directly. In this way, we essentially reduce the workload and virtually expand the resource of a smartphone.

The design of our framework leverages the emerging Cloud Computing and virtualization techniques. On the smartphone side, a client side synchronization module is deployed to collect smartphone user input data and transmit them to the mirror server, a powerful application server. With Cloud Computing techniques, the mirror server is capable of hosting hundreds of mirrors and each mirror is implemented as a VM. To keep loose synchronization between mirror and smartphone, the mirror server replays inputs to smartphone on its mirror with exactly the same order (Section 3).

To illustrate the feasibility and compatibility of deploying the proposed framework in 3G networks, we present a network architecture based on UMTS, a 3G network architecture currently used by T-mobile (Section 3.4). We show that only minor modifications are needed on the current UMTS 3G network. Moreover, we use multimedia service (MMS) and web browsing service as examples to illustrate our protocol design. To show the benefits of our framework, we present two applications (Section 4). One is the data caching application, which saves both the power consumption of the smartphones and network traffic of 3G networks by caching smartphones' data in a mirror server. Another application is virus scanning in a smartphone.

We have built a prototype of the mirror server using Dell PowerEdge 1900 (Section 5). By measuring its workload with different number of mirrors, we show that the scalability of our framework is acceptable. We also tested the power consumption of running synchronization client side module on Android Dev Phone 1. According to the result, the performance overhead in the smartphone is very small.

Scope of This Work: Given the complexity of our whole system, it is infeasible to describe every detail in this paper. From the system point of view, issues like how to implement and deploy the client-side software and how to design mirror servers efficiently are of great importance, but they require independent study. We believe with the advances of technology in Cloud Computing and VM techniques, such issues could be solved in the near future. Instead, in this stage of our work, we focus on describing the design from the networking point of view and demonstrating the feasibility of this framework by analyzing and evaluating its scalability and its benefit/cost. Note that due to the hardware diversities of existing commodity smartphones, building mirrors supporting hardware-specific features, such as accelerometer sensor, could be very difficult and costly. In this paper, we focus on common features, such as touch screen, key pad, and 3G connection. In other words, currently we do not provide mirroring services to applications whose functionalities rely on sensor inputs.

2 Related Works

The idea of leveraging cloud computing techniques to enhance smartphones has been discussed in recent years. Our framework distinguishes itself from two aspects. First of all, in our architecture, the cloud is located in TSP. A smartphone is connected with its mirror via 3G telecom network. Secondly, our mirror is for more general purposes, although it can certainly be used for specific applications. Smartphones and their mirrors are kept synchronized via a loose synchronization mechanism.

Satyanarayanan *et al.* proposed a cloudlet infrastructure [1], where a smartphone is connected with a cloudlet via high speed wireless LAN. In the proposed infrastructure, the mobile device delivers a small VM overlay to the cloudlet infrastructure. Smartphones provide input to launch VMs in the cloudlet. Launch VMs provide services based on the input and return the result. The focus of this work is to provide real-time interactive responses between smartphones and the launch VM, which are essential to such applications as augmented reality applications. It tries to connect service providers in the cloudlet with smartphones while making this process invisible and seamless to user. Different from this work, first in our proposed architecture, smartphones are connected with cloud via Telecom networks. A server in our case has much wider coverage than a cloudlet. On the other hand, the data transmission rate of existing 3G networks cannot compete with that of wireless LANs. Secondly, different from launch VMs, mirrors are always synchronized. Files exist on a smartphone also exist on its mirror.

Oberheide *et al.* extended the CloudAV platform [5] to a mobile environment [3]. In the proposed model, smartphones send suspicious files to a remote in-cloud antivirus network service, which provides a scanning service. The purpose of [3] is to move the mobile antivirus functionality to an in-cloud antivirus network service, so as to save the resources on the smartphones. The author claims that it is worthwhile to get the antivirus scanning service by paying for the file transmission cost. Different from [3], first we propose a generic framework which is not limited to the antivirus service. Secondly, because a smartphone and its mirror are always synchronized, scanning can be performed on-the-fly and directly on the mirror, reducing the file transmission cost in [3].

Zhang *et al.* suggested an elastic application model [4]. In this model, one application is partitioned into components called weblets. By partitioning, some workload can be outsourced from smartphones to cloud elasticity services. In our framework, no application partitioning is required.

Chun and Maniatis proposed a new architecture [2] to augment the capabilities of smartphones by moving, in whole or in part, the execution of resource expensive applications to smartphone clones at an external computational infrastructure such as a nearby PC. Their work focus on how to move parts of execution of applications to the augmented clones that are more powerful than smartphones. To get the clone of a smartphone, they proposed to perform synchronization via whole-system replication, which is expensive because of the power consumption in synchronization. Different from [2], we discuss and evaluate the feasibility of implementing a mirror for a smartphone in a real telecom network. We focus on saving power on smartphones and minimizing the network traffic between smartphones and telecom network.

3 System Design

Our system provides an architecture for shifting computing from smartphones to their mirrors in the telecom network. Fig. 1 illustrates a high-level overview of our system. On the smartphone side, a client-side synchronization module, called *Syn-Client*, is deployed within the smartphone operating system (OS) to collect smartphone input data, including user keyboard inputs, and transmit them to the mirror server for synchronization. The Syn-Client module is designed according to the specification provided by TSP and manufacturer. On the mirror server side, the mirror server is a powerful application server maintaining a set of VMs. Each VM is a mirror to one smartphone. To keep mirrors and smartphones synchronized, the server side synchronization module, called *Syn-Server*, updates mirrors based on the data provided by *Syn-Clients* and network traffic between smartphones and IP network, which are collected by the *Traffic Monitor* module. Next we will focus on location of mirror server, mirror design and synchronization. The detailed designs and implementations of the other modules are out of the scope of this paper.

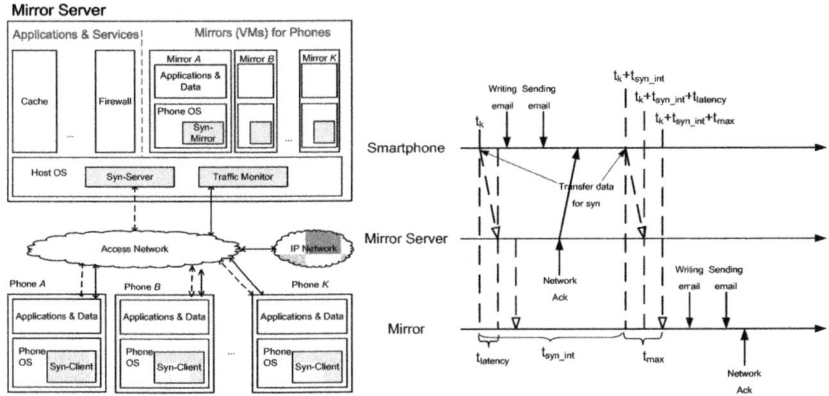

Fig. 1. The System Design   **Fig. 2.** Timelines for mirror synchronization

3.1 Location of Mirror Server

In the proposed architecture, the mirror server monitors all network traffic of smartphones. In our case, we consider traffic from both ISPs (e.g., Internet websites) and the TSP (e.g., AT&T). Bluetooth or WiFi connections are not considered in our discussion. We only consider the case when a smartphone is connected through a 3G network.

To deploy the mirror server, there are two options. One is to deploy it in ISPs, i.e. the server I shown in Fig. 3(a). The dotted line and dashed line represent the message flow

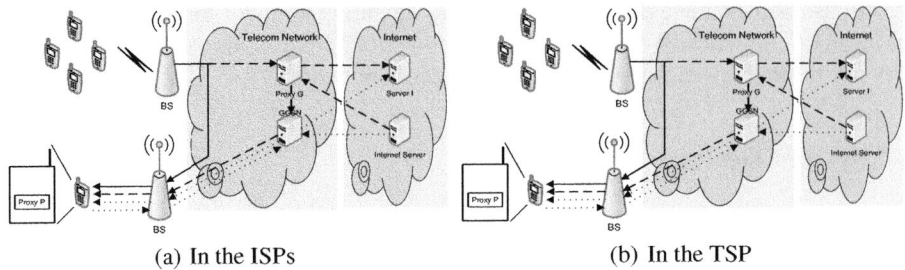

(a) In the ISPs (b) In the TSP

Fig. 3. Placing the mirror server

with proxy P and proxy G. And the other option is to deploy it in the TSP, i.e. the server T shown in Fig. 3(b). In this case, a copy of traffic from both TSP and ISP is forward to server T within telecom network. This forwarding requires modification to existing telecom network and we present details of modification in the next section.

In the systems proposed in [4], the Cloud assisting smartphones is deployed in ISPs and connected with smartphones via IP networks. However, in our architecture, if we want to deploy the mirror server in the ISPs, an additional proxy will be needed to collect incoming traffic of smartphones and forward it to the mirror server. As shown in Fig. 3(a), such a proxy can be deployed either besides GGSN in TSP or the smartphone, represented as proxy G and proxy P respectively. However, placing proxy in GGSN will also need to change the telecom network in order to forward messages, and then similar to placing mirror server in the TSP. Considering these concerns of placing mirror server in ISPs, we deploy the mirror server in the TSP.

3.2 Mirror Design

Compared with PC, creating mirrors for smartphones has many advantages. First of all, all smartphones of one model share the same hardware specifications, default factory settings, OS, and a lot of application software. Moreover, under the restriction of usage agreement with TSP, users are not allowed to modify the hardware or OS of their smartphones. Therefore, for one model, the mirror server only needs to keep one template of its factory default state.

On the other hand, we also notice that advanced smartphones have additional ways of system input, such as accelerometer sensors, which do not exist on PCs. These input methods introduce much difficulty to synchronization between a smartphone and its mirror. Currently, our mirror only supports general user interfaces, which are provided by most existing commodity smartphones, such as touch screen, keypad, and 3G network connection.

To build a new mirror for a smartphone, the mirror server creates a VM with exactly the same hardware/software specifications using the factory default template. If the smartphone is in its initial state, i.e. the first-time use, the mirror server can quickly update the mirror by supplying user information to the mirror. If the smartphone is not in its initial state, the mirror server copies software specifications and user personal data from the storage of the smartphone to the mirror. During the copying process, the state of the smartphone will be frozen and no operation from user is allowed.

3.3 Synchronization Mechanisms

To keep smartphone and its mirror synchronized through the telecom network, the mirror needs to be updated when the state of smartphone changes. The choice of synchronization mechanisms depends on applications which the mirror server supports.

Loose Synchronization. In this paper, we present one option for loose synchronization, which requires identical hardware specification, storage (e.g. SD card), OS, and installed application software. The loose synchronization is easier to achieve and is sufficient for applications like firewall, data caching applications, and virus scan applications. Indeed, because a smartphone rarely changes its hardware and OS, and user applications are not frequently changed either, in the first stage of our work, we mainly consider the issue of keeping storage consistency between a smartphone and its mirror.

An intuitive way for storage synchronization is data replication. That is, copying every new or updated data item to the mirror. This however is too expensive when the data activity is frequent in the phone. Instead, we propose a periodical synchronization mechanism by replaying smartphone inputs at its mirror periodically.

The first question for such a design is: why data storage is uniquely determined by inputs? We noticed that, the change of smartphone state is always triggered by user inputs (e.g. keyboard typing) or network inputs (e.g. arrival of data packets), and most applications are deterministic. That is, given a list of inputs, the application will generate deterministic outputs. In this case, if we apply the same list of inputs to a smartphone and its synchronized mirror, both will generate the same output and will be synchronized again after applying the inputs. Note that in our design, to save communication overhead, only the user inputs are forwarded to the mirror server. The mirror server caches the network inputs to a smartphone in the past synchronization period, so later it can directly feed in the cached data to the mirror. Also, the outputs from the mirror execution do not actually go out of the mirror server. We use the global clock of the telecom network to time stamping all kinds of inputs/outputs. For example, on the smartphone side, the user pushes a button to open an email client application, types an email, and sends the email by clicking an icon on the screen. On the mirror side, if we exactly replay the "pushing", "typing", and "clicking" actions in order, the mirror will send the same email and both will have the same archived email in the storage.

Based on this observation, we make an assumption that, by applying exactly the same inputs to a smartphone and its mirror in the same order, the storage of a phone and its mirror will be synchronized. This assumption may be untrue for certain applications, such as the application that generates a random number. However, it is suitable for most applications on current smartphones, such as email, IM, web browser, VoIP, MMS/SMS, and the present service. Therefore, we monitor all the inputs to the smartphone and later apply to the mirror in order.

The second question is: what are the pros and cons of periodical synchronization and the impact of the synchronization period on the system? Qualitatively speaking, the periodical synchronization improves the transfer efficiency and reduces much power consumption by sending the inputs in a batch. In normal cases, it does not cause problems to user applications, as long as the interaction between a mobile phone and the network need not involve mirror's cooperation. One counter example we can think of is related to our cache based file uploading application. In this application, when a user wants to share a file with a friend, to save power no file is actually sent out from the mobile; instead, the mirror will make the data transfer after it receives the send command through synchronization. Thus, no file transfer happens immediately.

Synchronization Timelines. Next we explain the detailed operations and timelines in our periodical synchronization mechanism. Let $t_{latency}$ be the network transfer latency between the mirror server and the smartphone, and t_{max} be the maximal network transfer latency. In 3G networks, t_{max} is a constant value, 100 ms [6]. Let t_{syn_int} be the synchronization interval and the last synchronization time point be t_k. Thus, the data input to the smartphone between t_k and $t_k + t_{syn\_int}$ are batched and transferred to its mirror at $t_k + t_{syn\_int}$. Due to the network transfer latency, it will arrive at the mirror server at $t_k + t_{syn\_int} + t_{latency}$. To make the replay interval even, the mirror server replays the network input data and keyboard input of the smartphone to the mirror at time $t_k + t_{syn\_int} + t_{max}$. As a result, the state of the mirror is $t_{syn\_int} + t_{max}$ later than that of the smartphone. Fig. 2 illustrates how this approach works with the email service by showing the time lines at the three parties: smartphone, mirror server and mirror. Note that in this figure the network acknowledgement is monitored and replayed by the mirror server to the mirror. Solid arrow lines stand for input data, and dash arrow lines stand for synchronization messages.

When a smartphone boots up, the initial time for the synchronization is set. At this time, on the smartphone side, the user can configure the value of t_{syn_int} for balancing the efficiency and reaction delay, while on the network side, the mirror server starts up this smartphone's mirror. When a smartphone is turned off, the mirror server stops this smartphone's mirror and saves its states.

3.4 Network Design

The mirror server stays in 3G networks as an application server, which supports such applications as security, storage and computation delegation services. It is easy to *configure* the nodes of 3G networks to route all incoming/outgoing messages (such as session control messages, data packets, and other 3G signaling protocols) of a smartphone to traverse the mirror server without any interaction with the smartphone. We do not need to modify any protocols or software in network nodes.

Taking UMTS as example, for a user who has subscribed to our mirror service, we will update his/her user profile in the Home Subscriber Server (HSS) node, as well as registering user policy and mirror service profile to the CSCF. HSS periodically updates the SGSN with the changes of user information; thus, SGSN becomes aware of the mirror service for this user. When SGSN receives a message from/to this user, it will either forward the message to CSCF, or forward it to the mirror server, subject to whether the message belongs to a service with session control.

Due to the limit of space, please refer to the technical report [7] for the details of deploying the mirror server in UMTS network. In this technical report, we present two sample message flows of typical telecom services to show that the mirror server works compatibly with UMTS network with only minor changes.

4 Applications

The proposed framework can support many applications. In this section, we identify two of them.

4.1 Data Caching Application

Recently, more and more people use their smartphones to share multimedia data with friends. Although data downloading/uploading speed between smartphone and BS has been significantly increased in recent years, receiving and sending bulk data still take time and consume a lot of power.

The current file sharing on smartphones has many weaknesses. For example, if the receiver runs out of battery or has very slow downloading speed in the current location (e.g. weak signal), the user may have to give up the file downloading. Also, if a user sends the same file to different people in a short time period, for each send, the smartphone has to do a separate file uploading, wasting lots of bandwidth and power.

Basing on the proposed framework, we present a data caching application. It is deployed on the mirror server which brings more flexibility to file downloading and significantly saves battery and time in file uploading.

Bulk Data Downloading. For bulk data downloading, the data caching service provides a temporary storage service. When the user wants to download a file from the Internet, the caching service will first download the file to its temporary storage. Once the downloading is complete, the caching service sends a "file-cached" notification to the user via the synchronization message. This file will be kept for a period of time depending on the TSP policy. When the user decides to download the cached file, he can download it from the mirror server directly. Moreover, according to the synchronization mechanism, once the smartphone starts to download a file from the mirror server, its corresponding mirror will also download the file from the mirror server. Therefore, no extra storage synchronization is needed.

For the file sender, the caching service provides an illusion that the file transfer has been completed. For the file receiver, he is able to choose a time which is suitable for the actual downloading. For example, when he moves to a place with better downloading speed or has the access to battery charging.

Bulk Data Uploading For bulk file uploading, the data caching service utilizes the mirror of the smartphone on the server. The mirror has the same storage as its corresponding smartphone and can replay all actions performed on the smartphone. Thus, if a user wants to send a bulk file to multiple people, he does not need to upload the file multiple times through the wireless link. Instead, the data caching service can send the file directly from the mirror to multiple people and save the wireless bandwidth.

To achieve this, the caching service registers the "upload" event in the synchronization modules. When the Syn-Client module intercepts the uploading action on the smartphone, instead of uploading file immediately, it replaces the default uploading action by a "fake" action without actual uploading. When the next synchronization is performed, the mirror sends the same data. In normal cases, the data sent from a mirror will be dropped by the mirror server. In the file uploading service, the data sent from the mirror will be delivered to the receiver pretending that it is sent by the smartphone. For both the file sender and the receiver, the caching application provides an illusion that the file is sent from the smartphone. Compared with uploading from smartphone, performing synchronization costs much less battery power, especially when the file size is large.

Note that the caching service can be used for all file uploading and downloading activities on the smartphone, such as IM client and web browser, etc. To reduce the delay caused by the synchronization, the user should have the flexibility to activate/deactivate the data caching service. For example, a user may want to send small files from the smartphone immediately, but use the caching service for files larger than 500KB.

4.2 Antivirus Scanning Service

Recent research [8] [9] shows there is an incoming number of mobile phone malware such as worm and virus in 3G networks. Virus scan is one of the most common functionalities provided by almost all antivirus software. In virus scan, the scanner reads files, compares them with some virus signatures, and returns a result of "virus found" or not to the user. Obviously, the scan process is CPU and I/O intensive. In case of smartphone, performing such scans adds inconvenience to users and drains the battery quickly.

In the proposed framework, an antivirus scanner can be deployed as a service on the mirror server, and the scanner can access the file system on the mirrors. Further, a hook application is installed on the phone. When a user launches the hook application on the smartphone, it sends a request with parameters to the mirror server. Instead of scanning the real files on the smartphone, the scanner service performs scanning on the synchronized mirror. The result will be sent back to the smartphone. As the smartphone and its mirror are synchronized, scanning on the mirror generates the same result as scanning on the smartphone. The feasibility of building an antivirus service outside of protected VM has been proved in the VMwatcher [10] which can be deployed directly on our framework.

Shifting antivirus scans are threefold. First, it saves battery power on smartphones. Second, as the CPU and I/O intensive workloads are shifted to the mirror server, the scanning will not affect the common usages of smartphones. Finally, the scan speed on mirror will be much faster than that on the phone due to its limited hardware capabilities.

5 Evaluation

In this section, we first describe our prototype. Then, we evaluate the synchronization cost and the scalability of our framework with this prototype.

5.1 The Prototype

Because the UMTS 3G network is very complicated with multiple protocols and many different nodes, currently only big telecom equipment vendors, such as Nokia,

Alcatel- Lucent and Ericsson have real prototypes. In order to evaluate the performance, scalability and feasibility of our framework, we setup a simplified prototype. In this prototype, the Android Dev Phone 1 is used as the smartphone, while Dell PowerEdge 1900 is used as the mirror server which has two Quad Core Intel Xeon 1.60GHz CPUs, 8GB Memory, 320GB Disk and installed Ubuntu 8.04 Desktop OS. The mirror server is connected to the IP network and can be accessed by the smartphone through T-mobile 3G network. In this prototype, for the synchronization feature, because we can not route or collect the incoming and outgoing messages of the smartphone in T-mobile 3G networks, we only replay the user input data on the mirror server so that the mirror takes the same tasks as the smartphone.

The Smartphone. We use Android Dev Phone 1 as our smartphone. It is a SIM-unlocked and hardware-unlocked device that is designed for advanced developers. We implement a syn-client module on it to monitor the user keyboard input. The user keyboard input can be got from system program "/dev/input/event2" in real time. Whenever the user enters a key, this program generates 32 bytes data which includes the key pressing time, key releasing time and key id. We develop a syn-client module to collect the input data from "/dev/input/event2" in real time and transfer it through 3G network to our mirror server periodically.

The Mirror Server. We use Android emulator as the mirror of Android Dev Phone 1. The size of the whole running Android emulator is 165MB. The Android OS is based on the simplified Linux kernel OS. Many virtualization techniques can be applied to our mirror server, such as OpenVZ [11] and QEMU. However, the current OpenVZ does not support Android OS. Android emulator basing on QEMU is simple and totally virtualization of the Android system.

3G Networks. To measure the uplink and downlink speed of T-mobile EDGE and 3G networks, we develop a test program, which sets up a socket connection between the smartphone and the server. Then it keeps downloading UDP data from the server to the smartphone or uploading data to the server. The result is measured in State College as shown in Table 1. The 3G bandwidth is much faster than the EDGE, but currently the EDGE has more coverage than 3G, especially in a suburban area.

To measure the average time for uploading or downloading files in EDGE networks, we transfer several different size of files. As shown in Fig. 4, to transfer same amount of data, the downloading takes less time than uploading. In addition, it takes several minutes to transfer 5 MB data which is around the normal video file size.

Table 1. The download and uploads speeds for EDGE and 3G

	Uploading (kbps)		Downloading (kbps)	
	Advertised	Tested	Advertised	Tested
T-mobile EDGE	75-135	140-160	75-135	170-190
T-mobile 3G	500-1200	708	700-1700	1259

5.2 Synchronization Cost

In the proposed framework, the Syn-Client module inside the smartphone will send the input information to the mirror server periodically. The battery cost of this periodical sending is determined by two factors. One is the frequency of inputs and the other is the time interval between two sending operations. More frequent inputs will generate more data to send. A shorter time interval means a more timely synchronization but will lead to more frequent sending operations.

To measure the power consumption with different input frequencies and sending time intervals, we have measured the power consumption of periodical sending in two cases. In the first case, the smartphone is very active and generates two events every second. In the other case, the smartphone only generates one event every five seconds. Each event is encoded as a 32-byte synchronization message, and the sending is through a UDP connection. In both cases, we have measured the total battery cost with different sending time intervals in 6 hours.

The battery consumption is shown in Fig. 5. As can be seen from the figure, with the same frequency of event generation, the shorter sending interval will bring more battery cost. Also, with the same frequency of sending operations, more input events cause more data to be sent thus cost more battery power. For example, in case the user generates two input events per second, if the synchronization interval is greater than 180 seconds, the total battery cost for sending synchronization information on the smartphone will be less than 4%.

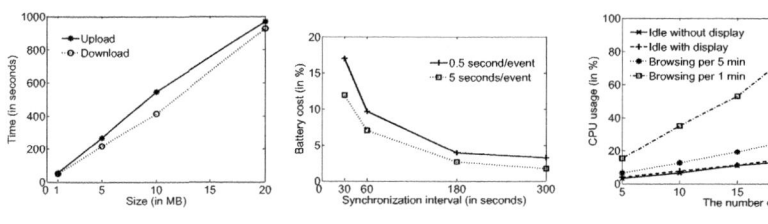

Fig. 4. The average time for uploading/downloading files in EDGE networks

Fig. 5. The Synchronization cost for 6 hours with different interval

Fig. 6. The scalability for running multiple emulators on a server

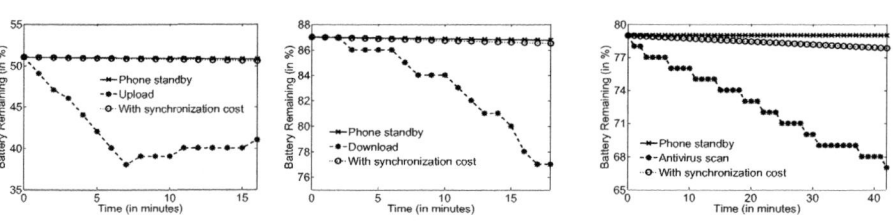

Fig. 7. The comparison of power consumption uploading

Fig. 8. The comparison of power consumption for downloading

Fig. 9. The comparison of power consumption

Table 2. The time and battery cost for performing antivirus scanning

Folder Size (MB)	SMobile on smartphone		Symantec on PC
	Time (sec)	Battery (%)	Time (sec)
10	120	2	12
50	600	3	31
100	1140	6	57
200	2340	10	110

To show the benefit in downloading/uploading service, we also compare the battery cost for synchronization and file downloading/uploading using ftp with 20MB file. When synchronizing periodically, we choose a case in which the user generates 0.5 event per second in average and the smartphone synchronization period is 60 seconds. Fig. 7 and Fig. 8 show the power consumption of a smartphone in standby or when it synchronizes periodically with the mirror server. When in standby, the average power consumption is around 0.09% per minute in average. In this case, the average power consumption is around 0.117% per minute.

To show the benefit in antivirus scan applications, we install the SMobile Virus-Guard [12] on an Android Dev Phone 1 and use it to scan folders with different sizes. All folders are filled with Android Package (APK) files which are used for application installation. As shown in Table 2, the time and battery needed for scan increases as the folder size increases. When the folder size is 200 MB, it takes more than 39 minutes and 10% battery to complete the scan, as shown in Fig. 9. During the scan, the user can explicitly feel the slow down of smartphone. Table 2 also shows much less time cost for scanning the same folders on PC using Symantec Endpoint Protection software.

5.3 Scalability

To evaluate the scalability of our framework, we study how many mirrors the server can support with different workload. In this experiment, we use web browsing service as an example to evaluate the scalability.

Fig. 6 illustrates the CPU usage versus different number of emulators, together with different workloads. In general, the mirror server can support at least 20 mirrors with heavy workload (browsing web every 1 minute) or more than 30 mirrors with light workload (open a webpage every 5 minutes). In this case, the CPU usage increases almost linearly with the number of mirrors.

6 Conclusion and Future Work

In this work, we propose a new framework, which uses Cloud Computing and VM techniques to shift the workload from smartphones to a computational infrastructure in telecom networks. It can greatly reduce the workload and virtually expand the resources of smartphones. We show the feasibility of deploying such a framework in telecom networks by protocol design, synchronization study and scalability test. A great number of issues are not addressed yet at this stage, especially those related to synchronization and actual system implementation. In our future work, we will study different synchronization mechanisms and find out what mobile applications can benefit from our framework. We will continue building the whole system and seek to evaluate its performance in a real 3G network. Finally, we will investigate the situation with additional incoming traffic through bluetooth and WiFi connections.

References

1. Satyanarayanan, M., Bahl, P., Caceres, R., Davies, N.: The case for vm-based cloudlets in mobile computing. IEEE Pervasive Computing (2009)
2. Chun, B.-G., Maniatis, P.: Augmented smart phone applications through clone cloud execution. In: Proc. HotOS XII (2009)
3. Oberheide, J., Veeraraghavan, K., Cooke, E., Flinn, J., Jahanian, F.: Virtualized in-cloud security services for mobile devices. In: Proc. MobiVirt (2008)
4. Zhang, X., Schiffman, J., Gibbs, S., Kunjithapatham, A., Jeong, S.: Securing elastic applications on mobile devices for cloud computing. In: Proc. CCSW (2009)
5. Oberheide, J., Cooke, E., Jahanian, F.: Cloudav: N-version antivirus in the network cloud. In: Proc. Security Symposium Conference, SS (2008)
6. 3GPP Specification TS 23.107: Quality of Service (QoS) concept and architecture, Std., Rev. Rel. 8, http://www.3gpp.org
7. Zhao, B., Xu, Z., Chi, C., Zhu, S., Cao, G.: Mirroring smartphones for good: A feasibility study. Tech. Rep. (2010), www.cse.psu.edu/~bzhao/mirror_report.pdf
8. Zhu, Z., Cao, G., Zhu, S., Ranjan, S., Nucci, A.: A social network based patching scheme for worm containment in cellular networks. In: Proc. IEEE INFOCOM (2009)
9. Zhao, B., Chi, C., Gao, W., Zhu, S., Cao, G.: A chain reaction dos attack on 3G networks: Analysis and defenses. In: Proc. IEEE INFOCOM (2009)
10. Jiang, X., Wang, X., Xu, D.: Stealthy malware detection through vmm-based "out-of-the-box" semantic view reconstruction. In: Proc. CCS (2007)
11. Openvz, en.wikipedia.org/wiki/OpenVZ
12. Smobile virusguard for google android, http://www.smobilesystems.com

μC-SemPS: Energy-Efficient Semantic Publish/Subscribe for Battery-Powered Systems

Davy Preuveneers and Yolande Berbers

Department of Computer Science, K.U. Leuven
Celestijnenlaan 200A, B-3001 Leuven, Belgium
{davy.preuveneers,yolande.berbers}@cs.kuleuven.be
http://www.cs.kuleuven.be

Abstract. In this paper, we present our lightweight semantic publish/subscribe system μC-SemPS that is targeted towards battery-powered micro-controllers and wireless sensor nodes in ubiquitous computing environments. The key challenge that we address is to minimize the overall energy consumption for subscription matching and delivery. Our system relies on an efficient representation of semantic subscriptions and favours computation over the more expensive wireless communication for the semantic matching and routing of events. When compared to more conventional pub/sub routing implementations, experimental results from network and energy simulations with MiXiM, an OMNeT++ modeling framework for wireless mobile networks, show that our approach for routing semantic events significantly reduces energy consumption.

Keywords: energy-awareness, publish/subscribe, semantic routing.

1 Introduction

The publish/subscribe (pub/sub) communication paradigm [4] provides a many-to-many anonymous event communication model that is decoupled in time and space, and as such an ideal distributed notification platform for ubiquitous computing systems. Traditionally, an event is a collection of attributes as key-value pairs, structured according to an a-priori known event schema, and communicated by means of push/pull interactions between possibly mobile publishers and subscribers. An event notification service acts as the mediator to notify subscribers about published events that match the constraints in at least one of their subscriptions. For scalability reasons, the event notification service is often implemented as an overlay network of event brokers [7] that match and route events/subscriptions to other brokers and clients.

The major challenge that we address in this paper is the energy-efficiency of semantic pub/sub systems for wireless and battery-powered embedded systems. This class of pub/sub systems [2,10,3,9] is gaining importance given the fact that (1) emerging computing paradigms like the Internet of Things will generate ever-increasing amounts of data, and that (2) participants in loosely-coupled heterogeneous systems will use a variety of terminologies. The fact that all events

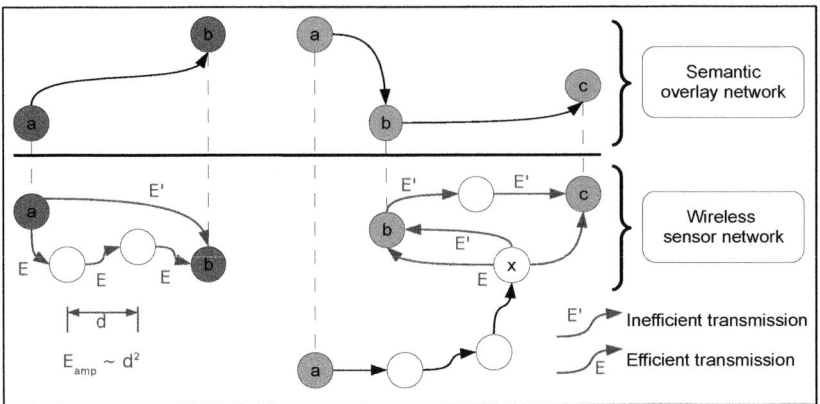

Fig. 1. Mapping of a semantic overlay structure onto a peer-to-peer network

must be unambiguously interpreted by the publishers and subscribers is the main motivation why semantics must become a first class entity of the pub/sub system.

In this paper, we investigate how a pub/sub system can be optimized for battery-powered wireless sensor nodes and embedded microcontrollers (μC), by matching and routing semantic events in such a way that reduces computation and communication energy usage. Merely implementing a semantic pub/sub system as a mobile overlay structure on top of a (power efficient) wireless peer-to-peer network can still be inefficient from an energy perspective. A direct unicast between two nodes may be more expensive than sending over intermediate nodes, because the energy consumption of the wireless signal amplifier is quadratically proportional to the distance. This is illustrated in Fig. 1. On the left, a message is transmitted with direct unicast from a to b requiring E' nJ/bit, or through $m - 1$ intermediate nodes requiring $m.E$ nJ/bit. For transmission over larger distances, the latter is often more efficient (depending on the number of intermediate nodes, the distance between them, and the energy dissipated in the transmitter/receiver/amplifier electronics). However, the decision is not clear cut. If the batteries of the intermediate nodes are almost depleted, a direct unicast can improve the network lifetime. On the right, a direct mapping of the routing in the semantic overlay network can lead to inefficient routing paths in the physical network. The event routing from b to c can be avoided by a multicast/broadcast at node x to reach nodes b and c.

In this paper, we present μC-SemPS, a semantic event pub/sub system for battery-powered micro-controllers in mobile and ubiquitous computing environments that aims to optimize the network lifetime. The main contributions are:

- An event subscription model for efficient semantic pub/sub matching
- A semantic-based clustering approach for efficient event routing
- A significant energy usage reduction vs. more conventional pub/sub schemes

The paper is structured as follows. After discussing related work in section 2, we present in section 3 our semantic subscription model. The semantic pub/sub middleware and algorithms are explained in section 4. We elaborate on the MiXiM-based experimental results in section 5. A concluding overview of our contributions and opportunities for further work are highlighted in section 6.

2 Related Work

Current solutions attempt to build semantic-awareness with a simple hierarchical topology of concepts (taxonomies of classes and properties). S-ToPSS [10] extends the conventional key-value pair-based systems with methods to process syntactically different, but semantically equivalent information. S-ToPSS uses an ontology with synonyms, a taxonomy and transformation rules to deal with syntactically disparate subscriptions and publications. The Ontology-based Publish/Subscribe system (OPS) [14] represents subscriptions as RDF graph patterns and uses a subgraph isomorphism algorithm for efficient matching. Follow-up research on semantic pub/sub systems of some of the authors resulted in the Semantic Publish/Subscribe System (SPS) [9], which leverages more of the expressive power of OWL Lite but relies on a centralized broker with a powerful server to do the semantic matching. Chand et al. present containment-based and similarity-based proximity metrics [2] for peer-to-peer overlays of pub/sub networks to tackle the lack of expressiveness in many subscription languages.

Many of the above semantic pub/sub systems go beyond the capabilities of embedded devices. The DSWare [8] middleware offers publish/subscribe services to distributed wireless sensor networks and supports the specification and detection of patterns of compound events. Resource-aware publish/subscribe in wireless sensor networks is studied in [13]. The authors propose a protocol that aims to extend the network's lifetime by offering tradeoffs between fixed event dissemination paths that increase communication efficiency, and resource-awareness that provides freedom for event routing. Network hops are used as a measure for the delivery cost. With energy efficiency being the main concern, hops are not a good measure for quantifying energy usage. Energy consumption of a wireless amplifier increases quadratically with the hop distance, so that transmitting over smaller distances (i.e. with more hops) could be more optimal. In [6], Demeure et al. present an energy-aware middleware for MANETs of handheld devices which includes a publish/subscribe event system. The middleware is made energy-aware by providing policies that specify adaptations in the middleware for a variety of battery thresholds. These adaptations may trigger the use of more energy-friendly encryption algorithms or change communication by using a non-acknowledged protocol to reduce network activity. The authors do not immediately focus on optimizing the event matching and routing. For an overview of energy-aware routing in wireless sensor networks, we refer to the survey of Akkaya et al. [1].

3 Semantic Event Subscription Model

In this section, we will briefly discuss the basic concepts of our event and subscription models as used in our μC-SemPS system.

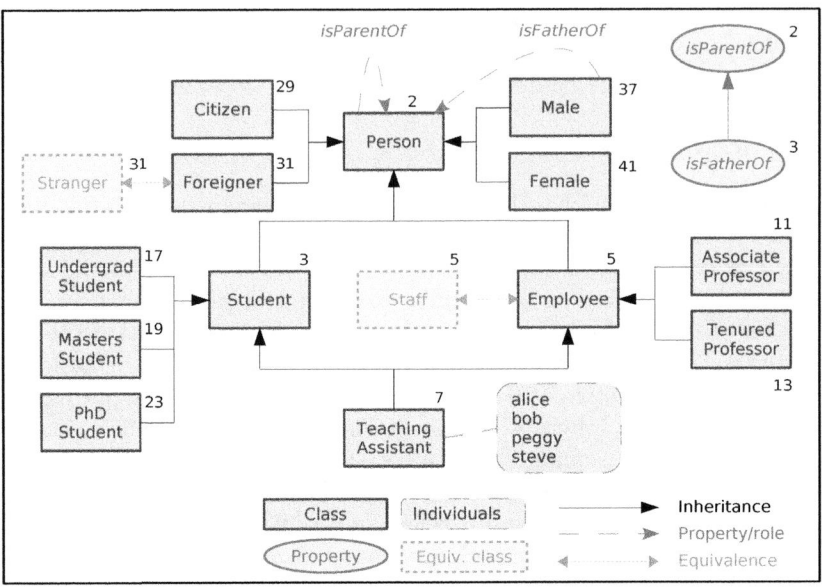

Fig. 2. Inheritance of ontology classes and properties with prime number assignment

3.1 Concept Model

Concept-based semantic publish/subscribe systems describe events and subscriptions at a higher level of abstraction than simple key-value pairs. The semantic relationships are defined in ontologies for an unambiguous interpretation of the event structure. In Fig. 2, we illustrate this with a familiar example often used for educational purposes. The concept model to define the semantic meaning of a subscription or event consists of the following constructs:

1. **Inheritance hierarchy of named classes:** A class (or concept) can have multiple parent classes and subclasses. Any individual (entity or instance of a class) of a class is also a member of its ancestor classes:

$$Student \subseteq Person$$

2. **Inheritance hierarchy of properties:** An (object or datatype) property defines a role of a class by pointing to another class. A similar inheritance relationship can also be defined for properties (or roles):

$$isFatherOf \subseteq isParentOf$$

3. **Equivalence of classes and properties:** Classes and properties are defined semantically equivalent if they represent the same set of individuals. For example, the class *Car* could be a synonym for the class *Automobile*:

$$Car \equiv Automobile$$

Synonyms and the two types of inheritance relationships are typically found in most semantic pub/sub systems. The acyclic, reflexive and transitive inheritance relationship is in the literature sometimes referred to as *containment*, *type inclusion* or *subsumption*.

3.2 Event Model and Subscription Language

The event model specifies the organization of data inside the events and provides an expressive subscription language for subscribers to define their interests in certain events. Rather than using a directed labeled graph representation for the semantic relationships of an event of interest, we use the user-friendly compact Manchester syntax for OWL [5] to express semantic subscriptions. For example,

$$\text{Student or Person that hasAge all xsd:integer}[<= 25]$$

would represent with the usual precedence rules an interest in students and persons that are 25 years or younger. Rather than relying on subgraph isomorphisms, we make use of an efficient encoding of semantic relationships based on the properties of prime numbers, as explained in the following subsection.

3.3 Semantic Event Matching

The encoding that we use in our system counters one of the major disadvantages of semantic matching with ontologies: the reasoning engines supporting the semantic modeling with ontologies are not designed with the limited computational resources of an embedded system in mind. Our prime number-based encoding algorithms were developed with the aim to reduce the memory and processing complexity. See our matching algorithms in [12,11] for more details.

The encoding of a multiple inheritance hierarchy $\chi = \{Person, Student, Employee, ...\}$ is based on the prime numbers assigned to each class as illustrated in Fig. 2. For example, class *Teaching Assistant* inherits the prime numbers 2, 3 and 5 from its ancestors and is assigned 7 as its own prime number. The encoding of this class becomes $\gamma(Teaching\ Assistant) = 2 \times 3 \times 5 \times 7 = 210$. As class *Person* has no ancestors, its encoding $\gamma(Person)$ is solely based on its own prime number. Testing for a semantic match can be done with a simple division. For example, *Undergrad Student* is a semantic match for a *Student* because $\gamma(Undergrad\ Student) = 102$ can be divided by the prime number of *Student* = 3. In several cases, the division is not even necessary as many heuristics can be used to rule out subsumption [12,11]. The degree of compaction solely depends on the prime number assigned to each class. For the very simple hierarchy in Fig. 2, the original OWL ontology file is more than 3000 bytes long, whereas our prime number encoding generates an equivalent representation of less than 200 bytes. The advantages of our semantic encoding are:

1. A class (or property) can be identified by a simple prime number rather than a possibly long string that represents its URI.

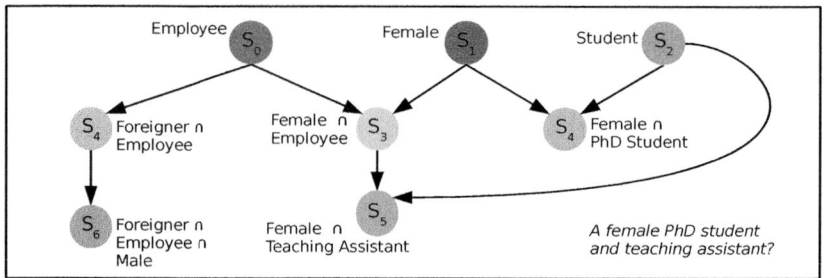

Fig. 3. Organizing nodes according to the specificity of their subscriptions

2. All the relevant semantic relationships are embedded in the encoding of each class and property.
3. Partial imports of an ontology are possible because you only need the class and property encodings of the terminology you are interested in.

These characteristics have an immediate effect on the message length of events that have to be routed, as textual descriptions are being replaced with a simple prime number. Message size reductions of a factor of 50 or more were not uncommon in our experiments.

4 Publish/Subscribe for Low-Power μC Systems

Given that the most energy savings can be gained from optimizing the wireless communication, our μC-SemPS pub/sub subsystem for microcontrollers will favor spending more effort in the semantic event matching if it can further reduce the energy impact of semantic event routing.

4.1 Semantic Event Subscription Polyhierarchy

The event notification routing protocol aims to reduce the delivery of events to nodes without a matching subscription, as well as the amount of undelivered events for nodes with a matching subscription (i.e. maximize precision and recall). Note that reducing energy consumption does not necessarily mean less messages between peers (as illustrated in Fig. 1). The challenge is to organize the peers according to their interests in events (i.e. their subscriptions) to improve the event routing efficiency (both in terms of event delivery accuracy and energy usage for communication).

Our event routing protocol will first establish semantic routing tables by organizing the peers into a polyhierarchy according to the specificity of their subscriptions. Nodes that have an a subscription for a very specialized concept will be at the bottom of the tree, whereas nodes interested in very general concepts are at the top of the tree. If a node has more than one subscription, it appears multiple times in the tree. See Fig. 3 for an example of a subscription polyhierarchy of named classes and anonymous classes (intersections of named classes).

Algorithm 1. BuildSubscriptionHierarchy(**in:** fromPeer, subscription)

1: (subscriptionRelevant, subscriptionForwarded) = (false, false)
2: **if** (*AlreadyReceived*(subscription)) **then**
3:   *FeedbackSubscription*(fromPeer, DUPLICATE, subscription)
4: **else**
5:   *MarkReceived*(fromPeer, subscription)
6: **if** (*IsMatchingAncestor*(subscription)) **then**
7:   subscriptionRelevant = true
8: **else**
9:   *LabelSubscription*(subscription, OUT_OF_INTEREST)
10: **if** (subscription.hopsLeft > 0) **then**
11:   subscription.hopsLeft = subscription.hopsLeft - 1
12:   **for each** Peer p **in** *ForwardFilter*(adjacentPeers, subscription) **do**
13:     subscriptionForwarded = true
14:     *ForwardSubscriptions*(p, *MatchingSubscriptions*(subscription))
15: **if** (**not** subscriptionForwarded) **then**
16:   **if** (**not** subscriptionRelevant) **then**
17:     *FeedbackSubscription*(fromPeer, OUT_OF_INTEREST, subscription)

The algorithm for constructing this polyhierarchy (and semantic routing tables) is partially shown in Algorithm 1. For every subscription a node receives, it basically checks if it already received the subscription via other routes and whether it has a matching local subscription (equivalent or subclass). If so, it forwards its own matching subscriptions to its adjacent peers for which it did not receive a matching out of interest message before. The purpose of the algorithm is to ensure that any event matching the subscription of a node always matches those of its ancestors, and that an out of interest event will not match the subscriptions of the node's descendants.

4.2 Energy Based Clustering and Event Routing

In principle, routing events from the root nodes towards the leaf nodes (until they stop matching the subscriptions) is a way to minimize sending events to nodes that do not match their subscriptions. However, there are a few drawbacks with strictly following this routing scheme:

- The root nodes with broader interests will have a lot more events to forward than the nodes at greater depths in the polyhierarchy.
- Strictly following the routing path may not be efficient from an energy point of view depending on the distance between the nodes involved.

Cluster-based data routing [15] is a way to improve energy consumption. To improve traffic confinement we cluster subscribers with similar interests in such a way that a subscriber that receives an event can broadcast it to all the nodes of the cluster it is part of. Our clustering algorithm uses a percentage P of the nodes (here 5%) to become cluster heads:

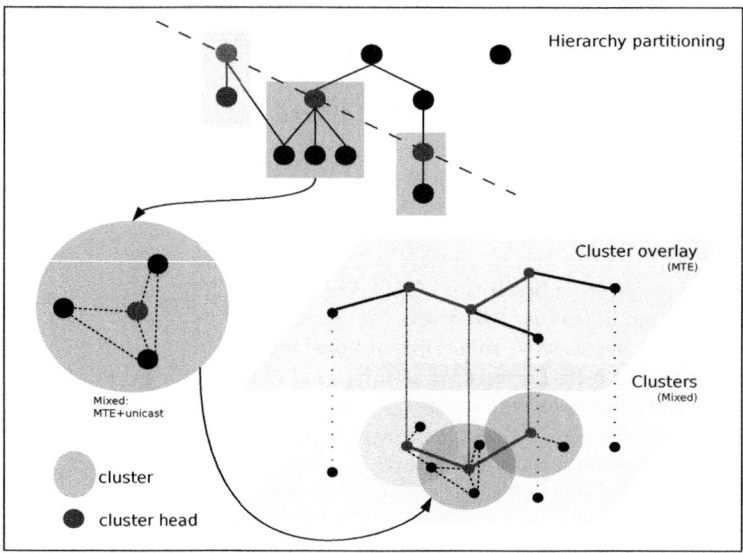

Fig. 4. Grouping into clusters of nodes with similar subscriptions

1. Cluster heads are self-selected at the beginning of a cycle. In the r-th cycle, a node that has not become a cluster head during the previous $1/P$ cycles decides to become a cluster head with probability $P/(1 - P(r \bmod 1/P))$.
2. Other nodes join all clusters for which the cluster head is an ancestor in the subscription polyhierarchy.

The first step rotates cluster heads in such a way that current cluster heads have a lower probability to become cluster heads in the next cycle. The second step groups nodes with similar interests. It is possible that a node does not find a matching cluster, especially if these nodes have subscribed to very general concepts (located near the top in the subsumption polyhierarchy). See Fig. 4 for an illustration of this grouping into semantically similar clusters. So rather than having to route events in the network according to subscription polyhierarchy of Fig. 3, we can now route on two different levels:

- **Clusters**: within a cluster, we use a mixture of unicasting and minimum transfer energy to maximize the lifetime of the cluster.
- **Cluster overlay**: between cluster heads and nodes without a cluster, we route events with minimum transfer energy (MTE) strategy, because it increases the lifetime better than unicasting and event matches are more likely for nodes with very general subscriptions anyway.

This approach basically breaks down into a partitioning of the subscription hierarchy into two parts: (1) the subhierarchies imposed by the cluster heads and their descendants that will form the clusters, and (2) the hierarchy imposed by the ancestor classes of the cluster heads and the nodes without a cluster. We use

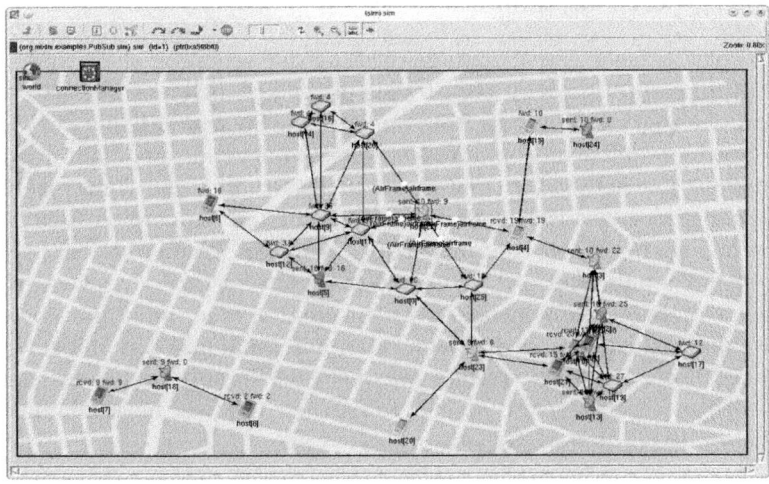

Fig. 5. Experimental results with the MiXiM wireless network modeling framework

two different routing strategies in the two partitions, and by reselecting the cluster heads we balance the energy load amongst all nodes. The efficient encoding and semantic matching makes the overhead of managing the clusters low. This is important for mobile nodes that connect to a variety of nodes and for nodes that frequently change their subscriptions. However, theoretical situations exist where the majority of communication is spent on regrouping messages.

5 Experimental Performance Evaluation

In this section, we will analyze both energy usage as well as the publish/subscribe effectiveness of the event delivery (in terms of event routing through non-subscriber nodes). To test the effectiveness of the semantic routing part of our μC-SemPS pub/sub system on larger setups (500 nodes), we conducted simulations with MiXiM[1], an OMNeT++ modeling framework (see Fig. 5) created for mobile wireless and ad-hoc sensor networks with built-in energy framework. We simulated a ZigBee wireless network using energy characteristics of a IEEE 802.15.4 low-power digital radio.

5.1 Energy Cost of Semantic Matching

Regarding the computational complexity, we estimated based on [12] that verifying for a semantic match between two concepts in a very large ontology with more than 25000 concepts requires on average about 50 instructions on an ARMv4 CPU, which at 2 nJ/instruction means something like 100 nJ.

---

[1] http://mixim.sourceforge.net/

5.2 Energy Cost of Semantic Routing

In Fig. 6, you can see the advantage of using our semantic clustering approach with cluster head rotation on the lifetime of the network. For this particular experiment, the routing protocols from our μC-SemPS semantic publish/subscribe system are more efficient compared to direct unicasting and minimum energy transmission. The energy for computation (the semantic matching) is quite negligible compared to the energy dissipated due to communication (less than 10% for some nodes). For nodes that were not transmitting themselves, either because they were not a publisher or because they did not forward any event, there was still some communication energy spent for creating and maintaing the clusters, but the significant difference between communication and computation was less outspoken. Note that these results are very specific for the experiment. If the average distance between the nodes grows, then this will have an effect on the total energy dissipation for all protocols.

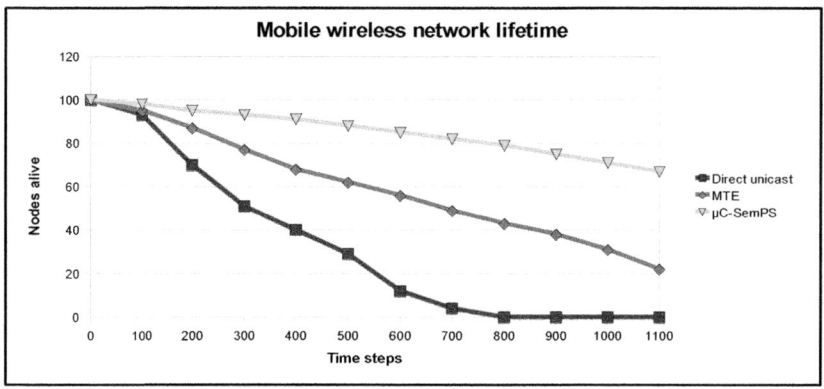

Fig. 6. Measuring the network lifetime

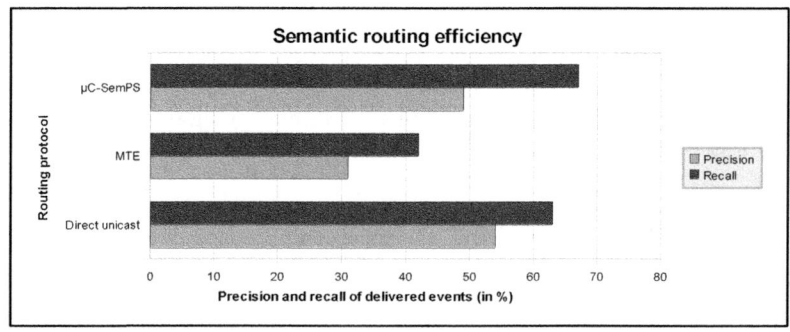

Fig. 7. Measuring the delivery of out-of-interest and missed events

5.3 Pub/Sub Efficiency of Event Routing

We compared in another experiment in which we measured how many out-of-interest events were delivered (counting the events received by a node that did not have a matching subscription) and missed events (events that should have been delivered but were not). These metrics boil down to precision and recall from an information retrieval point of view. See Fig. 7 for the results. Due the fact that MTE uses more intermediate nodes, it is clear that its precision is smaller. The max hop count for delivery also had an effect on its lower recall. The direct unicast performed better, but our μC-SemPS system performed even better due to the semantic clustering approach.

6 Conclusions

As the publish/subscribe community expands towards new computing paradigms like the Internet of Things, it is paramount that semantic awareness and energy consumption is taken into account in the matching and routing of events. While many energy-efficient routing protocols have been proposed for wireless sensor networks, they only focus on optimal delivery paths and do not consider optimizing the routing paths for precision and recall of the pub/sub system.

This paper addresses energy-efficient semantic matching and routing for low-power wireless and mobile devices. We presented our μC-SemPS semantic publish/subscribe for battery-powered microcontrollers and sensor nodes. Rather than relying on a centralized broker for ontology-based semantic matching, we use an encoding scheme that represents semantic relationships in an efficient compact representation that helps reduce the packet size of the network messages. We have demonstrated our clustering technique that takes both semantic similarity as well as energy awareness into account for the efficient delivery of events to subscribers. We have shown that our approach is significantly better in terms of energy consumption and network lifetime while maintaining a high precision and recall of delivered events.

A key direction of our future work, will be to investigate the effects of node mobility and churn to maintain the subscription polyhierarchy, and to create a larger testbed with many more energy-efficient wireless routing protocols. While none of them focus on semantic publish/subscribe systems, it will be interesting to see how well they fare in terms of precision and recall. This could provide interesting insights into how energy efficiency can be traded for more accurate routing to increase precision and recall for the semantic events.

Acknowledgments. This research is partially funded by the Interuniversity Attraction Poles Programme Belgian State, Belgian Science Policy, and by the Research Fund K.U. Leuven.

References

1. Akkaya, K., Younis, M.F.: A survey on routing protocols for wireless sensor networks. Ad Hoc Networks 3(3), 325–349 (2005)

2. Chand, R., Felber, P.A.: Semantic peer-to-peer overlays for publish/subscribe networks. In: Cunha, J.C., Medeiros, P.D. (eds.) Euro-Par 2005. LNCS, vol. 3648, pp. 1194–1204. Springer, Heidelberg (2005)
3. Cilia, M., Bornhövd, C., Buchmann, A.P.: Cream: An infrastructure for distributed, heterogeneous event-based applications. In: Chung, S., Schmidt, D.C. (eds.) CoopIS 2003, DOA 2003, and ODBASE 2003. LNCS, vol. 2888, pp. 482–502. Springer, Heidelberg (2003)
4. Eugster, F., Guerraoui, K.: The many faces of publish/subscribe. ACM Comput. Surv. 35(2), 114–131 (2003)
5. Horridge, M., Patel-Schneider, P.F.: Manchester syntax for owl 1.1. In: International Workshop OWL: Experiences and Directions, OWLED 2008 (2008)
6. Demeure, I., et al.: An energy-aware middleware for collaboration on small scale MANets. In: Proceedings of the Autonomous and Spontaneous Networks Symposium (ASNS 2008), Paris, France (2008)
7. Jafarpour, Hore, Mehrotra, Venkatasubramanian: Ccd: efficient customized content dissemination in distributed publish/subscribe. In: Middleware 2009: Proceedings of the 10th ACM/IFIP/USENIX International Conference on Middleware, pp. 1–20. Springer-Verlag New York, Inc., New York (2009)
8. Li, S., Lin, Y., Son, S.H., Stankovic, J.A., Wei, Y.: Event detection services using data service middleware in distributed sensor networks (2003)
9. Ma, J., Xu, G., Wang, J., Huang, T.: A semantic publish/subscribe system for selective dissemination of the rss documents. In: GCC 2006: Proceedings of the Fifth International Conference on Grid and Cooperative Computing, pp. 432–439. IEEE Computer Society, Washington, DC (2006)
10. Petrovic, M., Burcea, I., Jacobsen, H.A.: S-topss: semantic toronto publish/subscribe system. In: VLDB 2003: Proceedings of the 29th International Conference on Very Large Data Bases, pp. 1101–1104. VLDB Endowment (2003)
11. Preuveneers, D., Berbers, Y.: Prime numbers considered useful: Ontology encoding for efficient subsumption testing. Technical Report CW464, Department of Computer Science, Katholieke Universiteit Leuven, Belgium (October 2006)
12. Preuveneers, D., Berbers, Y.: Encoding semantic awareness in resource-constrained devices. IEEE Intelligent Systems 23(2), 26–33 (2008)
13. Taherian, S., Bacon, J.: A Publish/Subscribe Protocol for Resource-Awareness in Wireless Sensor Networks. In: Proceedings of the International Workshop on Localized Algorithms and Protocols for Wireless Sensor Networks. LNCS, vol. 4549, pp. 27–38. Springer, Heidelberg (2007)
14. Wang, Jin, Li: An ontology-based publish/subscribe system. In: Middleware 2004: Proceedings of the 5th ACM/IFIP/USENIX international conference on Middleware, pp. 232–253. Springer, Heidelberg (2004)
15. Wang, H.-L., Chao, Y.-Y.: A cluster-based data routing for wireless sensor networks. In: Hua, A., Chang, S.-L. (eds.) ICA3PP 2009. LNCS, vol. 5574, pp. 129–136. Springer, Heidelberg (2009)

Collaborative Algorithm with a Green Touch

Luciana Oliveira, Djamel Hadj Sadok, Glauco Gonçalves,
Renato Abreu, and Judith Kelner

Federal University of Pernambuco (UFPE), Recife, Brazil
{lpo,jamel,geg,rra3,jk}@cin.ufpe.br

Abstract. Discovery and announcement processes are regular tasks in ubiquitous scenarios, since they are important to support devices' self-adaptation. However, devices enable a multitude of protocols by default, including many not always required by users or networks. For example, the NetBIOS protocol is rarely used in some networks, but it is continuously executed by several Desktops. Consequently, today's networks waste energy processing protocol messages that should be turned off. In this paper, we analyze such energy cost, and we propose devices' collaboration and sharing of traffic knowledge to save energy. Firstly, we suggest that devices should view protocols in the same way of users through a social analogy. Skype, MSN, ARP and other protocols are seen as societies with specific languages, which devices can learn and teach, similarly to societies. Finally, simulations showed that our proposal is able to reduce energy consumption for all network's devices.

Keywords: Energy efficiency and awareness, Architectures, systems and applications.

1 Introduction

The ubiquitous scenario brought new opportunities with regard to sharing services between users, providers and autonomous devices. In such environment, providers and users frequently use several protocols to announce their services, but such behavior have been increased the energy consumption of users' devices and providers' network devices. For instance, researchers from Bell Labs found that current networks use 10,000 times more energy than the absolute minimum required to transport data bits across them. For this reason, researchers have been studying solutions to save energy and how efficiently to use energy resources.

According to [1], approaches to reduce consumption span several levels from applications and operating systems to hardware. For example, work in [4] introduces changes to the Bitorrent application to reduce its energy consumption and to encourage P2P and grid networks to save energy in data centers. At the hardware level, [18] introduced optical technologies to save energy. It approached the optical switching to avoid the use of electronic forwarding and switching of IP routers that have high consumption of energy. The research in [5] and [6] are examples of efforts for finding hardware solutions to build devices that consume renewable energy.

Authors in [1] and [19] also point to the network infrastructure as the key issue to control energy consumption. The proposal in [19] identified the consumption of conventional devices in a 3G infrastructure and proposed changing the conventional devices to optical in the access network architecture. Work in [7] presents several factors that should be analyzed to reduce the energy consumption such as collisions at the MAC layer, amount of switching between sending and receiving data, communication modes, computation of transmission schedule and others.

Future networks must consider the use of energy-aware devices, applications and protocols. Such networks, one may call, as green networks [8]. Green networks may have membership policies whereby, only devices that turn off some network protocols, or those that implemented certain operation modes for energy saving, etc., would be able to join the network.

In this work, a communication device is seen as a mean for a user to join in some groups or services which we refer to as societies. Societies are distinguished by their inherent set of protocols and languages (messages). As a result of the above, a green network could define and enforce, within its policies, the sets of societies (protocols) that devices must participate (turn on) or must reject (turn off) in order to joining the network.

As these policies differ between networks, it is important that devices implement a mechanism to recognize societies' information in order to change their behavior to suit the rules of a given network before joining it. To achieve this goal, this work develops a flexible and sharable mechanism for network elements to learn and teach each other knowledge about traffic and consequently allows devices to self-configure while minimizing their energy consumption.

Formally, our approach slices the communication in societies modeled to be recognized by automata. An automaton sharing algorithm is developed for devices to learn and teach others about the identification of societies, providing a flexible and dynamic solution for acquiring knowledge by observing traffic. Using wireless trace data, we show how sharing of automata allows devices to build dynamically a green network and reduce the energy consumption.

The remainder of this paper is organized as follows. Next section introduces the relationship between the standard concepts of societies and network protocols. Section 3 introduces the proposal for a collaborative algorithm to automatically share automata between network devices in order to recognize societies and provide the functionality of management for users and networks. Section 4 presents an evaluation of these algorithms in a scenario of green networks. Finally, a discussion about related works, conclusion of this paper and future works are presented respectively in Sections 5 and 6.

2 Why and How Should We Recognize Societies?

Our work in [2] showed that relationships between machine and humans are towards socially inspired structures, embracing new networking contexts with new business models, services and technologies. We also showed that overlay networks are an interesting model to exemplify societies constituted of elements. Here, we address protocols as societies and protocol messages as the vocabulary for such societies. Thus, we see the vocabulary as a syntactic pattern to extract the society knowledge, i.e., the set of know addresses and messages within a given interval of time.

We suggest that such co-existence among vocabulary, messages and addresses can describe higher level information such as business relationships (representing customers or services of some enterprise) as well as low level or less abstract information such as network protocols (ARP, LLC, IP, NetBIOS and others). Although our proposal is towards modeling syntactically each society with its own vocabulary (pattern to express addresses and messages), this networking case study maps societies to the protocols or low level information that make them up.

We assumed that each society has a specific vocabulary or a syntactic pattern consisting of address information and message symbols. Then automata theory is a promising mathematical model to express a machine to recognize such vocabulary constituted by a set of words that identify a society. Such model is extensively used by compilers and linguistics research to describe the precise and correct vocabulary (formal grammar) of any programming language.

Conceptually, an automaton changes its current state according to read symbols. When the symbols of message are completely consumed the last state indicates if the input were accepted as part of the language. Formally, an automaton s_i is defined by $s_i = \{Q, \sum, \delta, q0, F\}$, where Q is a set of states. $\sum$ is a finite set of symbols a.k.a. the alphabet of the language whereas we will refer to as the vocabulary of a society. δ is a function such that $\delta: Q \times \sum \rightarrow Q$. q0 is the initial state, that is, the state in which the automaton is when no input has been processed yet, where $q0 \in Q$. Moreover, an automaton can have one or more states represented by F that is a set of states of Q called accept states. Each transition function of δ receives an input symbol and goes to some state or stays in the same state. For example, δ_a or $\delta(q,a)$ represents a transition in state q to q', where $\delta_a:Q \rightarrow Q$ and $q' \in Q$. Hence, a pair of symbols can be a unique function $\delta':Q \rightarrow Q$, for example, $\delta_{ab}:Q \rightarrow Q$, or denotes a composition function processed sequentially δ_a and after δ_b, or defined by $\delta_a \circ \delta_b:Q \rightarrow Q$.

Consequently, a society is completely known when the automaton s_i identifies all possible exchanged messages among its elements, i.e., when all combination of vocabulary $\sum^*$ are known. Hence, a society is identified by an automaton $s_i = \{Q, \sum, \delta, q0, F\}$, when $s_i = \{w \in \sum^* \mid \delta'(q0,w) \in F\}$. The messages identified by s_i of a given device constitute its knowledge $\mathbf{k_i}$.

Further the most intrinsic social mechanism, - knowledge sharing by learning and teaching - needs to be executed by networking devices. We suggest that sharing automata, users, service providers and any network may control the traffic.

Before giving details of our proposed algorithm for sharing automata, let us emphasize that the specific creation of automata is outside the scope of this work. We assumed that such task could be manually performed by any individual using any mechanism such as L7[1] or tcpdump/libpcap(T/L)[2] filters support to identify a society. Our focus was on the design of a generic algorithm that saves energy due to automata share. L7-filter deal regular expression that can be derived to translate it into finite automaton and T/L has a specific expression to identify societies.

These tools are able to receive an expression and translate it into a finite automaton. L7-filter is able to translate a file containing a regular expression. For example, Figure 1(a) shows an automaton used to identify the NetBIOS society.

---

[1] http://l7-filter.sourceforge.net/
[2] http://www.tcpdump.org/

Please note that such filter is limited to identifying societies only in the application layer. Differently, the T/L can identify any society in any layer of the OSI architecture. For comparison, Figure 1(b) shows how to do an automaton to recognize NetBIOS using T/L filter.

\x81.?.?.[A-P][A-P][A-P][A-P][A-P][A-P][A-P][A-P][A-P][A-P][A-P][A-P][A-P][A-P][A-P][A-P][A-P][A-P][A-P][A-P][A-P][A-P][A-P][A-P][A-P][A-P][A-P][A-P][A-P][A-P][A-P][A-P]	eth.dst and eth.src and eth.type == 0x0800 and ip.dst and ip.src and udp.dstport == 137 and nbns.id
(a)	(b)

Fig. 1. Recognizing NetBIOS by different filters: (a) L7-filter and (b) T/L filter

Due to the use of different mechanisms to identify societies, we generalized our solution to associate the regular expression of L7-filter, or the tcpdump filters or any other new identification mechanism for protocol/societies. This includes the representation of applications that change frequently theirs patterns and require often automata update. For example, we identified that, if a Skype user executes a login, a logout immediately after and it executes again the login after at least 15 minutes, the user machine executes the following steps:

1 Trigger NAT-PMP [20] to define the UDP port used by such protocol;
2 Send several messages using the same port configured by NAT-PMP. We identified that some UDP messages and NAT-PMP was inside ICMP messages;
3 Send DNS messages to find the IP that corresponds to the name containing skype.com and 4publishers.com;
4 Send TCP messages using IPs from Autonomous Systems (AS) associated to messages described in items 1, 2 and 3.

Using such procedure, we recognized more than 90% of Skype traffic executed in three distinct scenarios: a small network with NAT (3 machines); a medium network with firewall and NAT (40 machines); and a machine with direct Internet access. Note that such evaluation is not within the scope of this work. It was described to contextualize the creation of a regular expression to recognize Skype society as can be seen in Figure 2. However, we will need to change *tcpdump/libcap* to filter a list of AS numbers. Additionally, such list and port configured by NAT-PMP should be dynamically updated according new values found by sequence 1-4.

In this context, our central observation and also main contribution is that the use of an automata sharing algorithm can be efficient in the dynamic collaboration of devices to identify and consequently manage societies, enabling power-aware control of its resources. In the following sections the automata sharing algorithm is described.

(udp.srcport = port_identified_pmp or udp.srcport = port_identified_pmp) or (list_ASN contains ip.dst or list_ASN contains ip.src)

Fig. 2. Extension of tcpdump/libpcap to recognize Skype

3 Collaborative Algorithm to Identify and Acquire Knowledge

We assume that the user's device has a set of automata ($S^u = \{s_1, s_2 \ldots s_{n-1}, s_n\}$) used to to identify one or more societies. Thus, the device could then bind the semantic context of the traffic and acquire knowledge (store address information and messages) for the societies and their relationships. Such knowledge provides adequate semantic for users to select and announce routes, store data, block traffic and perform other network management tasks. As result of a lack of contextual knowledge, devices are forced to adopt the extreme behavior: block all traffic or unblock all traffic; cache all or cache nothing, enable all protocols, and so on.

In order to turn on or off such societies and avoid a lack of user control over its own device configuration, a dynamic learning mechanism is used. It is based on automata sharing used by devices to acquire $s_i \notin S^u$ and consequently to obtain new knowledge according to the social interests of each user. First, the following considerations need to be made.

User device (node) traffic can be classified according to two message types as showed in Figure 4: received messages (α) or sent message (β).

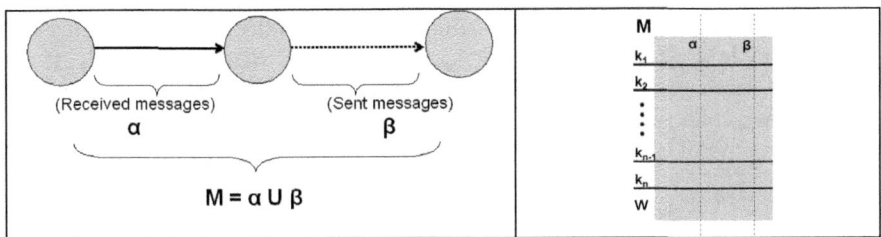

Fig. 3. Flow of user's messages   **Fig. 4.** Set of all user's messages

The set of all messages received and sent by each node is expressed by M as shown in Figure 5. They are subdivided in two sets: recognized or unrecognized messages, respectively expressed by K and W. Where $K = k_1 \cap k_2 \cap \ldots k_{n-1} \cap k_n$, $|S^u| = n$, and k_i is the set of messages recognized by the used automaton $s_i \in S^u$. W is the set of the messages unrecognized to all automata $s_i \in S^u$. Also, $M = (K \cup W) = (\alpha \cup \beta)$. Additionally, we state that $k_{i,\alpha} = k_i \cap \alpha$ and $W_\alpha = W \cap \alpha$, and β-type messages follow these same conventions.

Each device or node has a knowledge base (B) which can store the following set according to user's interests: S^u (the set of all automata that device has learned) and M (the set of messages stored along the communication). Hence, $B = <S^u, M>$, where $S^u = \emptyset$ when node has no automaton; $M = \emptyset$, when the device is configured to discard all the recognized messages or when the algorithm to share automata did not start yet.

Moreover, each device has a mechanism which frequently collects user traffic, generating a set of gathered messages in a given instant t expressed by G^t, where $G^t = \{G^t_1, G^t_2, \ldots G^t_{n-1}, G^t_n, G^t_w\}$. The gathered messages that were recognized by automaton $s_i \in S^u$ is expressed by G^t_i, and the gathered messages that failed to be recognized by any of the $s_i \in S^u$ is defined by G^t_w. Additionally we have: $G^t_{i,\alpha} = G^t_i \cap \alpha$ and $G^t_{w,\alpha} = G^t_w \cap \alpha$. Again, β-type messages follow these same conventions.

Given that the input and output messages for a node are common information to all nodes in any network and that each type of information in Figures 1 and 2 provides a basic and semantic information, even when the node does not have any automata, we propose an algorithm $A_{share\_automata}$ whose behavior depends on recognized and unrecognized, received and sent messages.

By approaching these variables with social semantics, and considering that there is an equilibrium generated by cooperation, we design an algorithm $A_{share\_automata}$(M, $v_{collect}$, v_{store}, λ_{learn}, $\bigcup_{i}^{|S^u|} \lambda_{i,teach}$, k, k_α, k_β) to share automata as shown in Figure 5.

Variables λ_{learn} and $\lambda_{i,teach}$, respectively, express numerically the node's motivation to learn automata and their capability to teach a specific automaton s_i; M is the set of messages already sent and or received that may contain nothing when the algorithm is started; the parameter $v_{collect}$ is a Boolean configuring the node to collect or not messages; the parameter v_{store} configures the node to store or not gathered messages into M and the other parameters (k, k_α, k_β) are values set according to user interests. They establish when a node must learn and/or teach by setting the rate of knowledge, of messages types α and β for each given automaton $s_i \in S^u$.

The algorithm to share automata is subdivided in four steps as seen in Figure 6. The algorithm $A_{collect\_data}(v_{collect})$ is the simplest one, it collects data according to the Boolean variable $v_{collect}$. The algorithm $A_{store}(v_{store}, M, c1, c2)$ is also simple, it stores the gathered messages from G^t into M according to Boolean the flag v_{store} once the algorithms A_{learn} and A_{teach} are finalized (c1 = true and c2 = true).

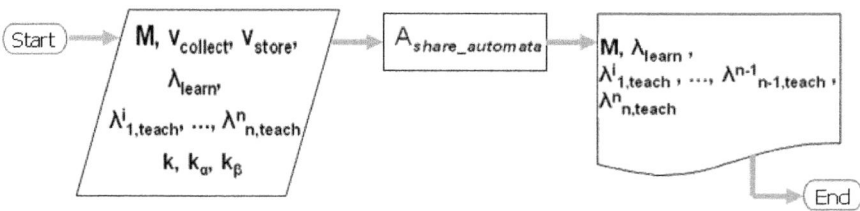

Fig. 5. Flowchart of algorithm to share automata

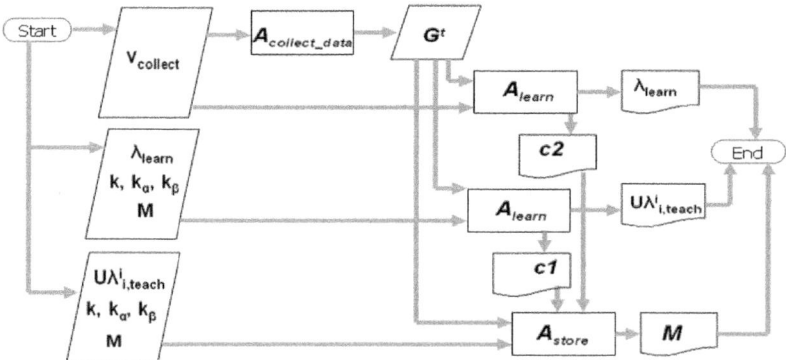

Fig. 6. Flowchart of entire algorithm to share automata

The algorithms $A_{learn}(M, \lambda_{learn}, G^t, k, k_\alpha, k_\beta)$ and $A_{teach}(M, \bigcup_i^{|S^u|} \lambda_{i,teach}, G^t, k, k_\alpha, k_\beta)$ are the head processes for a node to share an automaton. These were designed following the pseudo-code given in Figures 7 and 8 to increase or decrease the node's capability to learn and teach once the process of message gathering has finalized. We set the learning process to increase the node's motivation to learn (λ_{learn}) when:

1. The node has a low rate of recognized messages;
2. The node receives a high rate of unrecognized messages from the node or to the node;

Moreover, we also encourage that a node increases its capability to teach ($\lambda_{i,teach}$) the automaton s_i to other nodes, when:

1. The node has high rate of recognized messages;
2. The node frequently recognized received and sent messages;

$A_{learn}(M, \lambda_{learn}, G^t, k, k_\alpha, k_\beta)$	$A_{teach}(M, \bigcup_i^{\|S^u\|} \lambda_{i,teach}, G^t, k, k_\alpha, k_\beta)$

if ($\lambda_{learn}=\infty^+$ or $\lambda_{learn}=\infty^-$)
 then $\lambda_{learn}=1$
 if ($\|K\|=0$ or ($\|K\|\neq 0$ and $\|K\cap G^t\|/G^t < k$))
 then $\lambda_{learn}=\lambda_{learn}+1$
 if ($\|W_\alpha \cap G^t_{w,\alpha}\|/\|G^t_{w,\alpha}\| > (1-k_\alpha)$)
 then $\lambda_{learn} = \lambda_{learn} + 1$
 if ($\|W_\beta \cap G^t_{w,\beta}\|/\|G^t_{w,\beta}\| > (1-k_\beta)$)
 then $\lambda_{learn} = \lambda_{learn} + 1$
 $\lambda_{learn} = \lambda_{learn} - 1$
	if ($\lambda_{i,teach}=\infty^+$ or $\lambda_{i,teach}=\infty^-$)
	 then $\lambda_{i,teach}=1$
	 if ($\|K_i \cap G^t_i\|/\|G^t_i\| > k$)
	 then $\lambda_{i,teach} = \lambda_{i,teach} + 1$
	 if ($\|K_\alpha \cap G^t_{i,\alpha}\|/\|G^t_{i,\alpha}\| > k_\alpha$)
	 then $\lambda_{i,teach} = \lambda_{i,teach} + 1$
	 if ($\|K_\beta \cap G^t_{i,\beta}\|/\|G^t_{i,\beta}\| > k_\beta$)
	 then $\lambda_{i,teach} = \lambda_{i,teach} +$
	 $\lambda_{i,teach} = \lambda_{i,teach} - 1$

Fig. 7. Algorithm A_{learn} **Fig. 8.** Algorithm A_{teach}

When theses algorithms finalize, the result is a tuple constituted by the node's capability to learn and teach each automaton of node. When $\lambda_{learn} \in Z^+$, the device will accept to request and receive any automaton, otherwise the device will not accept any, neither could it request any. In same way, when $\lambda_{i,teach} \in Z^+$ then the device will agree to announce and send the automaton to nodes, otherwise the device will not announce it, neither would it send this to another node. However, two nodes n_A and n_B only share a automaton s_i, when λ_{learn} from n_A and $\lambda_{i,teach}$ from n_B are positive values or when λ_{learn} from n_B and $\lambda_{i,teach}$ from n_A are positive values.

4 Evaluation

The objective of this evaluation is to demonstrate qualitatively the benefit of using the proposed algorithm for automata sharing in a scenario where policies may be defined to reduce the energy consumption through the rejection of societies with high levels of energy consumption at their network devices.

The scenario was evaluated in a local area condominium network made up of four buildings and a total of 288 apartments. Along seven hours, we captured packets exchanged among the network devices using the wireshark[3] tool. We identified 418 devices and analyzed the traffic in this environment in terms of dynamicity, number of messages and devices in order to measure the energy consumption in Section 4.1.

For running our simulations, we assumed that the condominium's devices could belong to the same network, where equipments may collaborate and share automata and dynamically turn off some protocols to save energy as detailed in Section 3.

Unlike some existing energy saving approaches that rely on new hardware or its replacing by optical technologies, the present proposal adopts a simple automata based technique. Although there are security and trust issues with the exchange of automata, we leave these outside the scope of the current work. The proposed scheme is scalable as it does not require wire-speed processing as captured packets may be processed subsequently. The algorithm may run continuously or periodically depending on the amount of traffic passing through and the dynamicity of the societies. In a deployment scenario, automata sharing may be policy based as in [9], a proposal to negotiate policies considering a social approach. For simplification, we considered that policies to selectively turn off the recognized societies after the learning and teaching processing, would be enforced by all devices to reduce their energy consumption.

4.1 Analysis of Real Data from Wireless Network

Initially the wireless interface of a fixed notebook was set up in promiscuous mode to capture packets and to identify societies (protocols) and their traffic characteristics including packets size, number of received and sent messages seen here as are important parameters to measure the energy consumption. According to [13], we use the four linear models (a), (b), (c), and (d) to measure energy consumption in terms of broadcast and direct (peer-to-peer – p2p) communication, where size corresponds to the packet size:

$$E_{broadcast\text{-}send} = 1.9 \times size + 266 \qquad (a)$$

$$E_{broadcast\text{-}recv} = 0.5 \times size + 56 \qquad (b)$$

$$E_{p2p\text{-}send} = 1.9 \times size + 454 \qquad (c)$$

$$E_{p2p\text{-}recv} = 0.5 \times size + 356 \qquad (d)$$

Packet capture was terminated after seven hours. The collected traffic of three main protocols will be used in our case study. NetBIOS was chosen, because it is very rarely used by a network to support file and printer sharing and it is very insecure [10]. The ARP also was selected for turning off, because devices could use static instead of dynamic ARP in order to protect the network against ARP spoof attacks [11]. Broadcast of ICMP messages is another society selected to turn off, to mitigate ICMP Smurf attacks which can create a Denial of Service in one or more machines in the network. This decision is unlikely to provoke a problem to the network, because it is a setup already used by some operational systems. For instance, a device running Microsoft's Windows-XP with Service Pack2 security software is designed to drop ICMP packets by default [12].

---

[3] http://www.wireshark.org/

The NetBIOS Name Service (NBNS) and the NetBIOS Session Service (NBSS) protocols of NetBIOS were considered as two different societies in addition to ARP and ICMP, hence reaching a total of four societies. The *libpcap* filter was used to define the automata correspondent to each societies. The ARP society was identified by expression "eth.dst and eth.src and eth.type == 0x0806". The ICMP society was recognized by "eth.dst and eth.src and eth.type == 0x0800 and ip.dst and ip.src and ip.proto == 0x01". The NBNS society was identified by "eth.dst and eth.src and eth.type == 0x0800 and ip.dst and ip.src and udp.dstport == 137 and and nbns.id". The NBSS society was recognized by expression "eth.dst and eth.src and eth.type == 0x0800 and ip.dst and ip.src and udp.dstport == 137 and and nbss.type".

The total energy consumption by these protocols corresponded to 7% of all energy consumed by network along these seven hours. Figure 9(a) details consumption for each society, where the total energy consumption of NBSS was very low and for such reason it is not seen in the graph.

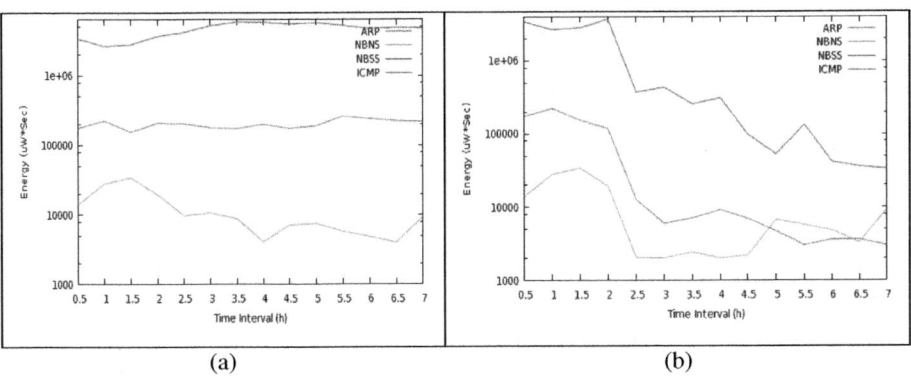

Fig. 9. Energy consumption by protocol: (a) without using automata sharing algorithm; (b) using automata sharing algorithm

4.2 Simulation

We defined a packet as a message in the form <src_address,dst_address,prot>, where src_address and dst_address are the source and destination MAC addresses, and the "prot" field is the protocol/society which can be one of ARP, NBNS, NBSS, or ICMP. Please note that this message structure was chosen for simplicity and that each society must define its message representation.

For this trace-based simulation, we assumed that all devices in the network ran the $A_{share\_automata}$ algorithm with the same parameters $k=0.5$, $k_\alpha=0.5$ and $k_\beta=0.5$. The set M is initialized without messages. Also, the sampling time of the algorithm was set to 30 minutes. Thus, at each time t the set G^t of each device is populated with messages sent/received by the node, which feed the algorithms A_{learn} and A_{teach} that control the parameters λ_{learn}, $\lambda_{ARP,teach}$, $\lambda_{NBNS,teach}$, $\lambda_{NBSS,teach}$, and $\lambda_{ICMP,teach}$. The initial 196 devices start the simulation with all automata and through the lambda parameters these devices may learn and teach automata to new devices. After learning a new automaton, the device turn off broadcasts messages for such automaton.

Figure 9(b) presents the behavior of energy consumption when the automata sharing algorithm was enabled. Although the energy consumption of society NBSS has not been showed, it uses the same energy than in the previous experiment (5,749.6 uW*sec). That occurred because the number of similar messages for each interval of 30 minutes was not enough for the algorithm increase λ_{learn} and $\lambda_{NBSS,teach}$.

The total energy consumption of the condominium reached 15023237.75 uW*sec. It corresponds to a reduction of 77.58% (51990376.52 uW*sec) without changing the conventional devices with new equipments based on renewable energy or other technologies. So, our approach could be seen as an intermediary solution before investing into new equipments.

5 Related Work

Authors in [14] used the Turing Machine as fundamental model to understand and reduce computation time and identified the similarities between research about how to reduce energy consumption and computation time in a device. However, they did not find any theoretical models to compute the energy complexity and for such reason propose to augment the Turing Machine (a variation of Automata Theory) to determine the energy consumption of algorithms.

Similarly to this approach, we proposed a mechanism also based on theoretical model to reduce energy. However, we did not propose to modify any aspect of Automata Theory, we used it as the mechanism to recognize societies in order to turn off some and consequently to reduce energy consumption. The principal contribution was a mechanism to learn, teach, and the recognition of societies by automata sharing.

Work in [17] dynamically adapts the network based on the exploitation of the resources for power management. For such approach, some policies were presented to configure and manage the operating frequency and voltage of devices maintaining a certain performance. Our proposal focuses on application level, providing a mechanism for users to dynamically detect and control network resources. So, our proposal is a more flexible and extensible algorithm than [3] and [15] that are likely to need additional mechanisms to support dynamic policies. The research in [3] studied the traffic of a cellular network and proposed to temporarily inactivate some cells when the traffic is not high. The work in [15] proposed a specific algorithm to turn off some devices, links, protocols and application of the network.

Our approach is believed to be closer to that in [16], given that both allow hosts to participate directly in network management and traffic engineering. Moreover, the automata sharing and [16] provide mechanisms for devices to distinguish among network applications, deal differently with the traffic of each one independently of port-based filters at routers. The difference to [16] is mainly in the mechanism adopted. The approach in [16] is based on exception handlers - the network administrator defines the combination of condition and actions, while our proposal is based on sharing automata with an explicit cooperation and synchronization between hosts. The first proposal needs predefined static policies (action and conditions) and our proposal is to slice the communication in societies which can be recognized by automata to detect societies' messages (protocol). Instead of having predefined automata, these can be shared and dynamically built by devices that learn and teach these to others including the network management space.

Although [16] and our proposal were evaluated in different scenarios, we believe that our proposal could be evaluated to control the network resources in term of firewall services, load-balance and link failures that are scenarios studied by [16].

6 Conclusions and Future Works

In this paper, we suggest that hosts should be able to control their own traffic in order to accept flexible policies and consequently collaborate with other nodes in the network. Devices were shown to be able to dynamically identify and turn off some protocols in order to save energy's device as well as to reduce the energy consumption of all network.

Core to the proposal are three points: the society's analogy to provide a flexible description of policies, the handling by automata to identify societies and automata sharing algorithm by devices that are able to dynamically obtain knowledge about the network traffic. However, the main objective was to demonstrate the handling by automata sharing in the context of energy saving.

Future work includes the building of automata and applying of policies in terms of high level information (i.e. business model information). The policies should reflect future green SLAs (service level agreement in term of energy) in order to reduce energy and its cost based on learning and teaching of societies.

Other studies and applications of the present solution are planned. Automata sharing will be evaluated in terms of load-balancing work, link failure recovery and new routing algorithms in disconnected networks. Future network devices, making use of our mechanism, will dynamically coordinate themselves to take rapidly decisions such as: store data (as a proxy) or forward data (as routing algorithms) or block traffic (as a firewall) according to its policies.

This work has shown that there is space for intermediary simple solutions that, when introduced into current networks, may offer important energy cost savings without the need for immediate hardware renewal. While our approach does not require wire-speed traffic analysis it suffers from security and trust problems which have been left out of this initial study.

Although the results are promising there is urgent need to conduct further experiments to determine how best the adopted parameters k, k_α and k_β can be tuned. Moreover, we intend to design the control messages of automata sharing algorithm and add such information to be executed and account by our simulation in order to measure the energy consumption of our proposal mechanism to save energy.

References

1. Baldi, M., Ofek, Y.: Time for a "Greener" Internet. In: Proceedings of GreenComm 2009 (2009)
2. Sadok, D., Oliveira, L., Kelner, J.: New Routing Paradigms for the Next Internet. IFIP AICT, vol. 1, pp. 190–201 (2010)
3. Ajmone Marsan, M., Chiaraviglio, L., Ciullo, D., Meo, M.: Optimal Energy Saving in Cellular Access Networks. In: Proceedings of the GreenComm 2009 (2009)

4. Blackburn, J., Christensen, K.: A Simulation Study of a New Green BitTorrent. In: Proceedings of the GreenComm 2009 (2009)
5. Hande, A., Polka, T., Walkera, W., Bhatia, D.: Indoor solar energy harvesting for sensor network router nodes. The Journal of Microprocessors and Microsystems (2006)
6. Chiu, H.-J., Yao, C.-J., Lo, Y.-K.: A DC-DC converter topology for renewable energy systems. The International Journal of Circuit Theory and Applications 37(3), 485–495 (2009)
7. Jones, C.E., Silvalingam, K.M., Agravawal, P., Chen, J.C.: Survey of Energy Efficient Network Protocols for Wireless Networks. Wireless Networks 7(4) (July 2001)
8. The Climate Group Smart (2020), Report: http://www.theclimategroup.org/assets/resources/publications/Smart2020Report_lo_res.pdf
9. Molina, B., Pileggi, S.F., Esteve, M., Palau, C.E.: A social framework for content distribution in mobile transient networks. Journal of Network and Computer Applications 31(5), 1000–1011 (2009)
10. Wool, A.: A quantitative study of firewall configuration errors, vol. 37(6), pp. 62–67. IEEE Computer Society, Los Alamitos (2004)
11. Chomsiri, T.: Architecture and Protocols for Secure LAN. IJCSNS International Journal of Computer Science and Network Security 8(7) (July 2008)
12. Kumar, S., Azad, M., Gomez, O., Valdez, R.: Can Microsoft's Service Pack2 (SP2) Security Software Prevent SMURF Attacks? In: International Conference on Internet and Web Applications and Services/Advanced International Conference, pp. 89–89 (2006)
13. Feeney, L.M., Nilsson, M.: Investigating the energy consumption of a wireless network interface in an ad hoc networking environment. In: Proc. IEEE Conf on Computer Communications, Infocom 2001 (2001)
14. Javin, R., Molnar, D., Ramzan, Z.: Towards a Model of Energy Complexity for Algorithms. In: Wireless Communications and Networking Conference. IEEE, Los Alamitos (2005)
15. Chiaraviglio, L., Mellia, M., Neri, F.: Energy-aware Backbone Networks: a Case Study. In: Proceedings of the GreenComm 2009 (2009)
16. Karagiannis, T., Mortier, R., Rowstron, A.: Network Exception Handlers: Host-network Control in Enterprise Networks. SIGCOMM Computer Communication (2008)
17. Bolla, R., Bruschi, R., Davoli, F., Ranieri, A.: Energy-aware performance optimization for next-generation green network equipment. In: Proc. of Workshop on Programmable Routers for Extensible Services of Tomorrow, pp. 49–54 (2009)
18. Harai, H.: Optical Packet & Path Integration for Energy Savings toward New Generation Network. In: Proc. of the International Symposium on Applications and the Internet (2008)
19. Etoh, M., Ohya, T., Nakayama, Y.: Energy Consumption Issues on Mobile Network Systems. In: Proc. of the International Symposium on Applications and the Internet (2008)
20. Cheshire, S., Krochmal, M., Sekar, K. N.: Port Mapping Protocol (NAT-PMP), http://tools.ietf.org/html/draft-cheshire-nat-pmp-03

Evaluating Mobile Phones as Energy Consumption Feedback Devices

Markus Weiss[1], Claire-Michelle Loock[2], Thorsten Staake[2],
Friedemann Mattern[1], and Elgar Fleisch[2]

[1] Institute for Pervasive Computing
[2] Information Management, ETH Zurich

Abstract. With smart electricity meters being widely deployed, data on residential energy usage is increasingly becoming available. To make sensible use of these data, we investigated the suitability of mobile phones as an interface to provide feedback on overall and device-related energy consumption. Based on the results of a user survey, we implemented the most highly valued feedback features on an iPhone that communicates with a smart meter. In a follow-up user study, we evaluated how users perceive the experience of such energy consumption feedback and how they rate the importance of different functionalities. Our work confirms the suitability of energy feedback delivered on a mobile phone. It outlines that a clear and easy to explain use case scenario is key and that knowledge-increasing functionalities as well as those functionalities from which monetary savings can be directly implied are perceived as most important. To address technophobe users, action-guiding feedback that goes beyond displaying aggregated information is required.

Keywords: Advanced metering, energy conservation, feedback systems, behavioral change, energy use.

1 Introduction

Information and communication technology (ICT) can help in many ways to conserve energy. On the one hand it does so by optimizing various industrial processes (and thus increasing their energy efficiency), on the other hand it can increase transparency, create energy awareness, and support individuals to make informed decisions that contribute to a more efficient energy use. Besides the industrial and the transport sector, residential and commercial buildings are a major consumer of resources. Their share is increasing, and they now account for approximately 73% of total electricity use in the U.S. The residential sector alone consumes 37% of electricity end use in the U.S., whereas it accounts for 29% in the EU[1].

It is widely accepted that electricity consumption in residential buildings is highly dependent on the habits of the occupants [16]. However, feedback on energy usage is typically only provided by a monthly (if not yearly) utility bill and thus remains rather vague

---

[1] See www.eea.europa.eu/data-and-maps/figures/final-electricity-consumption-by-sector-eu-27-1 and www.eia.doe.gov/aer/

and opaque to most consumers. Most people could reduce their energy consumption (and thus save money and contribute to a reduction of greenhouse gas emissions), but few know how much they consume and even fewer know for what purposes they use how much energy and which of their appliances in the house consume the most energy.

"Smart" ICT now provides opportunity to change that situation. With smart electricity meters being widely deployed and the broad availability of technologies from the ubiquitous computing domain (such as cheap sensors, low-power CPUs, wireless proximity communication, spontaneous networking, touch screen displays, mobile internet connectivity, etc.), it becomes possible to provide real-time energy use feedback on the spot, with almost no additional hardware equipment.

The main contribution of this paper is the goal-driven development of a mobile phone application[2] for residential energy monitoring based on consumption data acquired by a smart meter. It considers general experience on feedback systems described in the psychological literature as well as results obtained from a survey of potential users. The resulting prototype on an iPhone shows how mobile phones can help users to monitor and control their energy consumption and allows evaluating which functionalities are regarded as most important, also with regard to individual differences concerning technological affinity.

The structure of the remainder of this paper is as follows: We first discuss related work before we outline in section 3 different feedback features that are important to get "users in the loop" when it comes to electricity conservation. In section 4, we describe the survey we conducted to gather an indicative basis for our prototypical development. Section 5 describes the experimental setting we used to evaluate the suitability of electricity feedback on mobile phones. In section 6, we describe the user study and report on selected results, before we summarize our main findings and conclude with an outlook in section 7.

2 Information Provisioning on Mobile Phones

With the rise of ubiquitous computing, data about real-world events is being captured at an increasingly detailed level. Together with the rapid growth of the mobile phone market and mobile internet access, this has led to a large number of mobile applications which aim to support users' daily life in a wide range of areas. To name a few, this ranges from insurance claims assistance [3] over shopping assistance [2] to emergency response [13].

Other work focuses on providing information about the personal environmental impact of travel, shopping, and residential resource consumption. Ecorio (ecorio.org) and Carbon Diem (carbondiem.com) for example allow for tracking the personalized carbon footprint with the help of the smart phone's GPS sensor. The greenMeter (http://hunter.pairsite.com) aim at reducing the fuel consumption and resulting cost by using the mobile phone's internal accelerometer to measure forward acceleration and calculate fuel economy as well as carbon footprints. The Carbon Tracker (www.clearstandards.com/carbontracker.html) application serves a similar purpose, but bases the calculation mainly on self-reporting. The authors of [10] developed a mobile application prototype that semi-automatically senses and reveals information about personal transportation behavior and motivates users to choose green transportation.

---

[2] Technical details about the infrastructure and its components can be found in [22].

Information provisioning with respect to residential resource consumption has received considerable attention lately. There exist numerous mobile phone applications that allow users to monitor the electricity consumption of individual household devices. These solutions are often based on smart power outlets, like Tendril (tendril.com) or the Energy UFO (visiblenergy.com). Once installed, they measure the attached load and are capable of transmitting the measurement data wirelessly to a remote user interface (UI). However, these products lack the possibility to aggregate the consumption of multiple sensors and to fuse the different data into a comprehensive picture. To surpass this limitation, other work has focused on developing systems that combine multiple sensors. The authors of [23] built a system that enables the integration of commercially available smart power sockets that transmit their measurements via Bluetooth or Zigbee. A gateway is responsible for the discovery of the smart sockets within wireless communication distance. The approach facilitates functions such as remote on/off switching, and offers local aggregates of device-level services (e.g., the accumulated consumption of all sockets). While the concept is interesting and helps to provide important findings for future work, deploying a large number of sensors in a residential environment is often too cumbersome and expensive. In contrast to solutions which provide consumption feedback on device level, Peterson et al. [18] use a circuit breaker box that has to be attached to the fuse box to acquire the electricity data per circuit. Other work has been conducted by Björkskog et al. [4] who developed a mobile prototype targeted to establish a more playful access to energy consumption data to better address non-engineering-related user groups.

Our mobile phone application differs from those applications as it is fully integrated into a backend architecture that is based on smart meter technology which is going to be installed in large numbers in the U.S. and Europe over the next years. This allows us to combine both, feedback on the entire electricity consumption and – with a simple yet powerful functionality – on the consumption on device level. We also emphasize on concepts to present and preprocess information in a way that motivates users to become engaged into their energy consumption.

3 Energy Consumption Feedback

We expect that automation and ambient, autonomous systems will play a major role in increasing energy productivity and energy efficiency. However, in addition to systems which do not require user involvement, approaches that explicitly get users engaged into a sustainable lifestyle can considerably help to achieve today's ambitious saving targets. User-induced saving effects mainly result from two factors: First, the energy demand of many loads (including heating, air conditioning, ventilation, warm water systems, driving habits, etc.) is highly dependent on how we operate them. Virtually identical households (same buildings, same number of habitants, identical age, same location etc.) can vary by a factor of 2.6 in energy usage [17]. Second, the decision to invest in efficient devices and energy saving technologies (including ambient systems, thermal insulation, etc.) is largely up to the consumer. Therefore, awareness and willingness to take action are crucial. Both aspects can be addressed by ICT.

The existence of many unnecessary loads can be attributed to a lack of transparency in energy consumption [5]. This leads, at least partly, to lost saving potentials, because residents lack knowledge about their energy consumption in general as well as about the pool of devices used at home. In fact, they have rather limited possibilities to investigate their household's efficiency with simple measures [20]. However, feedback has been shown to be one of the most effective strategies in reducing electricity usage in the home [12]. With the advent of low-cost sensing technologies and advances in machine learning, we now have the potential to provide personal, relevant feedback in real time for a variety of consumption activities [11].

It is generally expected that with detailed and immediate feedback, 5% to 15% of the residential electricity consumption can be conserved [6]. However, to maximize saving potentials, technology itself is not sufficient, nor is the pure visualization of consumption values in some "obscure" electrical measurement unit. In fact, feedback on energy consumption is often presented in a rather technical and non-interactive way on somber devices that lack the ability to motivate users. We tried to improve on that: By connecting a smart electricity meter with a mobile phone application, our electricity feedback prototype is not only particularly easy to use (and features a nice interface on an iPhone), but it realizes those features that seem to be most promising in terms of energy feedback. Following the literature [9], feedback in the context of energy monitoring has to feature several key characteristics, which we outline below.

Reduce usage barrier. Many available energy monitoring systems require either complex installation around the central fuse box or the use of many electricity sensors. These systems typically induce a high usage barrier because the wiring around the fuse box is – at least across Europe – only accessible to technicians, and because equipping appliances throughout the house with a dedicated sensor is costly and rather burdensome[3]. However, since energy monitoring is a low involvement topic for many people, systems should be designed to allow for easy interaction.

Strong integration into daily life. Integration of feedback into users' daily life is important for long term energy conservation. Trials have shown that when using an additive battery-dependent display for feedback, in 50% of all cases users do not replace the battery once it is depleted [21]. This indicates a loss of interest after the initial curiosity has been satisfied. Thus, since not being integrated into daily life, these additive displays seem not capable motivating for a longer time.

Timeliness of informational support. Feedback should be provided frequently, in real time, and at hand when needed allowing users to relate feedback to a certain behavior or device usage [1]. Continuous feedback has been proven to be most effective. The authors of [20] investigated the effects of continuous versus monthly feedback. The results show that people confronted with continuous feedback save more (12 %) than those who had received monthly feedback (7%). In addition, only feedback that is at hand when needed is able to satisfy spontaneous curiosity.

Breakdown of the entire energy consumption. It is important to provide a possibility to disaggregate the overall consumption. A breakdown (e.g., allocation to specific rooms, appliances, or times of the day) is a powerful way of establishing a direct link between action and result. This considerably improves the intensity of reflection and interpretation of a measure [9].

---

[3] A comparison of different energy monitoring solutions can be found in [16].

The above-mentioned characteristics are believed to be important for energy feedback systems to be effective. Hence, we further investigate them in the next sections. We specifically consider how to get "users in the loop", to attract users' attention, and to increase the added value for users.

4 The User Survey

Before designing and developing an energy consumption feedback prototype, we conducted a survey to provide us with an idea what functionalities users would expect. The survey design was developed in three steps. First, we initiated and led a discussion in a focus group[4] in order to identify two applications where the participants were confident that these were easy to explain in a paper-based survey and offered varying degrees of the above-mentioned characteristics. Applications which were found to be suitable were:

1) A mobile phone application that utilizes a smart metering infrastructure that allows users to get feedback on the consumption of individual appliances (timeliness of informational support, suitable for investigation of specific loads, high degree of interaction, and portability).

2) A washing machine with a simple display which provides feedback on energy consumption of specific programs and information on the energy that is saved by choosing eco-programs (low timeliness of information, low degree of interaction, and no portability).

The approach of contrasting two different applications allowed us to draw conclusions based on the "inverse" application, and we also expected the study design to reveal the degree to which user expectations achieved a mature state, i.e. a state where they can be used as a stable basis for a requirements analysis. In a second step, we extended the question catalogue by a set of constructs to evaluate general functionalities of consumption feedback. We used established constructs taken from the Technology Acceptance Model (TAM) [7], constructs on the word of mouth, and questions concerning the willingness to pay. To evaluate the validity of the reported willingness to pay, we used a technique called framing. Framing means that information can be presented in different contexts which affect the perception of the information [14]. In a third step, we evaluated the comprehensibility by reviewing the constructs of the questionnaire with non-experts.

The survey was conducted at lively points throughout the city center of Zurich., Switzerland. 185 persons participated in the survey (50.3% male) with all age groups evenly represented. The sample was slightly biased as respondents with a higher education degree and an above–average income were overrepresented. However, we do not expect this to considerably reduce the validity of the findings. In the following, selected results of the survey are presented. Regarding the general attitude towards conserving energy, roughly 50% of the participants think it is rather cumbersome to save energy and not fun. In doing so, 89% like to be supported by innovative technology.

The left side of Fig. 1 depicts the comparison of the perceived usefulness and the intention to use of the mobile phone application with the washing machine display.

---

[4] 3 experts from academia, 4 industry experts, and 4 employees of consumer organizations.

The figure shows the mean as well as the standard deviation (as black error bars). The ratings for both applications are above average, with the mobile phone application performing worse over all the ratings. It was surprising to us that the washer application performed better than the mobile device with regard to the likelihood to recommend the application to a friend, the likelihood to use the application, and the expectation that the information would lead to energy savings. A potential explanation for this unexpected outcome is that the washer application had an inherent use case (the application domain "washing" and saving by selecting an "eco-program" was clear), while the innovative phone application had a more general and not very obvious use case ("you can follow your energy consumption and measure how much energy your devices consume, which can help you save").

In addition, we investigated the self-reported willingness to pay for a mobile application that allows for measuring the electricity consumption only. Knowing of the difficulty to obtain a reasonable indicator for a sales price with this method, we used the findings to get an indication for the relative price range, and for developing a better understanding on how stable the perceived value is. For that, we used two different, slightly modified versions of the questionnaire and distributed them half-half amongst the participants. The first version asked people for their willingness to pay, but not indicating the possible saving potential of such a technology measure. The second version indicated the monetary amount of potential savings (85$), thereafter asking for the amount people would spend. While the mean values do not vary significantly between both versions, the median increases from 16$ to 30$ when presented the version with frame. Thus, by providing an annual saving value, participants were willing to pay a higher price for the application. In addition, the standard deviation decreases significantly. This effect is referred to as framing effect and highlights that the expectations are only vague. From these results we can conclude that the price regarding such a mobile phone application is not yet set. Participants have a general understanding that mobile phone applications are low-priced. On the other hand, the results also show that it is hard to determine the price for an innovative product that is not yet touchable and the price can be varied according to the context the application is presented.

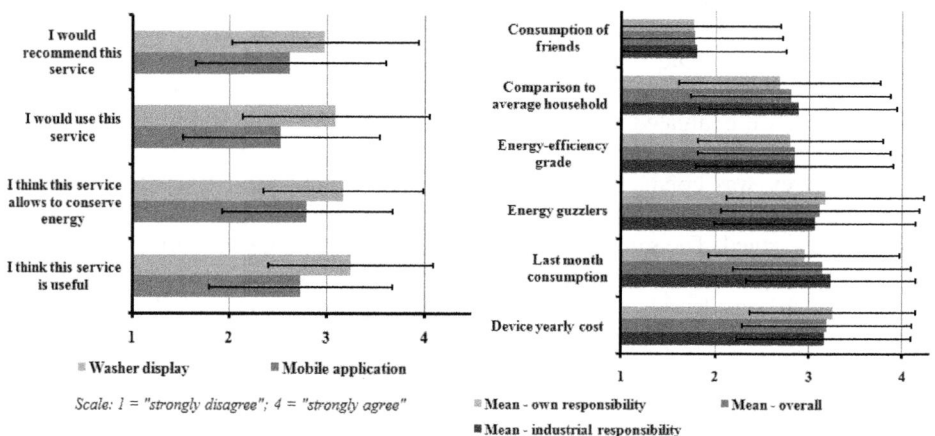

Fig. 1.

In order to conserve energy, it is important to be aware of how much particular appliances consume as well as of effective measures that increase the energy-efficiency. Thus, we asked users with what information they would like to be provided on a mobile phone application. The right part of Fig. 1 shows the participants' assessment of the different functionalities offered. Besides depicting the overall mean of the whole sample and the standard deviation in error bars, the figure provides a more in-depth view on the mean of participants that regard energy conservation as their own responsibility (N=65) and those that believe it is industry's responsibility (N=120). Those who strongly or slightly disagreed on the statement "The industry is primarily responsible for saving energy" (1 or 2 points on the four point Likert scale) belong to the first group. The second group consists of people who slightly or strongly agreed with the statement mentioned above (3 or 4 points). Overall, we find that participants prefer to be provided with the yearly cost of single appliances followed by the last month consumption and the biggest energy guzzlers of the household. However, users do not want to compare their consumption to the one of their friends or family. A closer look on the attitude of the participants reveals that participants who regard saving electricity as their own responsibility rate the identification of the biggest energy guzzlers significantly higher than others. We assume this goes along with their higher involvement and interest for both, their personal energy efficiency and defined measures that allow for conserving energy.

The survey results served as an indicative basis for our prototypical development. It encouraged us to offer a simple use case that makes the value of an innovative energy consumption feedback application clear even to individuals who are not familiar with the system. As an easy to explain use case, we identified a measurement function that can be explained as "learn how much a device consumes by just switching it on or off". On that basis yearly costs can be calculated and the biggest energy guzzlers can be identified.

However, the results on willingness to pay, perceived usefulness, and the intention to use the system also showed us that innovative services seem to go beyond the imagination of users. Thus, it is necessary to investigate possibilities of electricity consumption feedback on mobile phones in a controlled environment, ideally in a user study where users can get in touch with the application and where the understanding of the application's functionalities can be guaranteed.

5 Experimental Setting

In this section, we present the setting we used to evaluate the suitability of electricity consumption feedback on mobile phones. Our prototype [16, 22] consists of a back-end architecture and an UI on the iPhone, and implements the above mentioned key feedback features. Being based on smart meters, which are currently rolled out in large numbers in Europe and in the U.S., the system requires no further modifications by the user. Thus, the necessary effort is limited to downloading an application, which leads to a low usage barrier.

By providing feedback through a mobile phone application, the prototype features both: feedback on a device that is already integrated in users' daily life, as well as the opportunity to provide immediate feedback that is at hand when needed. Through an interactive measurement functionality, the application not only allows users to monitor, measure, and compare the energy consumption of the entire household, but also the electricity consumption of individual devices. It thus allows for users to link specific actions to their effects and helps them to identify the biggest energy guzzlers, which is key to take effective energy conservation measures [19].

The System Architecture of our prototype consists of three independent, loosely coupled components: The *mobile phone application* is linked through *a gateway* to *a smart meter* which measures the total electricity consumption of all household devices. The gateway's function is to manage and provide access to the data acquired by the smart electricity meter, while the mobile phone application allows users to interactively monitor, measure, and compare their energy consumption in real time. The system is based on the REST (Representational State Transfer) paradigm [8]. It enables easy and seamless integration of physical resources to the Web and makes them available through URLs. Through providing a web-based interface to the smart meter, we allow for easy interoperability with other applications, such as the mobile phone application.

Compared to traditional electricity meters used today, a smart meter contains a communication interface that is intended to enable remote meter readings for billing purposes. In order to achieve real-time feedback, we extended the meter's capabilities to send out all available consumption data every second to a gateway, which parses and stores the received data in a SQL data base. A small Web server provides access to the gateway's functionalities and the meter readings through URLs. It further handles the incoming requests from the UI. A more detailed view on the system and the benefits gained by using embedded Web technologies can be found in [22].

The User Interface. The mobile phone application, realized on an iPhone 3GS, is used to provide electricity feedback. The UI consists of four main views (Fig. 2) – current power consumption, history, device inventory, and measurement – on which we elaborate in the following.

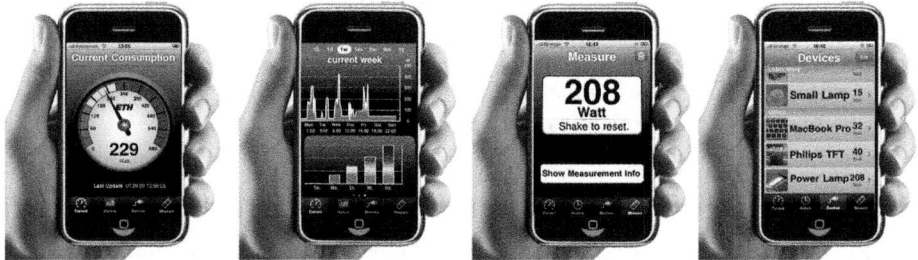

Fig. 2. The UI (left to right): Consumption view, history view, measurement view, and device inventory view

The *consumption view* is used to determine the real-time electricity usage of the entire electricity demand of the household. Once the view is active, the scale is auto-updated every second to provide instantaneous feedback. The color-coded intervals help users distinguish how high their current consumption is compared to their historic values. Red indicates a high, yellow a normal, and green a low consumption. Since an equidistant partition of the color ranges leads users to believe their consumption is seldom high, we tried experimenting with different interval sizes and ended up with a partition where the intervals correspond to the quantiles. The blue part of the scale depicts the level of the household standby electricity consumption (i.e., base load). It is determined by a weighted moving average of the consumption values measured between two and four o'clock in the morning. The weight for each new value taken into account corresponds inversely proportional to the difference between this value and the previous average consumption value.

The *history view* combines historical consumption feedback with budgeting and projections. The view's upper half depicts the historical consumption of the household in form of a line graph. Users can choose between different predefined time periods ranging from one hour to one year. The lower part of the screen can be changed to either display aggregated consumption or budgeting and projections. For both the scale can be swapped between kWh and cost. Per default it shows the aggregated consumption of the last five selected periods (Fig. 2) and provides users with an easy way to identify at what times their accumulated consumption peaked. The bar scale is color-coded, providing users with feedback on how their consumption compares to a typical household with same characteristics (e.g., size, location). When changed to budgeting and projections, users can see their remaining budget on a bar that drains as budget runs out together with a projection of where their consumption is going to cease in case it remains at the current level. All calculations are done based on values users enter on the application's first startup.

The *measurement view* allows for users to interactively breakdown the entire electricity usage and to detect how much power switchable appliances consume in operation (and by unplugging the device also in standby). To perform a measurement, users simply activate the process by pressing the start button and thereafter switching the device under measurement on or off. The corresponding result is shown in two to ten seconds. Besides considering the increase or drop in real power, the measurement algorithm [24] also takes the different electric circuits and other physical variables, such as apparent power and power factor, into account. This allows determining on which circuit the switching event has occurred, and can be used for failure detection.

The *device inventory view* provides users with an overview of the previously measured devices as illustrated in Fig. 2. After measuring a device, users can personalize the measurement. The UI offers the possibility to take a picture of the appliance, adjust utilization on which incurred cost calculation is based, enter a location, and assign a device category. In case a category is selected, further category-specific information and guidance on how to save energy can be obtained.

6 The User Study and Selected Study Results

In the following, we first focus on the user study we conducted to evaluate the suitability of mobile phones as energy feedback devices, before we report on selected study results with respect to the application and the perceived value of different feedback functionalities.

The User Study. To evaluate which of our implemented functions are perceived as most valuable on a mobile phone, we conducted a user study with 25 participants of different background from students over marketing and sales persons to industry experts. Twelve of them were male (48%). We covered all age groups: 32% were between 18 and 25 years old, 28% between 26 and 35 years, 36% between 36 and 49 years, and 4% between 50 and 70 years.

The user study was divided into two parts: The first part aimed at familiarizing participants with the functionalities of the application. First, the consumption as well as standby consumption of five appliances (a light bulb, an energy saving lamp, a kettle, a game console, and a flat screen monitor) had to be measured by each participant (Fig. 3) to provide an idea of how easy it might become in the future to determine the consumption of different devices – compared to the rather cumbersome solutions that exist today. This also allowed us to validate the measurement functionality of the iPhone application. After that, we used a guided interview to explore and discover the different functionalities and their meanings. For that, each user had to accomplish different tasks which involved various implemented features (e.g., determine highest historical consumption or current consumption, how current consumption compares to historical, standby consumption, etc.) and aimed at gaining a solid understanding of the application.

The second part of the study consisted of a questionnaire that aimed at a general evaluation of the application and at assessing the functionalities that are perceived most valuable from a user perspective. The questionnaire was anonymously completed in an unobserved environment. We asked the participants to rate the importance of implemented as well as possible future functionalities. Moreover, users had to rate the complexity, usefulness, ease of use, ease of learning, and satisfaction of the mobile phone application. The latter four factors are taken from the USE, an established questionnaire for measuring the usefulness, satisfaction, and ease of learning of a UI by Lund [15]. Additionally, we asked the participants to indicate their intention to use (once a week, once a month, or never), and the willingness to tell their friends (word of mouth). All items were rated on a five-point Likert scale.

Selected Study Results. The general evaluation of the mobile applications shows that participants had understood the UI and the underlying functionalities. On a scale from one (lowest) to five (highest), participants rated the ease of use (4.04), ease of learning (4.04), and satisfaction with the application (4.16) all significantly above average (means in brackets). Taking into account that we covered a wide age range (18 to 51 years) and only five participants were iPhone users, we regard this as a very positive response for a prototypical application. General results also indicate that the feedback latency was perceived as more than satisfactory, the measurement functionality as easy to handle, and the individual views as easy to understand (Fig 3 right). Determining the consumption of individual devices, over all conducted measurements participants achieved an accuracy of ±5% compared to the real consumption we had previously verified.

The left of Fig. 4 illustrates the results for the mobile application assessment in terms of usefulness, intention to use, and word of mouth. Besides depicting the overall mean of the whole sample and the standard deviation in error bars, the figure provides a more in-depth breakdown in terms of technological affinity. It was assessed based on four items: 1) my friends and colleagues often ask me what I think about new telecommunication technologies, 2) my friends and colleagues are better informed about new technologies than me (inversed), 3) I am always up on the latest technologies in my area, and 4) I think it is fun to test new technologies. We calculated the average response for the four items after inverting question two. People

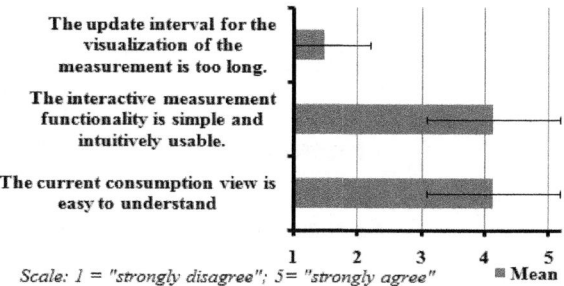

Fig. 3. User measuring the power consumption of a device (left) and evaluation of the iPhone application (right)

with a scale mean of 3.64 or less were grouped to "not technological affine" (N=10), whereas people with a higher mean were assigned to the group "technological affine" (N=15). The evaluation on perceived usefulness, intention to use, and word of mouth was only marginally affected by the technology affinity (except frequency of use where the technophile users received higher scores). The application reached high scores especially for positive expectations towards saving effects and knowledge gains. Not surprisingly, technophile people indicate they would use the application more often than others. The word of mouth effect is significantly high. It thus offers potential for utilities or smart meter manufactures to positively influence their image providing such an application. A large part of the participants agrees that the application is useful. The prime use of the application – especially with technophobe participants – is seen in increasing the knowledge about the electricity consumption of individual devices. In consistence with the survey results, the claimed external social motivation ("demonstrate good behavior to others") to use such a mobile application remains low. Users in general do not feel the need to express their proenvironmental behavior to others (or are not willing to admit it), but technophile people would rather do it than non-affine.

In order to evaluate which functions with respect to energy feedback are perceived as most valuable on a mobile phone, we asked the participants to indicate their impressions on the following functionalities: Real-time visualization of the total consumption; visualization of the household's standby consumption; comparison of the current consumption with the historic consumption; costs of recent months; consumption of individual devices; consumption of recent months; projections of yearly cost on device level; efficiency grade of appliances; overview of biggest energy guzzlers; comparison of the consumption with the one of friends; possibility to show others my appliance pool; possibility to set a saving target.

Fig. 4 right provides an overview of the assessment of mobile electricity consumption feedback features. It depicts an in-depth view on the rating per functionality sorted in an ascending order according to the overall mean value (shown on the right). Overall, we find that participants value at-a-glance-feedback on their most prominent energy guzzlers most (mean of 4.72), followed by those functionalities that increase the knowledge about consumption or cost. The real-time view of the entire consumption achieves similar ratings with a mean of 4.6. For both, 96% of the participants

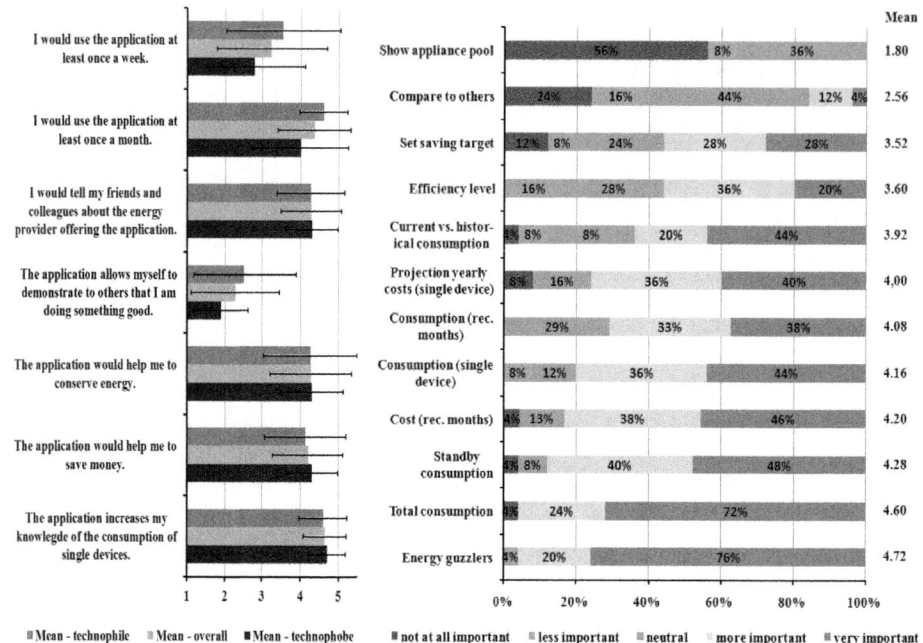

Fig. 4. Mobile application assessment: Usefulness, intention to use, and word of mouth (left). Mobile electricity feedback: Functionalities (right).

indicated the importance of these functionalities. Down to a mean value of 4.16, still 80% of the participants perceive functionalities such as standby consumption and consumption of an individual device important. All these functionalities have in common that they provide an action-guiding feedback from which users can directly draw effective measures to lower their electricity usage. Surprisingly, cost of the recent months receives a high importance ranking of 84%, although the feature itself is not action-guiding. Below a mean value of 4.16 the picture changes. Those functionalities that present aggregated information from which people cannot imply a direct action (e.g., consumption of recent months, comparison of the current vs. historical consumption) receive significantly lower ratings and reside in the bottom half. Functions aiding users through motivational support (e.g., set a saving target) are not perceived as important, nor are those that deal with social aspects (e.g., compare to others). They reside in the bottom third and especially the latter two receive a low importance rating of 16% and 0% respectively. A closer look on the technological affinity reveals that in general technophiles rate the functionalities higher. However, technophobes value features that present action-guiding (e.g., energy guzzlers) and device-level information (e.g., consumption of an individual device) over aggregated information.

7 Discussion and Future Work

Our contributions are threefold: First, we could confirm that the existence of a clear and simple to explain use case behind energy consumption feedback systems is a key success factor, whereas when left to the imagination of the potential users, energy consumption feedback applications receive only medium ratings. Second, we developed a system that implements several promising feedback features, and which succeeded to receive positive ratings from users. The findings can serve as a starting point for further application development in this field. And third, we tested the benefits and capabilities and confirmed the suitability of such a mobile phone application to serve as an energy feedback system in a user study. We also identified the functionalities that are perceived most valuable by users, in general and with regard to individual differences concerning technological affinity.

We consider the applied methodology to develop a prototype application based on preliminary interviews and a survey with diametric application scenarios to be well suited for application development at an early stage. It helped us to critically assess the user requirements and to extend the application's functionalities. Exposing people to a functioning prototype was crucial for us to gather experience with the application, while at the same time participants better understood the usefulness of the application. In order to get users in the loop, we implemented the most promising feedback features and evaluated the different functionalities with our application. We found that the knowledge-increasing functionalities as well as those from which monetary savings can be directly implied are perceived as most useful. In contrast, functionalities that present aggregated information receive lower scores. In addition, the survey results as well as the user study indicate that social motivation is so far not an important factor in terms of energy consumption feedback. In order to address different target groups, we focused on the individual difference between technologically and non-technologically oriented people. To target the latter ones, a closer look revealed that it is important to implement simple, easy to understand, and action-guiding feedback that goes beyond aggregated information, such as a list of energy guzzlers.

Another important aspect is the interactivity of energy consumption feedback that is introduced through the mobile phone application. We believe that this is key to get users involved into energy conservation. The measurement functionality, which allows users to determine the consumption of individual devices, is a good example how interactivity can be used in this context. It easily enables users to familiarize with their energy consumption. However, providing feedback to get users in the loop is just the first step. As our results show, participants are more likely to use the application once a month rather than once a week without additional measures. Thus, it will be crucial to develop concepts that aim at keeping users in the loop. Besides further analyzing which functions users expect, our future work will investigate engagement strategies (e.g., competitions, alerts, and bonus points) to further involve users after their initial curiosity has been satisfied. Overall, we believe that the user-in-the-loop-paradigm together with automated energy saving systems can contribute considerably to mitigate energy consumption.

Acknowledgements. The authors would like to thank the anonymous reviewers, the study participants, our industry partners Landis + Gyr, Illwerke VKW, and EWZ as well as David Abdurachmanov, Fabian Aggeler, Adrian Merkle, and Wolf Roediger for their help.

References

1. Abrahamse, W., Steg, L., Vlek, C., Rothengatter, T.: A review of intervention studies aimed at household energy conservation. J. of Environmental Psychology 25(3), 273–291 (2005)
2. Adelmann, R.: Mobile Phone Based Interaction with Everyday Products on the Go. In: Proc. NGMAST 2007 (2007)
3. Baecker, O., Ippisch, T., Michahelles, F., Roth, S., Fleisch, E.: Mobile claims assistance. In: Proc. MUM 2009 (2009)
4. Björkskog, C., Jacucci, G., Lorentin, B., Gamberini, L.: Mobile implementation of a web 3D carousel with touch input. In: Proc. MobileHCI 2009 (2009)
5. Chetty, M., Tran, D., Grinter, R.: Getting to green: understanding resource consumption in the home. In: Proc. UbiComp 2008 (2008)
6. Darby, S.: The effectiveness of feedback on energy consumption. A review for DEFRA of the literature on metering, billing, and direct displays (2006)
7. Davis, F.D.: Perceived usefulness, perceived ease of use, and user acceptance of information technology. MIS Quarterly 13(3), 319–340 (1989)
8. Fielding, R., Taylor, R.: Principled Design of the Modern Web Architecture. ACM Trans. Internet Technology 2(2), 115–150 (2002)
9. Fischer, C.: Feedback on household electricity consumption: a tool for saving energy? Energy Efficiency 1(1), 79–104 (2008)
10. Froehlich, J., Dillahunt, T., Klasnja, P., Mankoff, J., Consolvo, S., Harrison, B., Landay, J.A.: UbiGreen: investigat-ing a mobile tool for tracking and supporting green transportation habits. In: Proc. CHI 2009 (2009)
11. Froehlich, J., Everitt, K., Fogarty, J., Patel, S., Landay, J.: Sensing opportunities for personalized feedback technology to reduce consumption. In: Proc. CHI Workshop on Defining the Role of HCI in the Challenge of Sustainability (2009)
12. Geller, E.S., Winett, R.A., Everett, P.B.: Preserving the Environment: New Strategies for Behavior Change. Pergamon Press Inc., Oxford (1982)
13. Landgren, J., Nulden, U.A.: A study of emergency response work: patterns of mobile phone interaction. In: Proc. CHI 2007 (2007)
14. Levin, I.P., Schneider, S.L., Gaeth, G.J.: All frames are not created equal: A typology and critical analysis of framing effects. Organizational Behavior and Human Decision Processes 76(2), 149–188 (2002)
15. Lund, A.M.: Measuring Usability with the USE Questionnaire. STC Usability SIG 8(2) (2001)
16. Mattern, F., Staake, T., Weiss, M.: ICT for Green – How Computers Can Help Us to Conerve Energy. In: Proc. e-Energy 2010 (2010)
17. Parker, D., Hoak, D., Cummings, J.: Pilot Evaluation of Energy Savings from Residential Energy Demand Feedback Devices. FSEC, Rpt: FSEC-CR-1742-08 (2008)
18. Petersen, D., Steele, J., Wilkerson, J.: WattBot: a residential electricity monitoring and feedback system. In: Proc. CHI 2009 (2009)
19. Prudenzi, A.: A neuron nets based procedure for identifying domestic appliances pattern-of-use from energy recordings at meter panel. Proc. IEEE Power Engineering Society Winter Meeting, USA 2, 941–946 (2002)

20. Van Raaij, W.F., Verhallen, T.M.M.: A behavioral model of residential energy use. J. of Economic Psychology 3(1), 39–63 (1983)
21. van Rensburg, L.: Energy demand research project: Review of progress 2008. In: TR Ofgem 2009 (2009)
22. Weiss, M., Graml, T., Staake, T., Mattern, F., Fleisch, E.: Handy feedback: Connecting smart meters with mobile phones. In: Proc. MUM 2009 (2009)
23. Weiss, M., Guinard, D.: Increasing Energy Awareness Through Web-enabled Power Outlets. In: Proc. MUM 2010 (2010)
24. Weiss, M., Staake, T., Fleisch, E., Mattern, F.: PowerPedia – A Smartphone Application for Community-based Electricity Consumption Feedback. In: Proc. Smartphone 2010 (2010)

Developing Pervasive Systems as Service-Oriented Multi-Agent Systems

Jorge Agüero, Miguel Rebollo, Carlos Carrascosa, and Vicente Julián

Departamento de Sistemas Informáticos y Computación,
Universidad Politécnica de Valencia,
Camino de Vera S/N 46022 Valencia (Spain)
{jaguero,mrebollo,carrasco,vinglada}@dsic.upv.es

Abstract. The development of *Pervasive Systems* is an emerging research topic due to the high heterogeneity of involved technologies and the changing nature of the existing platforms/devices, which make it hard to develop this kind of systems. This work presents a Model Driven Development approach to develop *agent-based* software for Pervasive Environment in order to design and implement application prototypes in an easy and productive way. Our approach provides a method for the specification of *Pervasive Systems*, which allows to face the development of such systems from a higher abstraction level. The deployment over different execution platforms is achieved by means of automatic transformations among models that described entities and the environment (UML-like). The result is a simplified and homogeneous deployment process for *Agent-Based Pervasive Systems*.

Keywords: Multi-Agent Systems, Pervasive Systems, Model Driven Development.

1 Introduction

Pervasive Systems is a paradigm in which technology is virtually invisible in our environment. *Pervasive Systems* are very common nowadays and will be even more in the next years. This is due to the appearance of *new objects* (of daily usage) with different technological capabilities, because they incorporate different electronic devices [14]. So, it is easy to think that this paradigm requires, from a designer point of view, the development of applications in different software and hardware platforms depending on the diversity of the objects in the environment. This raises big challenges.

In this way, the development of *Pervasive Systems* is a complex task, with multiple actors, devices and different hardware environments; where it is difficult to find a compact view of all components. The requirements of this kind of systems are very different[15]. However some of them are basic: (i) integration of external devices and software systems, the services that are provided by *Pervasive System* can be supplied by physical devices and also by existing software systems, and it is essential that the system supports these issues; (ii) the isolation of the technology and the manufacturer-dependent devices, in order to facilitate the development of this kind of systems, the manufacturer dependent devices must be well encapsulated in independent and generic functionalities.

Software engineering based on Multi-Agent Systems (MAS), particularly Open MAS, has the capability to fulfill these requirements[4]. This approach supports the integration of highly heterogeneous platforms, where agents work together to support complex tasks, in a collaborative and dynamic way, with the ability to adapt, coordinate and organize each other[20]. Therefore, it seems appropriated a MAS-based development process to facilitate the orchestration and integration of the functionalities of physical devices.

Moreover, Model Driven Development (MDD) approach can facilitate and simplify the design process and the software quality in the development process of a MAS-based *Pervasive System*. It allows to re-use software and the transformation between models[23]. This methodology can be applied in the development of embedded agents for *Pervasive Systems*, where different technologies and execution platforms coexist. That is, to design applications automatically, where the toolkits guide the developer in the design process, using unified models to apply transformations allowing to obtain specific code for the deployment platforms.

This work proposes to use a MDD approach to facilitate the development process of *Agent-Based Pervasive Systems*, providing the user with a set of abstractions that ease the implementation of *Pervasive Systems* and the deployment of a platform for their execution. To sum up, this work presents: (i) an Environmental meta-model, which allows to incorporate in the modeling process the different environment devices of a *Pervasive System*; and (ii) a layered deployment architecture. This allows to design pervasive applications using high-level abstractions, avoiding the low-level implementation details and, after that, the *Pervasive System* deployment (with embedded agents and devices) is generated by using automatic transformations. In this way, a non-expert programmer will be able to develop *Agent-Based Pervasive Systems*, reducing the gap between the design and the implementation phases.

This document is structured as follows. Section 2 explains the MDD approach and some related works. Section 3 explains how we use MDD for *Agent-Based Pervasive Systems*. Section 4 presents the proposed Environment meta-model. Section 5 shows how to implement the MDD approach in order to develop *Pervasive Systems* and explains the deployment architecture. Finally, some conclusions are presented in section 6.

2 Model Driven Development

The purpose of MDD is to create models legible by computers that can be understood by automatic tools to generate code templates and proof models, integrating them in multiple platforms[5]. Model driven methods make a clear distinction between the problem space (centered on what the system is) and the solution space (centered on how it is implemented as a software product).

2.1 MDD for MAS

From the viewpoint of the design of agent-oriented systems, application development consists of how to obtain the agent code that could be executed in different platforms. That is, to concentrate the development of the application from a unified agent meta-model and, after that, to apply different transformations to obtain implementations for different platforms.

In MAS literature, researchers have formulated a set of typical meta-models that guide the process of MAS development using a model-driven approach. Some works have concentrated their efforts on creating a generic unified model for analyzing and modeling the system using different methodologies. Some of the most significant proposals are: INGENIAS[17], PIM4AGENT[18] FAML[9], Agent UML(AUML)[7] and AML[13]. However, although these proposals use similar components in their meta-models, none of them focus in the development of the MAS as a *Pervasive System*. Those approaches have a limited scope in order to design how the agents perceive/act on the physical world.

The difference of our proposal with respect to the existing approaches is that we propose specific Environment meta-model which provides access to physical devices located in the real world. Furthermore, some of the proposals (FAML, AUML and AML) just define high-level meta-models, and they do not arrive to the implementation phase, and difficulting enormously the developer work when trying to obtain executable code. Moreover, even in those methodologies that include the implementation phase, there exist a gap between design and implementation models. Since there exist notable differences between the high-level agent definition and the implemented agent class.

2.2 MDD for Pervasive Systems

Although the application of model-driven approaches has no been widely adopted in the Pervasive System field, some heterogeneous efforts can be identified that follow this development paradigm. The most important proposals include: CML[19], pervML[22], VRDK[21], among others [11][19]. These approaches using a modeling language to model *Pervasive Systems*, focused on the description of context-aware data (or framework system), and to provide visual tools to facilitate the system development. Some proposals (such as CML and VRDK) focus directly on the device as the main system component and the implementation of the functionality using this device.

Our proposal provides high-level abstractions to decouple the dependencies of the devices with the functionality that the device provides. So, it is based on the encapsulation of the device in an independent service controlled by agents. Similarly as pervML and other[11], we have the aim to decouple the system functionality from the implementation software, using a Service Oriented Architecture (OSGi environment).

The OSGi[1] framework, (Open Service Gateway Initiative) is a standard technology that defines an environment for the execution of services and their

---
[1] http://www.osgi.org/

life-cycle management. The OSGi is originally focused in networked devices market, using services that provide access to different existing device networks (EIB, UPnP, X10, etc.). These features are very interesting for developing *Pervasive Systems*. The OSGi proposal uses a component called *Bundle*, which is Java-base technology that provides a mechanism for releasing and deploying applications.

In summary, our approach allows agents to manage, interact and control physical worlds and thereby create a versatile *Pervasive System* model, since the agents (and developer) interacts with the environment using independent services, which can be combined, composed, and so on.

3 πVOM Approach

Models is a MDD approach are defined through meta-models[23]. One fundamental challenge when defining a meta-model is selecting which concepts or components will be included in order to model the system. To achieve this objective, some of the most well-known approaches in the area of MAS and *Pervasive Systems* were studied (mentioned in Section 2). The purpose of this analysis was to extract the common features from the methodologies studied and adapt them to the current proposal, specifying a generic platform-independent meta-model of an *Agent-Based Pervasive System*.

This set of meta-models is created by the detection of common concepts in an iterative cycle consisting of a bottom-up analysis. Common elements in existing MAS and *Pervasive Systems* methodologies, have been identified and incorporated to the Computation Independent Model (CIM) level (see Figure 1). These models can be adjusted as MDD models that specify the concepts of the system, as roles, behaviors, tasks, environment, interactions or devices. The models can be used to describe an *Agent-Based Pervasive System* without focus on platform-specific details and requirements, as a Platform Independent Model (PIM). After that, it is possible to transform PIM models into Platform Specific Models (PSM). Figure 1 shows relationships between the concepts of different MDD models and their transformations.

The proposed set of meta-models integrate different MAS modeling approaches, and it mainly focuses on the integration of Services and MAS techniques for supporting dynamical and open MAS societies[4]. This set of meta-models is called πVOM (*Platform-Independent Virtual Organization Model*). The main views of πVOM are the *Structure, Functionality, Normative, Agent,* and its *Environment*. Therefore, to model the characteristics of these components in our approach, five key concepts are used: *Organizational Unit, Service, Environment, Norm,* and *Agent*[4]. According to this, πVOM is structured in five meta-models or views, which cover the above mentioned key aspects.

These five elements describe those members (entities) that form the organization: the topology of the organization; the services and features that the organization offer; the evolution of the organization over time; the environment where the organization is situated; and the rules about the behavior of members respectively. These five elements are described in more detail in [3]. In this paper, an extension of the Environment meta-model is described in detail. This

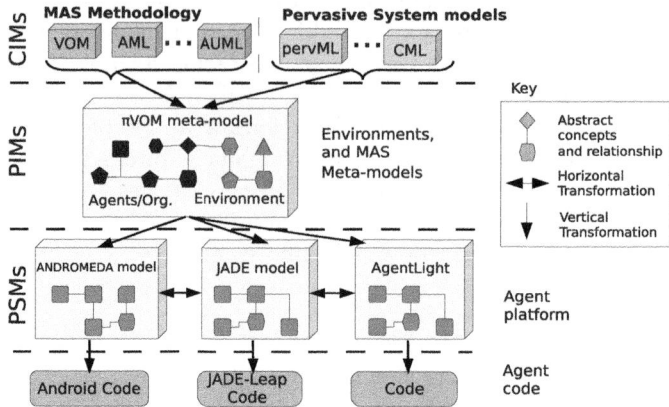

Fig. 1. MDD for MAS

meta-model allows agents to easily interact with the different physical devices in the environment, which enables agents to manage, perceive and be located in a *Pervasive System*. This new Environment meta-model is able to capture the details and requirements of a *Pervasive Systems* (whereas the old meta-model no). Next section explains in detail this Environment meta-model.

4 Environment Meta-model: An Ubiquitous Meta-model

This new meta-model is focused on describing environmental components, representing perceptions and acts of the devices. Moreover, it defines permissions for accessing them. The proposed Environment meta-model is showed in Figure 2. The *Environment* concept represents the physical world where the agent is located. The *Environment* can be perceived using the *Perceive* relationship. The *Environment* is a recursive model, which allows the creation of sub-worlds (workplaces) contained one inside another. These sub-worlds may be connected with other neighboring worlds using the relationship *neighborhood*.

The *Environment* consists of *Resources* and *Devices*. A *Resource* is a software-oriented environment component and its access is performed by a standard protocol, not requiring an adjustment of the device drivers. A *Device* represents a physical component, which access is made directly through a physical interface (sensor or actuator). In this case, it is required to bind the low-level driver (firmware) to a software component.

The *EnvironmentService* concept allows to use the functionalities of the physical devices through a service, decoupling the low-level abstraction. *Resources* and *Devices* are accessed through an *EnvironmentPort*. An *entity* (agent or organization) is in charge of managing the access permissions to these elements using the *Port* abstraction. Each *Port* is controlled by an entity and it can be used by one or more roles (played by the entities). A *Port* represents a point of interaction between the *Entity* and physical devices, and it serves as an interface

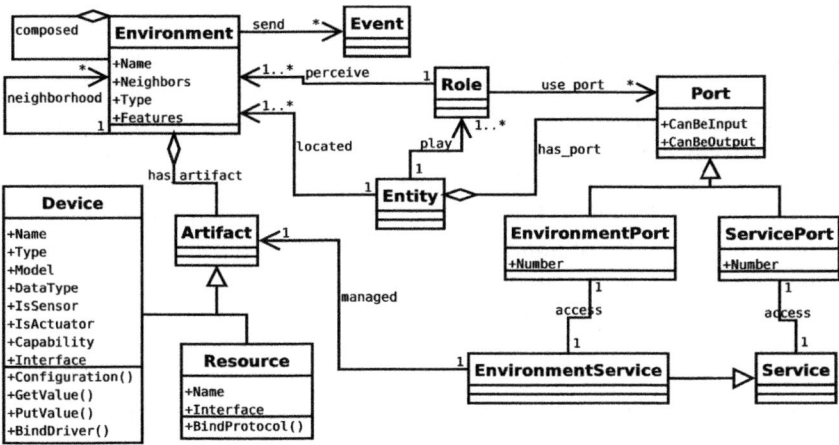

Fig. 2. Environment meta-model of πVOM

Table 1. Main concepts used in the Environment meta-model

πVOM concepts	Description
Entity	Specification of something that has an individual existence in the MAS.
Role	Specification of a behavioral pattern expected from some MAS entities.
Service	A single activity (or complex block of activities) that represents a functionality of the agents/devices/resources.
Environment	Physical world, workspace where the agents are situated.
Resource	Specification of a software artifact, that has a reasonable representation in the environment, which can be perceived and shared using data protocols.
Device	Specification of a hardware artifact, which we can perceive and act through low-level interfaces (using proprietary firmware).
Port	This abstraction is an interface to the service, that allows the input/output of data.
EnvironmentPort	Access point to interact with the environment (with the physical world).
ServicePort	Access point to use an agent-based service.
EnvironmentService	High-level functionality, which decouples the protocol or firmware of the environmental artifacts.

to the physical world. Finally, all these basic concepts and relationships will enable agents (and users) to create new high-level representations of the *Pervasive Systems*, as context-aware models[6], Service discovery/composition models[10], and adaptation models[24]. Table 1 summarizes the main concepts used in the Environment meta-model.

Finally, in order to illustrate the use of this meta-model, a typical case study of a *Pervasive System* is analyzed. The case study presents a little office where some pervasive services are provided. The office is structured in three locations: the reception desk, the head office and meeting room. Every location provides a lighting service which is activated when is detected in the space the presence of a user. The head office and the meeting room provide services for multimedia content (audio, video, and presentations). When employees make a presentation or use multimedia devices in the meeting room, the light intensity is decreased and the blinds are automatically closed to improve the visibility. The ubiquitous office provide others services, such as: security and recording; but, due to the space limitations of the paper, these functionalities are not described.

In order to model the ubiquitous office, the developer must specify the different components which model the different devices, resources, environments, services and agents, of the *Pervasive System*. Figure 3 shows the partial model (focused on meeting room) of the office using πVOM Environment meta-model. Figure 3 shows that the ubiquitous office uses different devices, such as: smart bulbs (using X10 protocol), gradual light bulbs and blinds automatic, are used to control the lighting room. Also, the ubiquitous office uses infrared sensors that are motion detectors and cameras. Furthermore, the developer can create new services such as Security and Recording, using the basic services (*EnvironmentService*) offered by each device.

5 Implementing Agent-Based Pervasive Systems

This work proposes to use a homogeneous and unified model for implementing *Agent-Based Pervasive Systems* permitting its translation into different execution MAS platforms through MDD, in which agents act/perceive about the environment and thereby manage/control the *Pervasive System*. This means that the user can design a *Pervasive Systems* with a unified, intuitive, visual model, i.e., with a high level of abstraction. Then, the user can get the agent code automatically using MDD with minimal user intervention. Finally, the drivers or firmware must be added to support environmental devices. These drivers will be encapsulated within a service (an OSGi service), to export their functionality as a high-level abstraction (which will be managed by agents). Finally, this code should be compiled for execution over an OSGi framework, as is showed in Figure 4.

5.1 Development Process

The design process starts trying to modelize the agents and the environmental devices using the abstract components of the proposed meta-models (commented in the previous section). The *Pervasive System* design process is formed by a set transformations that finally will obtain the OSGi-Java code. In order to do these steps, a set of tools, which support the process, are required. The tools used at each stage of the design can be summarized as follows (see Figure 4):

Step 1: At the beginning, the developer must create the different diagrams (using the EMFGormas toolkit[16]) which model the different devices, resources or services of the agents. To perform this step, an Eclipse IDE with a set of *plugins* is employed[16]. These plug-ins are mainly *EMF, Ecore, GMF* and *GEF*, which allow the user to draw the models that represent the *Pervasive System*.

Step 2: Once the model has been developed, it is necessary to select in which platforms the user wants to execute the agents. This phase corresponds with the PSM model definition of each agent. To do this, it is necessary to apply a model-to-model transformation (PIM-to-PSM). This is done using the Eclipse IDE and the ATL *plug-in* incorporating the appropriated set of transformation rules. It

Table 2. Transformation rules between agent meta-model and JADE-Leap model

Rule	Concept	Transformation
1	Agent	πVOM.Agent $\Rightarrow$ JADE.Agent
2	Behaviour	πVOM.Behaviour $\Rightarrow$ JADE.ParallelBehaviour
3	Capability	πVOM.Capability $\Rightarrow$ JADE.OneShotBehaviour
4	Task	πVOM.Task $\Rightarrow$ JADE.Behaviour

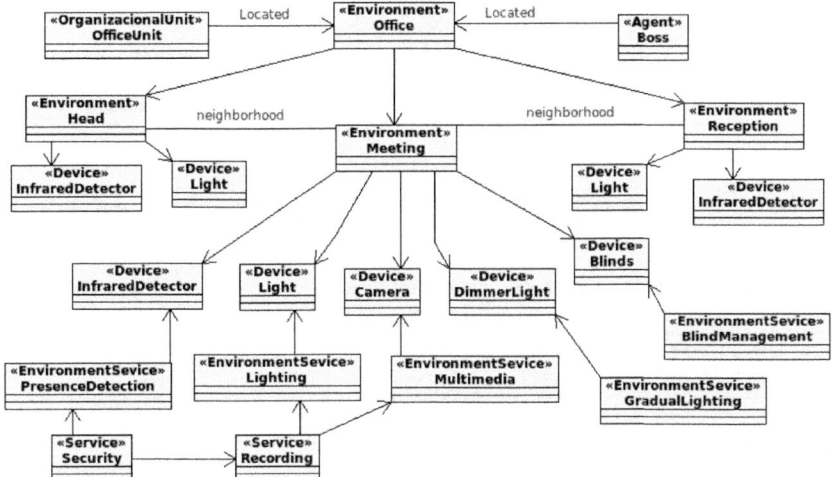

Fig. 3. Partial view of the office using Environment meta-model of πVOM

is important to remark that the same agent model can be transformed into different specific agent platforms. Table 2 illustrates the agent transformations, from agent meta-model of πVOM to JADE-Leap[8]. These rules are a subset of the transformation rules needed in this phase, which are explained in detail in [2]. In this way, agent concepts are mapped from source models to target models, and agent components are transferred or changed from one model to another.

Step 3: After the second step, the developer must apply a transformation to convert the models into the MAS code (OSGi-based). To do this, we must use a PSM-to-code transformation. In this case, we use *MOFScript* which is an Eclipse plug-in that uses templates to do the translation. These templates have been developed for two MAS platforms: JADE-Leap and ANDROMEDA[1]. Figure 5 illustrates how one rule is implemented using *MOFScript*. Part of the code of the rule shows the transformation of the *agent* concept.

Step 4: Finally, the process finishes by adding the necessary drivers for the different environment devices needed in the *Pervasive System*. All the functionality of physical devices are encapsulated as OSGi services, which allow agents to use them without worrying about low-level features. However, it is necessary to provide the driver/firmware/protocol of the new devices that are not in the OSGi *bundle* library. This application has a *bundle* library, which store the *EnvironmentService* (service devices) for frequent use or re-use.

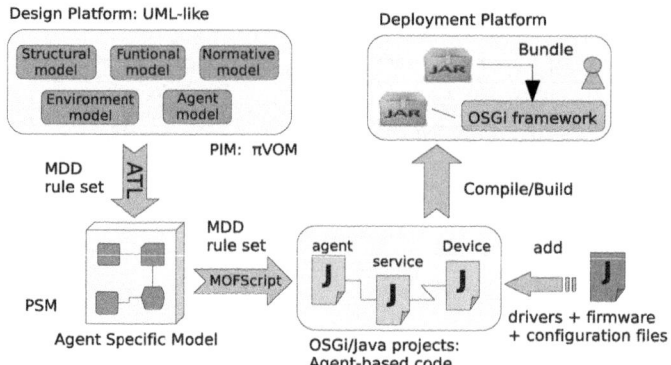

Fig. 4. Implementing Agent-based Pervasive System using MDD transformations

```
texttransformation AGENT2ANDROMEDA (in myAgentModel:uml2)
//Rule1: Agent transformation
uml.Package::mapPackage () {
  self.ownedMember->forEach(c:uml.Class)
     if (c.name != null)
        if (c.name = Agent) c.outputGeneralization()   }
uml.Class::outputGeneralization(){
  file (package_dir + self.name + ext)
  self.classPackage()
  self.standardClassImport ()
  self.standardClassHeaderComment ()
  <% public class %> self.name <% extends Agent { %>
     self.classConstructor()
     <% // Attributes   %>
     self.ownedAttribute->forEach(p : uml.Property) {
         p.classPrivateAttribute()
     } newline(2) ...
```

Fig. 5. *Agent* translation using MOFScript

5.2 Deployment Platform

Implementing *Pervasive Systems* is a challenging and exciting task, since a solid background knowledge about how to implement this kind of systems do no exist. Many research efforts are currently being developed on prototype implementations[15]. However, some *Pervasive System* prototypes share a common architectural style, which correctly fits the requirements of this systems.

As discussed above, the requirements of this kind of systems are basically two: (i) is essential that the *Pervasive System* must support the integration of services provided by external devices and software systems (as another services); (ii) the isolation of the low-level abstraction of the devices (manufacturer dependent), the environmental devices must be well encapsulated in independent and generic functionalities. Therefore, an architecture style that meets these requirements is the well-known layered architecture[15]. By means of this architecture, the sys-

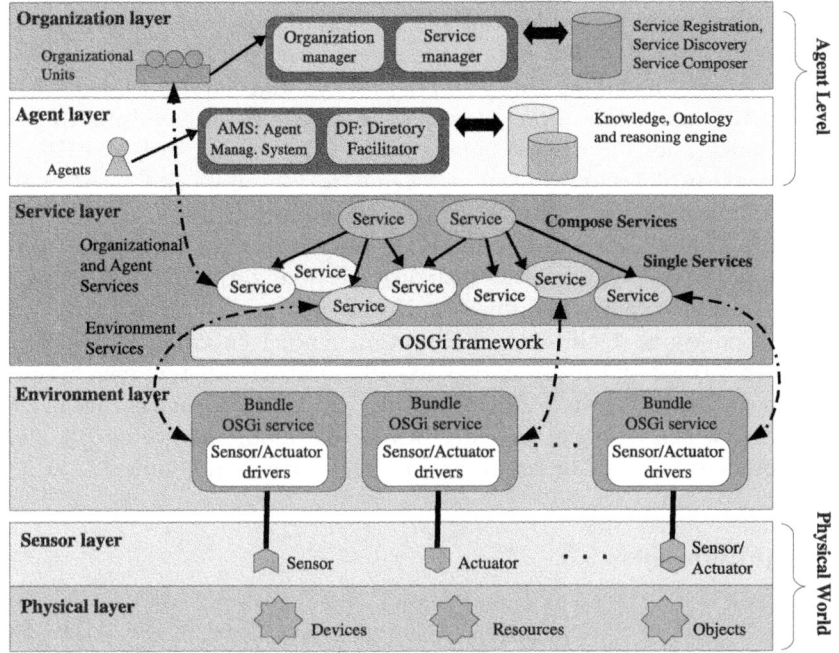

Fig. 6. Proposal architecture of Pervasive System

tem elements are organized in different levels with well-defined responsibilities. Our proposal follows this architecture style. Figure 6 shows our deployment platform, which is a framework based on OSGi technology. The main layers of this deployment platform are:

Physical layer, has the resources/devices that are perceived in the environment. Fully represents the real world, where the *Pervasive System* is located.

Sensor layer, has the responsibility of accessing to physical devices, through actuators and sensors, which allow to change or read the state of the devices.

Environment layer, has the responsibility of encapsulating the manufacturer dependent technology of the environment devices. The drivers that conform this level directly export their functionalities through a *bundle*. The *bundles*, which manage similar devices of software systems from different technology or vendors, are implemented as a common interface, in order to provide a uniform way of communicating within the environment devices.

Service layer, provides the system functionality, offering services that the *Pervasive System* must supply. The services are provided by the devices located in the physical world, and by the MAS entities (agent or organizational unit). Also, at this level the services can be *single* or *composite services*, which are formed by the composition of other services.

Agent layer, supports the mechanisms in order to register, de-register and discovery of agents. In this layer the agents work together through different interactions to support complex tasks in a collaborative and dynamic way. This layer also supports the information management (*knowledge*) and the needed *knowledge models*, including the *reasoning engine* and *ontologies* needed by the agents. Furthermore, this layer provides the necessary mechanisms to support the communications and needed languages used by agents, such as FIPA ACL.

Organizational layer, is used as a regulatory framework for the coordination, communication, and interaction among different computational entities. This layer is formed by a set of individuals and institutions that need to coordinate resources and services across institutional boundaries. This layer supports high level interoperability to integrate diverse information systems in order to share knowledge and facilitate collaboration among entities. This layer is an open system formed by the grouping and collaboration of heterogeneous entities. From a technical view, these functionalities are obtained using the THOMAS platform[12], which consists basically of a set of modular services that enable the development of agent-based organizations in open environments.

6 Conclusions

This work presents the application of the ideas proposed by the MDD for the design of *Agent-Based Pervasive Systems*. Although the use of MDD refers primarily to methodologies of object-oriented software, it was verified that the approach can be adopted in the development of *Agent-Based Pervasive Systems*.

The application of these technologies to the development of *Pervasive Systems* can provide many benefits: (i) the *Pervasive System* can be adapted to new devices technologies in an easy way, because the approach uses high-level abstraction that are not so hardware dependent; (ii) the development of a *Pervasive System* is more intuitive using this model-driven method than using classical approaches, because the developer models the *Pervasive System* as an UML-like model avoiding technical details; (iii) the implementation of *Pervasive System* over a service-based framework (as OSGi) using agent-oriented software engineering, allows agents to manage the services, the context, adapting the environment, giving the possibility to create more advanced and powerful models.

Future work of this research will focus on developing an explicit support to Context-Awareness specification, using ontologies to describe the contextual information, allowing new knowledge in the environment to be inferred.

Acknowledgment. This work was partially supported by TIN2009-13839-C03-01 and PROMETEO/2008/051 projects of the Spanish government and CONSOLIDER-INGENIO 2010 under grant CSD2007-00022.

References

1. Agüero, J., Rebollo, M., Carrascosa, C., Julián, V.: Towards on embedded agent model for Android mobiles. In: Proceedings of Mobiquitous 2008, pp. 1–4 (2008)

2. Agüero, J., Rebollo, M., Carrascosa, C., Julián, V.: Agent design using Model Driven Development. In: 7th Int. Conf. on PAAMS 2009, vol. 55, pp. 60–69 (2009)
3. Agüero, J., Rebollo, M., Carrascosa, C., Julián, V.: MDD for Virtual Organization design. In: Trends in Int. Conf. on PAAMS 2010, vol. 71, pp. 9–17 (2010)
4. Argente, E., Julian, V., Botti, V.: MAS Modeling Based on Organizations. In: Luck, M., Gomez-Sanz, J.J. (eds.) AOSE 2008. LNCS, vol. 5386, pp. 16–30. Springer, Heidelberg (2009)
5. Atkinson, C., Kuhne, T.: Model-driven development: a metamodeling foundation. IEEE Software 20(5), 36–41 (2003)
6. Baldauf, M., Dustdar, S., Rosenberg, F.: A survey on context-aware systems. International Journal of Ad Hoc and Ubiquitous Computing 2(4), 263–277 (2007)
7. Bauer, B.: UML Class Diagrams Revisited in the Context of Agent-Based Systems. In: Wooldridge, M.J., Weiß, G., Ciancarini, P. (eds.) AOSE 2001. LNCS, vol. 2222, pp. 101–118. Springer, Heidelberg (2002)
8. Bergenti, F., Poggi, A.: LEAP: A FIPA Platform for Handheld and Mobile Devices. In: Meyer, J.-J.C., Tambe, M. (eds.) ATAL 2001. LNCS (LNAI), vol. 2333, pp. 436–446. Springer, Heidelberg (2002)
9. Beydoun, G., Low, G., Henderson-Sellers, B., et al.: FAML: A Generic Metamodel for MAS Development. IEEE Trans. on Software Engineering, 841–863 (2009)
10. Brønsted, J., Hansen, K., Ingstrup, M.: A survey of service composition mechanisms in ubiquitous computing. In: RSPSI 2007 at Ubicomp (2007)
11. Cano, J., Madrid, N., Seepold, R., Aguilar, F.: Model-driven development of embedded systems on OSGi platforms. In: FDL 2007, pp. 1–6 (2007)
12. Carrascosa, C., Giret, A., Julian, V., Rebollo, et al.: Service oriented multi-agent systems: An open architecture. In: AAMA 2009, pp. 1–2 (2009)
13. Cervenka, R., Trencansky, I.: The Agent Modeling Language – AML. Whitestein Series in Software Agent Technologies and Autonomic Computing (2007)
14. Davidsson, P., Boman, M.: Distributed monitoring and control of office buildings by embedded agents. Inf. Sci. Inf. Comput. Sci. 171(4), 293–307 (2005)
15. Endres, C., Butz, A., MacWilliams, A.: A survey of software infrastructures and frameworks for ubiquitous computing. Mobile Inf. Syst. 1(1), 41–80 (2005)
16. Garcia, E., Argente, E., Giret, A.: A modeling tool for service-oriented Open Multi-agent Systems. In: Yang, J.-J., Yokoo, M., Ito, T., Jin, Z., Scerri, P. (eds.) PRIMA 2009. LNCS, vol. 5925, pp. 345–360. Springer, Heidelberg (2009)
17. Garca-Magario, I., Gómez-Sanz, J., Fuentes, R.: INGENIAS Development Assisted with Model Transformation By-Example. In: PAAMS 2009, pp. 40–49 (2009)
18. Hahn, C., Madrigal-Mora, C., Fischer, K.: A platform-independent metamodel for multiagent systems. In: AAMAS 2008, vol. 18(2), pp. 239–266 (2008)
19. Henricksen, K., Indulska, J.: Developing context-aware pervasive computing applications: Models and approach. Pervasive and Mobile Comp. 2(1), 37–64 (2006)
20. Huhns, M., Singh, M., Burstein, M., et al.: Research directions for service-oriented multiagent systems. IEEE Internet Computing 9(6), 65–70 (2005)
21. Knoll, M., Weis, T., Ulbrich, A., Brändle, A.: Scripting your home. In: Location and Context-Awareness, pp. 274–288 (2006)
22. Munoz, J., Pelechano, V., Fons, J.: Model driven development of pervasive systems. In: International Workshop MOMPES 2004, pp. 3–14 (2004)
23. OMG: Object management group. MDA guide version 1.0.1 (June 2008), http://www.omg.org/docs/omg/03-06-01.pdf
24. Poladian, V., Sousa, J., et al.: Task-based adaptation for ubiquitous computing. IEEE Trans. on System, Man, and Cybernetics 36(3), 328–340 (2006)

Adaptation Support for Agent Based Pervasive Systems

Kutila Gunasekera[1], Shonali Krishnaswamy[1],
Seng Wai Loke[2], and Arkady Zaslavsky[1,3]

[1] Faculty of Information Technology, Monash University, Australia
{kutila.gunasekera,shonali.krishnaswamy}@monash.edu
[2] Department of Computer Science & Computer Engineering, La Trobe University, Australia
S.Loke@latrobe.edu.au
[3] Luleå University of Technology, Sweden
arkady.zaslavsky@ltu.se

Abstract. Pervasive computing systems execute in dynamic highly variable environments and need software that are context-aware and can adapt at runtime. Mobile agents are viewed as an enabling technology for building software for such environments due to their flexibility, migratory nature and scalability. This paper presents a novel approach which aims to further enhance this advantage by building compositionally adaptive mobile software agents that are also context-driven, component-based and have the ability to exchange their components with peer agents. We present the formal underpinnings of our approach and a decision making model which assists agent adaptation. We also describe our current implementation and experimental results to evaluate the benefits of the proposed approach.

1 Introduction

As we move towards the era of *invisible computers* envisioned by Weiser [1], computers are becoming smaller and increasingly pervasive in day-to-day environments. Pervasive computers now have to execute in diverse and rapidly changing environments where low powered devices and wireless communication media are *de rigueur*. The miniaturization and portability of devices have also contributed to these devices having lower computing capacity in comparison to their stationary counterparts.

Developments in pervasive computing have been primarily driven by advances in hardware and communication technologies, whereas application development for such environments has lagged behind considerably. Pervasive applications need to be adaptive and versatile to survive rapidly changing environments and requirements. How to build such applications is a current challenge in pervasive computing research [2]. Self-adaptive software, autonomic computing and mobile software agents [3] are amongst some of the approaches seen as attractive options for pervasive application development. Flexibility, scalability and ability to reduce complexity by delegation are some of the desirable features that mobile agents bring to pervasive computing applications [4]. Pervasive applications are also likely to require mobility - a salient feature of mobile agents – in order to support mobile users as well as to migrate when faced with intermittent connectivity and device resource constraints. The ability of a statically defined agent to survive in an uncertain environment is limited. Thus, adaptive agent systems are being investigated to overcome this limitation [5].

Our research explores the use of compositionally adaptive software agents to build applications for pervasive environments. In our proposed approach, agents are lightweight, mobile and can autonomously adapt based on contextual input [6]. We aim to provide adaptations in agent capabilities in a manner that accommodates dynamic changes, with flexibility that is unprecedented compared to earlier work. This we see as important in pervasive environments where actual resources available at a particular location cannot be determined *a priori*, so that dynamic uptake and exchange of capabilities become useful. Context-awareness is another major requirement in pervasive applications [2] as they should be able to sense changes in the environment and adapt accordingly. For example, an application may need to monitor the power level of its current device and migrate to another node if the level falls below a threshold. Adaptation decisions are affected by multiple criteria such as resource availability, application semantics and user preferences. This paper discusses a multi-attribute cost model to assist agents in making dynamic adaptation decisions based on contextual and user given criteria. We formally define our cost model and present preliminary evaluation results which illustrate the benefits of the model. In particular, we show that an agent, using our cost model, and having access to contextual information about parameters of the model, can make accurate estimates of performance with different agent capabilities, and so, can make effective adaptation decisions at run-time.

The rest of the paper is structured as follows. In section 2 we present a conceptual overview of our compositionally adaptive agents. Section 3 describes the formal underpinnings of the approach and the adaptation cost model. A brief introduction to our prototype implementation is given next, followed by experimental evaluations in section 5 and some related work in section 6. Finally, we conclude in section 7.

2 Conceptual Overview

We propose a novel approach to develop smart pervasive applications through the use of dynamic compositionally adaptive mobile software agents [6]. The proposed VERSAG (VERsatile Self-adaptive AGents) agents are context-driven, adapt by acquiring new software components at runtime, and execute on dynamic heterogeneous environments. An agent's component-based structure allows it to have different architectures embedded within during its lifetime, and its useful functionality is provided in the form of reusable software components termed *Capabilities*. Two salient features are the ability of agents to acquire new behaviours from peer agents without depending on designated component providers, and an agent's ability to adapt itself based on contextual input. An agent's high-level task is to execute an *itinerary* assigned to it where an itinerary specifies a list of *locations* it has to traverse and activities to execute at each location. To carry out an activity, the agent may need multiple capabilities. It decides when and from where the necessary capabilities are acquired and may load necessary capabilities in advance, or load them at a later location based on criteria such as capability availability, number of locations a particular capability is required at, network cost and resource constraints at locations.

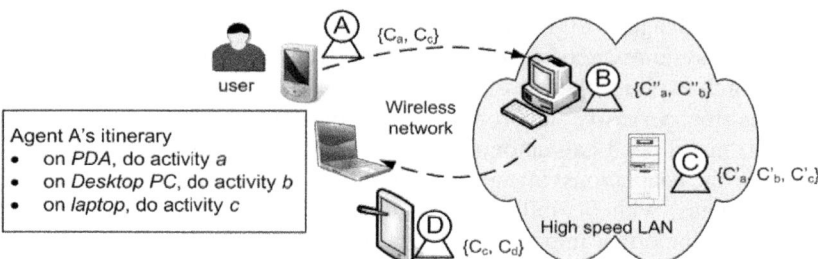

Fig. 1. A hypothetical scenario showing VERSAG agents deployed in a pervasive environment. The capabilities contained in each agent are shown beside it within braces, and the subscript corresponds to an activity that the capability can carry out.

Figure 1 illustrates a hypothetical scenario of how this approach can be useful in a pervasive computing environment. A human user requests his personal agent A, which resides on his PDA, to do a task. This task is converted into an itinerary as shown. The agent first does activity a using capability C_a. Then, discards its capabilities and migrates without carrying anything. On the desktop PC, it asks nearby agents for capabilities to do activity b. Agents B and C respond with C''_b and C'_b. A selects C''_b which is nearer, executes it, and then discards it before moving to the laptop. Once on the laptop, A searches for C_c to do activity c. D and C respond and D's response is selected because it is more suitable for resource-constrained devices.

If the agent has a choice of capabilities to select from, it needs to be able to select those which minimize execution *cost*. Time required to carry out the activity, generated network load and reliability, memory/CPU requirements, component accuracy, security, probability of reuse, monetary costs and user preferences are some of the many possible criteria which make up the cost. While many of these criteria are interrelated, it is neither necessary nor realistic to consider each one before a decision making step. Thus, this research limits itself to supporting the following main criteria: time, network load, maximum memory needs, CPU usage and level of accuracy of execution. Which costs should be minimized for a given situation can be explicitly specified by the user or inferred from environmental conditions.

Specific adaptation actions of an agent include stopping and starting a capability, discarding a capability, acquiring a new capability, migrating to a different location or terminating itself. Two other forms of adaptation an agent could undertake are changing its base implementation to support migration to a different platform or changing the order of visits in its itinerary. These two forms of adaptation are outside the scope of this research and are therefore deferred for further investigation.

3 The Formal Model

Having provided an overview of our proposal, in this section we formally define the key concepts that underpin our approach. We also describe how an agent executes an activity at a particular location and the adaptation cost model.

Definition 1: Adaptive agent
Let I be an agent's itinerary, $O = \{move, terminate, get\_c, drop\_c, start\_c, stop\_c\}$ be the set of operations an agent can perform, with $*\_c$ operations performed on capabilities and, C be the set of capabilities that the agent is carrying. Then an agent A can be defined as a tuple of the form (I, O, C).

Definition 2: Capability
A capability is a central concept in our approach and provides an agent with application specific behaviours. Let U be a set of unique identifiers, F be the set of functionalities that the capability contains, $Y = \{primitive, compound\}$ be a set of capability types, E be the set of environments on which the capability can execute and M represent meta-data about the capability such as owner, version, security certificates, algorithms/units used and optimizations. Then, a capability is a tuple of the form (u, F, y, E, M) where $u \in U$ and $y \in Y$. Functionality of a capability could range from a simple database query execution to providing the agent with a new architecture such as the BDI model.

Definition 3: Capability Specification
A capability specification (spec) is used to identify capabilities that can fulfil a given set of functions and can execute in a given set of environments. Let F represent the set of functions the capability contains and E define the set of environments on which the capability can execute. A capability spec is then a tuple of the form (F, E).

Given a capability spec $cs = (F_s, Env_s)$ and a capability $c = (u, F, y, Env, M)$, capability c will *match* the spec cs if and only if $F_s \subseteq F$ and $Env_s \subseteq Env$. That is, for a capability to *match* a capability spec, the spec's full set of functionalities and supported environments must be supported by the capability.

Definition 4: Activity
An activity is the unit of work an agent has to carry out at a particular location. Multiple capabilities may need to be executed in sequence to carry out the activity. Let CS be a set of capability specs and O be a set of operations $\{get\_c, drop\_c, start\_c, stop\_c\}$. A tuple (cs, o) where $cs \in CS$ and $o \in O$ represents application of an operation on a capability. An activity A is defined as a finite sequence $(cs_1, o_1), (cs_2, o_2), ..., (cs_n, o_n)$ where $cs_i \in CS$, $o_i \in O$, $i \in \{1, 2, .., n\}$ and n is the number of capabilities in the activity.

A *location* is a place an agent can visit and contains the necessary runtime environment for the agent to exist. Various computational resources are available at a location. A capability can only use resources available at its current location. *Cost elements* are used to represent different types of costs incurred during itinerary execution and have integer values. *Constraints* place limits on cost elements and can be associated with a location, an itinerary, an itinerary step, a capability or an agent. *Relative constraints* are generally user specified and used to indicate relative weights of cost elements. *Absolute constraints* place limits on permissible value ranges for cost elements.

Definition 5: Relative constraint
A relative constraint r is a finite sequence of the form $(q_1, w_1), ..., (q_i, w_i), ..., (q_n, w_n)$ where $q_i \in Q$ the set of cost elements and $w \in \mathbb{Z}$ represents the weight of that cost element.

Definition 6: Absolute constraint
An absolute constraint places an upper and lower limit on a cost element. It is a tuple of the form *(q, min, max)* where $q \in Q$ the set of cost elements, *min < max* and *min, max* $\in \mathbb{Z}$ represent the lower and upper limits of the cost element. For example, *(time, 0, 10)* means the "time" cost element must have a value between 0 and 10.

Definition 7: Itinerary
An itinerary consists of a sequence of *itinerary steps* the agent has to carry out. An itinerary step is a tuple *(l, A, R)* where *l* is a location, *A* an activity and *R* a set of constraints. Let *L* be the total set of locations and *A* the set of activities. Then, an itinerary is defined as a finite sequence $(l_1, a_1, R_1), ..., (l_i, a_i, R_i), ...,(l_n, a_n, R_n)$ where $l_i \in L, a_i \in A$ and $i \in \{1, 2, .., n\}$. Activity $a_i \neq \emptyset$ while constraints can be empty.

While an itinerary is defined as above, it is expected that users would be issuing high-level commands to the system, which would be converted to detailed itineraries for agents to execute.

We next describe an agent's activity execution process. The following discussion assumes that the agent first migrates to the relevant location and then searches for and acquires capabilities required at that location. Therefore the activity fails if suitable capabilities could not be found. Figure 2 provides an algorithmic description of this process.

```
1 Let CS = {cs₁, cs₂..., csₘ} be the set of capability specs which can
  be used to carry out activity a
2 Search for capability instances matching CS
3 Combine capability instances to form groups of capabilities
  G = {g₁, g₂ ..., gₚ} where each gᵢ can fulfil activity a
  If G = ∅
    Fail
  End if
4 Build set of constraints CONS = {r₁, r₂ ..., rₙ} from explicit and
  implicit constraints identified through context sensing.
5 Sort G in increasing order of cost By applying the adaptation cost
  model.
6 For each gᵢ in G
    Acquire capability instances of gᵢ by applying operation get_c()
    If acquisition successful
        Break loop
    End if
  End for
  If a completed group is not available
    Fail
  End if
7 Execute capabilities in correct sequence to fulfil activity
```

Fig. 2. Activity execution algorithm of an agent

Step 1: Identify sequences of capability specifications that can accomplish the given activity. It is possible that more than one such sequence exists. Activity to capability spec group mappings may be contained in the itinerary, stored with the agent or available from an external source.

Step 2: For each capability spec, search for matching capability instances in the agent's local repository and from peer agents. The search mechanism itself is implementation specific and is expected to be changeable.

Step 3: Combine received capability descriptions to build groups that can accomplish the given activity. If no groups can be formed, the activity has to fail. Group formation needs to take into account limitations of capability instances and their compatibility with each other.

Step 4: The agent senses environmental conditions at the current location and identifies applicable implicit constraints. These are combined with the explicitly specified constraints to build the complete set of constraints.

Step 5: Apply the *adaptation cost model* to sort the capability groups according to cost. Constraints, capability group details and available capability descriptions are used as input to the selection process. Groups that don't meet absolute constraints are removed.

Step 6: Select the least cost capability group and acquire the relevant capability instances (if they are not already with the agent). If acquisition fails due to any reason, acquire the next best capability group.

Step 7: Execute the acquired capabilities in correct sequence.

The adaptation cost model of step 5 above is illustrated in Figure 3.

```
1    Let Q = {q₁, q₂, … q_k} be the set of cost elements
     relative constraint {(q₁, w₁), … (q_m, w_m)} where m ≤ k
     Cons = {con₁,con₂,… con_n} be absolute constraints where n ≤ k
     G = {g₁, g₂, … g_p} be the set of alternative groups
2    For each con_i in Cons with cost element q_i
        For each g_j in G
            Estimate cost in terms of q_i for g_j
            If estimated cost does not satisfy con_i
                Remove g_j from G
            End if
        End for
     End for
     If G = ∅
        Fail activity
     End if
3    From relative constraint, build normalized priority vector P_{k×1}
4    For each g_i in G
        For each q_j part of a relative constraint
            If cost estimate of g_i not available
                Estimate cost in terms of q_j for g_i
            End if
        End for
     End for
5    Build utility matrix U_{p×k} from group cost estimates
6    Calculate utility vector V_{p×1} ← U_{p×k} × P_{k×1}V
     Sort V in decreasing order of utility
```

Fig. 3. Adaptation cost model

The input to the cost model consists of cost criteria, relative constraint (provides weights of each criterion), absolute constraints and a list of available groups. Step 2 evaluates each capability group against the *absolute constraints* and discards groups that fail to meet the necessary constraints. If no groups remain at the end of this step, the process fails. Step 3 normalizes the weights from the relative constraint to build a normalized priority vector of the cost elements. In step 4, for remaining groups, costs are estimated for elements that are part of the relative constraints. Step 5 builds a utility matrix from these estimated costs. Given that there are p alternative groups remaining and k different cost criteria, the utility matrix is an $p \times k$ matrix where element u_{ij} is the reciprocal of the estimated cost value of group g_i for cost criterion q_j (except for "accuracy" criterion which is used as it is). The reciprocal of the cost value is used as an indicator of the benefit or "utility" that can be gained by selecting a particular group. Thus, lower the estimated cost, higher the utility gained. In step 6, by multiplying this utility matrix with the priority vector, which indicates the relative importance of each cost element, we obtain a vector which provides aggregate utilities for each group. The group with highest utility is the one that has lowest cost and is to be selected by the agent.

4 Implementation Details

This section presents the prototype implementation of the VERSAG concepts and theory previously discussed. It is built on top of the JADE agent toolkit [7] with an OSGi [8] based capability model. Figure 4 illustrates the structure of an agent. The agent, as previously mentioned is itinerary-driven at its core with the kernel driving this behavior. The capability repository stores the agent's application specific capabilities. The itinerary service holds the agent's itinerary and provides methods to interpret itinerary commands. The capability execution service provides the means to load, run and stop capabilities that are available in the repository. The capability exchange service (not in figure) fulfils the dual roles of a capability requestor and provider. In its provider role it listens for capability requests from peer agents and respond as appropriate. The requestor role gives an agent the ability to request capabilities from peers. The base JADE agent is the framework's point of contact with the underlying agent platform and provides access to services such as mobility and communication. Capabilities themselves are agent-platform agnostic and can

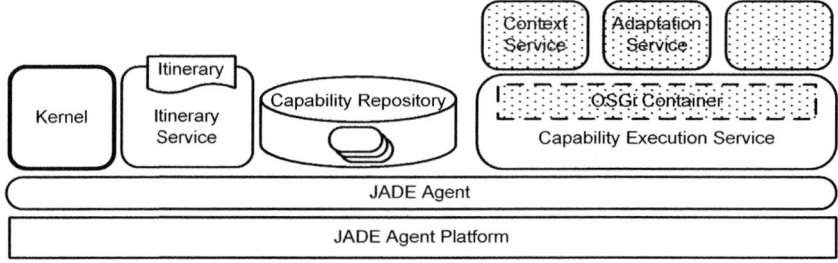

Fig. 4. Structure of a JADE based VERSAG agent

execute wherever a suitable execution environment is made available. Since the context service and adaptation service are implemented as capabilities, it is possible for a VERSAG agent to switch to different implementations of these services or be stripped of them to become a lightweight itinerant agent. An agent's reference architecture and more design details are described in [6].

This implementation limits itself to considering time, network load, maximum memory needs, CPU usage and level of accuracy as the cost criteria to be considered before adaptation. Itinerary execution *time* is often a crucial factor that needs to be minimized. It is also desirable to reduce *network load*, especially in wireless networks. Hence, these two are the primary cost elements supported. For a mobile agent, it is difficult to make design-time assumptions about the computational resources that would be available during its lifetime, as it traverses devices with different enabling opportunities. Thus, it is desirable if the agent can dynamically adapt itself to work with available computational resources. A VERSAG agent is able to achieve this by dynamically changing its constituent components. For example, an agent maybe required to execute an activity on a resource-constrained device, which has limited memory, processor and battery capacity. Available resources may be further limited by having to share them with a number of other agents. Memory and CPU cycles are two key resources that have to be thus shared. However, due to differences in computer architectures, hardware and system software, it is difficult to provide estimates of CPU or memory requirements of a software component in a useful platform-independent manner. Therefore, we use a *CPU usage ranking* which can be used as a relative indicator of a capability's processing needs. For memory, the maximum required *heap memory size* in the Java Virtual Machine is used as a criterion. In certain applications, it is desirable to generate an approximate result using fewer resources rather than a more accurate output which is considerably more resource-intensive. The cost element of *accuracy* is incorporated into the decision making model to reflect this requirement. Thus, an agent may decide to select a less accurate capability over a more accurate one in order to reduce overall costs.

5 Experimental Evaluation

This section describes experimental evaluations carried out to verify the performance of the agent adaptation cost model. The experiments were conducted with two main goals: to check whether an agent correctly selects the least cost alternative from among many, and to check whether the generated estimates are representative of the actual costs. We describe two sets of experiments: one using *network load* and the other using *time* consumed as the main cost criterion. No absolute constraints are specified in both cases. Three separate workloads representing a simple agent GUI, an information retrieval task and a file sorting task were used. Time based tests were conducted over a high-speed LAN and an IEEE 802.11g wireless network while a 3G wireless broadband link was also used for load based tests. The three test computers used each had comparable computational capacity and were running Windows XP, Vista and Ubuntu Linux. All had Java SDK 1.6.0 and JADE version 3.7 with the LEAP add-on was used. We first describe the formulae used by the prototype to estimate costs followed by the experimental results.

The total network load generated when executing an itinerary step consists of the load due to execution of the capabilities as well as the load generated when the agent searches for and acquires the relevant capabilities. Given that an alternative group has n capabilities, the total network cost of selecting that group is made up of the load generated during adaptation decision making (B^{search}), load generated during the actual acquisition of the capabilities (B_i^{acq}) and any network load generated during capability execution (B_i^{exec}). The total cost can be expressed as shown in equation 1.

$$B^{total} = B^{search} + \sum_{i=1}^{n}(B_i^{acq} + B_i^{exec}). \quad (1)$$

Further, B_i^{acq} is made up of request size (B^{creq}), capability size (B_i^{cap}) and the overhead of each request (B^{ovrhd}) as shown in equation 2. It is assumed that the size of a capability request and overhead of each request is constant for a given network.

$$B_i^{acq} = B^{creq} + B_i^{cap} + 2B^{ovrhd}. \quad (2)$$

The network cost of capability execution (B^{exec}) is functionality and implementation specific and it is expected that a fixed value or a formula to estimate the same will be provided with the capability's meta-data. For the experimental workloads used, network traffic generated during execution was nil.

For an alternative that has n capabilities; the time to sequentially acquire the capabilities and execute the itinerary step can be expressed as follows:

$$T = T^{search} + \sum_{i=1}^{n}\left(T_i^{cap\_req} + T_i^{cap\_res} + T_i^{start\_p} + Tp_i^{loc}\right). \quad (3)$$

Here T^{search} is the time taken for the decision making process, $T_i^{cap\_req}$ the time to request a capability, $T_i^{cap\_res}$ the time to receive a capability in response, $T_i^{start\_p}$ the time to start execution of the capability, and Tp_i^{loc} the time to execute it at location loc. The time T, to move B bytes over a network link with latency λ and bandwidth bw can be represented as $T = \lambda + B/bw$. Using this in equation 3 we get:

$$T = T^{search} + \sum_{i=1}^{n}\left(2\lambda_i + \frac{(B_i^{cap} + B^{creq} + 2B^{ovrhd})}{bw_i} + T_i^{start\_p} + Tp_i^{loc}\right). \quad (4)$$

Every prototype agent is equipped with a context-sensing capability which measures network parameters. While it could measure bandwidth and latency, for our experiments request size, overhead, load and time for decision making were estimated manually using trial runs and supplied to this context-sensing capability. For the wireless network, the context-sensing capability measured the average latency as 1ms and average bandwidth available to the agents as 13.5Mb/sec. For the LAN, latency was negligible and average available bandwidth was 81.3Mb/sec. Based on our trial runs it was concluded that B^{search} can be estimated as $(3900 \times no.\,of\,provider\,agents)$. Similarly, request size and message overhead were estimated to be 224 and 2000 bytes respectively while decision making time (T^{search}) was estimated to be 300 milliseconds. It is expected that in real-life situations the context-sensing capability makes these estimations. Of the needed capability meta-data, capability size was measured by the peer agent. Time to start a capability and execution time were also estimated in trial runs and added to its meta-data. During the experiments, Wireshark was used to measure the actual network load generated and code was instrumented to log actual time consumption.

Based on experimental data, the error between estimated and actual values was calculated as a percentage of the estimated value. For the load based experiments, mean percentage error was 7.17% with a maximum of 13.5%. For time based experiments, higher error percentages could be observed when the estimates were less than 2 seconds whereas it was less than 10% when the estimate was larger than 2 seconds. In figures 5 and 6 we compare the estimated and actual costs for the two sets of experiments. They clearly illustrate that our estimates are representative of the actual cost, an indication that the strategies used to estimate time and network load are accurate. Since they are used by the cost model to make adaptation decisions, it is essential that the estimates are accurate.

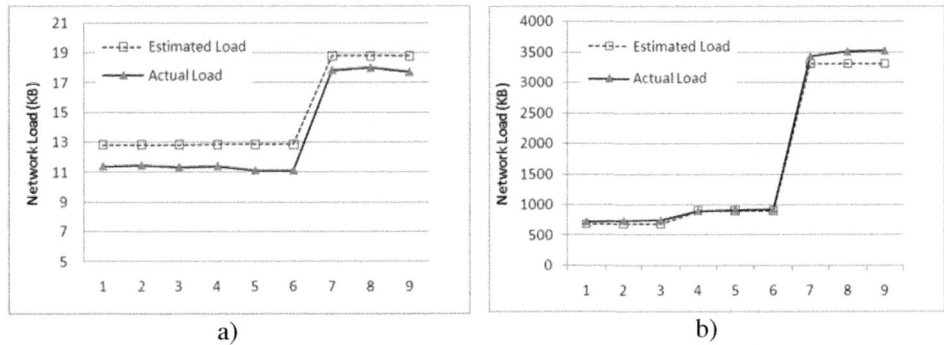

Fig. 5. Comparison of Actual and Estimated Loads for the tests, sorted in ascending order of estimate size. Two graphs are plotted for clarity, each graph containing 9 tests.

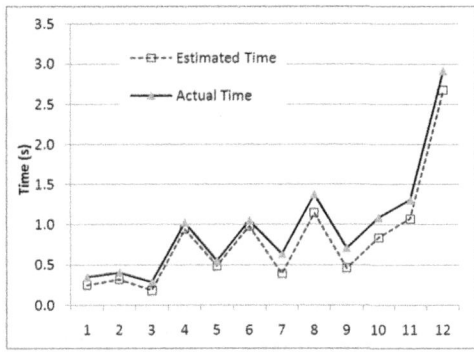

Fig. 6. Comparison of average Actual and Estimated time consumption. The first 6 tests were on a LAN environment and the following 6 conducted over a WLAN.

Table 1. Predicted utility values and actual costs for the three workloads, each with two available alternatives a and b are shown. Load and time values shown are averages.

Workload	Alternative	Network Load (KB) Utility	Network Load (KB) Actual	Time (ms) Utility	Time (ms) Actual
1	a	0.021	731.0	0.78	709
1	b	0.971	17.8	0.22	1089
2	a	0.214	3495.9	0.75	1306
2	b	0.786	913.5	0.25	2919
3	a	0.4995	11.2	0.90	638
3	b	0.5004	11.4	0.10	1378

In Table 1, we compare predicted utility values for the three workloads with actual measured values. Columns 3 and 4 are for the first set of experiments with *network load* as the constraint. For the first two workloads the agent correctly predicts higher utility values while in workload 3, the utilities are identical to three decimal places and the actual difference between the two workloads less than 200 bytes. In columns 5 and 6, *time* is the constraint and we see that all predictions are accurate. The results clearly indicate that the cost model is able to correctly select the least cost alternative for the agent to fulfil its goals.

While our experimental results validate the accuracy of the model, prediction accuracy is dependent on multiple factors. Primary among these are the meta-data provided by capability developer which includes time to start and execute a capability and network traffic generated by the executing capability. These are in turn likely to depend on other factors such as the runtime environment and data being processed. We expect that a capability's temporal and traffic generation behaviours will be tested on a standard environment and used to build the meta-data which in term could either be constant values or formulae for use by the adaptation cost model. These then need to be adapted by the agent to suit its current environment. Non-deterministic nature of the environment, especially the network, is another primary factor which can affect the prediction accuracy.

A major feature of our approach is the ability to simultaneously aggregate multiple criteria for decision making. The accuracy of cost aggregation has been separately tested using synthetic data and is not reported here. Our twin aims were to check whether the predictions are representative of the actual costs and whether the alternative with lowest cost is correctly selected. As per the experimental results, it is evident that the proposed approach achieves these two aims and can be beneficial for adaptation decision making in pervasive environments.

6 Related Work

We briefly list some previous research where mobile agents were applied to pervasive computing scenarios. GAMA (Generic Adaptive Mobile Agent architecture) [9] builds compositionally adaptive mobile agents from ground up for use in ubiquitous computing environments. The open source JADE agent platform [7] also provides rudimentary support for runtime adaptation of agent functionality. Preuveneers and Berbers [10] describe a non-agent solution to mitigate service disconnections in mobile ad hoc networks with dynamically migrating services. Further related research is listed in [6]. Our approach aims to overcome various limitations found in these and to provide a flexible component based solution to build intelligent adaptive agents suited for pervasive environments.

7 Conclusions

Building software for pervasive computing environments is a current challenge in pervasive computing research. Our goal was to build an approach to address this challenge while making use of desirable features provided by the agent paradigm. To this end, we proposed a novel context-driven component based mobile agent framework. As contributions of this paper, we presented the formal underpinnings of the approach and described a multi-criteria decision making model which assists the agents make context and user preference driven adaptation decisions at runtime. A prototype implementation of this novel approach was also described.

Experimental evaluations were carried out using three different workloads on wired and wireless networks. The results confirm that our estimation strategies for time and network load are accurate and allow agents to effectively select the least cost alternative to carry out its given task when multiple options are present. We are currently engaged in further evaluations and intend to verify the scalability of our approach in environments with large numbers of agents and alternatives. As future work, we also plan to implement a case-study scenario to demonstrate the benefits of our approach in real-life pervasive computing applications.

References

1. Weiser, M.: The Computer for the 21st Century. Scientific American (1991)
2. Niemelä, E., Latvakoski, J.: Survey of requirements and solutions for ubiquitous software. In: 3rd International Conference on Mobile and Ubiquitous Multimedia, pp. 71–78. ACM, College Park (2004)
3. Cardoso, R.S., Kon, F.: Mobile Agents: A Key for Effective Pervasive Computing. In: ACM OOPSLA 2002 Workshop on Pervasive Computing, Seattle (2002)
4. Zaslavsky, A.: Mobile agents: can they assist with context awareness? In: IEEE International Conference on Mobile Data Management, pp. 304–305. IEEE Press, Los Alamitos (2004)
5. Marín, C.A., Mehandjiev, N.: A Classification Framework of Adaptation in Multi-Agent Systems. In: Klusch, M., Rovatsos, M., Payne, T.R. (eds.) CIA 2006. LNCS (LNAI), vol. 4149, pp. 198–212. Springer, Heidelberg (2006)
6. Gunasekera, K., Krishnaswamy, S., Loke, S.W., Zaslavsky, A.: Runtime Efficiency of Adaptive Mobile Software Agents in Pervasive Computing Environments. In: ACM International Conference on Pervasive Services (ICPS 2009), pp. 123–132. ACM, London (2009)
7. Jade - Java Agent DEvelopment Framework, http://jade.tilab.com/
8. About the OSGi Service Platform: Technical Whitepaper. OSGi Alliance, pp. 1–19 (2007)
9. Amara-Hachmi, N., Fallah-Seghrouchni, A.E.: Towards a Generic Architecture for Self-Adaptive Mobile Agents. In: European Workshop on Adaptive Agents and Multi-Agent Systems, Paris (2005)
10. Preuveneers, D., Berbers, Y.: Pervasive Services on the Move: Smart Service Diffusion on the OSGi Framework. In: Sandnes, F.E., Zhang, Y., Rong, C., Yang, L.T., Ma, J. (eds.) UIC 2008. LNCS, vol. 5061, pp. 46–60. Springer, Heidelberg (2008)

Mining Emerging Sequential Patterns for Activity Recognition in Body Sensor Networks

Tao Gu[1], Liang Wang[1,2], Hanhua Chen[1,3],
Guimei Liu[4], Xianping Tao[2], and Jian Lu[2]

[1] University of Southern Denmark
{gu,wang,hhchen}@imada.sdu.dk
[2] Nanjing University
{txp,lj}@nju.edu.cn
[3] Huazhong University of Science and Technology
[4] National University of Singapore
liugm@comp.nus.edu.sg

Abstract. Body Sensor Networks offer many applications in healthcare, well-being and entertainment. One of the emerging applications is recognizing activities of daily living. In this paper, we introduce a novel knowledge pattern named Emerging Sequential Pattern (ESP)—a sequential pattern that discovers significant class differences—to recognize both simple (i.e., sequential) and complex (i.e., interleaved and concurrent) activities. Based on ESPs, we build our complex activity models directly upon the sequential model to recognize both activity types. We conduct comprehensive empirical studies to evaluate and compare our solution with the state-of-the-art solutions. The results demonstrate that our approach achieves an overall accuracy of 91.89%, outperforming the existing solutions.

Keywords: Body sensor networks, activity recognition, data mining.

1 Introduction

The recent advance of wireless sensing and the development of miniaturized sensors have led to the use of Body Sensor Networks (BSNs). A BSN consists of a number of sensor nodes, placed or implanted on a human body, which provide sensing and wireless communication capabilities. Such systems offer many promising applications in healthcare, assistive living, well-being, sports, and entertainment. One of the applications is recognizing activities of daily living. In such an application, user observations in the form of a continuous sensor data stream are collected from various sensor nodes and transmitted to a gateway device. Useful features will be first extracted from the sensor data to train an appropriate activity classifier, which can then be used to identify new observations.

Recognizing activities using BSNs has attracted many research interests from academic researchers and industry participants. Most of the existing work focus on recognizing sequential activities (i.e., one activity after another in a timeline) in different settings [1–5]. However, the situations in real life are more complex since people often multitask when performing their daily activities. Such multitasking can occur in an interleaved (i.e., switching between the steps of two or more activities) or concurrent (i.e., performing two or more activities simultaneously) manner. Little work has been done in addressing complex issues rise in recognizing sequential, interleaved and concurrent activities in a unified framework. Existing solutions, such as Interleaved Hidden Markov Model [6] and Factorial Conditional Random Field [7], rigidly model interleaved or concurrent activities using Markov chains which require proper training. However, in real life, there exists many different ways in which activities can be interleaved and performed concurrently, e.g., when performing an activity, one may start another activity any time interleavedly or concurrently. Hence, it may not be feasible, if not impossible, to construct the complete models for interleaved and concurrent activities through training.

In this paper, we investigate an efficient way to recognize both simple and complex activities. By analyzing the trace involving both simple and complex activities, we observe many unique feature sets for each sequential activity and there exists an inherent sequential order among these features. In addition, we observe that the sequential order of the feature sets will not be changed regardless of whether this activity is performed sequentially, interleavedly, or concurrently. Intuitively, such sequential orders correspond to the intermediate steps or subactions of an activity, e.g., in the *brushing teeth* activity, there exists a sequential order—*taking a toothbrush, squeezing toothpaste, brushing,* and *washing with water*. Discovering such sequential orders (complete or partial orders) provides an additional, useful discriminator to recognize the activities which are performed interleavedly or concurrently.

To address the aforementioned problem, we aim to eliminate the training process of interleaved and concurrent activities. To achieve this, we propose a novel knowledge pattern, named Emerging Sequential Pattern (ESP), to capture unique contrast *sequential* patterns among different activity classes, and build our activity models based on ESPs. While the sequential activity model requires a training process, our complex activity models are built directly upon the sequential model without training. In this way, our system has a great flexibility and applicability for real-life applications. We design and implement our activity models and recognition algorithms in our ESP-based Activity Recognizer (ESPar). Through comprehensive empirical and comparison studies, we demonstrate both the effectiveness and flexibility of our proposed system.

The rest of the paper is organized as follows. Section 2 discusses the related work. Section 3 describes ESP and the mining algorithm. We then present ESPar in Section 6. Section 7 reports our empirical studies, and finally, Section 8 concludes the paper.

2 Related Work

To classify activities in BSNs, probabilistic models are typically used due to the non-deterministic nature of human activity. Probabilistic models can be categorized into static and temporal classification. In the static model, features are first extracted from sensor readings, and then a static classifier is applied for classification. Typical static classifiers include naïve Bayes used in [1,2], decision tree used in [1,9], and k-nearest neighbor used in [1,2]. Multiple binary classifiers can be exploited to recognize interleaved and concurrent activities; however, this solution may not work properly because many activities share the common features.

In temporal classification, state-space models are typically used to enable the inference of activity labels. We name a few examples here: Hidden Markov Models (HMMs) used in [3,4] and Conditional Random Fields (CRFs) used in [4,6]. Recent work showed that Interleaved Hidden Markov Model [6] can be used to model interleaved activities and Factorial Conditional Random Field [7] can be used to model concurrent activities. However, they require a predicting instance must have its model presented in the training dataset. On one hand, this implies that the training dataset has to be large enough to build the complete models for interleaved and concurrent activities. Any partial model will result in loss of accuracy. On the other hand, in real life, there exists a great variety of ways in which activities can be interleaved and performed concurrently by different individuals. As a consequence, it may not be possible to obtain the complete models for interleaved or concurrent activities through training. Hence, the applicability and flexibility of these solutions are limited.

The solution presented in this paper is fundamentally different than the existing work by introducing a new knowledge pattern—ESP. We use ESP to build the models for interleaved or concurrent activities directly on the sequential activity model without training. In this way, our system is more easy to train. ESPs extend our earlier approach of Emerging Pattern based model [8] by taking the sequential order of observations into account, and mining unique contrast, sequential patterns to achieve better classification accuracy.

3 Emerging Sequential Pattern

In this section, we introduce ESP and present an algorithm to efficiently mine ESPs. ESP is motivated by the concept of Emerging Pattern [11] which is a kind of contrast patterns among unordered features/items. ESP greatly extends Emerging Pattern by considering the inherent sequential information among sequences and discovering the unique contrast sequential patterns.

3.1 Definitions

Let $I = \{i_1, i_2, \cdots, i_n\}$ be a set of items and $\mathcal{C} = \{C_1, C_2, \cdots, C_k\}$ be a set of class labels. An itemset is a subset of I. A sequence is an ordered list of non-empty itemsets, denoted by $s = \langle e_1, e_2, \cdots, e_m \rangle$, where e_i is a non-empty itemset and it is also called an element of s. Each sequence has a class label $C_i \in \mathcal{C}$. We use $label(s)$ to denote the class label of a sequence s. In the context of activity recognition, items are feature items, elements are feature vectors, sequences are segments of traces within a continuous period of time that belong to the same activities, and the class labels of the sequences are the activities they belong to.

Given two sequences $s_1 = \langle a_1, a_2, \cdots, a_n \rangle$ and $s_2 = \langle b_1, b_2, \cdots, b_m \rangle$, if there exists integers $1 \leq i_1 < i_2 < \cdots < i_n \leq m$ such that $a_1 \subseteq b_{i_1}, a_2 \subseteq b_{i_2}, \cdots, a_n \subseteq b_{i_n}$, then we say s_1 is a subsequence of s_2, and s_2 is a super-sequence of s_1, denoted as $s_1 \prec s_2$. If $s_1 \prec s_2$, we also say s_2 contains s_1.

Definition 1. *(Support of a Sequential Pattern)* Given a sequence database D, the support of a sequence s in D is defined as the number of sequences in D that contain s, denoted as $sup_D(s) = |\{s_i | s \prec s_i, s_i \in D\}|$.

Note that the support of a pattern is defined as the number of distinct sequences containing it instead of the total number of occurrences. The support of s in D with respect to a particular class C is defined as the number of sequences in D with label C that contain s, that is, $sup_D(s, C) = |\{s_i | s \prec s_i, s_i \in D, label(s_i) = C\}|$.

Definition 2. *(Frequent Sequential Pattern)* Given a user-specified minimum support threshold min_sup, if $sup_D(s) \geq min\_sup$, then we say s is a frequent sequential pattern in D.

We are interested in sequential patterns that can distinguish sequences of different classes. We expect such patterns to occur frequently in one class and rarely in other classes. The discriminative power of a sequential pattern s with respect to a class C is defined as follows:

$$dpower(s, C) = p(C|s) \cdot p(\overline{C}|\overline{s}) = \frac{sup_D(s, C)}{sup_D(s)} \cdot \frac{(|D| - sup_D(s)) - (sup_D(C) - sup_D(s, C))}{|D| - sup_D(s)}$$

where $p(C|s)$ is the probability of class C if s is present, and $p(\overline{C}|\overline{s})$ is the probability of other classes if s is absent. The range of $dpower(s, C)$ is [0,1]. The higher the $dpower$ value is, the more discriminative the pattern is. In the ideal case, all sequences containing s belong to class C ($p(C|s)=1$), and all the sequences that do not contain s belong to classes other than C ($p(\overline{C}|\overline{s})=1$), and we have $dpower(s, C)=1$.

Definition 3. *(Emerging Sequential Pattern)* Given a user-specified minimum support threshold min_sup and minimum discriminative power threshold ρ, if $sup(s) \geq min\_sup$ and $dpower(s, C) \geq \rho$, then we say s is an emerging sequential pattern with respect to class C.

Our task here is to enumerate the ESPs with $sup \geq min\_sup$ and $dpower \geq \rho$ from training sequences, and then use them to recognize the activities of new sequences. To mine ESPs, we use the pattern growth approach and propose an algorithm named ESPMiner presented in the next section.

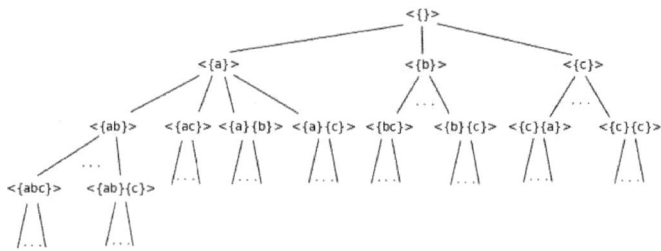

Fig. 1. Our mining framework

3.2 Mining ESPs

The search space of the sequential pattern mining problem can be represented as a prefix tree as shown in Fig. 1. In this tree, each node represents a sequential pattern, and it is a prefix of its children nodes. A sequential pattern $s = \langle e_1, e_2, \cdots, e_k \rangle$ can be extended in two ways: (1) element extension: one new item is added to the last element e_k; and (2) sequence extension: one new element containing only one item is added after e_k. For element extension, we sort the items in lexicographical order, and an item can be added to an element only if the item is larger than all the existing items in the element. For example, an element $e = \{b\}$ can be extended by item c but cannot be extended by item a. The purpose of this restriction is to avoid generating the same pattern more than once.

ESPMiner explores the search space in depth-first order, and it generates a sequential pattern from its prefix. The support of sequential patterns has the anti-monotone property. That is, the support of a sequential pattern is always no larger than that of its sub-sequences. ESPMiner uses this property to prune the search space. It starts from the empty pattern $\langle \rangle$ and recursively adds items to the pattern until the support of the pattern is below the minimum support threshold.

Algorithm 1 illustrates ESPMiner. When it is first called, $s=\langle \rangle$ and $D_s = D$. For each sequential pattern s, ESPMiner scans its projected database D_s to find its frequent element extensions and sequence extensions (line 1). An item x is a frequent element extension of $s = \langle e_1, e_2, \cdots, e_k \rangle$ if $s' = \langle e_1, e_2, \cdots, e_k \cup \{x\} \rangle$ is frequent. An item x is a frequent sequence extension of s if $s' = \langle e_1, e_2, \cdots, e_k, \{x\} \rangle$ is frequent. ESPMiner constructs the project databases of extensions of s from D_s (line 2). If a sequence in D_s contains $s' = \langle e_1, e_2, \cdots, e_k \cup \{x\} \rangle$ or $s' = \langle e_1, e_2, \cdots, e_k, \{x\} \rangle$), then the sequence is put into the projected database of s'. Next, ESPMiner outputs s' if $dpower(s', C) \geq \rho$ (lines 5-7, lines 12-14), and then extends s' recursively (line 8, line 15).

3.3 Mining ESPs for Activity Recognition

We then apply ESPMiner to mine the ESPs for activity recognition. Before the mining process, we first need to convert the data stream to a sequence of

Algorithm 1. The ESPMiner Algorithm

Input: $s = \langle e_1, e_2, \cdots, e_k \rangle$ is the frequent sequential pattern currently being processed;
D_s is the projected database of s;
min_sup is the minimum support threshold;
ρ is the minimum discriminative power threshold;
Output: Mined ESPs

1: Scan D_s to find frequent extensions of s, denoted as $F_{elm}(s)$ and $F_{seq}(s)$, where $F_{elm}(s)$ contains frequent element extensions of s, and $F_{seq}(s)$ contains frequent sequence extensions of s;
2: Scan D_s a second time to construct projected databases for super-sequences of s that extend s by one new item in $F_{elm}(s)$ or $F_{seq}(s)$;
3: **for all** items $x \in F_{elm}(s)$ **do**
4:     $s' = \langle e_1, e_2, \cdots, e_k \cup \{x\} \rangle$;
5:     **if** $dpower(s') \geq \rho$ **then**
6:         output s';
7:     **end if**
8:     DFSMine(s', $D_{s'}$, min_sup, ρ);
9: **end for**
10: **for all** items $x \in F_{seq}(s)$ **do**
11:     $s' = \langle e_1, e_2, \cdots, e_k, \{x\} \rangle$;
12:     **if** $dpower(s') \geq \rho$ **then**
13:         output s';
14:     **end if**
15:     DFSMine(s', $D_{s'}$, min_sup, ρ);
16: **end for**

observation vectors by concatenating all of the raw data in a fix time interval (set to one second in our experiments), and then we extract useful features as follows.

For acceleration data, we compute five common features—DC mean, variance, energy, frequency-domain entropy and correlation. The DC mean is the mean acceleration value in a time interval. Variance is used to characterize the stability of a signal. Energy captures the data periodicity, and it is calculated as the sum of the squared discrete FFT component magnitudes of a signal. Frequency-domain entropy helps to discriminate activities with similar energy values, and it is calculated as the normalized information entropy of the discrete FFT component magnitudes of a signal. Correlation between axes is especially useful for discriminating between activities that involve translation in just one dimension. It is computed for every two axes of each accelerometer and all pair-wise axes combinations of two different accelerometers. For RFID reading or location information, we use object name or location name as features. We then transform these *observation vectors* into *feature vectors*. A *feature vector* consists of many *feature items*, where a *feature item* refers to a feature-value pair. Feature vectors are indexed by a simple encoding scheme and will be used as inputs to the mining process.

The mining process leverages on sequential activity instances only. Specifically, for each sequential activity class SA_i, we mine a set of ESPs to contrast its instances, D_{SA_i}, against all other activity instances D'_{SA_i}, where $D'_{SA_i} = D - D_{SA_i}$ and D is the entire sequential activity dataset. After mining, we obtain a set of ESPs for each sequential activity. Table 1 presents an ESP subset for *making oatmeal*. Columns 3 and 4 show the corresponding values of support and dpower, respectively.

Table 1. An ESP Example for *making oatmeal*

ESP	Support	dpower
{$ACCEL\_BODY\_X\_MEAN@(-900.75 \sim -822.75]$, $ACCEL\_LEFT\_X\_MEAN@(214.39 \sim 352.87], LOCATION@kitchen$}; {$ACCEL\_LEFT\_X\_MEAN@(543.27 \sim 843.19]$, $ACCEL\_LEFT\_Z\_MEAN@(441.53 \sim 847.59]$}; {$ACCEL\_RIGHT\_Z\_MEAN@(467.36 \sim 701.17], OBJECT@tablespoon$}; {$LOCATION@kitchen$}; {$OBJECT@burner$}	75.0%	1.0

4 ESP-Based Activity Recognition

4.1 ESPar Overview

We first give an overview of ESPar. The input is the sensor data stream which will be first pre-processed into a sequence of feature vectors. ESPar operates in two phases—model training and activity recognition. In the training phase, a training dataset consisting of sequential activity instances will be used to train our activity models. In the recognition phase, given a sequence of feature vectors (i.e., S_t, $t = 0 \sim T$), we first segment its sequence using a sliding window (i.e., L_{A_i}) to obtain a test instance (i.e., $S_{t \sim t+L_{A_i}}$, $t = 0 \sim T$), and then we apply our recognition algorithm to label this sequence segment. This process will be performed recursively, and each sequence segment will be assigned with a candidate label. For each pair of consecutive sequence segments we label, we apply a boundary detection algorithm to detect the boundary, and adjust the length of each sequence segment according to the new boundary. The above recognition processes will be performed recursively until the end of the sequence. In the following sections, we first describe our activity models. We then present the ESPar algorithm.

4.2 Sequential Activity Model

We design the activity model for each sequential activity SA_i based on a set of ESPs (i.e., ESP_{SA_i}) we mined from sequential activity instances. Each set of ESPs contains many subsets in which a single ESP can sharply differentiate the class membership of a fraction of the test instance $S_{t \sim t+L_{SA_i}}$ that contains the ESP. To make use of each subset of ESPs to achieve good overall accuracy, we combine the strength of each ESP based on the aggregation method described as follows.

Suppose an instance $S_{t \sim t+L_{SA_i}}$ contains an ESP, X, where $X \in ESP_{SA_i}$, then the odds that $S_{t \sim t+L_{SA_i}}$ belongs to SA_i is defined as $dpower(X)$. The differentiating power of a single ESP is then defined by the odds and the fraction of the population of class that contain the ESP. More specifically, the differentiating power of X is given by $dpower(X) * sup_{SA_i}(X)$. The aggregated probability of $S_{t \sim t+s_{A_i}}$ belongs to SA_i is defined as follows.

$$aggr\_p(SA_i, S_{t \sim t+s_{A_i}}) = \sum_{X \subseteq S_{t \sim t+s_{A_i}}, X \in ESP_{SA_i}} dpower(X) * sup_{SA_i}(X) \quad (1)$$

where $sup_{SA_i}(X)$ is the support of X in SA_i. The ESP measurement of each activity are then normalized by dividing them using the median probability value in the training instances of that activity. Finally, the ESP measurement is defined as follows.

$$esp(SA_i, S_{t \sim t+L_{SA_i}}) = \frac{aggr\_p(SA_i, S_{t \sim t+L_{SA_i}})}{base\_p(SA_i)} \qquad (2)$$

where $base\_p(SA_i)$ is the median probability value of $aggr\_p(SA_i, S_{t \sim t+L_{SA_i}})$ in the training data.

In addition to the ESP measurement, we model activity correlation (i.e., when an activity SA_j has been performed, the probability of another activity SA_i being performed) in the design of our sequential activity model. We use condition probability to model correlations between activities. We define the activity correlation probability of SA_i as $P(SA_i|SA_j)$, which is the conditional probability of SA_i given SA_j. The activity correlation probability for each sequential activity can be easily obtained from the training dataset.

4.3 Interleaved and Concurrent Activity Models

The design of complex activity models is crucial. A common practice is to train both interleaved and concurrent activity models from complex activity instances. We design our complex activity model based on our sequential model only, eliminating the need for training. To illustrate, we denote CA_i as both interleaved and concurrent activities. We denote CA_i as SA_a & SA_b) for an interleaved activity and $SA_a + SA_b$ for a concurrent activity, where two single sequential activities SA_a and SA_b are involved in. We set the number of single activities involved in interleaved or concurrent activities to two for illustrations although in theory it can be more than two. We define the sliding-window length of CA_i as $L_{CA_i} = L_{SA_a} + L_{SA_b}$, and use L_{CA_i} to get the test instance $S_{t \sim t+L_{CA_i}}$. Since an instance of CA_i containing both ESP_{SA_a} and ESP_{SA_b} (i.e., some of the steps that belong to SA_a and SA_b respectively are interleaved or overlapped), the ESP measurement of CA_i can be computed as follows.

$$esp(CA_i, S_{t \sim t+L_{CA_i}}) = max[esp(SA_a, S_{t \sim t+L_{CA_i}}), esp(SA_b, S_{t \sim t+L_{CA_i}})] \qquad (3)$$

The computation of activity correlation probability for interleaved and concurrent activities can be quite complex. There are three cases: a sequential activity followed by an interleaved or a concurrent activity, an interleaved or a concurrent activity followed by a sequential activity, and an interleaved or a concurrent activity followed by another interleaved or concurrent activity. Given the rational that a higher condition probability implies a stronger activity correlation, we choose the maximum value of all possible condition probabilities for all these cases. To illustrate, given CA_j, where $CA_i = SA_a$ & SA_b or $CA_i = SA_a + SA_b$, the activity correlation probability of CA_i, where $CA_j = SA_c$ & SA_d or $CA_j = SA_c + SA_d$, can be computed as follows.

$$P(CA_i|CA_j) = max(P(SA_a|SA_c), P(SA_a|SA_d), P(SA_b|SA_c), P(SA_b|SA_d)) \quad (4)$$

The computation of $P(SA_i|CA_j)$ and $P(CA_i|SA_j)$ follows the same method.

In summary, our activity model for sequential, interleaved and concurrent activities is defined as follows.

Definition 4. *(Activity Model)* Given a time t and an activity A_j which ends at t, for each activity A_i, a test instance $S_{t \sim t+L_{A_i}}$ is obtained from t to $t + L_{A_i}$, the activity model of A_i is then defined as follows:

$$model(A_i, A_j, S_{t \sim t+L_{A_i}}) = esp(A_i, S_{t \sim t+L_{A_i}}) * P(A_i|A_j)$$

4.4 The ESPar Algorithm

We are now ready to classify activities based on the above model. We first use a sliding window based algorithm to segment the sequence. For each possible activity, we obtain the test *feature vectors* using its corresponding sliding window (the length is the average duration of this activity), and compute the probability based on Definition 4. We then assign the activity label with the highest probability to the *feature vectors*. The process can be applied recursively to the entire sequence. However, since the sliding-window length of each activity is an approximation of its actual length, the segmentation may not be accurate. To overcome this drawback, we use a boundary detection algorithm in [8] to detect and adjust the boundary between two adjacent activities so that the next sliding window can be applied from the correct boundary.

5 Empirical Studies

We now move to evaluate our proposed algorithm. In this section, we first describe the trace we use, and then present and discuss the results obtained from a series of experiments.

5.1 Trace and Methodology

We use a real-wrold trace collected in [8]. The dataset consists of 26 sequential activities, 11 interleaved activities and 13 concurrent activities with a total number of 532 instances. We use *ten-fold cross-validation* (a common technique to evaluate predictive models) for our empirical studies. Note that the training process involves sequential instances only.

We evaluate the performance of ESPar using time-slice accuracy which is a typical technique in time series analysis. It represents the percentage of correctly labeled time slices. The length of time slice Δt is set to 15 seconds as our experiment shows different Δt does not affect the accuracy much. This time slice duration is short enough to provide precise measurements for applications. The metric of the time-slice accuracy is defined as follows.

$$Time\_slice = \frac{1}{N} \sum_{n=1}^{N} [inferred(n) = true(n)] \quad (5)$$

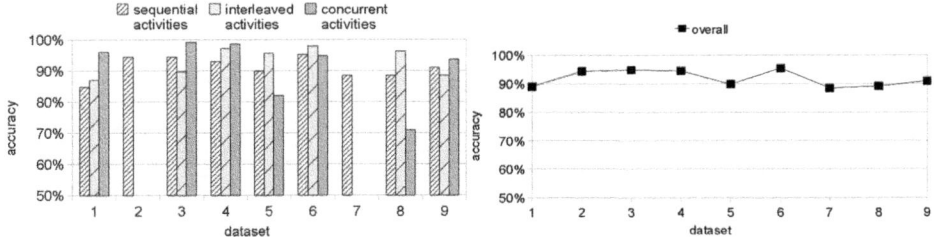

Fig. 2. Breakdown for type of activities   **Fig. 3.** Breakdown for overall accuracy

where $[inferred(n) = true(n)]$ produces 1 when true and 0 when false; $N = \frac{T}{\Delta t}$, i.e., the total number of time slices.

5.2 Accuracy

In this experiment, we evaluate the accuracy of ESPar. ESPar achieves an accuracy of 91.61% for sequential activity, 92.26% for interleaved activity, 92.32% for concurrent activity, respectively, and an overall accuracy of 91.89%. Figure 3 shows the breakdown for the overall accuracy, and Fig. 2 shows the breakdown for type of activities. One observation is that we obtain similar results for all the three types of activities. Most confusion takes place in the following three cases:

Case 1: A sequential activity is predicted as another sequential activity, e.g., for a sequential activity *washing face*, our result shows while 51.2% of its entire observation sequence is predicted correctly, 16.3% of them is predicted as *brushing teeth* and 14.1% of them is predicted as *brushing hair*.

Case 2: In a complex activity, only one of the sequential activities is detected and the other one is missed, e.g., for an interleaved activity *reading book/magazine & using phone*, while 76.1% of its entire observation sequence is predicted correctly, 19.1% of them is predicted as *reading book/magazine* (i.e., *using phone* is missed).

Case 3: A sequential activity is predicted as an interleaved or a concurrent activity, e.g., while 75.6% of the entire sequence of *reading book/magazine* is predicted correctly, 23.9% of them is predicted as an interleaved activity *reading book/magazine & using phone*.

To analyze the above results, we suggest a number of reasons for Case 1. Firstly, we observe many similarities among these activities such as similar hand movements and the same location. Secondly, RFID sensors may not work in some activities such as *washing face* since *face towel* is not tagged, or fail to detect tagged objects due to out of range. Thirdly, many tagged objects may be placed in close proximity. As a result, sensor noise may be introduced to the data stream. For Cases 2 and 3, the above possibilities may hold as well. In addition, for the two activities involved in a complex activity, ESPar seems to bias to the longer one. Possible solutions include adding more sensor modalities, which we leave for our future work.

Table 2. Comparison Results

Classifier	Accuracy				Remark
	Sequential Activity	Interleaved Activity	Concurrent Activity	Overall	
ESPar	91.61%	92.26%	92.32%	91.89%	pattern based model
HMM	70.99%	N.A.	N.A.	70.99%	temporal probabilistic model
CRF	86.93%	N.A.	N.A.	86.93%	temporal probabilistic model

5.3 Comparison Studies

In this experiment, we compare ESPar with temporal models (HMMs and CRFs). The results of the temporal models are based on sequential activities only since both models can not be directly applied to recognize complex activities. For a fair comparison, we use the same training and testing datasets for all the models. Table 2 summarizes the results.

For each dataset, an HMM was trained and the Viterbi algorithm was used to recover the state sequence; similarly for CRF. The result is shown in Table 2. ESPar outperforms both CRF (86.93%) and HMM (70.99%). This result can be probably explained as follows: First, the amount of training data is relatively small. In such circumstance, mining the differences between classes tends to be more effective for building a discriminative model. Although CRF is also a discriminative model, it focuses on mining the regularities in which a large amount of training data is typically required. This result reveals that ESPar tends to be more efficient with the same amount of training data. HMM, as a generative joint model, is least effective due to the known shortcomings such as over-fitting the training data and strong independence assumptions. ESPar also outperforms the Emerging Pattern based model [8], demonstrating that the ESP-based model is more efficient because it mines not only the differences between classes, but also the inherent ordering among activity sequences.

6 Conclusions

In this paper, we study the problem of recognizing sequential, interleaved and concurrent activities using a BSN. We propose a novel Emerging Sequential Pattern, and demonstrate that, leveraging on ESP, both simple and complex activities can be effectively recognized in a unified framework.

For our future work, we will integrate more sensors such as acoustic sensor and gyro sensor into our BSN. The more sensor modalities we use, and the stronger ESPs we obtain. We will also explore the concept of ESPs to other classification tasks.

Acknowledgement. This work was supported by the Danish Council for Independent Research, Natural Science under Grant 09-073281, National 973 program of China under Grant 2009CB320702, National 863 program of China under Grant 2009AA01Z117, Natural Science Foundation of China under Grants

60736015 and 60721002, Project for Core High Technology of the Ministry of Science and Technology of China under Grant 2009ZX01043-001-06, and Jiangsu Climbing Program under Grant BK2008017.

References

1. Bao, L., Intille, S.S.: Activity recognition from user-annotated acceleration data. In: Ferscha, A., Mattern, F. (eds.) PERVASIVE 2004. LNCS, vol. 3001, pp. 1–17. Springer, Heidelberg (2004)
2. Logan, B., Healey, J., Philipose, M., Tapia, E.M., Intille, S.: A long-term evaluation of sensing modalities for activity recognition. In: Krumm, J., Abowd, G.D., Seneviratne, A., Strang, T. (eds.) UbiComp 2007. LNCS, vol. 4717, pp. 483–500. Springer, Heidelberg (2007)
3. Patterson, D., Fox, D., Kautz, H., Philipose, M.: Fine-grained activity recognition by aggregating abstract object usage. In: Proc. IEEE Int'l Symp. Wearable Computers, Osaka (October 2005)
4. Vail, D.L., Veloso, M.M., Lafferty, J.D.: Conditional random fields for activity recognition. In: Proc. Int'l Conf. Autonomous Agents and Multi-agent Systems, AAMAS (2007)
5. van Kasteren, T.L.M., Noulas, A.K., Englebienne, G., Kröse, B.J.A.: Accurate activity recognition in a home setting. In: Proc. Int'l Conf. Ubicomp, Seoul, Korea (September 2008)
6. Modayil, J., Bai, T.X., Kautz, H.: Improving the recognition of interleaved activities. In: Proc. Int'l Conf. Ubicomp, Seoul, South Korea (September 2008)
7. Wu, T.Y., Lian, C.C., Hsu, J.Y.: Joint recognition of multiple concurrent activities using factorial conditional random fields. In: Proc. AAAI Workshop Plan, Activity, and Intent Recognition, California (July 2007)
8. Gu, T., Wu, Z., Tao, X., Pung, H.K., Lu, J.: epSICAR: An Emerging Patterns based Approach to Sequential, Interleaved and Concurrent Activity Recognition. In: Proc. IEEE Int'l Conf. on Pervasive Computing and Communications (Percom 2009), Galveston, Texas (March 2009)
9. Lombriser, C., Bharatula, N.B., Roggen, D., Tröster, G.: On-body activity recognition in a dynamic sensor network. In: Proc. Int'l Conf. Body Area Networks, BodyNets (2007)
10. Fayyad, U., Irani, K.: Multi-interval discretization of continuous-valued attributes for classification learning. In: Proc. Int'l Joint Conf. on Artificial Intelligence, San Francisco (1993)
11. Dong, G.Z., Li, J.Y.: Efficient mining of emerging patterns: discovering trends and differences. In: Proc. ACM Int'l Conf. on Knowledge Discovery and Data Mining, San Diego, CA, USA, pp. 43–52 (August 1999)

Indoor Cooperative Positioning Based on Fingerprinting and Support Vector Machines

Abdellah Chehri[1,*], Hussein Mouftah[1], and Wisam Farjow[2]

[1] School Information Technology and Engineering (SITE),
800 King Edward Avenue, Ottawa, Ontario, Canada, K1N 6N5
{achehri,mouftah}@uottawa.ca
[2] Mine Radio Systems Inc.
394 Highway 47, R.R. #1, Goodwood, 10C 1A0, Canada
wisam_farjow@mineradio.com

Abstract. For location in indoor environments, the fingerprinting technique seems the most attractive one. It gives higher localization accuracy than the parametric technique because of the existence of multipath propagation and fast fading phenomena that are difficult to model. This paper introduces a novel positioning system based on wireless the IEEE802.15.4/ZigBee standard and employs Support Vector Machines (SVMs). The system is cost-effective since it works with real deployed IEEE 802.15.4/ZigBee™ sensors nodes. The whole system requires minimal setup time, which makes it readily available for real-world applications. The resulting algorithm demonstrates a superior performance compared to the conventional algorithms.

Keywords: Location-based Services, Support Vector Machines, Radio Mapping, RSSI, Underground mines.

1 Introduction

Wireless sensor networks are becoming more and more popular in many areas such as agriculture activity, construction, industrial, manufacturing plants, large buildings, and so on. In these environments it is preferable to set up a wireless network rather than the traditionally wired sensor network because the flexibility and freedom that can provided. Based on the wireless infrastructure, a many new and promising applications are coming into our lives [1].

Even if the main applications of wireless sensor are security and environmental data monitoring, they can be used to locate and track people and objects. The technique is based on the received signal from a few anchors (node with known location) and some traditional location algorithm. This application is called the WSN positioning system [2].

Location tracking systems has brought tremendous benefits, especially to the mining industry; one of its benefits is to help keeping track of miners and equipment is a challenge in any the underground mines.

---

[*] Corresponding author.

To implement a successful WSN-based location application, the most important is to locate sensor node accurately, at least within an acceptable range, to pin-point the user's positioning.

In practice, two types of location techniques are used: the first one is based on received signal information [2]-[4] and the second is the fingerprinting technique [5]-[8].

In the first technique, an important factor is to find the mechanisms for physical distances/angles measurements [9]. Several techniques have been used. The most known are the angle of arrival (AoA), the received signal strength (RSS) and the time delay information (ToA).

However, the performance of this localization mechanism is largely determined by both physical distances measurements mechanism and radio propagation channel characteristics. As a result, this positioning algorithms may not provide optimum performance in indoor environments, which necessitates the design of new positioning algorithms for indoor geolocation systems

The second approach is fingerprinting technique. This technique assumed that the characteristics of the propagation signal are different at each location of the zone of interest. In other words, each location has a unique fingerprint or signature in terms of propagation characteristics [7].

The basic idea of the fingerprinting positioning algorithms is simple. Each location spot in the area of interest have a unique signature in terms of the RSS observed from different sensors. A fingerprinting technique determines the unique pattern features and then this knowledge is used to develop the rules for recognition.

This technique seems to be the most attractive particularly for indoor environments and small coverage areas. In fact, a judicious sensor network planning can significantly reduce the estimation errors of the location metrics caused by the both NLOS and multipath propagation condition. Therefore, the structural information of the sensor network can be easily employed in the intelligent positioning algorithms [9].

In this paper we present a new ZigBee–based sensor networks for localization in underground environments. Amongst all of the possibilities of choosing methods of positioning, we focused on the received signal strength method along with fingerprinting. So, in the first time, a measurement campaign was conducted in an experimental underground mine to collect the RSS from two transmitter sensors nodes. We use RSS as a metric instead of trying to extract the ToA or AoA which is more challenging task at the physical layer. Secondly, we introduce localization in a cooperative technique when two transmitters' nodes share the responsibility of estimating the position of mobile node.

After building the database, we introduce a new algorithm which uses the support vector machine technique as a matching algorithm to localize. Then we present the performance of localization algorithm in terms of mean localization squared error is evaluated.

The remainder of the paper is organized as fellows. Section 2 provides an overview of localization systems. The importance of the fingerprinting method is also examined. Section 3 describes the testbed development and discusses on the SVM which is used as the matching algorithm for localization. Section 4 provides the details of measurement scenarios and discusses on the result. Finally, Section 5 summarizes the results and discusses the future work which can be done through this testbed measurement.

2 An Overview of Fingerprinting Localization Techniques

The technique using the RSS fingerprint was applied for the first time in the RADAR system [5] for a mobile user's location on a second floor of a 3-storey building. It opened the door for many different techniques to be applied for the localization problem.

For example, Nibble [10] is one of the first systems to use a probabilistic approach for location estimation. To date, Ekahau's Positioning Engine Software [11] claims to be the most accurate location system based on probabilistic model; they claim a one-meter average accuracy with a short training time.

Statistical learning theory [12] and neural networks [6], [7], [8], have also been investigated for localization. Some works [13], [14] also try to aggregate localization data from different technologies (e.g., Wi-Fi and Bluetooth) in order to achieve finer accuracy.

Another emerging approach that has better accuracy and potential is Ultra wideband (UWB) technology. The large bandwidth provides high time-domain resolution which in return provides better ranging accuracy [15].

In underground mines, the geolocation can be placed in two main categories: commercial, like task optimization, logistic, traffic management in galleries and the most important factor, miners' safety.

Accurately predicting the location of an individual or an object definitely can be a difficult task producing ambiguous results because of the harsh wireless environment.

Few works have been carried out in underground mining environments [2], [6], [7], [8]. The characteristics of underground mining environments differ from other traditional indoor environments since the roughness of its walls leads to scattering of the RF signal. Thus, underground mining environments have specific radio propagation characteristics.

On one hand, the harsh site-specific multipath environment in underground mines introduces difficulties in accurately tracking the position of objects or miners. The growing interest and demand for such applications dictates examining position estimation more carefully. The underground mines propagation channel poses a serious challenge to system designers due to the harsh multipath environment.

On the other hand, the structural information of the wireless network can be used in the intelligent positioning algorithms. The small (relatively) coverage of underground mines, as compared without outdoor environment, makes it possible to conveniently conduct extensive pre-measurements in mine gallery.

The main contributions of this paper are:

(1) Investigate more profoundly on the location problem in indoor environment and particularly in underground mine. This task is considered as important for many applications.
(2) Developing a testbed for RSS-based indoor positioning systems.
(3) Using ZigBee signal as reference for our localization system. The current trend for localization is based on IEEE 802.11standard. However the main limiting factor of localization systems for underground mines is the absence of line of sight (LOS). In an underground mine, the surface and the architecture are often irregular. The localization system based on a wireless sensor network can

solve this problem. Instead of employing a few relatively expensive 802.11 access points, we can use several inexpensive ad-hoc sensor nodes to estimate the position of an object. In addition, it is easier to exploit the small size of ZigBee nodes compared to a PC Laptop for example.
(4) We adopt cooperative localization technique where more than one transmitter nodes shares the responsibility of estimating the position of the receiver node in experimental underground mines.
(5) We present an efficient algorithm based on SVM technique for location. The proposed algorithm is practical and scalable. The choice of this point is motivated by the high performance of SVM compared to other classification technique such as Neural Networks.

3 Design of Indoors Positioning System

Concurrently, there has been an increasing deployment of wireless sensor networks in many domains including mining industry. The popularity of wireless sensor networks opens a new opportunity for location-based services.

In our localization system, mobile node measure RSSI from two fixed nodes. The collected data is compared by this mobile node with a known data base in under to estimate his location.

3.1 Characteristics of the Sensor Nodes

When choosing deployment of WSN in underground mine, it should be necessary to make a compromise between conflicting requirements. The priority is to insure a robust global network with battery-operated nodes.

Wireless communication is achieved with a transceiver compliant with the IEEE 802.15.4/ZigBee^{TM} standard. ZigBee^{TM} is a global standard for wireless network technology that addresses remote monitoring, environmental data measurements and control applications. ZigBee^{TM} is an open specification that enables low power consumption, low cost and low data rate for short-range wireless connections between various electronic devices.

3.1.1 Hardware Description
The Silicon Laboratories 2.4 GHz 802.15.4 Development Board (DB) provides a hardware platform for the development of 802.15.4/ ZigBee^{TM} networks. The DB includes a Silicon Labs 8051-based MCU, a Chipcon CC2420 RF Transceiver, a JTAG (Joint Test Action Group or IEEE 1149.1 standard) connector for in-circuit programming, an assortment of programmable buttons and LEDs and a USB interface for connecting to the host computer.

3.1.2 Software Description
The 2.4 GHz ZigBee^{TM} development kit contains all necessary files to write, compile, download, and debug a simple IEEE 802.15.4/ ZigBee^{TM} -based application. The development environment includes an IDE, evaluation C compiler, software libraries, and a several code example. The software library includes the 802.15.4 MAC and PHY layers.

The ZigBee™ demonstration provides a quick and convenient graphical PC-based application. The kit also includes an adapter for programming and debugging from the IDE environment.

A Network Application Programming Interface (API) contains all necessary network primitives to build a 802.15.4 network from a user-defined application. A software example illustrates the MAC API. This example builds an ad-hoc 802.15.4 network using the included MAC API software library [16].

3.2 Support Vector Machines

SVMs (Support Vector Machines) are a useful technique for data classification. SVM is considered easier and more powerful than the Neural Networks [17].

Let $S' = (x', y')$ be the mobile node location to be determined, $\Theta' = [R'_A, R'_B]$ are the observed RSS vector form sensor A and B respectively.

A set of fingerprint is the radio signature of many known location. For each geographic location (x, y) a known vector Θ represented by:

$$\Theta = [R_A, R_B] \quad (1)$$

with $R_A = [RSS_{left}, RSS_{midle}, RSS_{right}]$

where R_A is the received signal from transmitter node A at three adjacent positions (located in left, middle and right), ditto for node B. So each physical location is characterized by six parameters.

When the mobile node changes his location, the parameter of vector Θ changed. We can get data sample $D = [\Theta, S]$ by sampling $S(x, y)$ and $\Theta = [R_A, R_B]$ in grid. The input of D is RF signal (featured by RSSI) and output is location (coordinate).

Given training set (here $D = [\Theta, S]$), the support vector machines require the solution of the following optimization problem:

$$\min_{w,b,\xi} \frac{1}{2} \|w^T w\| + C \sum_{i=1}^{N} \xi_i \quad (2)$$
$$\textbf{Subject to } (w^T \phi(x_i) + b) \geq 1 - \xi_i$$
$$\xi_i \geq 0$$

Here training vectors x_i are mapped into a higher dimensional space by the function ϕ. SVM finds a linear separating hyperplane with the maximal margin in this higher dimensional space. C> 0 is the penalty parameter of the error term. Furthermore, $K(x, x') = \phi(x_i)^T \phi(x_j)$ is called the kernel function.

We can change linear inseparable problem of (2) in lower dimension space into linear inseparable problem in higher dimensional space by kernel function $K(x, x')$. The optimization result is as follows:

$$S' = f(\Theta') = \sum_{i=1}^{N} \alpha_i K(\Theta, \Theta'_i) + b \quad (3)$$

In localization phase, the input of system is the collected RSS vector $\Theta' = [R'_A, R'_B]$ of the mobile. The output of the system is the estimated location $S' = (x', y')$ of the mobile node.

We choose LS-SVM [18] to build model. LS-SVM is an improvement version of the standard SVM. It is simple-calculation, can improve convergence speed, and suitable for WSN for resource constraints. Here we chose the radial basis function (RBF) given by:

$$K(x, x') = \exp\left(\frac{|x-x'|}{2\sigma^2}\right) \quad (4)$$

The SVM-based localization mechanism is represented in figure 1.

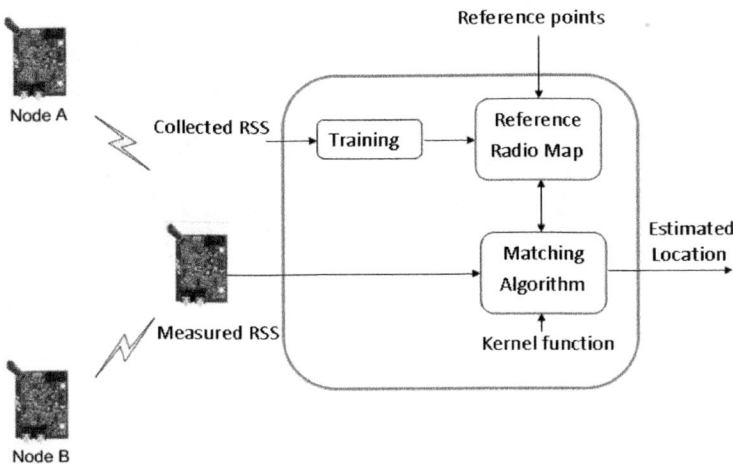

Fig. 1. Simulation procedure of sensor networks localization system

3.3 Closest Neighbor (CN)

Our algorithm will be compared with classical algorithm in pattern recognition class is called the closest neighbor (CN) or nearest-neighbor (NN) [19]. In this algorithm mobile sensor compares an observed RSS vector with all available fingerprints in the reference radio map and finds a reference point with the smallest Euclidean distance in signal space and reports that as the current location of the device. Suppose that MS observes $\Theta^{\wedge\prime}$. The Euclidean distance between this vector and the k-th reference point entry in the radio Θ is given by:

$$d = \left(\sum_{i=1}^{N}(\Theta' - \Theta_i)^2\right)^{1/2} \quad (5)$$

The CN algorithm maps the location of MS to an entry on the radio map.

4 Measurement Setup and Results

4.1 Experimental Testbed

The measurement campaign necessary to build the fingerprints' database has been carried out in an underground mining gallery. Figure 2 shows digital picture of the underground mine gallery. It has a width of approximately 2.5 m, a length of 72 m and is located at 70 m below the level of the ground. Two transmitter node (Tx node A and B) are located on the ceil of the mine gallery.

Fig. 2. Illustration of underground mine gallery and the deployment of the node (A)

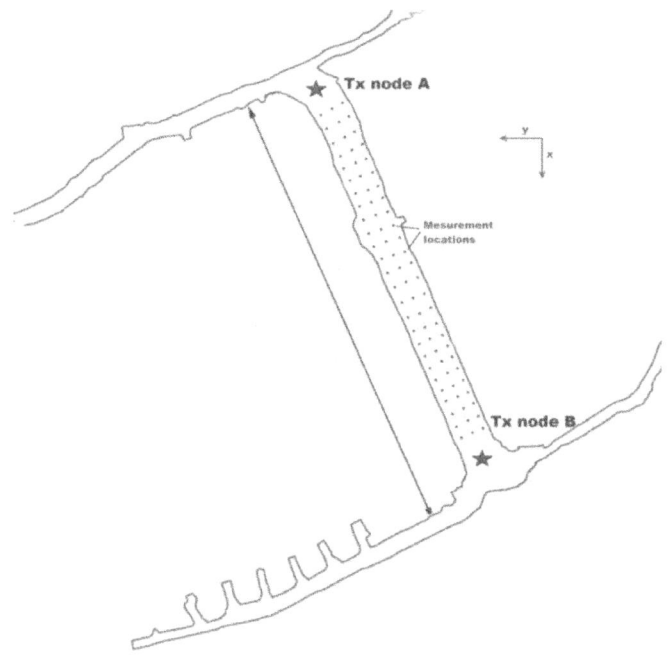

Fig. 3. Layout of the underground mining gallery used for measurements

The mobile node start form 1 m with regard to the node A (reference node) to 105 meter. The grid spacing between two adjacent mobile user's locations was set to 1 m along the x-axis and 0.8 m along the y-axis. At each location, the RSS value has been recorded in each three point (left, middle and right). A set of RSSs has been collected at 630 (105*3*2) different mobile user's locations. All the measurement locations are shown in Figure 3.

As for the pattern-matching algorithm, the SVM has been trained with the 530 location points leaving the remaining 100 points for localization purposes.

4.2 Reference Radio Map

In figure 4, received power measurements signatures are recorded follow the gallery long. The received power is recorded at uniform distance step from the source emitter following a route located at the gallery center, a route following the right lateral wall and the route following left lateral wall of the gallery. So for each separation distance three data were collected (middle, left, and right).

4.3 Separation of Location Fingerprint

The performance of indoor positioning systems depends greatly on the separation of location fingerprints. A location fingerprint corresponding to a location can be identified correctly if it is difficult to classify it (incorrectly) as another fingerprint by a pattern classifier [20].

Theoretically, a change in RSS is proportional to the logarithm of the distance between a transmitter and a receiver. Therefore, two different locations with different distances from the same transmitter node should have different average RSS values. However, in practice the RSS is a random variable that has its value fluctuating around the average value due to the dynamics in the environment. These fluctuating values can be grouped to get her as patterns of RSS at a particular location.

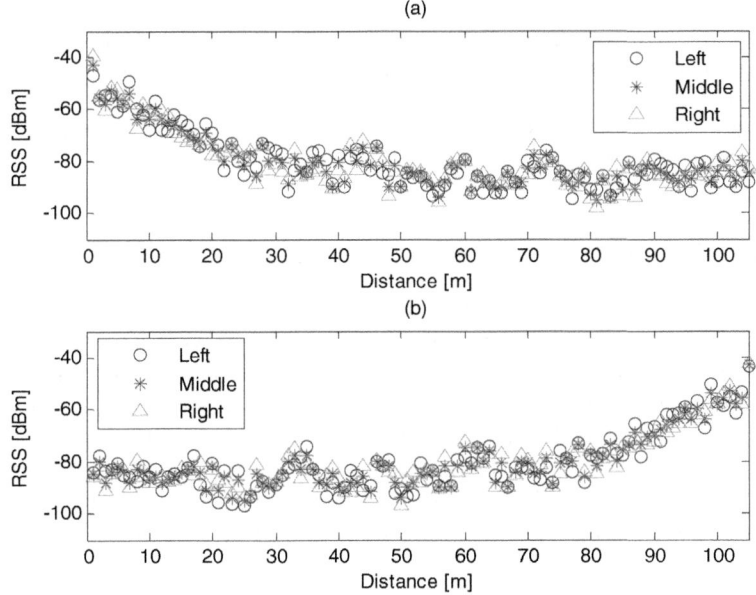

Fig. 4. The received power measurements signatures

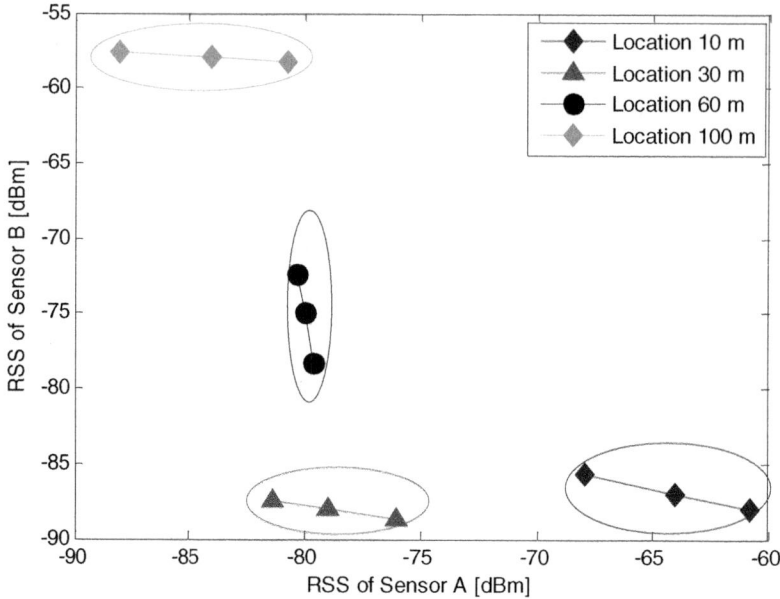

Fig. 5. Signature separation problem of four location

Figure 5 shows two-dimensional plots of patterns form node A (x-axis) and node B (y-axis). The group of patterns at each location can be called the location fingerprint of that particular location. From the plot, patterns of each location can be grouped together as a cluster.

This figure indicates that RSS's patterns can be separated by a separate cluster. Each signature is well distinct. This good indication is useful for localization mechanism.

4.4 Localization Results

The performance of the presented localization techniques will be evaluated using the CDF graph. The first two plots show the results of the SVM localization technique based on cooperative localization (node A and node B). The third plot represents the position errors when using SVM with only one transmitter node. The last plot shows the results of using the localization technique based on closet neighbor.

As shown in Figs. 6, closet neighbor algorithm provides the worst results. Form the results 3.03 meter of localization error is needed to cover 90 % of the testing data. The positioning error when only one transmitter is used is around 2.2 m for 90% of the testing data. However, when two nodes are used for localization (the cooperative localization), this error decrease to 1.45 m for 90 % of testing data. This can be explained by the size of signature location (six parameters instead of three when one transmitter is used).

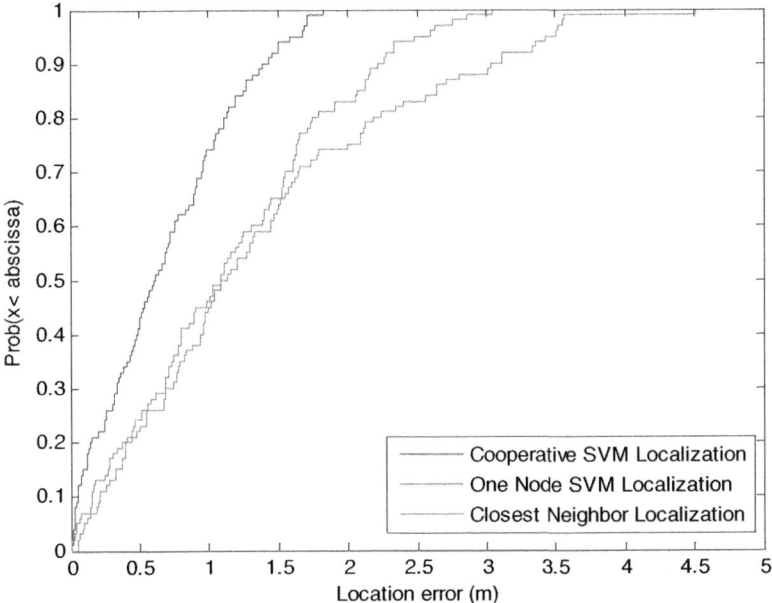

Fig. 6. CDF plots of the position estimation errors at a receivers' several localization techniques

5 Conclusion

This paper has described the implementation of a novel algorithm, based on a set of support vector machine (SVMs) applied to ZigBee RSS fingerprinting technique in an underground mining gallery. We investigated the performance of our system experimentally. The advantage of our technique is that it does not use coordinates, special hardware is not required, and it is simple and robust in dynamic environments. The performances in terms of localization accuracy of algorithm is evaluated and compared to a conventional algorithm. The experimental results showed that the percentage of measurement points where our system could localize the mobile node with a median distance error of 0, 67 m. The algorithm performs slightly better than the conventional closet neighbor algorithm (median error distance is 1. 4 m).

In the future our algorithm will be compared with other more sophisticated algorithm such as generalized radial neural network.

References

1. Holger, K., Willig, A.: Protocols and Architecture for Wireless Sensor Networks. John Wiley and Sons, Chichester (2005)
2. Chehri, A., Fortier, P., Tardif, P.-M.: UWB-based Sensor Networks for Localization In Mining Environments. Elsevier Journal Ad Hoc Networks (October 2008)

3. Bulusu, N., Heidemann, J., Estrin, D.: GPS-less Low Cost Outdoor Localization for Very Small Devices. IEEE Personal Communications Magazine 7(5), 28–34 (2000)
4. Priyantha, N.B., Chakraborty, A., Balakrishnan, H.: The Cricket location-support system. In: The 6th Annual ACM/IEEE International Conference on Mobile Computing and Networking (MobiCom 2000), Boston, MA, USA, pp. 32–43 (2000)
5. Paramvir, B., Padmanabhan, B.V.: RADAR: An In Building RF-based User Location and Tracking System. In: Proceedings of IEEE Infocom 2000, Tel Aviv (March 2000)
6. Nerguizian, C., Despins, C., Affes, A.: Geolocation in mines with an impulse response fingerprinting technique and neural networks. IEEE Transactions on Wireless Comm. 5(2), 603–611 (2006)
7. Wassi, G.I., Grenier, D., Despins, C., Nerguizian, C.: Radiolocation Using Fingerprinting Technique in an Underground Mining Environment. In: 1st International Workshop on Wireless Comm. In Underground and Confined Areas, Val-d'Or, Canada, pp. 163–166 (June 2005)
8. Dayek, S., Affes, S., Kandil, N., Nerguizian, C.: Cooperative Localization in Mines Using Fingerprinting and Neural Networks. In: Proc. of IEEE WCNC 2010, Sydney, Australia, April 18-21 (2010)
9. Pahlavan, K., Li, X.: Indoor Geolocation Science and Technogy. IEEE Communications Magazine (February 2002)
10. Castro, P., Chiu, P., Kremenek, T., Muntz, R.: A probabilistic room location service for wireless networked environments. In: Abowd, G.D., Brumitt, B., Shafer, S. (eds.) UbiComp 2001. LNCS, vol. 2201, pp. 18–34. Springer, Heidelberg (2001)
11. Ekahau, http://www.ekahau.com/
12. Youssef, M., Agrawala, A., Shankar, A.: WLAN location determination via clustering and probability distributions. In: IEEE International Conference on Pervasive Computingand Communications (PerCom), pp. 143–150 (2003)
13. Gwon, Y., Jain, R., Kawahara, T.: Robust indoor location estimation of stationary and mobile users. In: Proc. of the 23rd Annual Joint Conference of the IEEE Computer and Communication Societies (INFOCOM 2004), pp. 1032–1043 (2004)
14. Pandya, D., Jain, R., Lupu, E.: Indoor location using multiple wireless technologies. In: Proc. IEEE PIMRC, pp. 2208–2212 (2003)
15. Taok, A., Kandil, N., Affes, S.: Neural Networks for Fingerprinting-based Indoor Localization Using Ultra-Wideband. Journal of Communications 4(4) (2009)
16. http://www.silabs.com
17. Vapnik, V.: The Nature of Statistical Learning Theory. Springer, Berlin (1995)
18. Suykens, J.A.K., Van Gestel, T., De Brabanter, J., De Moore, B., Vandewalle, J.: Least Squares Support Vector Machines. World Scientific, Singapore (2002)
19. Hatami, A.: Application of Channel Modeling for Indoor Localization Using TOA and RSS. PhD thesis, Worcester polytechnic institute (2006)
20. Kaemarungsi, K.: Design of indoor positioning systems based on location fingerprinting technique. Ph.D. dissertation, University of Pittsburgh, Pittsburgh (2005)

Crowd Sourcing Indoor Maps with Mobile Sensors

Yiguang Xuan, Raja Sengupta, and Yaser Fallah

University of California, Berkeley
Berkeley, CA, U.S.A., 94720

Abstract. The paper describes algorithms required to enable the crowd sourcing of indoor building maps, i.e., where GPS is not available. Nevertheless to enable crowd sourcing we use the 3-axis accelerometers and the 3-axis magnetometers available in many smart phones and the piezometer in a Nike running shoe. Volunteers carry the sensors while walking around in buildings, and use some application on their smart phone to send the data to a mapping server. We present the algorithms to obtain walking trajectories from the data by dead reckoning, and to estimate indoor maps with multiple walking trajectories.

Keywords: crowd sourcing, indoor mapping, dead reckoning.

1 Introduction

Open map systems (e.g., openstreetmap.org, waze.com) are viable today due to the rise of crowd sourcing. Open map systems combine two major sources of data: GIS databases and motion trajectories contributed by the crowd. Any person with a GPS receiver/logger on an increasing number of consumer devices can record his/her own trajectory and send it in to help an open map system. However, this approach does not work when the GPS signal is poor or not available, for example, within buildings. Thus the objective of this paper is to explore ways to crowd source indoor maps without relying on GPS.

The literature shows how to navigate and map indoor environments with relative and absolute positioning technologies other than GPS. For example, [1] describes the use of laser to scan buildings from outside through windows and map the interior of buildings. The quality of the map depends on obstructions due to the windows. The work in [2] uses radio frequency (RF) ranging to measure the distance from a sensor worn by a person to multiple base stations, and then triangulate to localize the person, in a manner similar to GPS. The accuracy of the positioning depends on the severity of the multi-path effect. These mapping methods require the installation of a bases station infrastructure.

In the relative positioning literature, maps are built using inertial navigation systems (INS) and imaging sensors such as vision, lidar, sonar, or radar. A robot equipped with such sensors navigates through the environment to be mapped without any absolute position measurement [3, 4, 5, 6]. [7] describes the use of a compass and an accelerometer to locate personnel within buildings. The work is similar to ours, but requires the compass to be fixed with known orientation.

We focus on indoor mapping using smart phones (iPhone or G-Phone) and Nike shoes. This paper describes the algorithms developed to crowd source indoor maps. We envisage people carrying these phones and wearing piezometer equipped shoes installing an application and communicating data to the mapping server. The shoe automatically networks with the phone. Our algorithms utilize the 3-axis accelerometers and 3-axis magnetometers on the smart phone. These are fused with the piezometer in the Nike running shoe. We do not need prior information about the orientation of the sensors as in [7].

The rest of the paper is organized as below. Section 2 introduces the sensors involved; section 3 presents the methodology to produce walking trajectories with the sensors; section 4 describes how floor map is estimated; section 5 summarizes and discusses the result.

2 Sensors

The 3-axis accelerometer and magnetometer used in this study are integrated in a G-Phone. The G-Phone coordinate is defined as in Fig. 1a: +x direction extends along the short edge of the screen to the right; +y direction extends along the long edge of the screen to the front; +z direction extends perpendicularly out of the screen. The G-Phone measures its own accelerations and the magnetic intensity along the three axes at a frequency of about 46Hz. The bias in the magnetic intensity measurements is well documented [8, 9], and can be calibrated by simply rotating the G-Phone along its three axes.

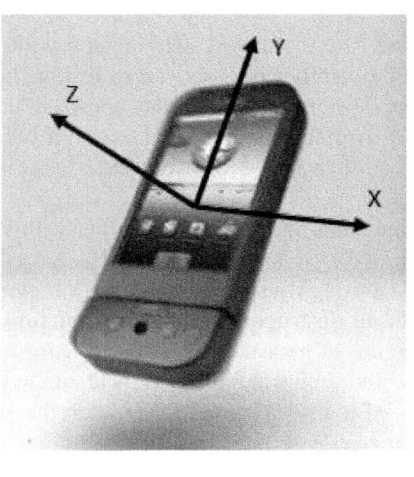
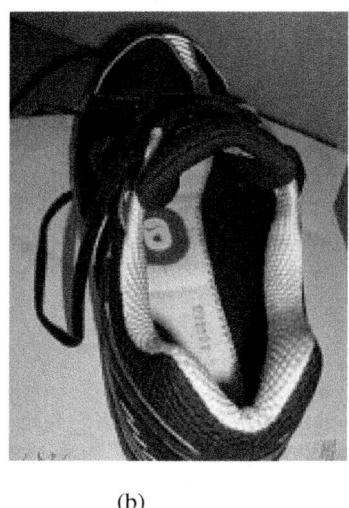

(a)   (b)

Fig. 1. The devices. (a) G-Phone with its three axes; (b) Nike sensor

The piezometer is mounted under the inner sole of the Nike shoe, as shown in Fig. 1b. The sensor measures the amount of time that the foot spends on the ground, and estimate pace based on the contact time [10]. The information is then sent to and recorded in an iPod or iPhone. The iPhone/G-Phone mix is only to facilitate data collection as a proof of concept. We envision that for a real application, data will be collected from a single platform.

During our study, volunteers walk around within the same building multiple times, carrying the G-Phone and wearing the Nike+ sensor. The logged data are used to map the building. We make the following assumptions to create an experiment simple enough for a proof of concept: (1) the map is restricted to one floor; (2) the hallways are straight, and are parallel or perpendicular to each other; (3) the orientation of the G-Phone is unknown (i.e., we do not know how people carry it during data collection), but is fixed or changes slowly with respect to the trunk of the volunteer (for example, as happens when carried in a chest pocket or backpack).

3 Navigation

The methodology for navigation is described in three steps. First, the orientation of the G-Phone is estimated with the measurements of the gravity and the magnetic north. Second, the walking direction is estimated with some human walking patterns. Finally, walking speed is estimated with the piezoelectric accelerometer. Then dead reckoning is applied to produce the walking trajectories.

3.1 Estimation of Phone Orientation

Besides the G-Phone coordinates, we define the world coordinate, with +E/+N/+U direction pointing to magnetic east/magnetic north/vertically up. Thus the orientation of the G-Phone can be described by a transformation matrix

$$\mathbf{T} = \begin{bmatrix} e_X & e_Y & e_Z \\ n_X & n_Y & n_Z \\ u_X & u_Y & u_Z \end{bmatrix},$$

where unit vectors along the +E, +N, and +U direction can be expressed as $\mathbf{e} = \begin{bmatrix} e_X & e_Y & e_Z \end{bmatrix}^T$, $\mathbf{n} = \begin{bmatrix} n_X & n_Y & n_Z \end{bmatrix}^T$, and $\mathbf{u} = \begin{bmatrix} u_X & u_Y & u_Z \end{bmatrix}^T$ respectively in the G-Phone coordinate. With this definition, $\mathbf{T}^T \mathbf{T} = \mathbf{I}$. Also, if a vector can be expressed as $\mathbf{V}_G = \begin{bmatrix} V_X & V_Y & V_Z \end{bmatrix}^T$ in the G-Phone coordinate and as $\mathbf{V}_W = \begin{bmatrix} V_E & V_N & V_U \end{bmatrix}^T$ in the world coordinate, then $\mathbf{V}_W = \mathbf{T} \mathbf{V}_G$ and $\mathbf{V}_G = \mathbf{T}^T \mathbf{V}_W$.

To estimate the orientation of the G-Phone, some prior knowledge is exploited: the gravity points into the −U direction; the magnetic north lie on the plane formed by the −U and +N directions. So we want to estimate $\mathbf{T}$, with the measurements of gravity $\mathbf{G}_G = \begin{bmatrix} G_X & G_Y & G_Z \end{bmatrix}^T$ and magnetic field $\mathbf{M}_G = \begin{bmatrix} M_X & M_Y & M_Z \end{bmatrix}^T$, knowing that $\mathbf{G}_W = \begin{bmatrix} 0 & 0 & -G_U \end{bmatrix}^T$ and $\mathbf{M}_W = \begin{bmatrix} 0 & M_N & -M_U \end{bmatrix}^T$. Interested reader can verify that the following procedure yield the estimate of $\mathbf{T}$:

$$\mathbf{u} = \frac{-\mathbf{G}_G}{\|\mathbf{G}_G\|},$$

$$\mathbf{n} = \frac{\mathbf{M}_G - \left(\mathbf{u}^T \mathbf{M}_G\right)\mathbf{u}}{\|\mathbf{M}_G - \left(\mathbf{u}^T \mathbf{M}_G\right)\mathbf{u}\|},$$

$$\mathbf{e} = \mathbf{n} \times \mathbf{u},$$

$$\mathbf{T} = \begin{bmatrix} \mathbf{e} & \mathbf{n} & \mathbf{u} \end{bmatrix}^T,$$

where × denotes cross product.

Note that the magnetic north is not the true north, and the magnetic declination angle (the direction of magnetic north with respect to the true north) can be looked up by location and by date [11].

3.2 Estimation of Walking Direction

Accelerometers have been widely used to predict human moving behavior [12, 13, 14]. Here, the accelerations along the 3 axes are used to predict the walking direction of the volunteer, with the knowledge of two human walking patterns. These walking patterns have been identified previously by researchers, but they have not been used to predict walking direction.

Define the volunteer coordinate: +AP (for anteroposterior) direction points to the front, +ML (for mediolateral) direction points to the left, and +V (for vertical) direction points up. We find that the acceleration along the ML axis always has the smallest root mean square (RMS) compared with the acceleration along the V/AP axis. For example, in one of our walking samples with speed 1.3 m/s, the RMS of the acceleration along the AP, ML, and V axes are 1.0 m/s^2, 0.6 m/s^2, and 2.4 m/s^2 respectively. Figure 4 of [15] confirms our finding, although the purpose of [15] is to estimate walking speed rather than to use this pattern to obtain the direction of walking. Since the average acceleration for motion with uniform speed is theoretically zero, the RMS can be interpreted as the intensity of motion. Thus the physical interpretation of this pattern is that there is less motion in the ML direction compared with AP and V directions.

Making use of this pattern, we carry out a principal component analysis (PCA) on (part of) the time series of the 3-D acceleration. PCA decomposes the 3-D acceleration into three principal components, so that the first principal component captures as much variance as possible, and the second principal component captures as much of the remaining variance as possible. Thus the third principal component is the estimate of the ML axis ($\overline{\text{ML}}$). Among the first and second principal components, the one closer to gravity is the estimate of the V axis ($\overline{\text{V}}$), and the other is the estimate of the AP axis ($\overline{\text{AP}}$).

Now we have identified the axes $\overline{\text{AP}}$, $\overline{\text{ML}}$, and $\overline{\text{V}}$ of the volunteer coordinate. But these axes are bidirectional and we need to decide whether the walking direction is in the $+\overline{\text{AP}}$ direction or the $-\overline{\text{AP}}$ direction.

We further find out that during walking the up and down movement along the V axis and the acceleration and deceleration along the AP axis are correlated. The acceleration along the AP axis is followed by the up movement along the V axis, then the deceleration along the AP axis, and then the down movement along the V axis. The process is shown with sample walking data in Fig. 2. [16] confirmed this pattern with similar experiments, though the purpose of their study is to estimate the energy related to walking, not to predict walking direction.

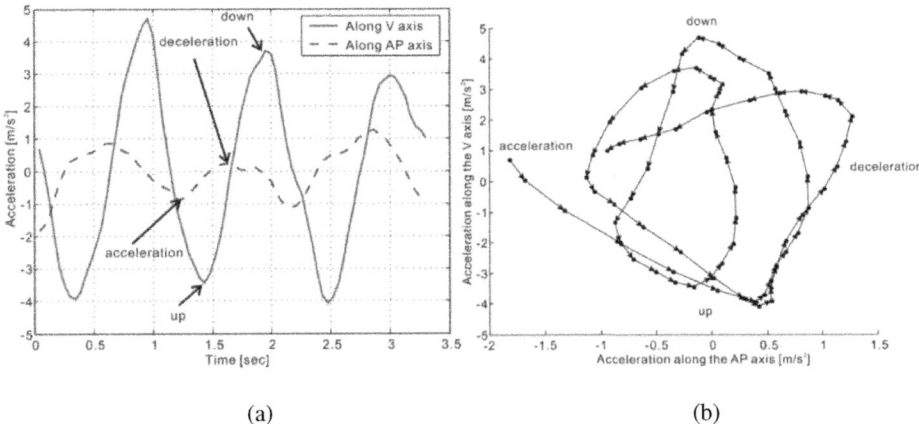

Fig. 2. Walking pattern: correlated V-axis component and AP-axis component of acceleration. (a) V-axis and AP-axis components vs. time; (b) V-axis component vs. AP-axis component.

Using this pattern, one can easily determine whether the walking direction is in the $+\overline{AP}$ direction or $-\overline{AP}$ direction, based on whether the gravity lies in the $+\overline{V}$ or $-\overline{V}$ direction. The estimated walking direction is then converted from the G-Phone coordinate into the world coordinate with the estimated transformation matrix **T** from Section 3.1.

3.3 Estimation of Walking Speed

The walking speed is estimated by measuring the contact time of foot on the ground with the piezoelectric accelerometer [10]. We do not have access to the contact time to calibrate the relationship between the contact time and the walking speed. Thus we use the calibration software available on the iPod or iPhone. The predicted speeds from the software are then calibrated against the true walking speeds measured from a treadmill with preset speeds. The result of the calibration is shown in Fig. 3. There are still errors in the speed estimation, and these errors are later diminished by aggregating multiple measurements.

The walking speed can also be estimated through the RMS of the acceleration (from the G-Phone). The rationale is that the RMS of the acceleration, which is an

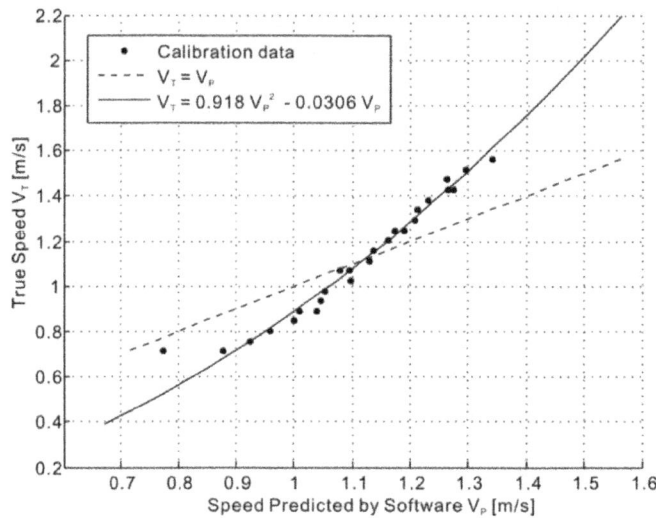

Fig. 3. Calibration of the speed predicted by iPod software against true speed

indicator of motion intensity, is correlated with walking speed. There have been studies like this [15, 17, 18]. But the motion intensity also depends on where the accelerometer (or the G-Phone) is located. Thus the relationship between the RMS of the acceleration and the walking speed will be different depending on where the G-Phone is located. This factor can introduce inaccuracy to the measurement of walking speed, and has been partially corrected by the proposed method here.

3.4 Post-processing

After obtaining the walking direction and walking speed in sections 3.2 and 3.3 respectively, dead reckoning is applied to obtain the walking trajectories as post-processing. During post-processing, we correct the trajectories, to assure that the hallways are straight and any two hallways are parallel or perpendicular.

The walking direction is represented by the azimuth angles, with 0, $\pi/2$, π, and $3\pi/2$ indicating true north, east, south, and west. Fig. 4a shows the azimuth angle versus time for a sample walking trajectory. First, we discretize the azimuth angle by clustering, with the assumption that hallways are straight. The center of the clusters are shown in Fig. 4a, while the discretized azimuth angle versus time is shown in Fig. 4b.

Note that the discretized azimuth angles can be neither parallel nor perpendicular due to distortion of the local magnetic field. To account for this problem with the assumption that any two hallways are either parallel or perpendicular, we define an independent orientation parameter to be the orientation of an arbitrary hallway. Then the orientation of all the hallways can be expressed by this orientation parameter plus or minus integer times of $\pi/2$. Thus we use the measured azimuth angles to estimate

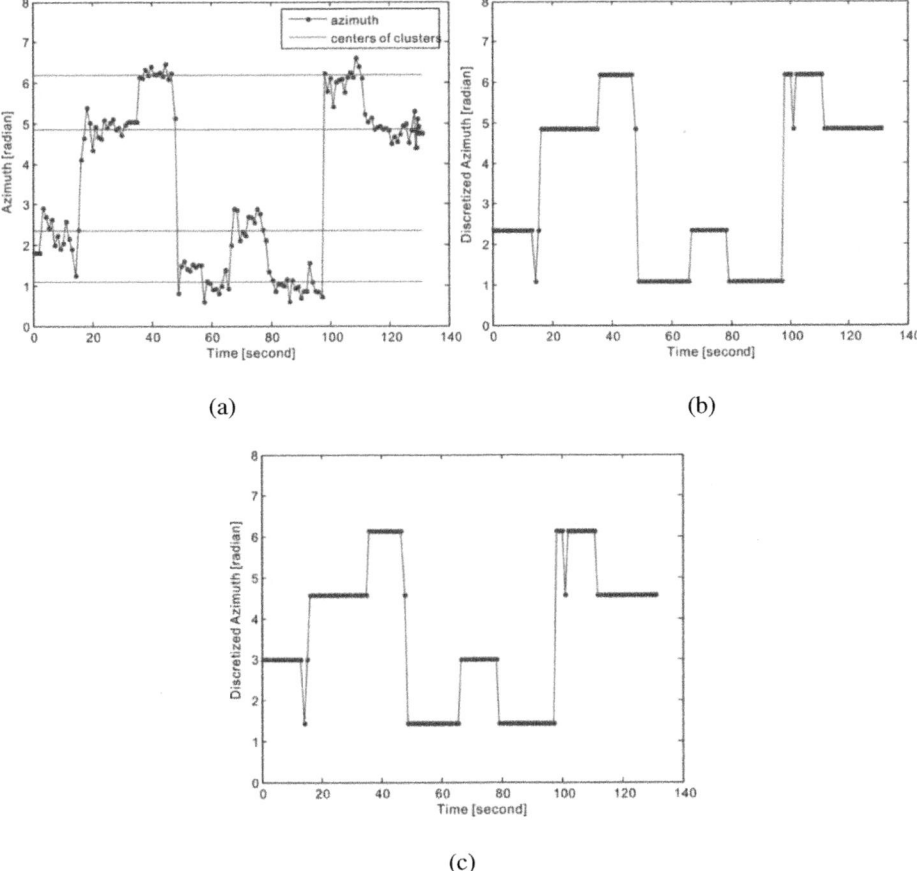

Fig. 4. Adjustment of azimuth angle. (a) clustering of azimuth angle; (b) discretized azimuth angle; (c) correction for distortion of the local magnetic field.

this parameter, and then express the orientation of all the hallways with the estimate of this orientation parameter. This adjustment is shown in Fig. 4c. If however, the magnetic distortion can be well corrected, then this assumption of parallel or perpendicular hallway is not needed.

A volunteer walked the hallways of the 5th floor of Davis Hall at the University of California, Berkeley, holding/wearing the aforementioned sensors. During walking, the G-Phone is held still with respect to the trunk of the volunteer, but the orientation of G-Phone is not used in the study. Shown in Fig. 5a is the center of the hallways, which is comprised of two side-by-side rectangles. The aforementioned method is used to obtain walking trajectories (we have no information about the walls though), two of which are shown in Fig. 5b and Fig. 5c. With discrete azimuth angles, it is easy to find out when the volunteer takes a turn. Note that the north shown in the figures is the true north, after correcting for the magnetic declination angle.

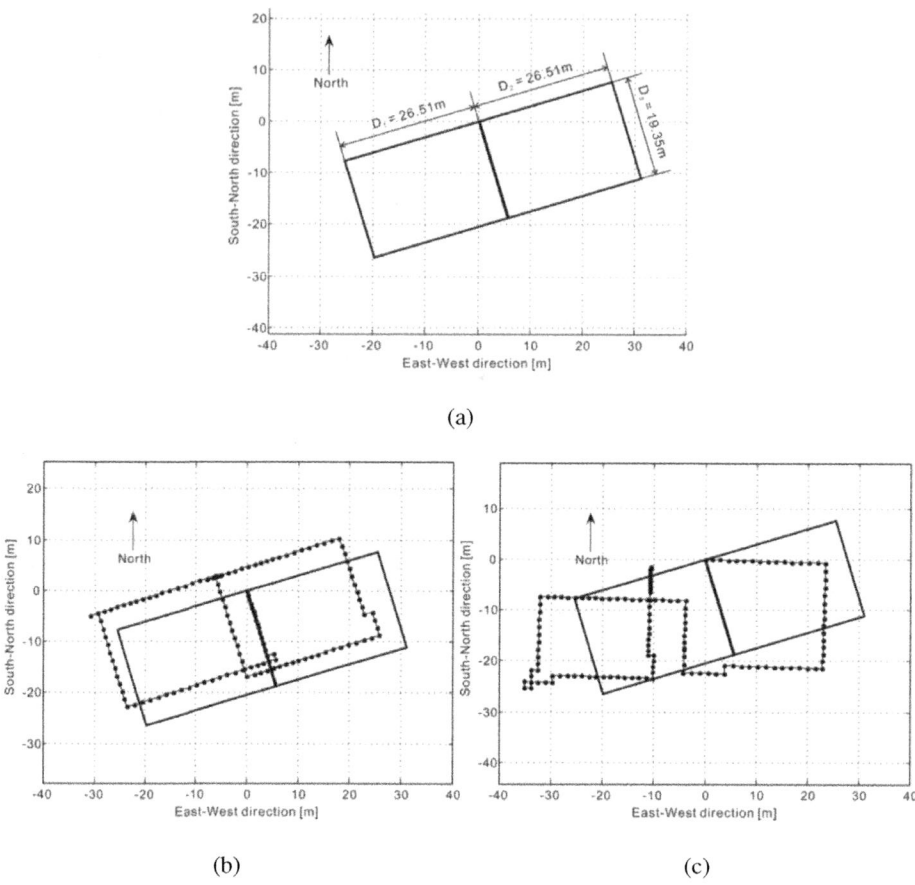

Fig. 5. True floor map and walking trajectories of the 5th floor of Davis Hall, UC Berkeley. (a) true floor map; (b) one sample walking trajectory; (c) another sample walking trajectory.

4 Estimation of Floor Map

One way to model a floor map is to represent each leg of the hallways with a length parameter and an orientation parameter. Not all parameters are independent. Since the hallways form loops, not all length parameters are independent. For example, only three length parameters are needed to determine the floor map shown in Fig. 5a. Also, all the hallways are either parallel or perpendicular, thus there only needs to be one parameter for orientation. We use the orientation of the northbound hallway.

Here we propose a method to estimate the dimension of floor maps with multiple walking trajectories, to diminish the inaccuracies in the estimation of walking speed. We assume the walking always starts and ends at the same location, e.g., the elevators or the entrance of buildings. To integrate the floor map into a geo-coded outside map, this location needs to be known in some global reference, e.g., from GPS or Wi-Fi hotspots.

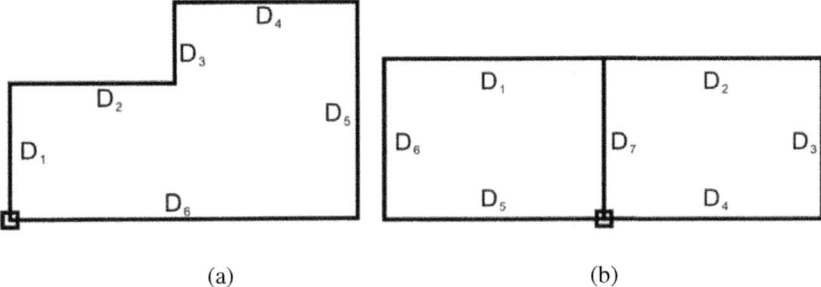

Fig. 6. Imaginary floor map.
(a) with only one loop; (b) with two loops (map of 5th floor of Davis Hall).

The floor map is divided into loops (Fig. 6a shows one such loop). Each leg of a loop is represented with unknown length parameters D_i ($i=1$ to n, n is the number of legs) and known azimuth angles a_i ($i=1$ to n). Let us assume that with trajectory j (from 1 to m_i, leg i is measured m_i times for each trajectory), the measurements of D_i is L_{ij}. Then the estimation of D_i's can be formulated as a constrained optimization problem so that the total sum of squared errors is minimized:

$$\{D_i^*\} = \arg\min \sum_{i=1}^{n}\sum_{j=1}^{m_i}(D_i - L_{ij})^2$$

$$s.t. \quad \sum_{i=1}^{n} D_i \cos(a_i) = 0, \quad \sum_{i=1}^{n} D_i \sin(a_i) = 0$$

Singularity problem may arise with quadratic optimization when the number of parameters is huge. But this is unlikely to happen because the number of parameters in this application is generally much less than the number of measurements.

For our example of the 5th floor of Davis Hall, as shown in Fig. 6b, it turns out that the sample mean is the best estimate for D_i, due to its simple geometry:

$$\{D_i^*\} = \arg\min \sum_{i=1}^{n}\sum_{j=1}^{m_i}(D_i - L_{ij})^2$$

$$s.t. \quad D_1 = D_5, D_2 = D_4, D_3 = D_6, D_6 = D_7$$

$$\Rightarrow \begin{cases} D_1^* = \dfrac{1}{m_1 + m_5}\left(\sum_{j=1}^{m_1} L_{1j} + \sum_{j=1}^{m_5} L_{5j}\right) \\ D_2^* = \dfrac{1}{m_2 + m_4}\left(\sum_{j=1}^{m_2} L_{2j} + \sum_{j=1}^{m_4} L_{4j}\right) \\ D_3^* = \dfrac{1}{m_3 + m_6 + m_7}\left(\sum_{j=1}^{m_3} L_{3j} + \sum_{j=1}^{m_6} L_{6j} + \sum_{j=1}^{m_7} L_{7j}\right) \end{cases}$$

Fig. 7a – Fig. 7c show how the estimates of D_1, D_2, and D_3 converge to their true values with increasing number of walking samples. With just six walking samples, the length parameters can be estimated with a relative standard error of 3%.

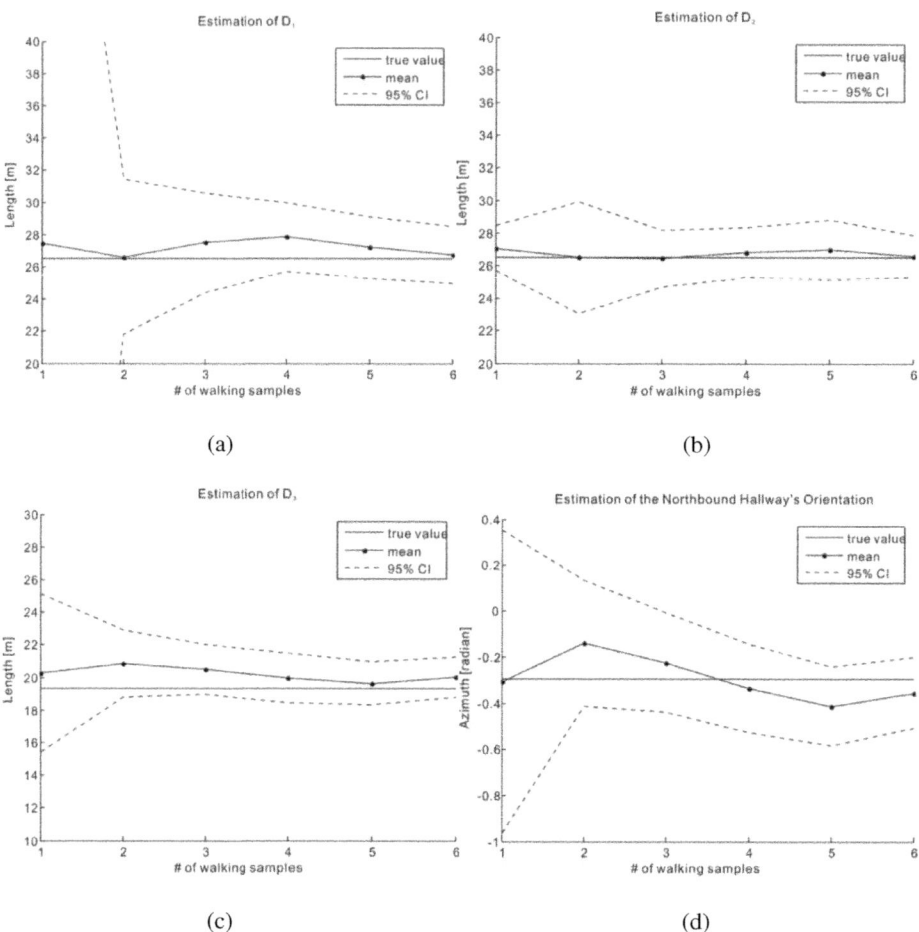

Fig. 7. Estimation of the length and orientation parameters vs. number of walking samples. (a) estimation of D_1; (b) estimation of D_2; (c) estimation of D_3; (d) estimation of the orientation of the northbound hallway.

Similarly, the orientation parameter is estimated with the measured azimuth angles as described in the previous section. The estimate of the orientation of the northbound hallway is shown in Fig. 7d. With six walking samples, the orientation parameter can be estimated with a standard error of 0.074 radian (about 4 degrees).

With just six walking samples, the estimated floor map has $\overline{D_1}$ = 26.75m, $\overline{D_2}$ = 26.58m, $\overline{D_3}$ = 20.02m. The estimated orientation of the northbound hallway is -0.3538 radian (about -20.3 degrees). The estimates are comparable with the true floor map with D_1 = 26.51m, D_2 = 26.51m, D_3 = 19.35m, and the orientation of the northbound hallway -0.2934 radian (about -16.8 degrees). The estimated and true floor maps are shown together in Fig. 8.

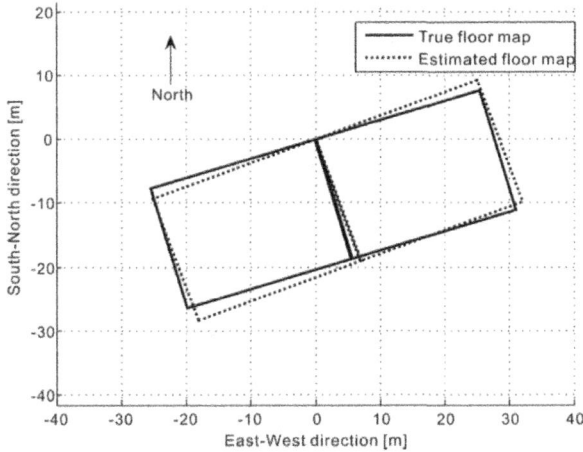

Fig. 8. Estimated vs. true floor map of the 5th floor of Davis Hall, UC Berkeley

5 Discussion and Conclusion

A successful project to crowd source indoor maps includes at least the following problems: the relative mapping, the global mapping, and the incentive problem. The relative mapping problem is to make a map with the correct proportions and orientation. This problem is the focus of the paper. The global mapping problem locates the relatively correct map in a common coordinate system such as the GPS Universal Transverse Mercator (UTM) system. We intend to address this problem next by using GPS readings at the entrances and exits of buildings (trajectory starting and ending point) and Wi-Fi hotspot readings. The incentive problem is about how to motivate the crowd to collect walking data and send them to a mapping server. Waze.com does an excellent job on this problem.

The contribution of this paper is a solution to the relative mapping problem. We proposed a methodology to estimate the lengths and orientations of the hallways. We then validate it by showing that we are able to estimate a floor map accurately with just a few walking samples. We compute the map for one floor requiring the correct estimation of three lengths and one orientation. The length parameters are estimated with a relative standard error of 3%, and the orientation parameter is estimated with a standard error of 0.074 radian (about 4 degrees), with six walking samples. The methodology should be generally applicable to other floors as long as the lengths and orientations of hallways can be estimated correctly. We expect the difference from floor to floor to lie in the number of samples required to obtain a correct map.

As future work, the proposed method needs to be tested on more than just one floor. Also, our study is restricted to identify hallways within buildings. They are relatively easy and not a lot of samples are needed. Looking forward, there are also open space and stairs within buildings, which are hard to identify, these are the cases when "crowd sourcing" will truly be needed.

References

1. Johnston, M., Zakhor, A.: Estimating building floor-plans from exterior using laser scanners. In: SPIE Electronic Imaging Conference, 3D Image Capture and Applications (2008)
2. Duckworth, J., Cyganski, D., Makarov, S., Michalson, W., Orr, J., et al.: WPI Precision Personnel Locator System – Evaluation by First Responders. In: The 20th International Technical Meeting of the Institute of Navigation Satellite Division (2007)
3. Thrun, S., Burgard, W., Fox, D.: A real-time algorithm for mobile robot mapping with applications to multi-robot and 3d mapping. In: IEEE International Conference on Robotics and Automation, vol. 1, pp. 321–328 (2000)
4. Naikal, N., Kua, J., Chen, G., Zakhor, A.: Image Augmented Laser Scan Matching for Indoor Dead Reckoning. In: International Conference on Intelligent Robots and Systems (2009)
5. Li, S., Kanbara, T., Hayashi, A.: Making a local map of indoor environments by swiveling a camera and a sonar. In: IEEE International Conference on Intelligent Robots and Systems, vol. 2, pp. 954–959 (1999)
6. Stanford Research Institute, Centibots (2004), http://www.ai.sri.com/centibots/index.html
7. Sharp, I.: Tracking method and apparatus. US patent application no. 10/547,238 (2004)
8. Graves, J.D.: Design and control of a vehicle for neutral buoyancy simulation of space operations. M.S. thesis, University of Maryland at College Park (1997)
9. Wang, J.H., Gao, Y.: GPS-based land vehicle navigation system assisted by a low-cost gyro-free INS using neural network. Journal of Navigation 57, 417–428 (2004)
10. Gaudet, P.J., Blackadar, T.P., Oliver, S.R.: Measuring foot contact time and foot loft time of a person in locomotion. US patent no. 6018705 (2000)
11. National Geographic Data Center, NOAA Satellite and Information Service, http://www.ngdc.noaa.gov/geomagmodels/struts/calcDeclination
12. Morris, J.R.W.: Accelerometry - A technique for the measurement of human body movements. Journal of Biomechanics 6, 729–736 (1973)
13. Godfrey, A., Conway, R., Meagher, D., Olaighin, G.: Direct measurement of human movement by accelerometry. Medical Engineering & Physics 30, 1364–1386 (2008)
14. Kavanagh, J.J., Menz, H.B.: Accelerometry: A technique for quantifying movement patterns during walking. Gait & Posture 28, 1–15 (2008)
15. Moe-Nilssen, R.: A new method for evaluating motor control in gait under real-life environmental conditions Part 2: Gait analysis. Clinical Biomechanics 13(4-5), 328–335 (1998)
16. Cavagna, G., Saibene, F., Margaria, R.: External work in walking. Journal of Applied Physiology 18, 1–9 (1963)
17. Schutz, Y., Weinsier, S., Terrier, P., Durrer, D.: A new accelerometric method to assess the daily walking practice. International Journal of Obesity 26, 111–118 (2002)
18. Menz, H.B., Lord, S.R., Fitzpatrick, R.C.: Acceleration patterns of the head and pelvis when walking on level and irregular surfaces. Gait Posture 18(1), 35–46 (2003)

Real Time Six Degree of Freedom Pose Estimation Using Infrared Light Sources and Wiimote IR Camera with 3D TV Demonstration

Ali Boyali[1], Manolya Kavakli[1], and Jason Twamley[2]

[1] VISOR (Virtual and Interactive Simulations of Reality),
Research Group,
Department of Computing,
[2] Physics Department,
Macquarie University
Balaclava Road North Ryde NSW 2109, Australia
{ali.boyali,manolya.kavakli,jason.twamley}@mq.edu.au

Abstract. The goal of this paper is to present the development of a tracking technology to interact with a virtual object. This paper presents the general procedures of building a simple, low cost tracking system by using Wiimote (a remote of Nintendo game console) and the Open source Computer Vision (OpenCV) software library as well as interfacing the tracking system with an immersive virtual environment (Vizard). We used an iterative position and orientation estimation (POSIT algorithm) which is optimized as an OpenCV function for extracting position parameters. We filter out the noise in the coordinate values using Kalman filters. The orientation and translation of the tracked system are then used to manipulate a virtual object created in the virtual world of Vizard. Our results indicate that it is possible to implement an inexpensive and efficient application for interacting with virtual worlds using a Wiimote and appropriate digital filters.

Keywords: WiimoteIR tracking, 6 DOF pose estimation, OpenCV, 3D TV visualization.

1 Introduction

Tracking of moving objects has a wide variety of application areas including medical science, interactive entertainment, gaming, control technologies and military applications. Motion tracking can be performed either by tracking a moving object with a fixed capturing device or by capturing movement of a camera using stationary objects. The information related to motion in both cases is extracted and used according to the requirements of the systems.

One useful technology for optical motion tracking are Infra-Red (IR) cameras which can report the 2D projected coordinates of IR light sources. The reported 2D projected coordinates of a known specific 3D arrangement of IR sources are then processed to estimate the position and orientation of an object in three dimensional space (six Degree of Freedom – 6 DOF) using appropriate computer vision algorithms.

Pose estimation and gesture recognition brings distinct advantages to Human Computer Interaction (HCI) in comparison to conventional input devices. These techniques enable users to interact with computers in a more intuitive and natural way. Commercially available tracking systems do not appeal to ordinary computer users due to their lack of affordability and reduced availability [1]. Wiimote is a game input device that allows the user to manipulate objects in a virtual environment, with the capability of sensing gestures through its accelerometer and optical sensors (Fig. 1). It can measure yaw, pitch and roll orientations via built-in acceleration sensors, and report the coordinates of infra-red light sources with a resolution of 1024x768 pixels.

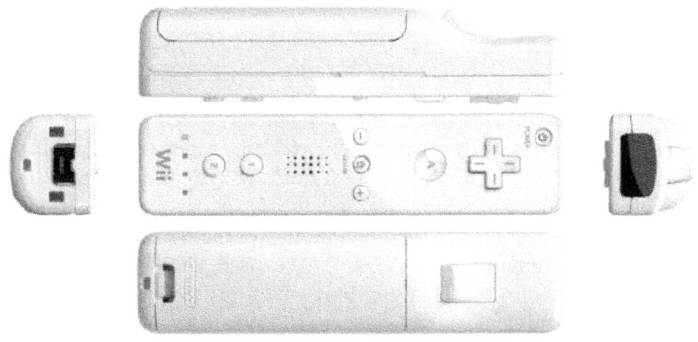

Fig. 1. Nintendo Wiimote controls [2]

Due to its unique features, affordability and availability, the Wiimote handset has attracted great attention in recent years. It has been used not only in computer games but also in HCI applications. For example, a head mounted IR LED arrangement can replace the ordinary mouse in order to enable people with limited mobility to surf the internet with minimal effort [3]. The click function can be realized with a pre-defined voice vocabulary in this system. In the study [4], Wiimote acceleration sensor readings are used to control a wheelchair with simple hand gestures. Low latency of the IR camera due to high refresh rates [5, 6] allows programmers to use Wiimote to control a virtual object [7], to navigate interactive maps [8], and build low cost interactive electronic whiteboards [9].

Our goal in this study is the development of an inexpensive 3DTV interactive environment for the user to interact with a virtual world, using head and ultimately, hand tracking.

We use a similar methodology for tracking IR light sources as presented in [5, 6], to demonstrate the potential use of such 3D tracking methods in medical imaging. The commonly used approach towards tracking objects using an IR camera and IR

LED (Light Emitting Diode) beacon is to extract orientation and translational movements from 2D image data. The structure of our tracked object is described in section 2 below. We present a method of pose estimation and its computer implementation in section 3. The software environment and coding are presented in section 4. We present the results of filtering in section 5. In section 6 we discuss our findings and directions for future studies.

2 IR Beacon Architecture

The Wiimote has a monochrome "camera" with a resolution of 128x96 pixels. On-camera proprietary Nintendo hardware post-processes the camera image and outputs the x-y coordinates of up to four infrared light sources at a refresh rate of 100 Hz. The coordinates of light sources are reported with a resolution of 1024x768 pixels via interpolation. There are scant details regarding the actual hardware makeup of the "camera" in the Wiimote (i.e. it is not clear whether it is an array device like a CCD chip in ordinary cameras), but it has a peak sensitivity in the IR though can also detect in the visible. In the Wiimote visible light is blocked with a plastic IR bandpass filter, which sharply attenuates light transmission below 900 nm. The combination of this IR filter and the spectral sensitivity of the "camera" results in the Wiimote being most sensitive to light around a wavelength of 940 nm [5, 6, and 10]. We conducted an experiment measuring the spectral properties of the IR bandpass filter as shown in Fig. 2.

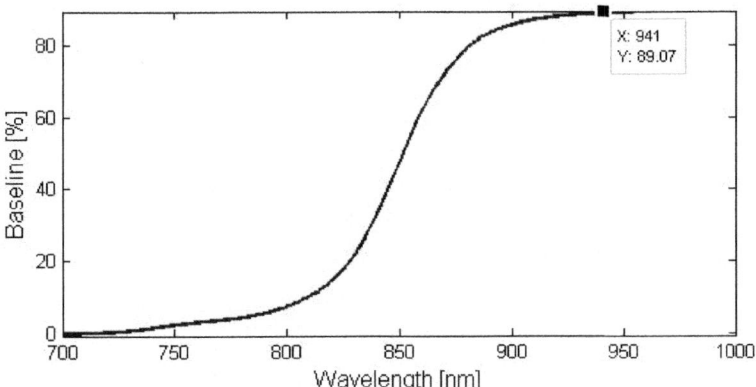

Fig. 2. Percentage transmission of light through the Wiimote's IR band-pass filter as a function of wavelength. We see a sharp attenuation below light of wavelength below 900 nm.

We used four LEDs which emit light at 940 nm wavelength in our LED beacon since the OpenCV pose estimation algorithm requires four non-planar points. A similar LED structure was used in the applications developed in [5] and [6] (Fig. 3, a, b and c).

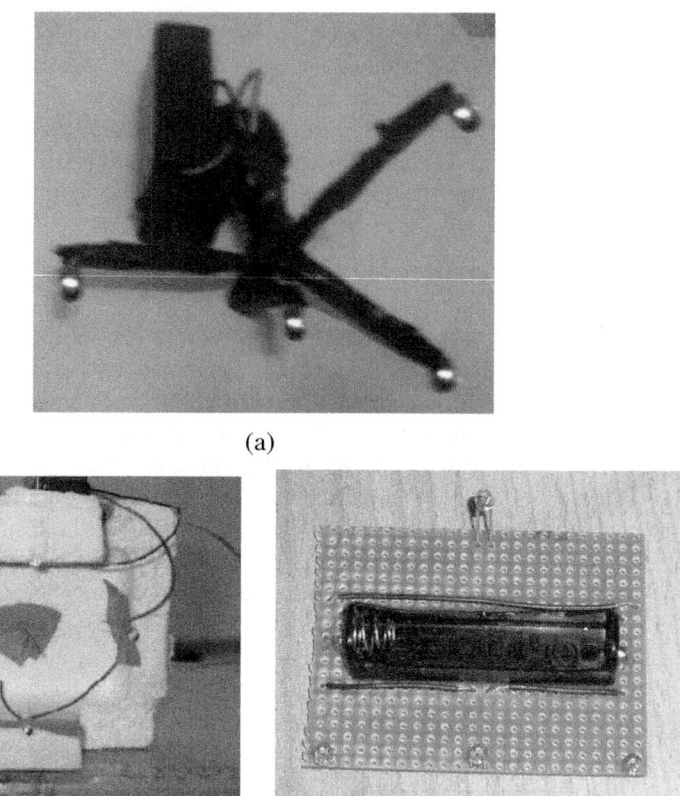

Fig. 3. (a) LED arrangement used in this study, IR Beacons used in [1] and [5] (b and c)

We used four resistances in the circuit to exceed the current limit (100 mA) for LEDs (Fig. 4). Resistance values are calculated using the rated forward current of these LEDs for a specified voltage bias and we found 3V batteries and 4 Ohms resistances where sufficient for our application.

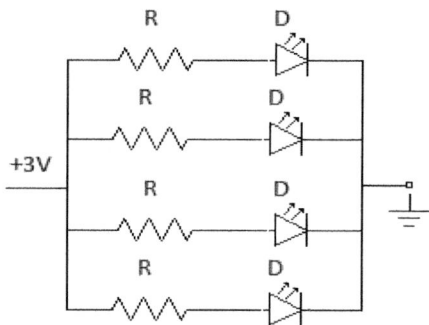

Fig. 4. Circuit diagram of our IR LED beacon where the LEDs (D) and associated resistances (R), are arranged in parallel and connected to a +3V voltage source

3 Methodology and Implementation

The Wiimote communicates bi-directionally via an on-board Broadcom 2042 Bluetooth driver chip [11]. This allows great flexibility in wireless communication between the device and the computer. There are freely available Wiimote libraries for different programming languages such as C#, Python, Matlab and C++. As OpenCV is a C++ based library we chose C++ as a development environment, and we used the Wiiuse library for the communication with Wiimote via Bluetooth.

There are a number of methods to estimate the orientations and positions of tracked objects, tagged by a constellation of tracking points [11, 12 and 13]. Among these methods the POSIT algorithm has some advantage since an initial pose estimation is not required [12]. However it suffers from a lack of accuracy when the tracking points lay on the same plane.

We used the POSIT function available in OpenCV to estimate the rotation matrix and translation vector of a tracked IR beacon. The OpenCV pose estimation method is also used in [5]. Alternatively one can solve a system of non-linear equations using the Levenberg-Marquardt minimization method to find the orientation matrix and translation vectors [6].

In the OpenCV C++ library, the cvPOSIT function takes the physical measurements of the configuration of lights on the object to be tracked (which are assumed to be pre-measured and fixed on the beacon to be tracked), the number of iterations, (or iteration sensitivity) as a termination criteria, and the camera-observed image points in 2D (as well as camera focal length in pixels). The function returns a rotation matrix and a translation vector. In nearly four or five iterations the process finds an estimate for the object's pose which is highly accurate [14, 15]. The use of extra non-coplanar points helps the algorithm to converge.

The camera focal length is an intrinsic parameter used to compute the z-coordinate of the 3D coordinate system and we model the optics using a pin-hole camera model. From this model the focal length can be found by using the camera calibration functions of OpenCV. A special IR beacon arrangement (a camera calibration pattern or a grid), whose geometry is known precisely (such as four LEDs positioned at eachcorner of a square), is required to use the camera calibration routines [14, 16]. However, camera calibration is beyond of the scope of this study and we used camera parameters as reported in [5, 6].

An important part of all pose estimation algorithms is the correct association between individual light points imaged by the camera and their respective sources on the object to be tracked. In [6] and [21], the Wiimote's linear accelerometer was used to obtain an initial point match and new poses were estimated successively with subsequent pose matchingbeing performed on a nearest neighborhood basis. In the study [5], three of the LEDs on the beacon were nearly collinear and thus the image points as seen by the camera lay also on a line in all poses. Thus the source and image point association can be achieved by applying co-linearity tests. We use a different geometrical approach to establish the association between the image and source points. Four different triangles can be constructed from four points (Fig 5). The triangle which has the biggest area excludes the point D, and this is true when viewed in any pose.

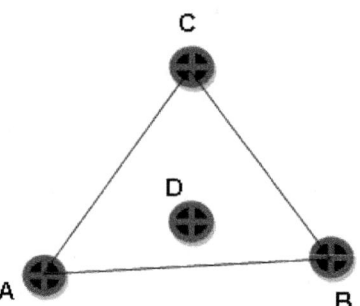

Fig. 5. The view of our IR LED arrangement from the z-axis direction

We can label point C since the vector DC constitutes the biggest angle in the combination of the vectors AD, AB and DC. Thus the point C can be distinguished from the others. Once the points, C and D are labeled, A and B can be labeled using a clockwise rotation check relative to point C.

The physical arrangements of IR LEDs on the beacon we use are given in Table 1. From our experiments the distances between the points are sufficient to track the device up to two meters away from the camera. The coordinates of point A are taken as the origin of the system. The other point's coordinates are then defined with respect to point A.

Table 1. Coordinates of the LEDs making up our IR beacon

Points/Coordinates [mm]	x	y	z
A	0	0	0
B	105	10	0
C	52	73	-10
D	54	13	7

3.1 Filtering Noisy Signal

Every digital measurement contains noise to some extent. Before processing the signals noise should be filtered out to increase the accuracy and to smooth the variation of the signal. Accuracy of six degree of freedom (6DOF) IR tracking with a single Wiimote IR camera was evaluated in [1, 6], by comparing the tracking performance with a more accurate 6DOF commercial tracking system.

In the study [1] the Wiimote tracking developed there had errors in estimating the coordinate and orientation parameters of the pose: x-y-z coordinates ~ 7-8 mm, yaw-pitch angles ~ 0.8 degree . Nearly same levels of accuracy are reported in [6]. The noise values as well as the offset in the y direction of the IR beacon are apparent in Fig. 6.

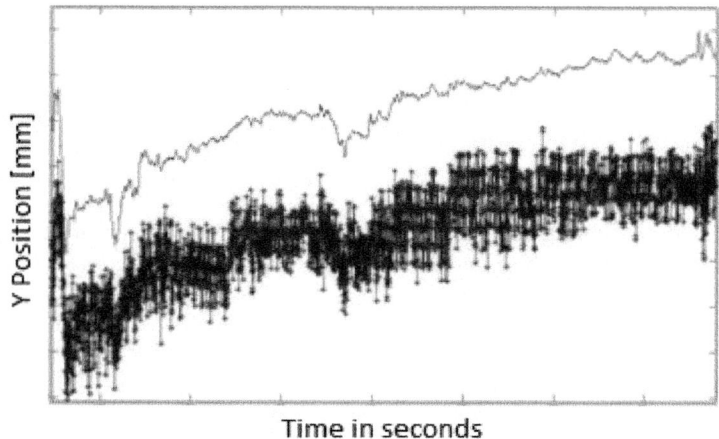

Fig. 6. Gauging the accuracy of the pose estimation. Graph plots the *y* coordinate in mm versus time at a stationary position taken from the study[6].

From these studies one observers that the Wiimote IR camera is not suitable for applications where precision is required, but it is sufficient for manipulating virtual objects or for use as a pointing device [1].

We can filter out the noise from the signal by applying a Kalman filter to the coordinates sent to the computer from the Wiimote. Kalman filtering is a powerful method to estimate the state of dynamical systems subject to random noise. It has been widely used in tracking systems to filter out noise from the signal [17]. Once the filter is initialized with initial state values and error covariance, the Kalman filter estimates the next state variables and error covariance in time by maximizing the posterior probability recursively and new measurements are used to update the estimation [14, 17].

Kalman filtering functions are available in the OpenCV library. We used these functions instead of writing the filtering class from scratch. Models of the system's dynamics and measurement are required for Kalman filtering.

Given a dynamic system at time step k, we can express the state and measurement models (1), (2) as follows

$$x_k = Fx_{k-1} + Bu_k + w_k \tag{1}$$

$$z_k = H_k x_k + v_k \tag{2}$$

Here, x_k and z_k represents the states and measured variables, u_k represents the control input, w_k and v_k represents the process and measurement noise respectively [14]. *F*, *B* and *H* are the matrices representing the state transition, control input and measurement matrices respectively.

State variables in our system for an individual LED position are

$$x_k = \begin{bmatrix} x \\ y \\ dx/dt \\ dy/dt \end{bmatrix}, \quad F = \begin{bmatrix} 1 & 0 & dt & 0 \\ 0 & 1 & 0 & dt \\ 0 & 0 & 1 & 0 \\ 0 & 0 & 0 & 0 \end{bmatrix} \quad (3)$$

Here dt is the sampling time (refresh rate). Since we only measure the position, our measured variables will be x and y.

$$H = \begin{bmatrix} 1 & 0 \\ 0 & 1 \\ 0 & 0 \\ 0 & 0 \end{bmatrix} \quad (4)$$

The random variables w_k and v_k are independent from each other and they have a normal probability distribution [14, 18].

$$P(w) \sim N(0, Q) \quad (5)$$

$$P(v) \sim N(0, R) \quad (6)$$

$$P_k = E[e_k e_k^T] \quad (7)$$

Q, R and P represent process noise covariance, measurement noise covariance and error covariance respectively. An initial state and initial error covariance matrix should be assigned to start the filtering process. We assigned the coordinates of the LEDs as an initial state as if they are estimated in a virtual preceding step by using the OpenCV Kalman filter functions' state_post method. We assumed that our tracking system is linear and our model reflects the dynamics of the system accurately.

The underlying aim of this assumption is to smooth the visualization of the object since high accuracy is not needed for our intended use (3D pointer). Thus we kept the measurement and process noise covariance values at very low levels.

4 Visualization and Virtual Environment / 3D TV

In the visualization stage, we used the Vizard Virtual Reality engine developed by Worldviz. By using the Vizard SDK we created a plugin that receives our tracking data and makes it available within the Vizard environment. We could then easily attach the 6 DOF location of our physical sensor to either a graphical 3D model or to the virtual camera position.

In order to display our test platform in stereoscopic 3D, we used a recently released 3D television (Samsung UA55C7000). Vizard is able to output to the display in various 3D standards, one of which is to use side by side stereo images, a left eye scene and a right eye scene displayed side by side horizontally, or one above the other vertically. The 3D TV can then be manually set to the relevant side by side 3D stereo mode.

We were then able to see the results of our tracking device, which being a 6 DOF device moves in 3 dimensions and 3 orientations, on a screen capable of displaying the result in full (Fig. 7).

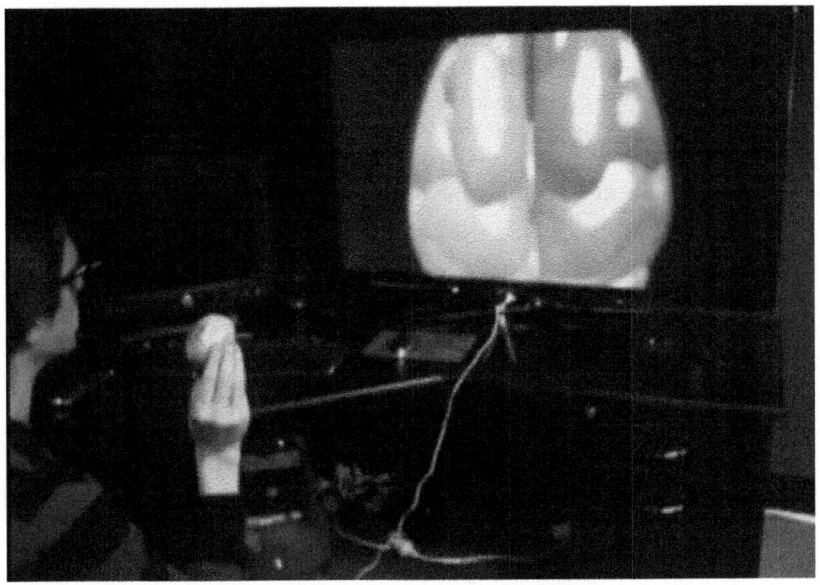

Fig. 7. Brain Model as a virtual object

5 Results

We used a Kalman filter to feed pose estimation model with a low level of noise, and not to allow the error level to increase after the use of applied functions. The effect of filtered coordinates on the Roll angle is seen in the Fig. 8.

The dotted line shows the roll angle computed using the Kalman filtered coordinates. There are eight coordinate values for four points. The coordinate values are integers and reported in the resolution of 1024x768 pixels. The errors occur

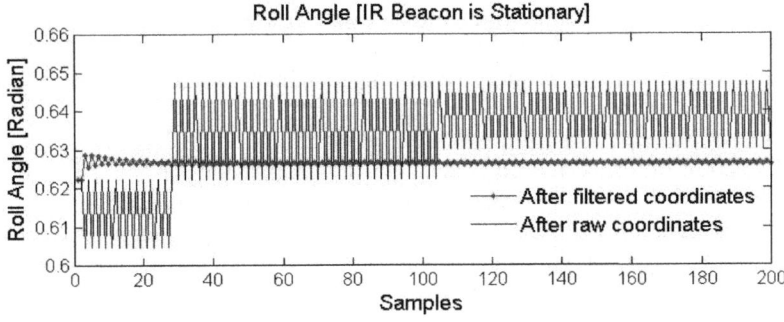

Fig. 8. Roll angle comparison computed using filtered and raw coordinates

jumping the coordinate values in the neighborhood of the LEDs' real position. The Kalman filter tries to estimate real position. The estimated coordinates have decimal values jumping between neighborhoods in small intervals.

Although the Kalman filter reduces error level in computed Euler angles, we still must use appropriate filters to smooth the visualization. In our experiments, we applied a moving-average filter to computed Euler angles for smoothing the visualization. The moving-average filter behaves as a low pass filter and attenuates the high frequencies. We continue to analyze the results, and design and test these and new filters for future phases of the project. A study of a first order low pass Butterworth filter on roll angle is given Fig. 9. As seen in the Fig. 9 Butterworth filter produces better results in smoothing the data.

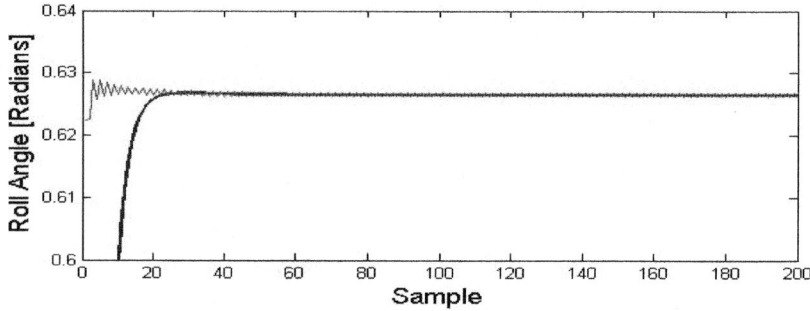

Fig. 9. Butterworth filtering after Kalman filter

6 Conclusion and Future Work

We demonstrated tracking and pose estimation of a moving IR beacon by using OpenCV and Wiiuse libraries. We built an IR LED beacon consisting of four non-coplanar LEDs and estimated the pose of this beacon using the OpenCV Posit function. Before calling cvPOSIT function to extract the rotation matrix and position vector of the device we applied a Kalman filter to smooth the visualization.

Movement of the manipulated object (in our study we used a 3D brain model) was very jittery before implementing the Kalman filter due to the very high level of noise associated with the Wiimote digitization hardware (also found by the studies [1, 12]). We suspect this noise arises from the actual low resolution of the Wiimote camera and the resulting hardware interpolation of coordinate values within the Wiimote camera.

Our results indicate that band pass filter algorithms can be used to achieve smooth pose estimation in virtual environments. Wiimotes provides an inexpensive and fast developing hardware platform for computer vision and gesture recognition studies, although they cannot be used in the application where precision and accuracy is essential. Future research will focus on developing greater pose accuracy, greater field of view and the ability to track many IR beacons simultaneously.

Acknowledgements. This project has been jointly sponsored through an Australian Research Council Discovery Grant, DP0988088 (Kavakli) 2009-2012 titled A Gesture-Based Interface for Designing in Virtual Reality and a Macquarie University Research Development Grant (Gilchrist, Twamley, Kavakli) 2008-2010 titled Immersive Virtual Scientific Collaboration. Special thanks to Richard Miller for the initial implementation of the IR tracking device, John Porte for the Vizard and 3D TV implementation and demonstrations and Peter Dekker for IR Filter measurements.

References

1. Chow, Y.W.: Low Cost Multiple Degrees of Freedom Optical Tracking for 3D Interaction in Head Mounted Display Virtual Reality. International Journal of Recent Trends in Engineering 1, 152–156 (2009)
2. Wronski, M.P.: Design and Implementation of a Hand Tracking Interface using the Nintendo Wii Remote, Dissertation, University of Cape Town (2008)
3. Azmi, A., Alsabhan, N.M., AlDosari, M.S.: The Wiimote with SAPI: Creating an Accessible Low-Cost, Human Computer Interface for the Physically Disabled. International Journal of Computer Science and Network Security 9(12), 63–68 (2009)
4. Duran, L., Carmona, M.F., Urdiales, C., Peula, J.M., Sandoval, F.: Conventional Joystick vs. Wiimote for Holonomic Wheelchair Control. In: Cabestany, J., Sandoval, F., Prieto, A., Corchado, J.M. (eds.) IWANN 2009. LNCS, vol. 5517, pp. 1153–1160. Springer, Heidelberg (2009)
5. Fleisch, T.: Wiimote Virtual Reality Desktop, http://www.vrhome.de/vr-applications
6. Vuong, P., Kurillo, G., Bajcsy, R.: Oliver Kreylos' Wiimote Tracking Algorithm and its Limitations, Summer, Research Report (2008), http://phongvuong.com/wordpress/wp-content/uploads/2009/11/wiimotePaper1.pdf
7. Dehling, E.: Using Multiple WiiMote Cameras to Control a Game, Student Project, http://hmi.ewi.utwente.nl/tkisis/persoon/Eike%20Dehling
8. Vural, G., Tekkaya, G., Erogul, C.: Using Head and Finger Tracking with Wiimote For Google Earth Control, Student Project, http://www.ceng.metu.edu.tr/~erogul/ceng701/project/.../paper-draft.pdf (accessed June 15 2010)
9. Wang, Z., Louey, J.: Economical Solution for an Easy to Use Interactive Whiteboard. In: Japan-China Joint Workshop on Frontier of Computer Science and Technology (2008)
10. Lee, J.: Hacking the Nintendo Wii Remote. IEEE Pervasive Computing 7(3), 39–45 (2008)
11. DeMenthon, D.F., Davis, L.S.: Model Based Object Pose in 25 Lines of Code. International Journal of Computer Vision 15(1-2), 123–141 (1995); Special issue: image understanding research at the University of Maryland
12. Daniel, G., Petersen, T., Kruger, V.: A Comparison of Iterative 2D-3D Pose Estimation Methods for Real-Time Applications. In: Salberg, A.-B., Hardeberg, J.Y., Jenssen, R. (eds.) SCIA 2009. LNCS, vol. 5575, pp. 706–715. Springer, Heidelberg (2009)
13. Campa, G., Mammarella, M., Napolitano, M.R., Fravolini, M.L., Pollini, L., Stolarik, B.: A comparison of Pose Estimation algorithms for Machine Vision based Aerial Refueling for UAVs. In: 14th Mediterranean Conference on Control and Automation, vol. 45(1), pp. 138–151 (2009); IEEE Transaction on Aerospace and Electronic System

14. Bradski, G., Kaehler, A.: Learning OpenCV: Computer Vision with the OpenCV Library. O'Reilly Media, Sebastopol (2008)
15. Open Source Computer Vision Library Reference Manual, Intel Corporation (2009)
16. Dorfmuller-Ulhaas, K.: Optical Tracking, From User Motion To 3D Interaction, PhD Thesis, Vienna University of Technology (2002)
17. Chui, C.K., Chen, G.: Kalman Filtering with Real-Time Applications, 4th edn. Springer, Hielderberg (2009)
18. Welch, G., Bishop, G.: An Introduction to the Kalman Filter, http://www.cs.unc.edu/~welch/kalman/kalmanIntro.html
19. Brindza, J., Szweda, J., Liao, O., Jiang, Y., Striegel, A.: WiiLab: Bringing Together the Nintendo Wiimote and MATLAB. In: Proceedings of the 39th IEEE International Conference on Frontiers in Education Conference, San Antonio, Texas, pp. 1373–1378 (2009)
20. Kavakli, M.: Gesture Recognition in Virtual Reality. Special Issue on Immersive Virtual, Mixed, or Augmented Reality Art of the International Journal of Arts and Technology (IJART) 1(2), 215–229 (2008)
21. Kreylos, O.: Research and Development Homepage, http://idav.ucdavis.edu/~okreylos/ResDev/Wiimote/index.html

VLOCI: Using Distance Measurements to Improve the Accuracy of Location Coordinates in GPS-Equipped VANETs

Farhan Ahammed[1], Javid Taheri[1], Albert Y. Zomaya[1], and Max Ott[2]

[1] School of Information Technologies,
The University of Sydney, NSW 2006, Australia
faha3615@it.usyd.edu.au
[2] NICTA, Australia, Australian Technology Park,
Level 5, 13 Garden Street, Eveleigh NSW 2015, Australia
max.ott@nicta.com.au

Abstract. Many vehicles rely on the Global Positioning System (GPS) to compute their locations. The inaccuracy of GPS devices means sometimes vehicles believe they are located in different lanes or roads altogether. Vehicular Ad Hoc Networks (VANETs) allow vehicles to communicate with each other using wireless means and thus connect them in a very dynamic wireless network. The algorithm **V**ANET **LOC**ation **I**mprove (VLOCI), proposed in this work, uses VANETs and distance measurements taken by each vehicle to improve the location estimates provided by all GPS devices. VLOCI is shown to perform efficient when erroneous distance measurements are present in the environment/computations.

Keywords: vehicular ad hoc networks, localization, GPS, distance measurements, location improve/refinement.

1 Introduction

VANETs are types of mobile networks—where the nodes are vehicles. The vehicles are equipped with wireless communication devices allowing them to transmit and share real-time information. With this information vehicles and drivers will have up-to-date information regarding the state of traffic, allowing them to avoid congested and other abnormally affected areas. VANETs are dynamic with vehicles travelling at speeds up to, and in excess of, 100 km/h. This leads to ever-changing wireless connections between vehicles resulting in some dense (on some city roads) and sparse (on country roads) areas which change over time (some city roads are dense only during certain hours of the day).

Many vehicles are nowadays equipped with GPS devices and it is quite possible that most, if not all, vehicles will have these devices as well in the future. GPS devices are accurate to within 10 metres [1]—more than the length of most family cars—resulting in situations where the GPS device incorrectly places its vehicles on the wrong road. Obtaining more accurate coordinates (position estimates) allows the vehicles to construct more precise models of their local traffic conditions.

With increased accuracy and better models, accidents can be prevented. Multi-car 'pile-ups' can be avoided if vehicles know immediately that other vehicles further in front are stopping suddenly or skidding. Although some sensors are already providing information about the vehicles directly in front and around the vehicle—VANETs can be used to provide information about vehicles further away. For example, when a vehicle detects a dangerous pot hole or other situation on the road, the exact co-ordinates of the problematic area can be immediately passed on to nearby vehicles. VANETs can be used to increase driver safety on the roads, but accurate coordinates is required for all vehicles—some drivers may incorrectly assume an accident or other incident is occurring on the wrong road.

This paper will look at using VANETs to improve on the position estimates provided by the GPS devices. Every vehicle can provide their position estimate to all vehicles within broadcasting range. It is also assumed every vehicle can measure the distance between them and other vehicles using already existing sensors/equipments [2,3]. When all vehicles combine the collected information, the algorithm LOCI can be used to adjust the GPS estimated position into a more accurate one.

An overview of previous work found in literature is presented in Section 2. Section 3 introduces the notation and defines the problem addressed in this paper. The method used to solve the problem is described and the VLOCI algorithm is presented in Section 4. Section 5 describes the simulations performed to test the devised algorithm. An discussion of the simulation results and concluding remarks are presented in Section 6 and 7, respectively.

2 Related Work

There does not seem to be much work in literature with the idea of improving location estimates in VANETs. There are algorithms designed to take advantage of some nodes that have GPS, or some other positioning, functionality to allow all nodes to compute their location.

Priyantha et al. proposed a technique called *anchor-free localization* (AFL) of providing localization to wireless sensor networks [4]. Their algorithm is decentralised where each node starts with a random initial coordinate assignment, and modifies its location estimates based on local distance measurements. The only information each node collects is the relative distance to its neighbouring nodes. With this information, the nodes construct a graph with the edges at the measured length/weight. A *mass-spring* based optimization is used to adjust the edge lengths of the graph. The edge lengths are adjusted based on the difference between the measured distances between neighbouring nodes and the corresponding computed distances in the constructed graphs.

Barani and Fathy [5] looked at the problem where not all vehicles are equipped with GPS devices or cannot receive signals from the GPS satellites. In their approach, the vehicles first attempt to find three neighbours within one-hop distance. If there is only one or two neighbours within one-hop, then distance information from neighbours within two-hop distance is used. They found when more than 40% of vehicles are equipped with GPS, most of the vehicles in the network will have at least three GPS-equipped neighbours. From these neighbours, methods such as trilateration [6] can be used to compute their location. Here the problem is not to improve GPS coordinates, but to provide localization to those unable to use the GPS.

Liu and Lin [7] improved location accuracy using the least-squares technique. Their technique is applied to cellular networks, where base transceiver stations are established and within communication range. Results obtained in field trials are compared to the co-ordinates computed by GPS devices. Consequently, these tests assume the GPS devices provide sufficient accuracy. The mean error produced, using their technique, is within the range of 200 metres: extremely large for use in VANETs.

Xu et al. [8,9] have developed an algorithm for use in wireless sensor networks where some of the sensors are located indoors—preventing detection of the signals from the GPS satellites. In their approach, only a subset of nodes are GPS-equipped, while other anchor nodes also know their true locations. The remaining nodes are able to compute their location based on distances between the GPS nodes and the GPS satellites, and distances between the GPS nodes and anchor nodes; as a result, not every node uses GPS to estimate their location. The DV-Hop algorithm is used by the remaining nodes for localization. Other techniques also exist that utilize the DV-Hop algorithm, where its accuracy is significantly proportional to the network density. Therefore, this cannot be used in VANETs where the network density continuously changes. Their tests results show the location error for the non GPS nodes achieving the same range as the nodes which use the GPS. However their algorithm does not improve further on this location error.

Some alternative solutions [10,11,12] are applied in situations where a subset of nodes, usually termed anchor nodes or base stations, have knowledge of their own positions. The remaining nodes then communicate with the anchor nodes to determine their locations as they have no other method to estimate their locations. Similarly, Benslimane [13] addresses the situation where not all vehicles are equipped with GPS devices, or some cannot obtain data from their GPS devices and need to collaborate with the GPS-equipped vehicles to determine their locations.

No work have so far been found in literature where all nodes are equipped with GPS devices and the problem addressed is finding ways to improve these estimates.

3 Problem Statement

The network in question is a VANET, thus the nodes are in fact vehicles. The set of all vehicles will be denoted V. The number of vehicles in the network is

$N = |V|$. Given any particular vehicle $n_i \in V$, any other vehicle that n_i can send and receive messages from are deemed its neighbour. The set of all neighbours of n_i is denoted $nbrs(n_i)$ and the number of neighbours of n_i is $m_i = |nbrs(n_i)|$.

Each vehicle n_i is located at $p_i = (x_i, y_i)$ (true location) and, due to inaccurate measurements, believes it is located at $\widehat{p}_i = (\widehat{x}_i, \widehat{y}_i)$ (computed/estimated location). The location error for n_i is the distance between its true and computed location $\delta_i = \|p_i - \widehat{p}_i\|$. It is assumed that every vehicle is able to take distance measurements between them and other vehicles. The true distance between vehicles n_i and n_j is denoted $d_{i,j}$, while the distance measured by n_i between itself and vehicle n_j is given as $\widehat{d}_{i,j} = \varepsilon \cdot d_{i,j}$, for some $\varepsilon \in \mathbb{R}$. Since each vehicle is assumed to have their own measuring device, it is also assumed that $\widehat{d}_{i,j} \neq \widehat{d}_{j,i}$ because $\widehat{d}_{i,j}$ is the distance measured by n_i while $\widehat{d}_{j,i}$ is the distance measured by n_j.

A metric used to gauge the performance of the localisation algorithm of a network is the *network location error* (Definition 1). For each vehicle n_i, its location error is already defined as δ_i. The network location error is the average of every vehicle's location error.

Definition 1 (Location Error). *Given a network of vehicles V, where each vehicle n_i believes it is located at $\widehat{p}_i$, while its true location is p_i. The* location error *of the network (E_V) is defined as*

$$E_V = \frac{1}{n} \sum_{n_i \in V} \delta_i = \frac{1}{n} \sum_{n_i \in V} \|p_i - \widehat{p}_i\|$$

Using the definition of network location error, the problem of localization can be formulated as shown in Definition 2.

Definition 2 (Localization). *Given a network of vehicles V, the goal of localization is for every vehicle $n_i \in V$, located at $p_i = (x_i, y_i)$, to compute its position $\widehat{p}_i = (x'_i, y'_i)$ such that the network location error is minimised.*

The problem of *Location Improvement* (Definition 3) is addressed in this paper. The aim is to find a method of adjusting every vehicle's current estimated position $\widehat{p}_i$ such that the location error is reduced.

Definition 3 (Location Improvement). *Let V be a network of vehicles. Assume that every vehicle n_i has computed its own estimated position $\widehat{p}_i$. The problem of* Location Improvement *is to find a function f which modifies some or all of the vehicle's estimated position such that the location error of the modified network has minimised.*

4 Method and Algorithm

4.1 Network Topology

In many situations, also addressed in literature, the topology of the network in question is modelled as a random topology with little or no restrictions on

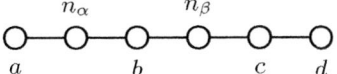

Fig. 1. The network topology within a single lane of a road. Node n_α has three neighbours $\{a, b, n_\beta\}$ and node n_β has four neighbours $\{n_\alpha, b, c, d\}$.

the nodes' locations: a non-realistic issue in VANETs sometimes. For example, within a lane of a road, the local network topology is already known and all vehicles in the same lane are lined up one behind another. Figure 1 depicts the 'lane topology.'

In this paper the vehicles are assumed to be travelling in one lane and in the same direction. In this model a co-ordinate scheme is used where $y_i = y_j$ for all $n_i, n_j \in V$. Each vehicle $n_i \in V$ has an estimate of their position $\hat{p}_i$ and this information is periodically sent to its neighbours. Using their distance-measuring device, each node measures the relative distance $\hat{d}_{i,j}$.

Road-side infrastructures (permanent, static nodes) can be used to provide vehicles the necessary linear transformation required to convert the local co-ordinates into their global counterparts.

4.2 Accuracy of Distance Measurements

The accuracy of the distance measurement devices refers to the values of $\hat{d}_{i,j}$ in the simulation and how far they differ from true distances $d_{i,j}$. Two techniques used to measure distance are time-of-arrival and received signal strength [14]. The accuracy of these techniques can be modelled such that as the distance to be measured increases, the accuracy of the measurements taken also decreases [15,16]. This is how distance measuring is modelled in this paper. The statement 'the accuracy is set to α metres' means that $\hat{d}_{i,j} = E \cdot d_{i,j}$ for all $n_i, n_j \in V$, where $E \sim N(1, \alpha^2)$; $N(1, \alpha^2)$ refers to the Gaussian distribution with mean 1 and standard deviation α. Thus, if a vehicle is $d_{i,j}$ meters away, then roughly 68% of the distance measurements $\hat{d}_{i,j}$ lie within the range $(1-\alpha)d_{i,j} < \hat{d}_{i,j} < (1+\alpha)d_{i,j}$.

Using this model, vehicles further away have a larger probability of producing more erroneous data. The further away a neighbouring vehicle is (i.e. the larger the value of $d_{i,j}$), the larger the error range of the measured distance. Measuring the distance to vehicles within close range is more accurate than measuring the distance to those further away.

4.3 Neighbourhood Size

Since the network is modelled as a single lane, the size of the network is straightforward. There are N vehicles, two are at the end points while the rest form the line. This leads to the definition of a vehicle's set of neighbours. Every vehicle has at most $2M$ neighbours (for some $M \in \mathbb{Z}$) they can communicate with; the M closest vehicles in front and the M vehicles in behind. That is, M defines the maximum number of neighbours in either direction. Figure 1 gives an example when $M = 2$. Here, every vehicle has at most 4 neighbours. Vehicle n_β has the maximum of four neighbours—two in front and two behind—while vehicle n_α only has three.

In this paper, the statement 'the half-neighbourhood size is M' means every vehicle can communicate with up to $2M$ vehicles.

4.4 Computing Position Estimates

Each vehicle receives messages from its neighbours containing their estimated positions. Additionally, each vehicle can measure the distance between itself and its neighbours. To counter the variance in erroneous distance measurements, multiple measurements can be taken for which the average can be used as the final distance measurement.

If vehicles n_j and n_i are neighbours, then n_i can obtain n_j's estimated position $\widehat{p}_j$ and the measured distance $\widehat{d}_{i,j}$. Using these two pieces of information, vehicle n_i can calculate its position assuming that n_j's estimated position is correct, i.e.

$$\widehat{p}_i^j = (\widehat{x}_i^j, \widehat{y}_i) = (\widehat{x}_j \pm \widehat{d}_{i,j}, \widehat{y}_j) \tag{1}$$

remembering that since in this model $y_i = y_j$, the vehicles can assume $\widehat{y}_i = \widehat{y}_j = 0$. Converting these co-ordinates into global co-ordinates requires a simple linear transformation. For generality define $\widehat{p}_i^i = \widehat{p}_i$ and $\widehat{d}_{i,i} = 0$. The reason for the plus/minus term in Equation 1 is that vehicle n_j may be relatively in front or behind vehicle n_i and so the value of $\widehat{x}_i^j$ should be adjusted accordingly.

Now given the set of neighbours $nbrs(n_i)$, vehicle n_i can construct the set $\{\widehat{p}_i^j \mid n_j \in \{n_i\} \cup nbrs(n_i)\}$. With this set of co-ordinates, an average can be computed to calculate a new estimate for $\widehat{p}_i$. A weighted average function $w : V \to \mathbb{R}$ is defined in this work to estimate $\widehat{x}_i'$ as follows:

$$\widehat{x}_i' = \frac{\sum_{\{n_i\} \cup nbrs(n_i)} w(n_k) \cdot \widehat{x}_i^k}{\sum_{\{n_i\} \cup nbrs(n_i)} w(n_k)} \tag{2}$$

This value becomes the new estimated x co-ordinate for vehicle n_i. Figure 2 shows graphically how an average estimate is used to compute the new estimated position from the cluster of points.

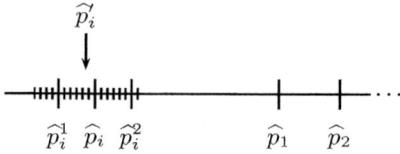

Fig. 2. The cluster of points from the computed set $\{\widehat{p}_i^i, \widehat{p}_i^1, \widehat{p}_i^2, \ldots, \widehat{p}_i^{m_i}\}$

This weighted average position becomes the new estimated position for vehicle n_i, and this is the coordinate transmitted to its neighbours in the next turn/iteration.

4.5 The Weight Function

To compute the weighted average, a weight function w needs to be defined. As the measured distances become more inaccurate the further away neighbouring vehicles are, it was decided the weights should be influenced by the measured distance $\widehat{d}_{i,j}$. For example, vehicles should put more weight to the estimated position constructed from data obtained from a vehicle that is 20 metres away, compared to a vehicle that is 400 metres away since there is more error involved in the calculations involving the latter case.

The weight function must therefore be inversely proportional to the measured distance. Two functions were considered: (1) inverse functions of the form $g(x) = A/x^B$ and (2) inverse exponential functions of the form $f(x) = Ae^{-x^2/B}$. In both cases, $x = \widehat{d}_{i,j}$. Thus the weight function takes the distance $\widehat{d}_{i,j}$ as the parameter and gives more weight to vehicles closer to n_i. The latter form f was chosen over the alternate inverse function because results have shown g generally decreases too quickly and better results were achieved using a function of the form of f.

4.6 VLOCI Algorithm Overview

The concepts explained in the previous section are used to design a location-improvement algorithm that is both (1) distributed as multiple vehicles collaborate to achieve the goal of improving every vehicle's position estimates, and (2) scalable as the number of vehicles able to participate is not restricted.

An iteration, performed by vehicle n_i is the process of receiving messages from its neighbours then updating its estimated position. The number of iterations hence determines the number of times every vehicle updates its estimated position as shown in Algorithm 3

1. Transmit a message containing the current estimated position ($\widehat{p}_i$).
2. Wait to receive messages from neighbouring vehicles. The received messages should contain the estimated position of the vehicle that sent the message.
3. For each vehicle a message was received from, measure the distance to it. The vehicle that is taking the measurements should take multiple measurements (D times) and use the average as the final measured distance. The value of D is based on the technique used to obtain distance measurements. Smaller values of D can be used with more accurate distance measuring devices.
4. Vehicle n_i now knows the values of $\{\widehat{p}_j, \widehat{d}_{i,j}\}$ for each of its neighbours. Equation 1 is then used to compute possible co-ordinates it could be located at. A set of these possible position estimates $\{\widehat{p}_i^i, \widehat{p}_i^1, \widehat{p}_i^2, \ldots, \widehat{p}_i^{m_i}\}$ is then constructed.
5. The weighted average of the set of possible position estimates (Equation 2) is calculated. This final co-ordinate becomes the new estimated position.

```
while iterationCount < I do
    transmitMessage(p̂ᵢ)   M contains the received messages
    // We have each nⱼ's computed position. Now measure the distance from
        them.
    // Wait for λ messages
    wait(M)  foreach Mⱼ ∈ M do
        // Use avg of D measurements.
        d̂ᵢ,ⱼ = (1/D) Σᴰₖ₌₁ takeDistMeas(nⱼ)
        if x̂ᵢ < x̂ⱼ then
        |   p̂ⱼᵢ = (x̂ⱼᵢ, ŷᵢ) = (x̂ⱼ − d̂ᵢ,ⱼ, 0)
        else
        |   p̂ⱼᵢ = (x̂ⱼᵢ, ŷᵢ) = (x̂ⱼ + d̂ᵢ,ⱼ, 0)
        end
    end
    // Now compute the weighted average of all the probable co-ordinates
        of nᵢ.
    x̂ᵢ′ = Σ w(nₖ)·x̂ᵢᵏ / Σ w(nₖ)     ŷᵢ′ = 0
end
```

Fig. 3. Algorithm VLOCI: updating position estimates (vehicle n_i)

4.7 Skewed Positions

Additional scenarios were devised where the initial estimated positions were predetermined. A vehicle n_i is said to be *skewed to the left*, if $\hat{x}_i < x_i$ and skewed to the right when $\hat{x}_i > x_i$. Figure 4 gives a examples of this concept. The number of vehicles initially skewed to the left is γ. The remaining $N - \gamma$ vehicles are then positioned skewed to the right.

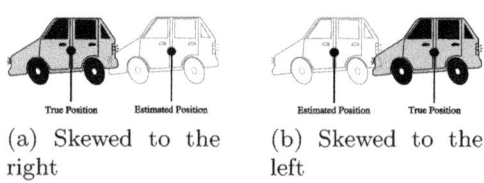

(a) Skewed to the right

(b) Skewed to the left

Fig. 4. Example of cars "skewed to one side"

4.8 Metric for Measuring Performance of VLOCI

The metric used to gauge the performance of VLOCI is the location error (Definition 1). The location error is the average distance between a vehicle's computed position and its actual position at the current point in time. The smaller the location error, the better the computed model of the vehicle's locations.

Assessing how the computed positions of all the vehicles would improve over time was the goal of the tests. The vehicles were simulated to receive GPS measurements only once and at the beginning. Afterwards, only VLOCI was used to further improve the position estimates.

5 Simulation Results

Simulations were performed to assess the effect of two variants in VLOCI: the accuracy of the distance measuring devices α (Section 4.2) and the neighbourhood size $2M$ (Section 4.3). The simulations were run with the parameters shown in Table 1. The network itself was static and the vehicles were set to be stationary. The weight function coefficients were chosen empirically. When vehicles were approximately 50m away the error in distance measurements were not detrimental to VLOCI's performance. The coefficients of the weight function were chosen to reflect this. It was the aim of the simulations to find how the neighbourhood size affects the rate of improvement of VLOCI and for what values of α does VLOCI still perform adequately.

Table 1. The simulation parameters used

Parameter	Value
Network size (N)	10
Half-neighbourhood size (M)	1–10
Distance measurement accuracy (α)	0–30%
No. distance measurements taken (D)	5
Distance between vehicles	20 metres
Weight function f coefficients	$A = 100$, $B = 550$
No. iterations (I)	10
No. tests per scenario	10 000
No. Vehicles skewed to left (γ)	0,1,2,3,4,5

5.1 Results

Figure 5a shows how the neighbourhood size effects VLOCI's performance, for half-neighbourhood sizes (M) of 1–5. To give a clearer picture of how the different curves compare after 7 iterations, Figure 5b shows the same graph for iterations 8–10 inclusive. The distance measurement error (α) was set to zero for these simulations to model errors only arising from the initial GPS measurements.

When the half-neighbourhood size ranges from 5–10, Figure 6 shows how the location error reduces, for iterations 8–10 inclusive. Again $\alpha = 0$. The best results occurred when the half-neighbourhood size is set to $M = 7$. This is the value M was fixed at when testing VLOCI's ability to handle errors in distance measurements.

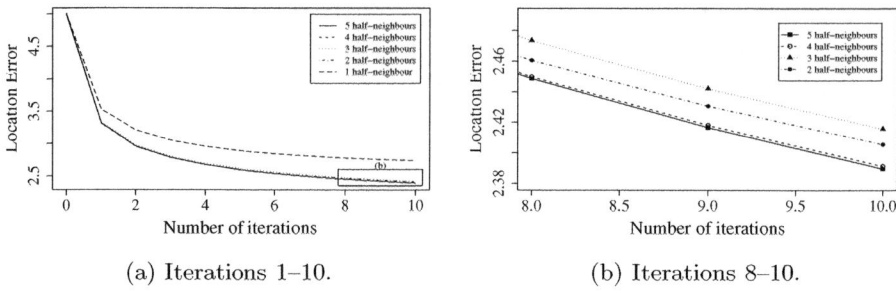

Fig. 5. The effects of local neighbourhood size on location error. Half-neighbourhood size 1–5. $\alpha = 0$.

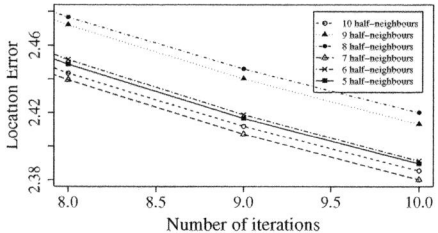

Fig. 6. The effects of local neighbourhood size on location error. Half-neighbourhood size 5–10. $\alpha = 0$. Magnified for iterations 8–10.

The graphs in Figure 7 show how VLOCI performs when errors are incorporated in distance measurements. For the value of $\alpha > 0.20$ the location error began to increase after the fifth iteration. When $\alpha < 0.20$ the location error still decreases during the first 10 iterations.

One of the reasons explaining why the location error begins to increase after some time is due to how the vehicles are skewed over time. Eventually too many cars believe they are located on the same side of their actual position. That is, too many cars are skewed on the same side. Figure 8 shows that once all the vehicles are on the same side (skewed to the right) the location error no longer improves. Here $\alpha = 0$.

The effectiveness of VLOCI when $\alpha = 0.1$ and 0.20 and with multiple cars skewed to one side is shown in figures 9a and 9b respectively.

6 Discussion and Analysis

The results show a definite improvement on the location error. Figures 5a, 5b and 6 show that with accurate distance measurements (i.e. $\alpha = 0$) VLOCI does indeed improve the average location error. When $\alpha = 0$ and the half-neighbourhood size is set to $M = 7$, after 10 iterations the average location error reaches a value of 2.38 meters—an improvement of 52%. With the accuracy set

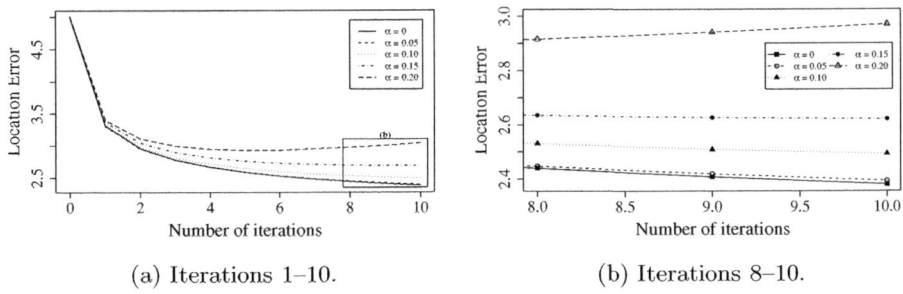

(a) Iterations 1–10.   (b) Iterations 8–10.

Fig. 7. Effects of the accuracy of distance measurements on location error. Half-neighbourhood size 7.

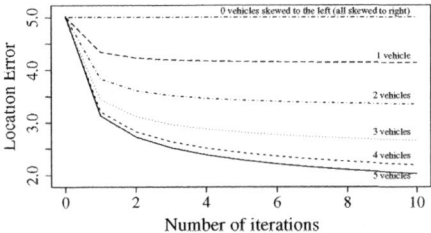

Fig. 8. Effects of skewness in initial positions. Half-neighbourhood size 7. $\alpha = 0$.

to $\alpha = 0.20$, the average location error improves by 41%, before increasing from the fifth iteration. This is still acceptable as there is an overall improvement on the location error after 10 iterations and due to the mobile nature of VANETs, the vehicles sometimes cannot perform time-consuming operations restricting the number of iterations possible.

Figure 7 shows that VLOCI still improves the location error when inaccurate distance measurements are introduced. For values of $\alpha < 0.20$ the location error, after 10 iterations, has improved to values less than 2.7 metres. If VLOCI is set to iterate 10 times, this value of α is an upper bound on the average location error.

Figures 8, 9a and 9b shows how VLOCI is affected when the initial position estimates of the vehicles, relative to their actual positions, is set to have a portion of vehicles skewed to one side (Section 4.7). When all the vehicles are skewed to one side, the location error does not noticeably increase or decrease within the first 10 iterations. This is because while the vehicles have adjusted their estimated positions such that the distance between them is consistent with the true distance, they are still skewed to the same side. There needs to be at least one vehicle skewed to the other side to 'pull' the other vehicles from one side of their true position to the other. As expected, the best results occur when approximately half the vehicles are skewed to one side (and the other half skewed to the other side). Even when only one vehicle is skewed to one side, the average location error still improves.

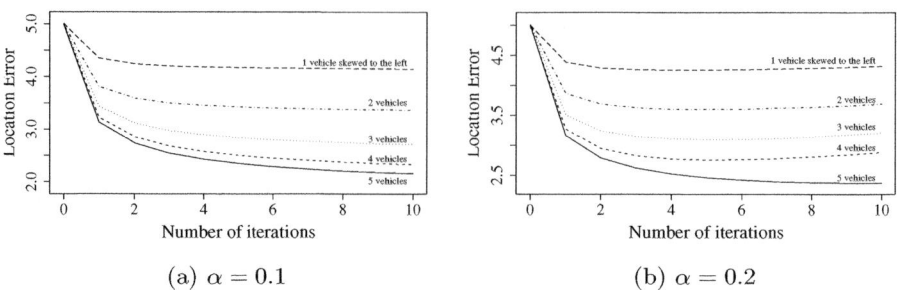

Fig. 9. Effects of skewness in initial positions. Half-neighbourhood size 7. $\alpha = 0.1$.

7 Conclusions

The VLOCI algorithm presented in this paper is shown to improve every vehicle's initial position estimate. Assuming the vehicles are connected via a VANET and are equipped with a distance measuring device, along with a GPS device to provide the initial position estimate, VLOCI is still able to improve locations estimation when erroneous distance measurements are included in the computations. The effect of skewness is also shown in this paper. Only in the situation where all vehicles are skewed to one side does VLOCI not improve the average location error. For the remaining situations, VLOCI still reduces the average location error.

References

1. LaMarca, A., Chawathe, Y., Consolvo, S., Hightower, J., Smith, I., Scott, J., Sohn, T., Howard, J., Hughes, J., Potter, F., Tabert, J., Powledge, P.S., Borriello, G., Schilit, B.N.: Place lab: Device positioning using radio beacons in the wild. In: Gellersen, H.-W., Want, R., Schmidt, A. (eds.) PERVASIVE 2005. LNCS, vol. 3468, pp. 116–133. Springer, Heidelberg (2005)
2. Fleming, W.J.: Overview of automotive sensors. IEEE Sensors Journal 1, 296–308 (2001)
3. Fölster, F., Rohling, H.: Data association and tracking for automotive radar networks. IEEE Transactions on Intelligent Transportation Systems 6, 370–377 (2005)
4. Priyantha, N.B., Balakrishnan, H., Demaine, E., Teller, S.: Anchor-free distributed localization in sensor networks. In: 1st International Conference on Embedded Networked Sensor Systems, SenSys 2003 (2003)
5. Barani, H., Fathy, M.: An algorithm for localization in vehicular ad-hoc networks. Journal of Computer Science 6(2), 168–172 (2010)
6. Thomas, F., Ros, L.: Revisiting trilateration for robot localization. IEEE Transactions on Robotics 21, 93–101 (2005)
7. Liu, B.-C., Lin, K.-H.: Enhanced location accuracy for hyperbolic positioning technique based on sssd measurements in wireless cellular networks via least-square algorithm. In: Global Telecommunications Conference, GLOBECOM 2006, December 27, pp. 1–5. IEEE, Los Alamitos (2006)

8. Xu, L., Deng, Z., Ren, W., Wang, H.: A location algorithm integrating gps and wsn in pervasive computing. In: Third International Conference on Pervasive Computing and Applications, ICPCA 2008, vol. 1, pp. 461–466 (2008)
9. Zou, D., Deng, Z., Xu, L., Ren, W.: Seamless lbs based on the integration of wsn and gps. In: International Symposium on Computer Science and Computational Technology, vol. 2, pp. 91–96 (2008)
10. Savarese, C., Rabaey, J.M., Beutel, J.: Locationing in distributed ad-hoc wireless sensor networks. In: Proceedings of 2001 IEEE International Conference on Acoustics, Speech, and Signal Processing (ICASSP 2001), vol. 4, pp. 2037–2040 (2001)
11. Cheng, X., Thaeler, A., Xue, G., Chen, D.: TPS: a time-based positioning scheme for outdoor wireless sensor networks. In: Twenty-third Annual Joint Conference of the IEEE Computer and Communications Societies, INFOCOM 2004, 7–11, vol. 4, pp. 2685–2696 (2004)
12. Lazos, L., Poovendran, R., Čapkun, S.: ROPE: robust position estimation in wireless sensor networks. In: Fourth International Symposium on Information Processing in Sensor Networks, IPSN 2005, pp. 324–331 (April 2005)
13. Benslimane, A.: Localization in vehicular ad hoc networks. In: Proceedings of Systems Communications, pp. 19–25 (August 2005)
14. Boukerche, A.: Algorithms and Protocols for Wireless, Mobile Ad Hoc Networks. Wiley-IEEE Press (2008)
15. Bellusci, G., Janssen, G.J.M., Yan, J., Tiberius, C.C.J.M.: Modeling distance and bandwidth dependency of TOA-based UWB ranging error for positioning. Research Letters in Communications, 1–4 (January 2009)
16. Powell, S., Shim, J. (eds.): Wireless Technology: Applications, Management, and Security. Springer, Heidelberg (2009)

On Improving the Energy Efficiency and Robustness of Position Tracking for Mobile Devices

Mikkel Baun Kjærgaard

Aarhus University, Denmark
mikkelbk@cs.au.dk

Abstract. An important feature of a modern mobile device is that it can position itself and support remote position tracking. To be useful, such position tracking has to be energy-efficient to avoid having a major impact on the battery life of the mobile device. Furthermore, tracking has to robustly deliver position updates when faced with changing conditions such as delays and changing positioning conditions. Previous work has established dynamic tracking systems, such as our EnTracked system, as a solution to address these issues. In this paper we propose a responsibility division for position tracking into sensor management strategies and position update protocols and combine the sensor management strategy of EnTracked with position update protocols, which enables the system to further reduce the power consumption with up to 268 mW extending the battery life with up to 36%. As our evaluation identify that classical position update protocols have robustness weaknesses we propose a method to improve their robustness. Furthermore, we analyze the dependency of tracking systems on the pedestrian movement patterns and positioning environment, and how the power savings depend on the power characteristics of different mobile devices.

Keywords: energy-efficiency, positioning, mobile devices, power consumption, GPS, position update protocols.

1 Introduction

An important feature of a modern mobile device is that it can position itself. Not only for use locally on the device but also for remote applications that require tracking of the device. Examples of such applications are geo-based information applications [2] or proximity and separation detection for social networking applications [9] just to mention a few. To be useful, such position tracking has to be energy-efficient to avoid having a major impact on the power consumption of the mobile device. Optimizing the operation of mobile devices for energy efficiency is an important issue and research is trying to address it from many angles as surveyed in [5], for instance, by trying to lower the impact of network traffic on power consumption [8] or by optimizing the execution at the operating system level [1]. Furthermore, tracking has to be robust in order to deliver position updates within limits when faced with changing conditions such as delays due to positioning and communication, and changing positioning accuracy.

As a basis for this work we divide the responsibility of remote tracking into *sensor management strategies* that on the device decides how to use available position sensors to estimate the current position and *position update protocols* that controls the interaction between the device and remote services. Such a division enables us to analyze the different combinations of sensor management strategies and position update protocols. Position update protocols has previously been studied, e.g., by Leonhardi et al. [11]. We will denote the combination of a sensor strategy and a protocol with *Strategy:Protocol*.

To quantify the impact of remote position tracking on power consumption, we have emulated the power consumption of a Nokia N95 phone in four different setups using the emulation tools and residential neighborhood dataset presented in Kjærgaard et al. [7]. In the first setup *(Periodic:Simple)* a periodic sensor management strategy every T_{period} seconds positions the phone using the built-in GPS receiver and then uses a simple protocol that immediately sends the position data using UMTS to a remote service hosted on an internet-connected server[1]. In the second setup *(Default:Distance)* a default sensor strategy positions the phone continuously by the update rate of the built-in GPS receiver (1Hz) and then uses a protocol that tries to minimize the number of position updates by only sending position data when the phone has moved more than a distance threshold T_{dist} meters from the last reported position. The third setup *(Default:Dead)* also uses a default strategy to position the phone by the update rate of the built-in GPS receiver and then uses a dead-reckoning protocol that sends an update when the distance between the current position and a server-side predicted position from the last reported heading, speed and position becomes greater than a given threshold T_{dist} meters. The fourth setup *(Dynamic(EnT):Simple)* uses a dynamic sensor strategy implemented by the EnTracked system [7] that tries to minimize the needed GPS fixes based on an accuracy limit T_{acc} meters and then uses a simple protocol that immediately sends the position data to the remote service.

The average power consumption for each setup with different accuracy threshold parameters are plotted in Figure 1 together with a robustness plot of the percentage of time the distance between the real position and the server known position is greater than the threshold. Comparing *Default:Distance, Default:Dead* and *Dynamic(EnT):Simple* we can notice that all three are able to lower the power consumption with between 560mW to 734mW compared to *Periodic:Simple* for the same accuracy threshold. The EnTracked system both minimize GPS and radio consumption whereas the distance-based and dead-reckoning protocols only save on radio consumption. Therefore we hypothesize that more power can be saved by combining EnTracked with either a distance-based or a dead-reckoning reporting protocol. However, a problem with either protocols is that they are less robust than *Periodic:Simple* and for most accuracy thresholds also

---

[1] 10 m/s is used as a conservative upper bound on the speed of pedestrian movement.

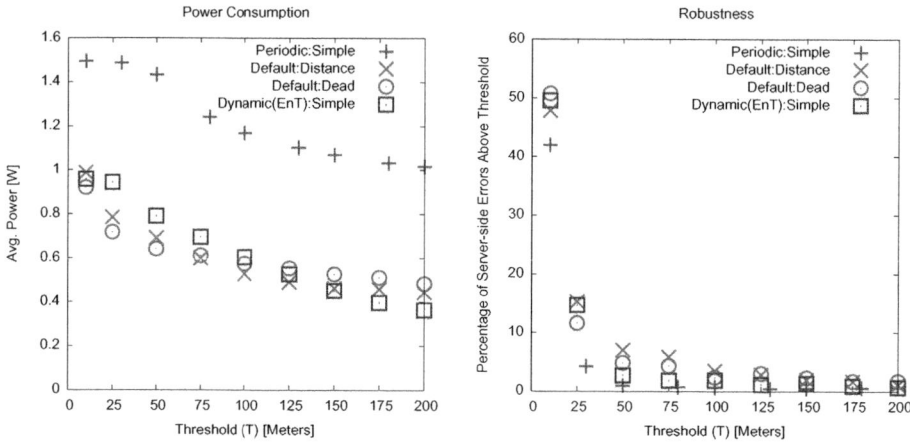

Fig. 1. Comparison of average power consumption and robustness for Periodic:Simple ($T = 10m/s * T_{period}$), Default:Distance ($T = T_{dist}$), Default:Dead ($T = T_{dist}$) and Dynamic(EnT):Simple ($T = T_{acc}$).

Dynamic(EnT):Simple as shown on Figure 1. This drawback of the protocols has been overlooked by previous work on position update protocols as they did not consider the ground truth accuracy [3,11,10].

We make the following contributions in this work: First of all, we propose a responsibility division for position tracking into sensor management strategies and position update protocols and combine the sensor management strategy of EnTracked with position update protocols which enables the system to further reduce the power consumption with up to 268 mW extending the battery life with up to 36% compared to the original system presented in [7]. Secondly, we propose a solution for improving the robustness of distance-based and dead-reckoning position update protocols that only marginally increases the power consumption. Thirdly, we evaluate how the power savings and robustness depend on the movement characteristics of pedestrian targets and the environment. Fourthly, we analyze how the power savings and the optimal system setup depend on the power characteristics of different mobile devices by deploying the system on the newer Nokia N97 phone which has significantly different parameters and power consumption levels.

2 Related Work

Previous work such as [3,10,11] has studied position update protocols to minimize communication and to minimize the load on server nodes by lowering the number of position updates. Leonhardi et al. [11] study time-based and distance-based protocols that takes a constant positioning accuracy and target speed into account. They study by simulation the number of updates each protocol produces and the average and maximum uncertainty of the server-known position.

They have later extended this work to consider dead-reckoning protocols [10]. Systems that tries to minimize the number of position updates for a specific application such as GeoPages have also been proposed [2].

A later work focusing both on sensor management strategies and position update protocols is Farrell et al. [4]. They propose strategies and protocols that take into account a constant positioning delay, target speed, and stress the importance of the fact, that it is not energy-free to use the GPS constantly as assumed by earlier work. Their solutions have been evaluated by simulation, where they can save around 50% energy in the evaluated scenarios. They have later extended this work for area-based tracking where they also take constant position accuracy and communication delays into account. For an indoor sensor network setting, You et al. [12] propose strategies and protocols that take into account a constant positioning accuracy and delay, target speed and acceleration to detect if the target is moving or not. They evaluate the techniques by emulation for IEEE 802.15.4 signal-strength-based indoor positioning and one of their results is that considerable energy savings can be gained from the use of an accelerometer to detect if the target is stationary or not. In our previous work Kjærgaard et al. [7] we proposed the EnTracked system that take into account dynamically estimated position accuracy and delays, communication delays, power constraints, target speed and acceleration (to detect if the target is moving or not). Furthermore the techniques was evaluated both by emulation and in real-world deployments.

In comparison, in this work we extend EnTracked with position update protocols to further reduce the energy consumption. Furthermore, we propose a method to improve the robustness of position update protocols and analyze the dependency of tracking systems on the pedestrian movement patterns and positioning environment, and how the power savings and optimal system setup depend on the power characteristics of different mobile devices.

3 Overview of Strategies and Protocols

As introduced earlier we divide the task of remote tracking into sensor management strategies and position update protocols. Sensor management strategies decide how to use available position sensors to estimate the current position. Sensor strategies could be implemented considering relevant properties such as position accuracy, power consumption, the availability of positioning in different environments (e.g., outdoor versus indoor) and privacy (e.g., WiFi positioning reveals a target's existence). Position update protocols control the interaction between the device and remote services which have to consider relevant properties such as server-side position accuracy, power consumption, data carrier availability and privacy.

In this paper we focus on outdoor GPS positioning of pedestrian targets and strategies and protocols for this setting. Figure 2 gives an overview of the considered strategies and protocols. A basic sensor management strategy is the default strategy that delivers position updates with the rate of a position sensor. This strategy was implicitly assumed by prior work such as Leonhardi et al. [11]. The second strategy is a periodic strategy that with a frequency $T_{frequency}$ Hz requests a new GPS position fix. The third strategy is a dynamic strategy that dynamically changes the sampling rate depending on requests and availability. A system implementing a dynamic strategy is the EnTracked system [7] which consists of several elements but in this work we will consider it as a whole and refer the reader to the evaluation of the individual elements presented in [7]. The system consists of the following elements: distance-based scheduling, device-aware power minimization and movement awareness. The distance-based scheduling estimates when the next GPS fix is needed according to an error model that takes into account the positioning accuracy and requested accuracy. The device-aware power minimization uses a power minimization algorithm implemented using dynamic programming to predict when sensors has to be turned on and off. The algorithm uses a profiled device model to ensure that the system will correctly minimize the consumption and take into account, e.g., the delays associated with powering on and off the GPS and the radio. The movement awareness enable the system to switch between GPS and sensing motion using accelerometer readings. If the system can sense that a mobile phone is not moving, there is no reason to update the position on the server and the GPS can be switched off. But as soon as motion is sensed, the system switches the GPS back on.

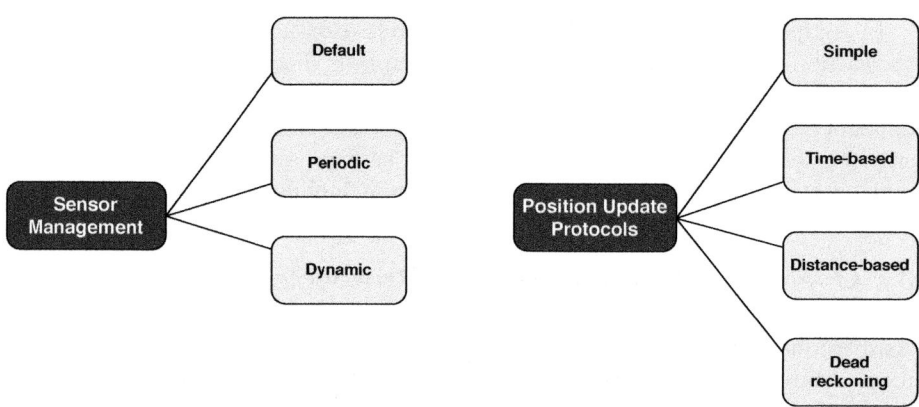

Fig. 2. Overview of sensor management strategies and position update protocols

In terms of position update protocols we restrict ourselves to protocols for a pedestrian scenario and device-controlled reporting protocols. Leonhardi et al. [11] list the following four types of reporting protocols that applies to the pedestrian scenario. Simple reporting which sends an update each time a position sensor provides a new position fix. Time-based reporting which sends an update each time a certain time interval of T_{period} seconds has elapsed. Distance-based reporting which sends an update when the distance between the current position and the last reported position becomes greater than a given threshold $T_{distance}$ in meters. Dead-reckoning which sends an update when the distance between the current position and the server-side predicted position from the last report position extrapolated with the reported speed and heading becomes greater than a given threshold $T_{distance}$ in meters. We refer the reader to [11] for a more detailed discussion and analytical analysis of these protocols in terms of their accuracy guarantees and communication efficiency. In this paper we will focus on the power consumption and robustness of these protocols on actual mobile devices. We will not consider Time-based combinations as any *Default:Time* combination is more efficiently implemented as a *Periodic:Simple* combination because this will lower both the GPS and radio usage.

The above discussion might indicate that out of the box any sensor strategy can be combined with any protocol, however, one have to take care of implementation pit falls. An example is the dead-reckoning protocol which assumes that the server can extrapolate the position as long as it does not receive new updates from the mobile device. In the classic protocol the threshold is tested continuously because a default strategy is implicitly assumed. The problem is what to do when a movement-aware dynamic strategy avoids to provide new updates because the device is detected not to move. In this case the server will continue to extrapolate the position which might violate the threshold. To address this issue we have extended the dead-reckoning protocol to test periodically if the server predicted position is about to violate the threshold and in this case send an extra position update with the last reported position and zero speed to stop the extrapolation.

A problem when implementing distance-based and dead-reckoning protocols is that they have robustness problems because they might not be able to keep the maximum error below $T_{distance}$ due to delays and positioning errors. Previous research such as Leonhardi et al. [11] did not observe this problem as they only considered data collected with a highly accurate differential GPS with a dedicated antenna in good signal conditions. In our work we focus on the more common case of less accurate GPS receivers found in mobile phones with smaller embedded antennas and in the non-optimal signal conditions found in urban and residential areas. To improve the protocols' robustness we propose to use the GPS receiver's estimates of it's current accuracy a_{gps} in meters and take this into account when evaluating if a threshold has been passed, e.g., for distance-based reporting the threshold equation then become: $d_{traveled} + a_{gps} < T_{distance}$ where $d_{traveled}$ is the distance between the last reported position and the current estimated position. This is an optimistic solution to strike a balance with power consumption as the pessimistic solution would be to also include the estimated accuracy of the last reported position because both the current and the last will be effected by GPS positioning errors.

4 Improving EnTracked Using Position Update Protocols

To evaluate the extension of EnTracked with position update protocols we will consider several datasets in the following sections. This section considers the residential neighborhood dataset presented in Kjærgaard et al. [7] which was recorded on Nokia N95 phones for three pedestrian targets walking a 1.7 km tour in a residential neighborhood with several stops. The dataset consists of ground truth positions and 1 Hz GPS and 35Hz acceleration measurements collected from the built-in sensors. The ground truth was collected manually by walking a known route and collecting timestamps on a mobile device when reaching known points on the route. As mentioned earlier we will denote the combination of a sensor strategy and a protocol with *Strategy:Protocol*. For the dynamic strategy we will add "(EnT)" to mark that it is the EnTracked system that is used and the protocols that implements the accuracy extension have a "+" attached to their name, e.g., *Distance+* or *Dead+*.

The results for combinations of sensor strategies and protocols are shown in Figure 3. From the figure one can observe how the results for *Default:Distance+* slowly approaches it's lower limit which is equal to the background power consumption (62mW) plus the GPS consumption (324mW) because it only saves radio consumption. When comparing *Dynamic(Ent):Simple* with *Dynamic(Ent):Distance+* we can see, as hypothesized, that the distance extended version is able to further decrease the power consumption with between 95mW to 268mW for the evaluated thresholds which equals a 17% to 36%[2] increase in battery lifetime. To highlight that *Dynamic(Ent):Distance+* both save GPS and radio usage we have split up the consumption for the 200 meter threshold. For this threshold the average power consumption of the radio was 45mW, for the GPS it was 176mW, and for the background, CPU and accelerometer it was 63 mW. We have also evaluated *Dynamic(Ent):Dead+* which also provides an improvement comparable to *Dynamic(Ent):Distance+*. The main difference is that the dead-reckoning version is a few mW better for thresholds smaller than 100 meters and a few mW worse for larger thresholds. One reason behind the negligible improvement over the distance-based protocol is that if one compares with ground truth the average accuracy for the speed and the heading estimates are 0.35 m/s and 52° and therefore the server predictions will often be extrapolated in a non-optimal direction. It can also be linked to the movement style of a pedestrian which often take turns. For the protocol part previous work [10] has in terms of communication efficiency for a pedestrian movement style also observed only a small decrease in the number of position updates specially for large thresholds.

A problem with the original position update protocols as discussed in the introduction is that they might not satisfy the given thresholds if the magnitude of GPS errors are significant compared to the length of the threshold. Therefore we have evaluated our proposal of extending the protocols to take the estimated accuracy into account. Figure 3 shows a robustness plot of the percentage of time the distance between the real position and the server known position is greater than the threshold. In most cases the *Dynamic(Ent):Simple* protocol has the lowest values, often below two percent. For the ten and twenty-five thresholds the percentage is higher because the GPS errors alone often are enough

---

[2] The largest percentage is for the 200 meter threshold.

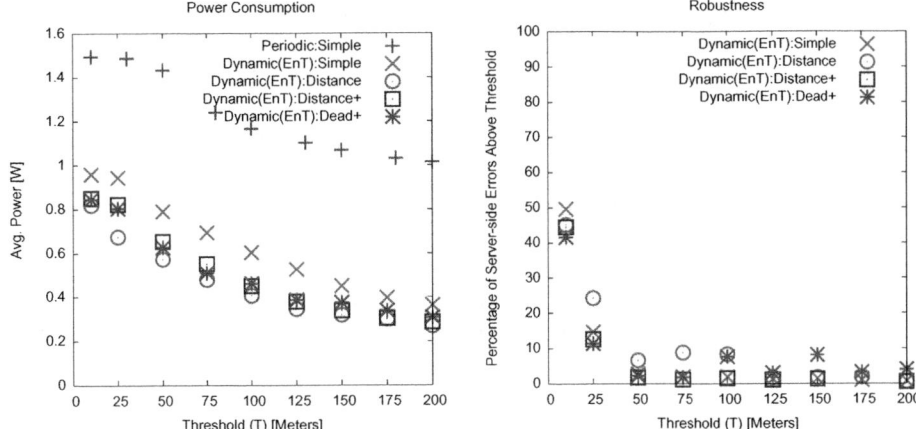

Fig. 3. Comparison of average power consumption and robustness for combinations of sensor strategies and position update protocols for the residential neighborhood dataset

to violate the smaller thresholds as the average GPS error for the dataset is 11.8 meters. Comparing *Dynamic(Ent):Distance* with *Dynamic(Ent):Distance+* the proposed extension is able to lower the percentage of violations with between six to eighteen percentage points. The *Dynamic(Ent):Dead+* generally performs similar to *Dynamic(Ent):Distance+* except for two cases where it performs worse.

5 Power Consumption and Mobility

The previous section presented results that provided evidence that the combination of EnTracked with position update protocols can lower the power consumption and that the proposed extension can improve the robustness. In this section we would like to consider how the combination performs given urban positioning conditions and a pedestrian movement pattern with no stops. To test this we have collected a dataset with Nokia N95 phones for three pedestrian targets walking a 4.85 km tour in a urban environment with no stops. The dataset consists of ground truth positions and 1 Hz GPS and 35Hz acceleration measurements collected from the built-in sensors. The ground truth was collected at 4Hz with a high accuracy u-blox LEA-5H receiver with an dedicated antenna placed on the top of a backpack carried by the collector. The ground truth measurements were manually inspected to make sure they followed the correct route of the target. Using an urban setting instead of an residential setting tripled the magnitude of average GPS errors to 29,1 meters. That the dataset does not include any stops is adding to the difficulty because it means that EnTracked cannot save power using motion detection, it can only save power by distance-based scheduling and device-aware power minimization.

The results from running different combinations of sensor strategies and protocols is shown in Figure 4. Due to the fact that the dataset does not contain any stops the average power consumption for *Dynamic(Ent):Simple* is higher than for the residential dataset especially for smaller thresholds. The combination of EnTracked with distance-based or dead-reckoning protocols provides a decrease in power consumption between 102mW to 274 mW. The difference between dead-reckoning and distance-based is again insignificant. For the robustness the percentage of threshold violations are twice as high as for the residential dataset. This difference can be explained by the magnitude of GPS errors in the urban dataset but again the accuracy extended version is able to lower the percentage of threshold violations with ten percentage points except for the smallest threshold of ten meters. Therefore even in more difficult conditions the combination can provide savings while improving the robustness.

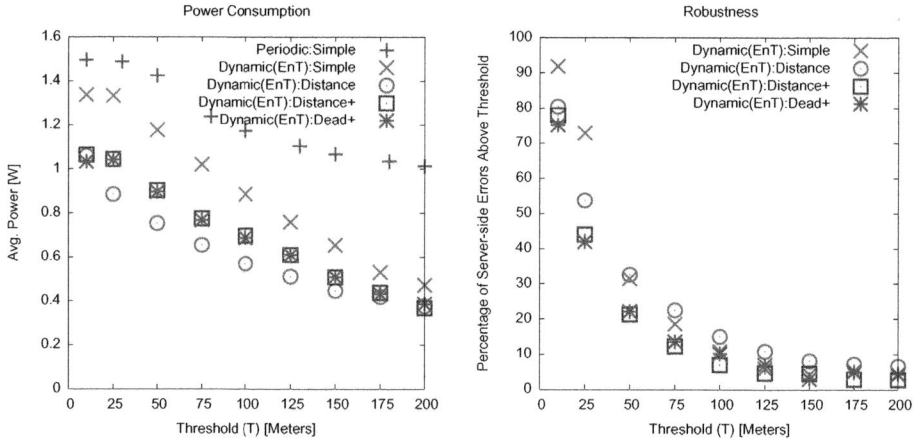

Fig. 4. Comparison of average power consumption and robustness for combinations of sensor strategies and position update protocols in the urban setting

6 Power Consumption and Hardware Characteristics

An interesting question is how much the preceding results depends on the characteristics of the specific device. To answer this we collected a dataset with Nokia N97 phones during the same walks as for the urban N95 datasets (the collector carried both N95 and N97 phones). For the emulation and for the parameters needed by EnTracked we have profiled the delays, power consumption and needed thresholds of the N97 which are quite different from the N95 as can be seen from Table 1 which list values for both phones. To illustrate the values as they impact the combination of *Periodic:Simple* with $T_{period} = 60s$ we have collected energy measurements on both phones which is shown in Figure 5. The plot clearly illustrates that the power consumption for using both the GPS and radio is lower on the N97.

Table 1. Comparison of parameters for N95 and N97

Power Consumption			Delays		
	N95 [mW]	N97 [mW]		N95 [s]	N97 [s]
GPS	324	255	GPS Off	30.0	1.00
Radio idle	466	-	Radio idle Off	31.3	-
Radio active	645	753	Radio active On	1.00	3.06
Accelerometer	50	51	Radio active Off	5.45	4.75
Idle	62	32			

Thresholds		
	N95	N97
$T_{Movement}$	1000	20
A_{norm}	3.71	2.53

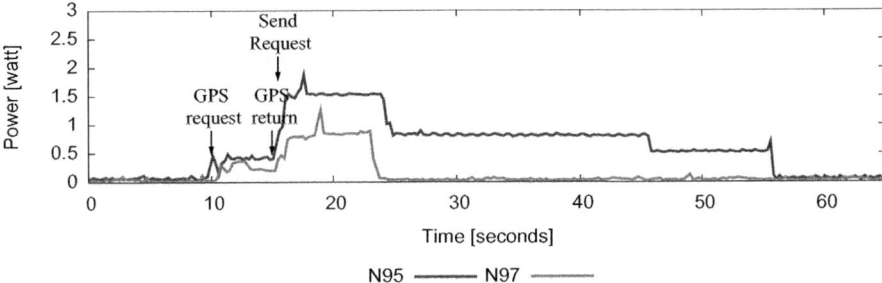

Fig. 5. Power consumption for N95 and N97 for *Periodic:Simple* with $T_{period} = 60s$

The new platform also significantly impacts the tracking results as shown in Figure 6. Generally the power consumption of all combinations are lower. The Periodic:Simple strategy is in this case a much better option than for the N95 data and is even better than Default:Distance which is limited by the GPS and background consumption as noted earlier. Considering the improvements of combining EnTracked with a distance-based or a dead-reckoning protocol there are significant savings for the thresholds below 125 meters, these savings are between 50mW-300mW and above between 20mW-30mW which extends the battery life with between 26% to 52%. Even though the savings are smaller in absolute numbers they are at least as significant as for the N95 due to the lower general power consumption. The robustness plot is given in Figure 6 for data collected in urban conditions with an average GPS error of 20.6 meters. The magnitude of GPS errors can again explain that values are higher for the ten meter threshold. The accuracy extensions are in this case able to lower the violation percentage with 2-7 percentage points for the distance-based combination. Therefore we can conclude that the combined system is able to improve the power consumption and robustness given another hardware platform.

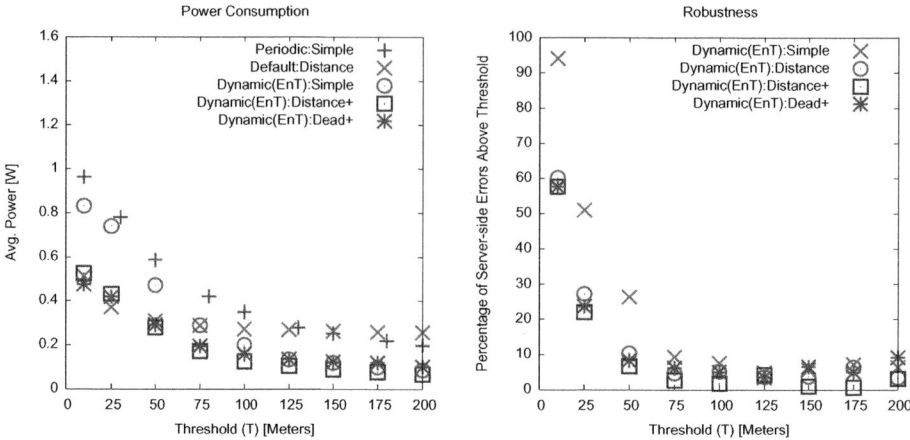

Fig. 6. Comparison of average power consumption and robustness for combinations of sensor strategies and position update protocols for N97 data

7 Conclusions

The primary contribution of this paper is the responsibility division of position tracking into sensor management strategies and position update protocols and the combination of the sensor management strategy of EnTracked with position update protocols which enables the system to further reduce the power consumption with up to 268 mW extending the battery life with up to 36% compared to the original system presented in [7]. Furthermore, we proposed a solution for improving the robustness of distance-based and dead-reckoning position update protocols that only marginally increases the power consumption. The experimental results also provided evidence that the system could save power and improve robustness for pedestrians with a high mobility level in urban positioning conditions and when deployed on a new hardware platform.

In our ongoing work we are trying to address several issues. These are: First, propose methods for automatically determine the parameters of our device model for new devices. Secondly, apply the proposed methods and findings to other positioning technologies such as location fingerprinting [6].

Acknowledgements. The authors acknowledge the financial support granted by the *Danish National Advanced Technology Foundation* under J.nr. 009-2007-2.

References

1. Anand, M., Nightingale, E.B., Flinn, J.: Ghosts in the machine: Interfaces for better power management. In: Proceedings of the Second International Conference on Mobile Systems, Applications, and Services (2004)

2. Cai, Y., Xu, T.: Design, analysis, and implementation of a large-scale real-time location-based information sharing system. In: Proceedings of the 6th International Conference on Mobile Systems, Applications, and Services (2008)
3. Civilis, A., Jensen, C.S., Pakalnis, S.: Techniques for efficient road-network-based tracking of moving objects. IEEE Trans. Knowl. Data Eng. 17(5), 698–712 (2005)
4. Farrell, T., Cheng, R., Rothermel, K.: Energy-efficient monitoring of mobile objects with uncertainty-aware tolerances. In: Proceedings of the Eleventh International Database Engineering and Applications Symposium (2007)
5. Kjærgaard, M.B.: Minimizing the power consumption of location-based services on mobile phones. IEEE Pervasive Computing
6. Kjærgaard, M.B.: A Taxonomy for Radio Location Fingerprinting. In: Proceedings of the Third International Symposium on Location and Context Awareness (2007)
7. Kjærgaard, M.B., Langdal, J., Godsk, T., Toftkjær, T.: Entracked: energy-efficient robust position tracking for mobile devices. In: Proceedings of the 7th International Conference on Mobile Systems, Applications, and Services, pp. 221–234 (2009)
8. Kjærgaard, M.B., Treu, G., Linnhoff-Popien, C.: Zone-Based RSS Reporting for Location Fingerprinting. In: Proceedings of the 5th International Conference on Pervasive Computing, pp. 316–333. Springer, Heidelberg (2007)
9. Küpper, A., Treu, G.: Efficient proximity and separation detection among mobile targets for supporting location-based community services. Mobile Computing and Communications Review 10(3), 1–12 (2006)
10. Leonhardi, A., Nicu, C., Rothermel, K.: A map-based dead-reckoning protocol for updating location information. In: Proceedings of 16th Int. Parallel and Distributed Processing Symposium (2002)
11. Leonhardi, A., Rothermel, K.: A comparison of protocols for updating location information. Cluster Computing 4(4), 355–367 (2001)
12. wen You, C., Huang, P., Chu, H.-H., Chen, Y.-C., Chiang, J.-R., Lau, S.-Y.: Impact of sensor-enhanced mobility prediction on the design of energy-efficient localization. Ad Hoc Networks 6(8), 1221–1237 (2008)

An ETX Based Positioning System for Wireless Ad-Hoc Networks

A.K.M. Mahtab Hossain, Preechai Mekbungwan, and Kanchana Kanchanasut

Internet Education and Research Laboratory (intERLab),
Asian Institute of Technology, Thailand
{mahtab,preechaim,Kanchana}@ait.ac.th

Abstract. RF-based localization has gained popularity because it offers low-cost positioning solution for ad-hoc networks. The Received Signal Strength (RSS) measured by a node has traditionally been used as a parameter to estimate location. However, RSS is not made readily available in the ad-hoc routing protocols like some other link quality indication parameter, e.g., *Expected Transmission Count (ETX)*. ETX predicts the number of transmissions required to deliver a packet over a particular link, including retransmissions. We reveal that ETX can be shown as a proximity indicator relative to an anchor node (i.e., node with known position), and thereby, could also be utilized as a location estimation parameter similar to RSS. We implement a localization plugin for the popular ad-hoc routing protocol, *Optimized Link State Routing (OLSR)* based on ETX. Our analysis and experiments show favorable results.

Keywords: Localization, Positioning System, Ad-hoc Networks, Expected Transmission Count (ETX), Optimized Link State Routing (OLSR).

1 Introduction

Ad-hoc network is generally characterized by a large number of unattended inexpensive nodes with varying capabilities devoid of any fixed infrastructure [1]. The mass production of cheap and low power sensors is making ad-hoc networks more ubiquitous where the sensors are used in health-care, habitat monitoring, inventory tracking, battle field surveillance, disaster management and forecasting, etc [2]. Many of these applications require the knowledge of the positions of the nodes.

Radio positioning techniques support virtually any wireless device, integrate well with existing data networks, and operate both outdoors and indoors. The Received Signal Strength (RSS) perceived by the transceiver has traditionally been utilized in calculating the position of such wireless device. The RSS measurements between a node and the anchors (i.e., nodes with known locations) are generally converted into range (i.e., distance) estimates using either empirical or existing RF propagation models [3], and multi-lateration algorithms are applied. This approach of localization is called range-based. On the other hand, a range-free scheme may discard the quantitative RSS values altogether. However, in this approach, the successful reception of radio messages from an anchor indicates whether the node is *connected* to that particular anchor or not (i.e., whether the node is close to the anchor in space). Subsequently, some connectivity-based algorithms like Centroid [4], APIT [5], etc. are applied for location estimation. Range-free schemes find more applicability in ad-hoc networks comprising of inexpensive nodes despite giving coarser accuracy. This is due to the fact that range-based schemes require complex hardware (e.g., directional antennae) in order to be accurate.

We aim to provide location as an add-on service for the wireless ad-hoc network that requires minimum or no modification at the node's protocol. The commercially accessible 802.11 (Wi-Fi) network interface cards (NICs) do not provide the RSS readings directly. Instead, a typical NIC only provides the Received Signal Strength Indicator (RSSI) parameter, in the form of an 8-bit unsigned integer which is intended for internal use by the NIC, e.g., to determine whether the channel is clear to send, or to decide whether it should attempt to roam. The 802.11 standard does not mandate how RSSI should be calculated from the sampled RSS. As a result, different vendors tend to have their own formulas or conversion tables for the mapping from RSS to RSSI, and vice versa [6]. Moreover, some NICs do not report RSS in user space of the operating system. The popular ad-hoc routing protocols (e.g., DSR [7], DSDV [8], OLSR [9], AODV [10], etc.) also do not make RSS available through their APIs. In order to integrate these protocols with location information, one has to make link layer system calls for retrieving the RSS values which may incur significant overhead for the protocol running on inexpensive sensors. Since we require a localization solution that could be transparently run over the various types of devices or sensors with little or no intervention, RSS option seems not too attractive.

The ad-hoc routing protocols may use some link quality metrics in order to discover efficient path. These metrics can easily be indicative of the *connectivity* of the node to its neighbors for the range-free algorithms of localization discussed previously. The better the link quality, the more *connected* the node is to its neighbor. *Expected Transmission Count (ETX)* is one such link quality indication parameter. ETX metric was first proposed in [11] to model the expected number of transmissions required to send a unicast packet over a link, including retransmissions. It is calculated by taking into account the successfully transmitted and received packets between a node and its neighbor within a certain time period. In this paper, we propose a positioning system for ad-hoc networks based on ETX. We analytically show ETX to be indicative of a node's *connectivity* or proximity to the anchors, and thereby, argue that it could be used to estimate its location. We also implement a localization plugin for the popular ad-hoc routing protocol, OLSR, based on ETX, and conduct experiments. Our experiments show favorable results which strengthen our claim that, ETX could be used as a location estimation parameter similar to RSS. We choose ETX as a location estimation parameter for ad-hoc networks for the following reasons: i) it is readily available as an extension to the popular ad-hoc routing protocols, e.g., DSR, DSDV, OLSR, etc [11,9], ii) it is an efficient proximity indication metric which will be elaborately discussed in Section 3.2, and iii) since it is a network layer metric, no interoperability issue across different NICs as discussed in case of RSS.

The rest of the paper is organized as follows. We provide a brief description of related works in Section 2. In Section 3, we present a brief overview of ETX, and analyze its suitability as a location estimation parameter. In Section 4, we discuss some localization algorithms used to obtain our results. We present experimental findings supporting our claims in Section 5. Finally, we depict in Section 6 the conclusions drawn, and our future work.

2 Related Work

Range-based localization techniques rely on specialized hardware (e.g., RF tags, ultrasound or infrared receivers, etc.) and extensive deployment of dedicated infrastructure solely for localization purpose [12,13,14]. They provide fine-grained location information, e.g., Active BAT [13] system is shown to have 2 cm average accuracy. On the contrary, range-free schemes [4,5,15] provide much coarser accuracy but require no infrastructure for positioning system. Therefore, they find applicability in localizing inexpensive nodes in ad-hoc networks. Generally, various link layer metrics, e.g., RSS [4], ordered sequence of RSSs [15], Signal Strength Difference (SSD) [16], Signal-to-Noise Ratio (SNR), signal quality, etc. are utilized as location estimation parameters. DV-Hop [17], Amorphous [18] and Self-Configurable [19] localization are proposed mainly for ad-hoc networks that provide coarse-level accuracy. They use number of hops (a network layer metric) to reach a node as an indication of its distance from it. To our knowledge, no work in literature has tried to utilize the network layer metric, ETX, as a location estimation parameter for ad-hoc networks so far.

There is also a third category of localization techniques which utilizes the correlation between easily measurable signal characteristics (e.g., RSS) and location. These location fingerprinting solutions try to build a positioning system on top of existing infrastructure (e.g., Wi-Fi networks) [20,21,16]. Some location-dependent signal parameters (e.g., RSS) are collected at a number of locations as location fingerprints in an "offline training phase". During the "online location determination phase", the signal parameter obtained is *compared* with those training data to estimate the user location. This family of localization generally entails a laborious training phase which may not be attractive for the low-cost easily deployable ad-hoc networks.

3 ETX Overview and Analysis

3.1 Overview of ETX

Expected Transmission Count (ETX) is a wired/wireless link quality indication parameter. ETX predicts the number of transmissions required to deliver a packet over a particular link, including retransmissions. The lower the value of ETX of a link, the better the link is. The ETX of a link is calculated by utilizing the forward and reverse delivery ratios of the link. The forward delivery ratio, $\mathcal{D}_f$, is the measured probability that a data packet successfully arrives at the recipient; whereas the reverse delivery ratio, $\mathcal{D}_r$, is the measured probability that the acknowledgement is successfully received. Whenever a data packet is received by a node, it sends an ACK packet to the sender to indicate its successful reception. The sender will retransmit the packet only if it does not receive the ACK packet. This incident occurs when

- the packet sent is lost, or
- the acknowledgement sent from the recipient is lost.

Consequently, it can be seen that a successful transmission (without retransmission) of a packet incorporates both the reverse and forward delivery ratio probabilities. The expected probability that a packet is successfully received and acknowledged is, $\mathcal{D}_f \times \mathcal{D}_r$. Since each attempt to transmit a packet can be considered a Bernoulli trial, the expected number of transmission is:

$$\text{ETX} = \frac{1}{\mathcal{D}_f \times \mathcal{D}_r} \tag{1}$$

In wireless networks, ETX link quality parameter has found practical application, where it is shown to be able to differentiate the high-throughput links from the lossy ones. The routing protocols (e.g., OLSR) of wireless ad-hoc networks make use of this metric in order to choose efficient routes. The ETX of a route is the sum of the ETX for each link in the route.

3.2 ETX as a Measure of Proximity

Proximity to an anchor node (i.e., node with known location) feature has been widely used in order to localize inexpensive sensors in a cost-effective way (i.e., without any infrastructure specifically deployed for localization). Centroid [4] uses the number of samples collected from an anchor node within a certain time period to indicate whether it is *close* to the anchor. Ecolocation [15] utilizes the signal strength measurements perceived by a node from two anchors, and picks the one with the stronger signal to be *closer* than the other. Based on some simplistic assumptions, we show that ETX can also be shown as a proximity indicator relative to an anchor node, and thereby, could be utilized as a location estimation parameter.

Suppose a node's communication range is, R, and the probability of a packet loss is, p. Subsequently, the probability that a packet is successfully received and acknowledged becomes, $(1-p)^2$.

Consider an example scenario where anchor node B is within the communication range of target node A. The ETX of the link AB can be denoted as,

$$\text{ETX}_{AB} = \frac{1}{(1-p)^2}. \tag{2}$$

Consider another anchor node C where it is inside the communication range, R, of node B but outside the communication range of A. This implies,

$$d_{AB} < d_{AC}, \tag{3}$$

where d_{AB} and d_{AC} represent the distances of node A from B and C, respectively. The ETX of the link BC can be formulated in a similar way as (2),

$$\text{ETX}_{BC} = \frac{1}{(1-p)^2}. \tag{4}$$

From the definition of ETX for a route, and utilizing (2) and (4), the ETX of the link AC (C is outside the communication range of A) can be written as,

$$\text{ETX}_{AC} = \text{ETX}_{AB} + \text{ETX}_{BC} = \frac{2}{(1-p)^2}. \tag{5}$$

Combining (2) and (5), we obtain,

$$\text{ETX}_{AB} < \text{ETX}_{AC} \tag{6}$$

For two nodes, B and C, that are inside and outside of A's communication range, respectively, both the constraints (3) and (6) are always satisfied. Therefore, ETX parameter could be used as a proximity measure for a multi-hop ad-hoc networks. Our analysis only verifies this considering two anchors (B and C) that are neighbors to each other. The analysis can easily be extended in a similar way for any two anchors that are separated by multiple hops. Furthermore, (3) and (6) could be merged into the following constraint:

$$\text{ETX}_{AB} < \text{ETX}_{AC} \Rightarrow d_{AB} < d_{AC}. \tag{7}$$

Note that, (7) is presented in the above form to depict ETX as a proximity indication parameter. However, (3) and (6) could actually be combined into the constraint, $\text{ETX}_{AB} < \text{ETX}_{AC} \Leftrightarrow d_{AB} < d_{AC}$ since for our particular scenario, $d_{AB} < d_{AC} \Rightarrow \text{ETX}_{AB} < \text{ETX}_{AC}$ as well.

Similar analysis holds if there is no packet loss, i.e., the communication channel is perfect ($p = 0$). In the analysis above, we consider two anchors where one is within the communication range of the target node, while the other one is not. In a practical scenario, both the anchors might be multiple hops away from the target node, or they both could be within the communication range of it. We consider both these scenarios in the following.

3.2.1 Anchors Multiple Hops Away

Here, we derive the constraint (7) if both anchors are multiple hops away from the target node. Suppose the nodes, B and C are m and n hops away from node A, respectively, where $m < n$. For simplicity of our analysis, we consider any node within the communication range, R, of the target node to be R distances away. In other words, all nodes within R are considered to be the same distance away from the target node. Subsequently, the nodes with the same hop count (i.e, k) turn out to be similar distances (i.e., kR) away from the target node.

This interpretation holds due to our simplistic channel model which states that any node within the communication range, R, is associated with the same packet loss probability, p. So, nodes B and C are mR and nR distances away from node A, respectively, thereby satisfying (3) since $mR < nR$. The ETX of the links AB and AC can be shown as, $\text{ETX}_{AB} = \frac{m}{(1-p)^2}$, and $\text{ETX}_{AC} = \frac{n}{(1-p)^2}$, respectively, yielding $\text{ETX}_{AB} < \text{ETX}_{AC}$ since $m < n$. Consequently, the constraint (7) is satisfied. The analysis in the next section differentiates between any two anchors that are same number of hops away from the target node (i.e., $m = n$) but might be at different distances from it.

3.2.2 Anchors within Same Communication Range

In this section, we consider a generic scenario where both the anchors, B and C are within range, kR ($k \geq 1$), of the target node A. However, their distances from the target node are not the same. Note that, $k = 1$ indicates the setting where both the anchors are within the communication range, R, of node A.

For this particular scenario, we drop our assumption that the channel is characterized by a constant packet loss probability, p. However, we assume same bandwidth and equal spectral efficiency for all the nodes within the range, kR. If the noise is flat, then the packet loss (i.e., bit error rate) probability between nodes A & B, and A & C, follows the relationship, $p_{AB} \propto \frac{1}{\text{RSS}_{AB}}$, and $p_{AC} \propto \frac{1}{\text{RSS}_{AC}}$, respectively. From path-loss model of RF propagation [3], we obtain, $\text{RSS}_{AB} \propto (\frac{1}{d_{AB}})^\beta$, and $\text{RSS}_{AC} \propto (\frac{1}{d_{AC}})^\beta$, where β is the path-loss exponent. If $d_{AB} < d_{AC}$, and the proportionality constant is the same, then, $\text{RSS}_{AB} > \text{RSS}_{AC}$. Subsequently, we obtain, $p_{AB} < p_{AC}$ which yields $\text{ETX}_{AB} < \text{ETX}_{AC}$. Therefore, the constraint (7) is satisfied.

4 Localization Algorithms

In this section, we briefly discuss the range-free localization algorithms that we have utilized in order to obtain our results.

4.1 Centroid

The Centroid scheme [4] defines a connectivity metric which indicates the closeness of a node to an anchor. During a certain time interval, all the anchors send a predefined number of beacons. The connectivity metric is defined as the number of beacons received by the node from a particular anchor to the number of beacons sent by it during a time interval. The final location estimate is the centroid of all the anchors for which, the connectivity metric is above a certain threshold. In our implementation, all the anchors are configured in similar way so that we know the number of beacons (RSS) or messages (ETX) sent by them during a particular time interval. We just capture the beacons/messages at the target node, and compare with the threshold to decide whether it is connected to a particular anchor or not.

4.2 Nearest Neighbor

Nearest Neighbor (NN) algorithm [22] returns the anchor's location from which the node receives the strongest signal (RSS) or perceives the best link quality metric (ETX). K-nearest neighbor (KNN) is a variant of the basic NN algorithm where K anchors' location entries are returned instead of returning only the best match. The final location estimate is obtained by averaging the coordinates of the K anchors' locations found. The value of K has usually been chosen empirically in the literature. In our implementation, we have chosen K to be 3.

4.3 Ecolocation

Ecolocation [15] uses ordered sequence of RSS measurements rather than the absolute RSSs. If $P(d_i)$ and $P(d_j)$ denote the RSSs from anchor$_i$ and anchor$_j$, which are at distances d_i and d_j from the node, respectively, then a constraint of the sequence is defined as,

$$P(d_i) > P(d_j) \Rightarrow d_i < d_j. \qquad (8)$$

First, the constraint set for each grid point of the localization area is calculated using the RHS of (8). During location determination phase, the ordered sequence of RSSs collected from the anchors is translated into the ordered sequence of distances using (8), and subsequently matched against the constraint set of each grid point calculated beforehand. The centroid of the grid points where the maximum number of constraints are matched is returned as the location estimate. Ecolocation can easily be adopted for ETX based localization as well using the constraint (7) derived in Section 3.2.

The discussion on fingerprint based localization algorithms, e.g., Bayesian Inference [21] and KNN in Signal Space [20] has been omitted for brevity.

5 Experimental Study

We first describe our experimental testbed and data collection procedure in Section 5.1. Then, we provide our results and findings in Section 5.2.

5.1 Testbed Setup and Data Collection

The experimental testbed is located inside our research lab which spans over an area of approximately 663 m^2. We have used six Mini Computer MicroClient Jr. (200 MHz thin client) to serve as anchors for our experiments. The choice of such devices is derived from our motivation to accommodate inexpensive devices or sensors. The locations of these anchors are shown in Fig. 1, marked as stars. The grids where data are collected are indicated by crosses. All our mini PCs run Puppy Linux 4.3.1 Linux distribution while our target node (an Asus Eee PC 901) runs Ubuntu 9.10. The target node has a Linksys WUSB54GC WLAN adapter attached to it while each mini PC is installed with Realtek

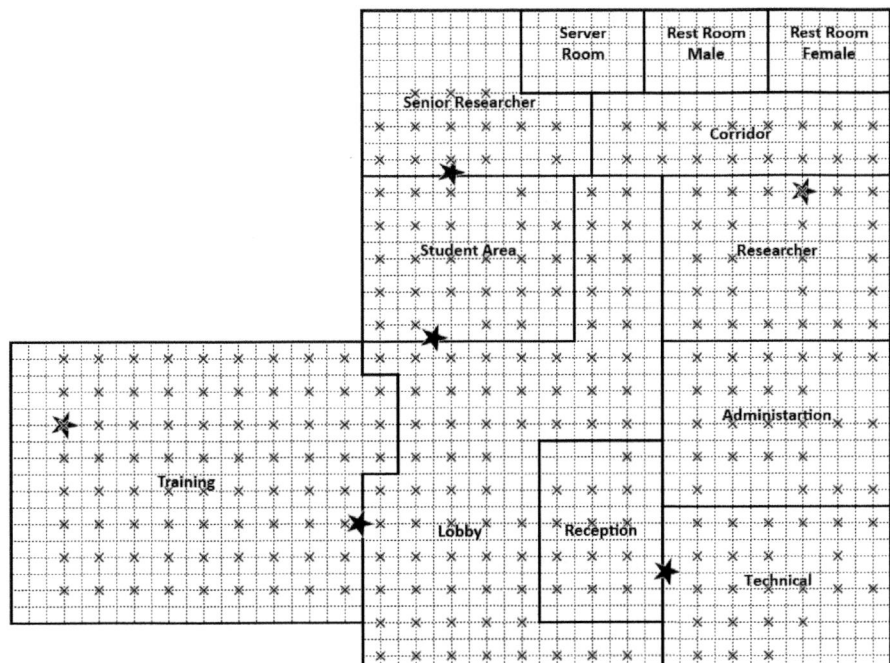

Fig. 1. Our experimental testbed inside our research lab – the six anchors are marked as stars, and all the data collection points are marked as crosses

8100B chip with an external antenna. All of them create an ad-hoc network, and run OLSR [23].

In our testbed, there are 281 training points or grids, marked as crosses in Fig. 1 where we collect both RSS and ETX measurements. For RSS, we have utilized "tcpdump" packet analyzer to capture the signal strength information at the target node stationed at a particular grid. We first put the node's NIC into *monitor mode*, and then run tcpdump where it snoops all the 802.11 packets from the air, and only retrieves the required RSS information from our desired anchors. For ETX, we consult the OLSR routing table at the target node periodically, and retrieve the ETX information only for the anchors. We have actually built a plugin for the OLSR daemon based on ETX which provides the location information of the node in our lab.

5.2 Experimental Results and Findings

5.2.1 Simplistic Localization Algorithms for Ad-Hoc Networks
We present the results of the localization algorithms discussed in Section 4 that are commonly applied for ad-hoc networks. We feed both RSS and ETX as location estimation parameter into the algorithms. We calculate the deviations (in meters) between actual and predicted locations for our 281 measurements,

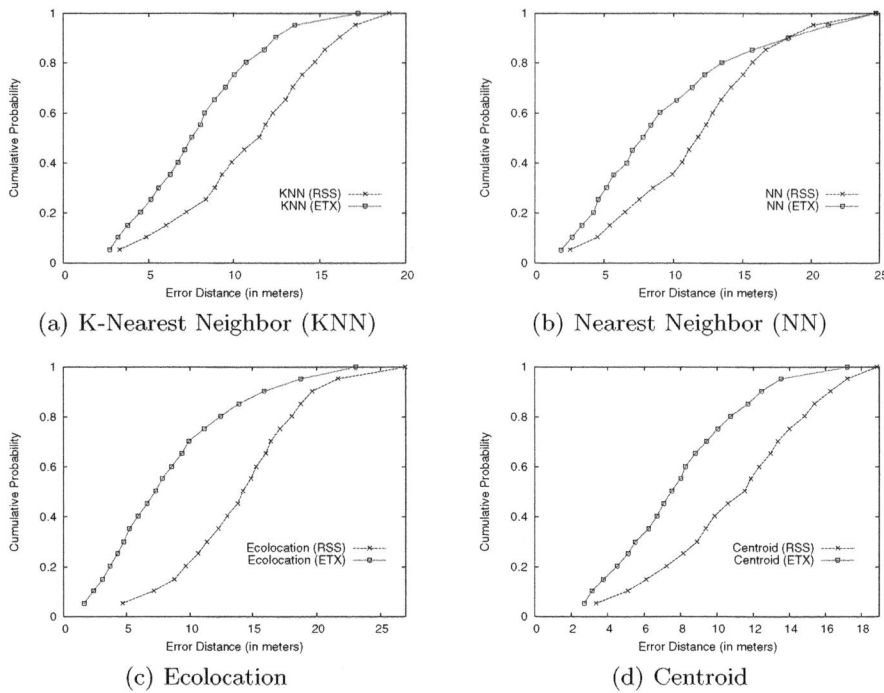

Fig. 2. Comparison of error performance for various simplistic localization algorithms seen in ad-hoc networks using both RSS and ETX

and draw the graphs of Fig. 2. The graphs reveal the percentage of locations (Y-axis) predicted correctly within a specific error distance (X-axis).

As can be seen from the graphs of all algorithms, ETX generally outperforms its counterpart RSS. These simplistic algorithms for ad-hoc networks is built upon the idea of "proximity measure" to a node with known location. ETX is shown to be an efficient proximity measure in the analysis of Section 3.2. Our experimental results here verify that. On the contrary, RSS is affected by reflection, diffraction and multi-path effect [3] which compels it to correlate poorly with distance. The performance gain of our ETX based Ecolocation compared to RSS based Ecolocation [15] is quite significant as depicted in Fig. 2(c). This is due to the fact that our ETX based constraint (7) follows quite nicely under some simplistic assumptions. On the contrary, RSS measurements generally do not represent distances accurately in the real world. Therefore, uncertainties could arise while using (8) as discussed in [15].

The numerical values (averages) of the experiments are listed in Table 1. It can be noted that all the accuracies reported by these simplistic algorithms are quite coarse which is in accordance with the findings of existing literature [4,5].

5.2.2 Fingerprint-Based Algorithms

In the previous section, we show ETX to perform better than RSS considering some simple localization algorithms designed specifically for wireless ad-hoc net-

An ETX Based Positioning System for Wireless Ad-Hoc Networks 183

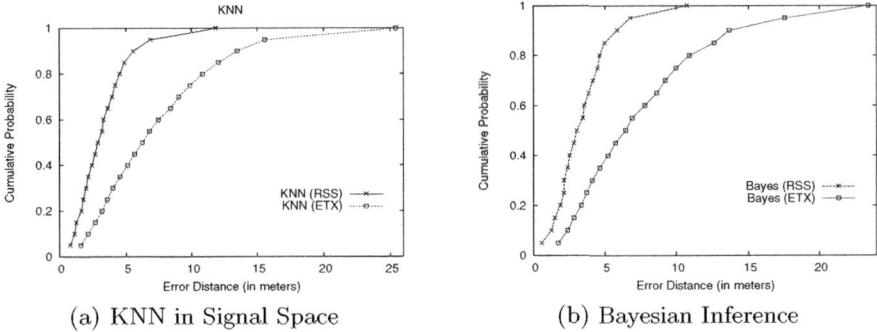

(a) KNN in Signal Space (b) Bayesian Inference

Fig. 3. Comparison of error performance for fingerprint-based localization algorithms using both RSS and ETX

works. Here, we investigate both RSS and ETX's performance regarding fingerprint-based algorithms even though such techniques may not be suitable for ad-hoc networks because of the burden of their deployment. Since our goal is to investigate whether ETX could be utilized as a location estimation parameter similar to RSS irrespective of the algorithms used, we consider these fingerprinting solutions for completeness.

For this experiment, in each trial, 100 testing samples are selected randomly from the 281 data points, and the rest 181 samples are kept as training database for fingerprinting algorithms [20,21]. We repeat this procedure for 101 times to obtain the graph of Fig. 3(a) and 3(b), and the averages with 95% confidence interval as shown in Table 2. The confidence interval is calculated using the results of 101 runs as a vindication that our results are not biased towards a particular separation of training and testing set.

From Fig. 3(a) and 3(b), it is evident that RSS based fingerprinting solutions comfortably outperforms the ETX based ones. The fingerprinting techniques involve collecting signal parameters (e.g., RSS) at places of interests, and subsequently building a radio-map [20] by correlating the signal parameters with locations. The signal parameters measured during the location determination phase is *compared* with the data of the radio-map collected beforehand, which is just *pattern matching* in signal space. Although RSS is reported to vary with many factors [15,24], fingerprinting techniques based on RSS is quite popular in existing literature [20,21,16] which provide reasonable average accuracy ($2 \sim 4\ m$). Our results here show similar trend (see Table 2). Apart form environmental surroundings, time of the day, etc., which are the common influencing factors for both ETX and RSS, ETX is also affected significantly by congestion, node density in ad-hoc networks, etc. A more detailed analysis can be found in [25]. These factors make ETX a poor candidate for location fingerprint since ETX stored at some place beforehand may hardly follow the same pattern at the same place later even though the proximity constraint (7) is satisfied in both occasions. Therefore, ETX based fingerprinting techniques is quite inferior to RSS based ones (see Fig. 3(a) and 3(b)) although its performance is slightly improved compared to the algorithms of Section 5.2.1.

Table 1. Average Errors (in meters) of the Simplistic Localization Algorithms for Ad-hoc Networks for the Experiments Conducted in Section 5.2.1

	KNN	NN	Ecolocation	Centroid
RSS	10.83	11.41	13.78	10.87
ETX	7.67	9.08	8.12	7.66

Table 2. Average Errors with Confidence Interval (in meters) of the Fingerprint-based Localization Algorithms for the Experiments Conducted in Section 5.2.2

	KNN	Bayes
RSS	3.19 ± 0.03	3.28 ± 0.02
ETX	7.11 ± 0.07	7.14 ± 0.07

6 Conclusion

In this paper, we propose an ETX based positioning system for wireless ad-hoc networks. We analytically show that ETX could be utilized as a proximity indication parameter and thereby, it can easily be adopted for the localization algorithms designed for wireless ad-hoc networks comprising of inexpensive nodes. Such ETX based systems are shown to outperform the RSS based ones in our experiments. However, ETX performed poorly compared to RSS in case of fingerprint based algorithms that are not suitable for wireless ad-hoc networks because of its exhaustive data collection procedure. In future, we plan to extend our work by considering other existing localization algorithms in addition to the popular ones that are considered here. We shall also investigate our ETX based system's performance in an anchor-free scenario utilizing collaborative localization approach [26].

Acknowledgement. This work has been supported by Intelligent Transportation System Cluster of the NSTDA, Thailand and the THNIC Foundation.

References

1. Kurose, J.F., Ross, K.W.: Computer Networking: A Top-Down Approach. Addison-Wesley Publishing Company, Reading (2009)
2. Akyildiz, I.F., Su, W., Sankarasubramaniam, Y., Cayirci, E.: A survey on sensor networks. IEEE Commun. Mag. 40(8), 102–114 (2002)
3. Rappaport, T.S.: Wireless Communications – Principles and Practice. Prentice Hall (1996)
4. Bulusu, N., Heidemann, J., Estrin, D.: GPS-less low-cost outdoor localization for very small devices. IEEE Personal Communications Magazine 7(5), 28–34 (2000)
5. He, T., Huang, C., Blum, B.M., Stankovic, J.A., Abdelzaher, T.: Range-free localization schemes in large scale sensor networks. In: Proc. ACM/IEEE Mobicom 2003, pp. 81–95 (2003)

6. Bardwell, J.: A discussion clarifying often-misused 802.11 WLAN terminologies, http://www.connect802.com/download/techpubs/2004/you_believe_D100201.pdf
7. Johnson, D.B.: Routing in ad hoc networks of mobile hosts. In: Proc. of the IEEE Workshop on Mobile Computing Systems and Applications, pp. 158–163 (1994)
8. Perkins, C.E., Bhagwat, P.: Highly dynamic destination-sequenced distance-vector routing (dsdv) for mobile computers. In: Proc. ACM SIGCOMM Conference (SIGCOMM 1994), pp. 234–244 (1993)
9. Jacquet, P., Mühlethaler, P., Clausen, T., Laouiti, A., Qayyum, A., Viennot, L.: Optimized link state routing protocol for ad hoc networks. In: Proc. of IEEE INMIC, pp. 62–68 (2001)
10. Perkins, C.E., Royer, E.M.: Ad-hoc on-demand distance vector routing. In: Proc. of the 2nd IEEE Workshop on Mobile Computing Systems and Applications, pp. 90–100 (1997)
11. De Couto, D.S.J., Aguayo, D., Bicket, J., Morris, R.: A high-throughput path metric for multi-hop wireless routing. In: Proc of MobiCom 2003, New York, NY, USA, pp. 134–146 (2003)
12. Want, R., Hopper, A., Falcão, V., Gibbons, J.: The active badge location system. ACM Trans. on Information Systems 10(1), 91–102 (1992)
13. Ward, A., Jones, A., Hopper, A.: A new location technique for the active office. IEEE Personal Communications 4(5), 42–47 (1997)
14. Priyantha, N., Chakraborty, A., Balakrishnan, H.: The Cricket location-support system. In: Proc. ACM MobiCom 2000, Boston, MA, pp. 32–43 (August 2000)
15. Yedavalli, K., Krishnamachari, B., Ravula, S., Srinivasan, B.: Ecolocation: a sequence based technique for RF localization in wireless sensor networks. In: Proc. ISPN 2005 (April 2005)
16. Hossain, M., Nguyen Van, H., Jin, Y., Soh, W.-S.: Indoor localization using multiple wireless technologies. In: Proc. IEEE MASS, Pisa, Italy (October 2007), http://www.ece.nus.edu.sg/stfpage/elesohws/mass07.pdf
17. Nicolescu, D., Nath, B.: DV based positioning in ad hoc networks. Journal of Telecommunications Systems (2003)
18. Nagpal, R.: Organizing a global coordinate system from local information on an amorphous computer. MIT A.I. Laboratory, Tech. Rep. A.I. Memo 1666 (1999)
19. Wu, H., Wang, C., Tzeng, N.-F.: Novel self-configurable positioning technique for multi-hop wireless networks. IEEE/ACM Transactions on Networking 13(3), 609–621 (2005)
20. Bahl, P., Padmanabhan, V.N.: RADAR: An in-building RF-based user location and tracking system. In: Proc. IEEE INFOCOM, Tel Aviv, pp. 775–784 (2000)
21. Youssef, M.A., Agrawala, A., Shankar, A.U.: WLAN location determination via clustering and probability distributions. In: Proc. IEEE PERCOM 2003 (2003)
22. Jain, A.K., Duin, R., Mao, J.: Statistical pattern recognition: A review. IEEE Transactions on Pattern Analysis and Machine Intelligence 22, 4–37 (2000)
23. olsrd, An adhoc wireless mesh routing daemon, http://www.olsr.org
24. Kaemarungsi, K., Krishnamurthy, P.: Properties of indoor received signal strength for WLAN location fingerprinting. In: Proc. MobiQuitous 2004, San Diego, CA, pp. 14–23 (2004)
25. Das, S.M., Pucha, H., Papagiannaki, K., Hu, Y.C.: Studying wireless routing link metric dynamics. In: Proc. of IMC 2007, NY, USA, pp. 327–332 (2007)
26. Priyantha, N.B., Balakrishnan, H., Demaine, E., Teller, S.: Anchor-free distributed localization in sensor networks. In: Proc. of SenSys 2003, pp. 340–341 (2003)

A Dynamic Authentication Scheme for Hierarchical Wireless Sensor Networks*

Junqi Zhang[1,2], Rajan Shankaran[1], Mehmet A. Orgun[1], Abdul Sattar[2], and Vijay Varadharajan[1]

[1] Department of Computing, Macquarie University, Australia
{janson,mehmet,rshankar,vijay}@science.mq.edu.au
[2] Institute for Integrated and Intelligent Systems, Griffith University, Australia
A.Sattar@griffith.edu.au

Abstract. Sensor networks offer economically viable solutions for a wide variety of monitoring applications. In surveillance of critical infrastructure such as airports by sensor networks, security becomes a major concern. To resist against malicious attacks, secure communication between severely resource-constrained sensor nodes is necessary while maintaining scalability and flexibility to topology changes. A robust security solution for such networks must facilitate authentication of sensor nodes and the establishment of secret keys among nodes In this paper, we propose a decentralized authentication and key management framework for hierarchical ad hoc sensor networks. This scheme is light weight and energy aware and reduces the communication overhead.

Keywords: Authentication, Wireless Sensor Networks.

1 Introduction

Ad hoc wireless sensor networks are self organizing wherein all nodes (either moving or stationary) can both provide and relay data. They provide solutions to a range of monitoring problems such as target tracking in battlefields, forest fire detection, medical monitoring and emergency response. However, this dynamic feature of wireless sensor networks poses security challenges which are aggravated not only due to the underlying peculiarities of sensor nodes such as small memories, weak processors, limited energy but also because they are prone to frequent topological changes with the topology being multi hop in nature. A robust security solution for such networks must facilitate authentication of sensor nodes and the establishment of secret keys among nodes.

Traditional authentication frameworks based on public key cryptography [7,18] and PKI [10] are not suitable for WSNs since the sensor network will ultimately consist of small, low-powered devices that are mobile and this necessitates alternatives to authentication based on central authorities and public key certificates. Due to the limited bandwidth and communication being the most expensive operation in terms of energy,

---

* This research has been supported in part by an Australian Research Council (ARC) Discovery grant (DP0452628) and a Macquarie University Research Development Grant (MQRDG).

messages should not be extended significantly in length when applying security services. Apart from this, a security service that is peculiar to sensor networks is Broadcast/group authentication wherein a sending node can broadcast/multicast a message to multiple nodes in an authenticated way. Some schemes such as [17] address this problem but are not scalable as the number of nodes increases.

An orthogonal problem in providing security solutions is to facilitate the provision of a key management infrastructure. Since the sensor nodes suffer from limited memory, battery and processing and communication capabilities traditional key management mechanisms such as those based on asymmetric cryptography are unsuitable for WSNs as they incur high computational overhead. A major limitation of these schemes is that most of them rely on a trusted third party (TTP), thus not fulfilling the self-organization requirement of an ad hoc sensor network. Some solutions based on random key pre-distribution [4,8,13] impose a limitation on the number of sensor nodes that can be compromised. Once a threshold is crossed, the entire network will be at risk of becoming compromised. Some schemes with a trusted intermediary [3,19] to establish key management have the problem of trusted intermediary being compromised.

In this paper, we address the problem of security by introducing the notion of a hierarchy in network topology wherein we divide the network into clusters, each of which consisting of a small group of nodes. The proposed model is distinguished with low power consumption, less computation workload and enhanced security and equipped with protocols that define how keys are distributed, added, revoked, and updated during the life time of the sensor network.

The remainder of this paper is organized as follows. Section 2 presents a discussion on security issues in hierarchical ad hoc wireless sensor networks. In section 3, we define a framework for authentication and key establishment protocols for such networks. This framework addresses issues that relate to group key generation, distribution and update. In section 4, we propose authentication protocols for both intra and inter cluster environments. In section 5, we compare our framework with similar works. Finally in section 6, we provide concluding remarks.

2   Security Issues in WSN

In this section, we provide an overview of the system constraints, security issues, and the security requirements in wireless sensor networks.

The constraints to be considered in sensor networks include two aspects: the network building (hardware) and network operating (software) [1]. The network infrastructure building aspect involves: infrastructure, deployment (location fixed) and mobility, network topology, density and network size, connectivity, and life time. The two common communication modes are the infrastructure based network and ad hoc network. In infrastructure based wireless sensor networks, the sensor nodes can only communicate directly with base stations. In ad hoc sensor networks, nodes can communicate directly with each other without any infrastructure.

The operating (software) aspects include self configuration, data aggregation and dissemination, node addressability, real time, reliability and security. WSNs must be self-organized to establish a topology to support communication. The sensor nodes will preclude manual configuration before deployment in networks. The networks are able to continuously and periodically to reconfigure themselves for dynamically changing nodes. Data aggregation is the summarization of the traveling data through the sensor network.

Security is a must for many applications of WSNs. There are a variety of potential attacks that breach security. These threats can be classified into four categories: changing message routing path attacks, injecting message attacks, disclosing message attacks and other attacks. Changing message routing path attacks includes sinkhole attack, wormhole attack, sybil attack, replay attack, selective forwarding attack, and non-replication or impersonation attack. In a sinkhole attack [12], an adversary tries to make all or some traffic from a certain area pass a compromised node. The attacker advertises a high quality link to the base station to change the message routing path. Sinkhole attack can enable other attacks such as privacy attack, and selective forwarding attack. In a wormhole attack, the attacker tunnels the captured data into a private link between two colluding nodes. The data can be dropped, forwarded or modified by malicious nodes [11]. In a replay attack, the attacker retransmits captured messages to disrupt or compromise the network. Without protection, the receiver node cannot distinguish a replayed message from the normal message [12]. In a selective forwarding attack, the malicious node will selectively drop some messages [12].

An inject message attack can be divided into inject false message attacks and injecting extra message attacks. Inject false message attacks include sybil attack, and non-replication or impersonation attack. In the sybil attack, the attacker employs a compromised node to masquerade as many other nodes. This can affect routing, data aggregation, and clustering [16] . The non-replication or impersonation attack is similar to the sybil attack. The difference is that the malicious node masquerades as an already existing node [6], which can lead to corrupted or misrouted data. Injecting extra messages attacks includes denial of service attacks, HELLO flood attacks and so on.

Disclosing message attacks include: traffic analysis attack and privacy attack. In the traffic analysis attack, the attacker locates an important node such as the base station so that it can be made unavailable or compromised. In the privacy attack, the attacker tries to discover the message by monitoring the network traffic and listening to the data.

3 Authentication in WSN

The security requirements for WSNs are similar to other networks. They may include the authentication, integrity, freshness, availability and confidentiality.

Authentication is the process of verifying the identity of someone or something. The three types of cryptographic functions used for authentication are hash functions, secret key functions, and public key functions. In traditional networks, the common way to authenticate someone is the use of public key functions. In WSNs, it is usually assumed that public key cryptography can not be used because of the elaborate constraints. This means that the two communicating entities must use secret key functions and hash functions. In WSNs, there are two types of authentication: device level authentication and group level authentication. The device level authentication means that a message is proved to originate from a certain device, whereas the group level authentication means a message is proved to originate from a certain group of devices.

Several authentication schemes have been proposed for WSNs. These schemes can be divided into three types: public key cryptography based, symmetric keys and hash functions, and one way key chain based on hash functions.

Public key based approaches include those based on the RSA public key cryptosystem and Elliptic curve cryptography. TinyPK uses the lower exponent variant of the RSA public key cryptosystem to implement authentication of an external party [20]. The external party is an entity that wishes to establish secure communication with the sensor network. The private part of the RSA is carried out at the certificate authority (CA). The nodes only need to implement the public parts, i.e., the data encryption and signature verification as this is much faster to perform than the private parts in RSA. The public key based approach can incur high computational overhead and network bandwidth consumption. Elliptic curve cryptography (ECC) can be implemented with a much smaller key size and memory usage than RSA. Blaβ and zotterbart give a software implementation of ECC on an Atme microcontroller [2]. ECC has the computational and memory size advantages, but it suffers from more complex arithmetic primitives and a large number of temporary operands [9].

In private keys and hash functions based schemes [21,23,22], each symmetric authentication key is shared by a set of sensor nodes. If an intruder compromises a sensor node, the shared key will be disclosed. Hence these approaches are not resilient to a large number of node compromises.

In one-way key chain type of schemes, the key hashed key chain and the technique of delayed disclosure of keys are used. μTESLA [17] and its variants [14,15,6] are such approaches. In μTESLA, a key chain with delayed key disclosure is used to create an asymmetry in time among the broadcasting source (sinks or users) and the receiver (sensor node) to emulate public key cryptography. Initially, sensor nodes are preloaded with $K_0 = h^n(x)$, where $h^n()$ is a hash function and x is the secret held by the sink (user). The sender(user or sink) sets up time intervals and in each time interval one key is used. During time interval I_1, $K_1 = h^{n-1}(x)$ is used to generate message authentication code (MAC) for all the broadcast messages sent. During time interval I_2, the sender (sink or user) broadcasts K_1, and sensor nodes verify $h(K_1) = K_0$). With K_1, the sensor nodes can verify the authenticity of the message received during the time interval I_1. The receiving sensor nodes need to verify that the key was not disclosed when it received the message. Therefore, loosely synchronized clocks between the sender (sink) and the sensor nodes are needed.

Recently a hierarchical wireless sensor network security protocol was proposed by [5]. This scheme employs hash functions, hash key chains and symmetric keys. Each sensor and the base station share a secret hash key chain. The sensor encrypts the data and sends it to the cluster head. The cluster head collects the data from the sensor nodes and then retrieves the secret keys from the base station. The cluster head decrypts the encrypted message and then sends these data to the base station. This scheme has several advantages. Firstly, it reduces the storage overhead, as each sensor node only stores three keys. Secondly, it reduces the probability for the guessing attack as the sensor nodes change keys once for each transmission. Finally, it uses two way challenge and response authentication method, so it can prevent replay attacks. However, this scheme has several disadvantages. Firstly, cluster heads can disclose all the secret keys of the sensor nodes in their cluster. A single compromised cluster head can affect a large number sensor nodes. Secondly, the cluster heads need to retrieve the sensor nodes secret key for every data transfer. This would cause communication overhead. Thirdly, the sensor nodes need to frequently change the secret keys for each time of data collection.

In order to mitigate these disadvantages, we propose a new authentication scheme which is similar to that in [5], but with more security and less computation and communication overhead.

4 Authentication Protocols for WSN

In the hierarchical wireless sensor network model, a wireless sensor network consists of a command node (or a base station), cluster heads and numerous sensor nodes which are grouped into clusters. The clusters of sensors can be formed based on various criteria such as capabilities, location and communication range, and usage of different cluster algorithms and strategies.

Each cluster includes the cluster head (or the cluster leader) and a set of distinct sensors. Each sensor has two main functions: sensing and relaying. Sensors probe their environment and gather data. They then transmit the collected information to the cluster head directly in one hop or by relaying via a multi hop path. Sensors transmit or relay data only via short-haul radio communication. A cluster head is in charge of its cluster. It is assumed that each cluster head can reach and control all the sensors in the cluster. Each cluster head receives data from different sensors, and then processes the data to extract relevant information, and sends it to the base station (command node) via long-haul transmission.

In the rest of this section, we first give the notations to be used, and then we describe the basic authentication protocols. We also discuss the the authentication protocols for dynamically moving sensor nodes.

4.1 Notation

The symbols and abbreviations used for the protocols are listed in Table 1. The base station stores the following information: two hash functions $H()$, $G()$, all the cluster head and sensors ID_{cl}, and ID_{si}, a shared secret key with each cluster head and sensors K_{bc_l}, K_{bsi}, and a shared secret group key for each cluster K_l, here $l = 1,...m$, $i = 1,..n$ for m clusters and n nodes in the sensor network. Each cluster head stores the following data: two hash functions $H()$, $G()$, all the sensor nodes ID_{si} in its cluster, a shared secret key with the base station K_{bc}, a session key for its cluster group K_{sk}, here $si = 1...p$ for a cluster with p sensor nodes. Each sensor stores the following information: two hash functions $H()$, $G()$, sensors ID_{si}, shared secret key with the base station K_{bc}, and the session key for the cluster group K_{sk}.

4.2 Basic Authentication Protocols

We consider the basic authentication protocols with three scenarios: the base station and cluster head(s), the cluster head and the sensors in the cluster, and the cluster head and a cluster head from a different cluster. In order to reduce the computation overhead, we employ symmetric key functions and hash chain functions in these protocols.

Table 1. Notations used in the protocols

	Symbol	Meaning
Base station	$H(), G()$	hash functions
	$ID_{cl}, (l = 1, \cdots, m)$	cluster ID list
	$K_{bc_l}, (l = 1, \cdots, m)$	shared secret with the cluster head
	$K_{sk}, (sk = 1, \cdots, m)$	session key
	$ID_{si}, (si = 1, \cdots, n)$	sensor ID list
	$K_{bs_i}, (i = 1, \cdots, n)$	shared secret with the cluster head
Cluster head	$H(), G()$	hash functions
	ID_{cl}	cluster ID
	$K_{bc},$	shared secret with the base station
	K_{sk}	session key
	$ID_{si}, (si = 1, \cdots, p)$	sensor ID list
Sensor node	$H(), G()$	hash functions
	ID_{cl}	cluster ID
	K_{sk}	cession key
	ID_{si}	censor ID
	$K_{bs},$	shared secret with the base station

Scenario one: the base station and the cluster head – This is a mutual authentication protocol between the base station and the cluster head in each cluster. We employ a hash chain to dynamically change the shared key between them. Hence guessing attacks can be prevented. The standard mutual authentication protocol can mitigate the reflection attacks. The authentication transfer protocol for the base station and the cluster head is shown in detail in Figure 1.

There are seven steps in our authentication protocol. The first three steps are for the cluster head to authenticate the base station and the next three steps are for the base station to authenticate the cluster head. We describe them as follows.

$Step$ 1. The cluster head sends the join message with its identity ID_{cl} and p_c (encryption of a nonce N_c and $IC_c l$) to the base station.

$Step$ 2. Upon receiving the message, the base station decrypts it, then update the shared key by rehashing it $K_{bc1} = H(K_{bc})$, and and then encrypts the nonce using the new shared secret key with this cluster $P_1 = E_{K_{bc1}}(R)$. Then the base station sends P_1 to the cluster head

$Step$ 3. The cluster head decrypts the message received from the base station, and then compares it with the original random number to verify the base station.

$Step$ 4. The base station chooses a nonce N_b, encrypts it with the shared key and sends it to the cluster head. The base station updates the shared key by rehashing it.

$Step$ 5. The cluster head decrypts P_b, and then encrypts the nonce N_b with the updated shared key with the base station $P_2 = E_{K_{bc2}}(N_b)$, and then sends P_2 to the base station along with its identity ID_{cl}.

$Step$ 6. The base station decrypts p_3 with the shared key, and then compares the nonce with the original one. If they are the same, this means the cluster head is authenticated. Then the base station chooses the session key for the cluster group and encrypts it with the dynamical shared secret key $P_3 = E_{k_{bc2}}(K_{sk})$. At the same time, the base

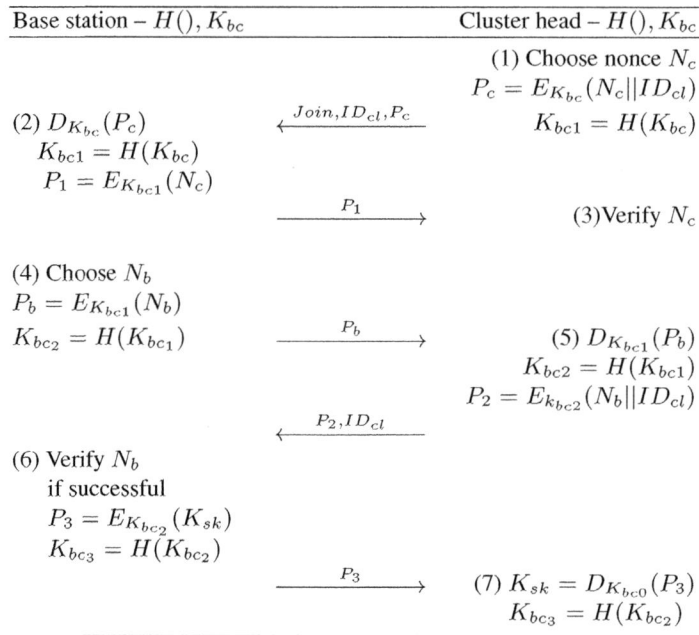

Fig. 1. The Authentication Protocol for the base station and the cluster head

station hashes the shared key $K_{bc3} = H(K_{bc2})$ and saves it for authentication next time. The base station then sends p_3 to the cluster head.

Step 7. The cluster head decrypts p_3 and then obtains the session key for the cluster group. Meanwhile the cluster hashes the shared key $K_{bc3} = H(K_{bc2})$ to achieve the dynamical shared key and save it for next time authentication.

Scenario Two: the base station and a sensor node – In our model, the base station shares a secret key with each sensor node, and the cluster head does not have a shared key with the sensor node. Therefore, the base station and the sensor node need to do a mutual authentication. Then the base station distributes the group key to the sensor. As the cluster head also has the group key for the same with all the sensors in its cluster, the cluster head then shares a group session key with all the nodes in the cluster.

The authentication protocol between the base station and the sensor node is similar to the one between the base station and the cluster head; we omit the details. The difference is that all the communication passes through the cluster head. The authentication transfer protocol for the base station and a sensor node is shown in Figure 2.

Scenario Three: two cluster heads – The authentication protocol between two cluster heads is similar to the mediated authentication with KDC (Key Distribution Center). The base station acts as a key distribution center. First, it generates the shared key for the two cluster heads, and then the two cluster heads mutually authenticate each other using this shared key. The protocol is shown in Figure 3. The steps in this protocol are described as follows.

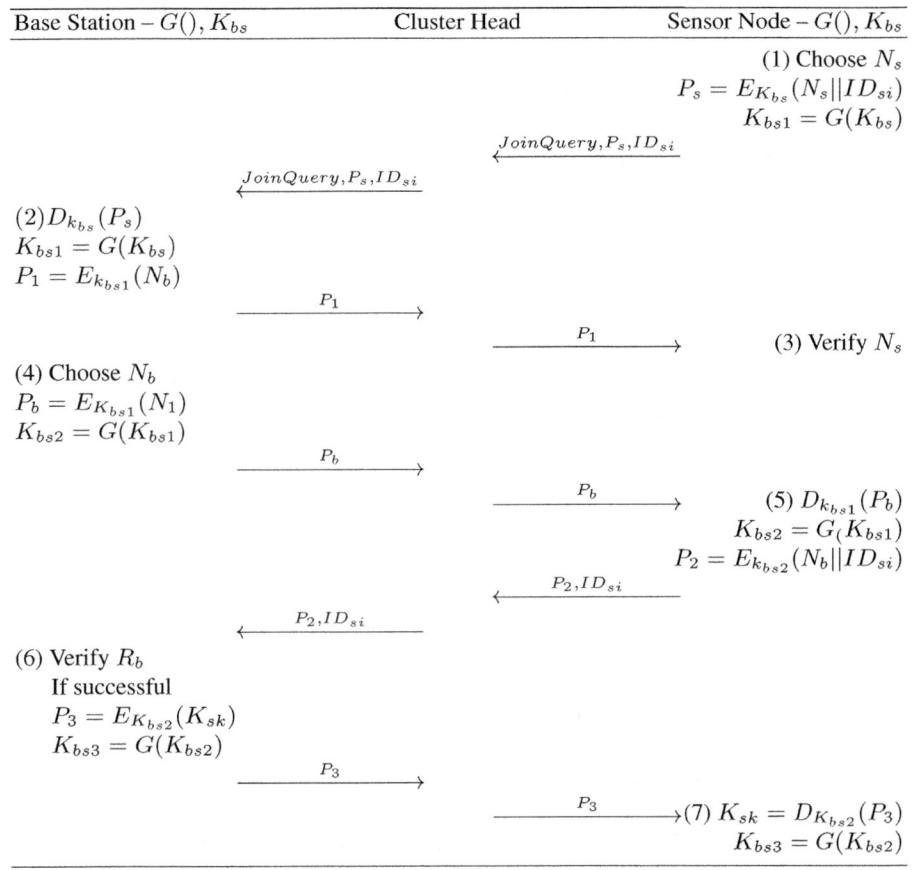

Fig. 2. Authentication Protocol for cluster head and Sensor Node

Step 1. The cluster head A sends the request for communicating with the cluster head B to the base station.

Step 2. The base station creates the session key K_{AB} shared by the cluster head A and the cluster head B. It then generates the ticket $ticket = E_{CB}(K_{AB})$, and encrypts the session key with the shared key with the cluster head A $F_0 = E_{CA}(K_{AB})$. The base station sends F_0 and $ticket$ to the cluster head A.

Step 3. The cluster head A decrypts the F_0 and obtains the session key K_{AB}, and then chooses a random number R_1. The cluster head A sends R_1 and the tickets to cluster head B

Step 4. The cluster head B decrypts the ticket with the shared secret key with the base station and obtains the shared session key K_{AB}. Then the cluster head B encrypts the random number R_1 with the shared secret key K_{AB} $F_1 = f(K_{AB}, R_1)$. The cluster head B sends F_1 to the cluster head A.

Step 5. The cluster head A decrypts F_1 and verifies the random number R_1.

Step 6. The cluster head B chooses a random number R_2 and sends it to the cluster head A.

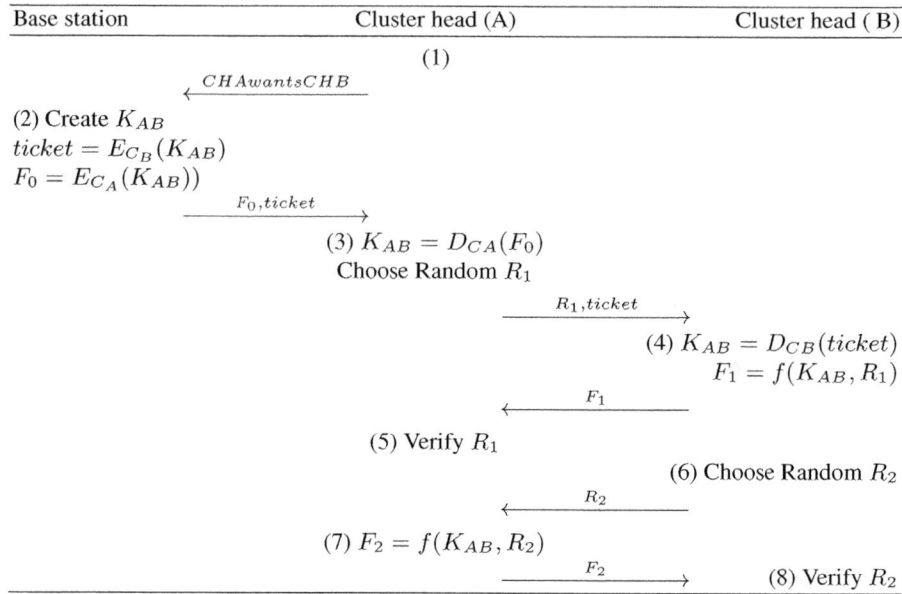

Fig. 3. Authentication Protocol for two cluster heads

Step 7. The cluster head A encrypts the random number R_2 with the shared session key K_{AB} $F_2 = f(K_{AB}, R_2)$, and then sends F_2 to the cluster head B.

Step 8. The cluster head B decrypts F_2 and verifies the random number R_2.

4.3 Authentication Protocols for Dynamical Movement of Nodes

In this section, we summarize the authentication protocols for the following three scenarios of dynamically moving nodes: (1) a sensor node moves from one cluster to another cluster, (2) one cluster is partitioned, and (3) two clusters merge.

Scenario One: A Node Moves from One Cluster to Another – A sensor node moves from cluster A to cluster B. There are two cases to consider: (1) an existing node leaves its cluster, and (2) an existing sensor node joins a new cluster. When one existing sensor node leaves a cluster, if the cluster does not want forward secrecy, then there is nothing to do. If the cluster wants forward secrecy, the new cluster group key must be redistributed. For the case that an existing sensor node joins a cluster, authentication can be done through the cluster head as the sensor nodes still share the dynamical secret key with the base station. If there is no requirement for backward secrecy, the cluster just sends the cluster group session key to the newly joined sensor node; otherwise, the sensor nodes in the cluster need a new session key to be distributed among them.

Scenario Two: Cluster partitioning – There are two cases for cluster partitioning: (1) some of the sensor nodes leave the cluster and organize a new cluster with a new cluster head, and (2) the sensor nodes are divided into two new clusters with two new cluster heads. In case one, the original part of the group only needs to distribute a new session key. The new cluster head needs to be authenticated, and then all the sensor nodes with the new cluster head form a new cluster and a new session key is generated. In case two, both of the new cluster heads need to be authenticated, and then all the sensor nodes within each new cluster form two new clusters and a new session key is also needed.

Scenario Three: Clusters merging – Cluster merging may include two cases: (1) the sensor nodes merge into an existing cluster, and (2) the sensor nodes merge with a new cluster head. In case one, the existing sensor nodes are authenticated through the cluster head, and a new session key needs to be distributed to all the sensor nodes in the cluster. In the second case, the new cluster head needs to be authenticated, and then all the sensor nodes form a new cluster; a new group cluster session key also needs to be distributed to all the sensor nodes.

5 Discussion and Analysis

In this section, we compare our proposed scheme with the DSKG scheme as both of them have the similar hierarchical architecture. We compare them on several aspects: communication overhead, memory overhead, computation overhead and security etc. The comparison table is shown in table 2

Table 2. Comparison of the DSKG scheme and the new scheme

	DSKG scheme	New scheme
Communication between base station and cluster head	much more	less
Communication for cluster heads and sensor node	roughly same	roughly same
Memory overhead for nodes	roughly same	roughly same
Memory overhead for cluster head	more	less
Cryptographic functions	hash & symmetric	hash & symmetric
Computation overhead for node	roughly same	roughly same
Computation overhead for cluster head	more	less

Communication overhead – For DSKG, communication is not efficient for several reasons. Authentication is required for each data transition. The cluster head needs to request the secret key from the base station for each message it obtains from the sensor node. In our new scheme, we use a group key, so there is no need to transfer the secret key for each message.

Memory overhead – In DSKG, a cluster head needs to store more data than our scheme because it stores lots of messages before it transmits them to the base station. In our scheme, the cluster head does not store much data. For sensor nodes, memory overhead is roughly the same.

Computation overhead – For DSKG, there are more authentication processes and more encryptions and decryptions in the cluster head. Hence there is more computation overhead than our new scheme.

Security – Our scheme employs the dynamical hash key chain technique and has the same advantages over the DSKG scheme. Firstly, it reduces the probability for guessing attacks as the sensor nodes exchange keys once for each authentication. Secondly, it uses two way challenge and response authentication method, so it can prevent reflection and replay attacks. One drawback of our scheme is that if the sensor nodes in one cluster change frequently, the group key will have to be changed. Therefore, our scheme will be have a better performance if it is applied to the relatively less changing clusters. In the real world, most applications may fall under this category.

6 Concluding Remarks

Wireless Sensor networks provide economically viable solutions for a wide variety of monitoring applications. When WSNs are deployed in an unattended or hostile environment, security is a major concern. In this paper, we analyzed WSNs security issues and classified them into three categories. We also reviewed the proposed authentication approaches . We proposed a dynamical key authentication scheme for hierarchical WSNs. This new scheme has several advantages over a recently proposed similar scheme.

References

1. Belinda, M.J.C.M., Dhas, C.S.G.: A study of security in wireless sensor networks. MASAUM Journal Of Reviews and Surveys 1, 91–95 (2009)
2. Blab, E.O., Zitterbart, M.: Towards acceptable public key cryptography in sensor networks. In: The 2nd International Workshop on Ubiquitous Computing (2005)
3. Chan, H., Perrig, A.: Pike: Peer intermediaries for key establishment in sensor networks. In: Proceedings of IEEE Infocom. IEEE Computer Society Press, Los Alamitos (2005)
4. Chan, H., Perrig, A., Song, D.: Random key predistribution schemes for sensor networks. In: Proceedings of the 2003 IEEE Symposium on Security and Privacy, p. 197. IEEE Computer Society, Washington, DC (2003)
5. Chen, C., Li, C.: Dynamic session key generation for wireless sensor networks. EURASIP Journal on Wireless Communications and Networking (2008)
6. Deng, J., Han, R., Mishra, S.: Countermeasures against traffic analysis attacks in wireless sensor networks. In: Proceedings of 1st IEEE Conference on Security and Privacy for Emerging Areas in Communication Networks (2005)
7. Diffie, W., Hellman, M.E.: New directions in cryptography. IEEE Transactions on Information Theory 22, 644–654 (1976)
8. Du, W., Deng, J., Han, Y.S., Varshney, P.K.: A pairwise key pre-distribution scheme for wireless sensor networks. In: Proceedings of the 10th ACM Conference on Computer and Communications Security, pp. 42–51. ACM Press, New York (2003)
9. Gaubatz, G., Kaps, J.P., Sunar, B.: Public key cryptography in sensor networks - revisited. In: Castelluccia, C., Hartenstein, H., Paar, C., Westhoff, D. (eds.) ESAS 2004. LNCS, vol. 3313, pp. 2–18. Springer, Heidelberg (2005)
10. The Open Group. Architecture for Public-Key Infrastructure, APKI (1999)
11. Hu, Y.C., Perrig, A., Johnson, D.B.: Packet leashes: a defense against wormhole attacks in wireless networks. In: Twenty-Second Annual Joint Conference of the IEEE Computer and Communications Societies, vol. 3, pp. 1976–1986. IEEE, Los Alamitos (2003)

12. Karlof, C., Wagner, D.: Secure routing in wireless sensor networks: Attacks and countermeasures. In: Proceedings of the 1st IEEE International Workshop on Sensor Network Protocols and Applications, Anchorage, AK, USA (2003)
13. Liu, D., Ning, P.: Establishing pairwise keys in distributed sensor networks. In: Proceedings of the 10th ACM Conference on Computer and Communications Security, pp. 52–61. ACM Press, New York (2003)
14. Liu, D., Ning, P.: Multi-level mtesla: Broadcast authentication for distributed sensor networks. ACM Transactions in Embedded Computing Systems (TECS) 3 (2004)
15. Zhu, S., Liu, S.J.D., Ning, P.: Practical broadcast authentication in sensor networks. In: Proceedings of Proc. of MobiQuitous, Mobicom 2001 (July 2005)
16. Newsome, J., Shi, E., Song, D., Perrig, A.: The sybil attack in sensor networks: Analysis & defenses. In: Proceedings of the 3rd International Symposium on Information Processing in Sensor Networks, pp. 259–268 (2004)
17. Perrig, A., Szewczyk, R., Tygar, J.D., Wen, V., Culler, D.E.: Spins: Security protocols for sensor networks. In: Proceedings of 7th Annual ACM International Conference on Mobile Computing and Networks (Mobicom 2001), Rome, Italy (2001)
18. Rivest, R.L., Shamir, A., Adleman, L.M.: A method for obtaining digital signatures and public-key cryptosystems. Communications of the ACM 22, 120–126 (1978)
19. Singh, K., Muthukkumarasamy, V.: A minimal protocol for authenticated key distribution in wireless sensor networks. In: Proceedings of the 4th International Conference on Intelligent Sensing and Information Processing, Bangalore, India (December 2006)
20. Watro, R., Kong, D., Cuti, S.F., Gardiner, C., Lynn, C., Kruus, P.: Tinypk: Securing sensor networks with public key technology. In: Proceedings of the 2nd ACM Workshop on Security of Ad hoc and Sensor Networks, Washington DC, USA (2004)
21. Ye, F., Luo, H., Lu, S., Zhang, L.: Statistical en-route filtering of injected false data in sensor networks. In: IEEE Infocom 2004 (March 2004)
22. Zhu, S., Setia, S., Jajodia, S.: Leap: Efficient security mechanism for large-scale distributed sensor networks. In: Proceedings of the 10th ACM Conference on Computer and Communication Security (CCS), Washinton DC, USA (2004)
23. Zhu, S., Setia, S., Jajodia, S., Ning, P.: An interleaved hop-by-hop authentication scheme for filtering false data in sensor networks. In: IEEE Symposium on Security and Privacy (2004)

Anonymity-Aware Face-to-Face Mobile Payment

Koichi Kamijo, Toru Aihara, and Masana Murase

IBM Research - Tokyo,
1623-14, Shimotsuruma, Yamato-shi, Kanagawa-ken, 242-8502, Japan
{kamijoh,aihara,mmasana}@jp.ibm.com

Abstract. On-line payments are increasingly popular in paying bills for Internet shopping, and payment-capable mobile phones support making purchases anytime and anywhere, without cash. However, mobile payments are rarely used for making face-to-face payments, with concerns about anonymity, security, and usability. This paper proposes a face-to-face mobile payment protocol that addresses these concerns. To address anonymity and security concerns, the proposed protocol uses unique information for the payment transaction, such as the location and the time, and introduces two procedures for optimizing the matching time slots and exchanging random numbers when needed, to secure the transactions without exposing the seller's or the buyer's personal identification. To address usability concerns, the proposed protocol optimizes the parameters for the two introduced procedures to match the seller-buyer pairs, depending on the number of the people involved in the mobile payments, the delays caused by human operations with the mobile phones, mobile communication, and so on. Experimental results prove that the proposed protocol is practical, solving the addressed concerns.

Keywords: Mobile payment, mobile phone, anonymity, security, usability.

1 Introduction

Mobile phones are changing the way we shop every day. While on-line payments are increasingly popular for paying bills for Internet shopping and public services by paying with a credit card, a bank account, or a prepaid on-line account, mobile phones are also enhancing our experiences by allowing us to pay anytime and anywhere, without carrying cash [1,2]. For example, Safaricom M-PESA [3,4,5,6,7] and similar services [8,9,10,11,12] are useful for sending payments within a country where the financial infrastructure is still immature and expensive. Such mobile payments are leading to new financial infrastructures, especially in African countries, where Automatic Teller Machines (ATMs) are scarce and people have difficulty in withdrawing money from banks. The service of M-PESA is based on Short Message Service (SMS), which supports a widely used application "Twitter", and most mobile phones have SMS capabilities, without regard to the brand or system software.

However, mobile payments have yet to make major inroads, especially for micropayments in face-to-face transactions. The Global System for Mobile communications Association (GSMA) [13] is promoting a pay-by-mobile [14] initiative to address this situation. The main inhibiting factors seem anonymity, security, and usability. Especially, anonymity is rarely addressed mainly due to identification of the sellers and the buyers are usually assumed with currently existing mobile payment technologies.

Cash payments are inherently anonymous. On the other hand, mobile payments may lack anonymity, being backed by information technology, and are often traced based on the security requirements. In fact, people, not only the buyers but the sellers, sometimes do not wish to disclose their identities, i.e. their names or telephone numbers, in some cases, e.g. at flea market, in charity bazaars, in street stalls in Asian countries or even though they are performing legal economic activities.

In this paper, we propose a face-to-face mobile payment (F2FMP) protocol with full anonymity, in which the anonymity of both the sellers and the buyers is guaranteed. We also briefly discuss anonymity levels and symmetry, relaxing the anonymity of either a seller or a buyer or the both. We mainly focus on the anonymity concerns, since anonymity is the problem not addressed in existing technologies, and security and usability can be solved accordingly once a proposal addressed to anonymity is finalized. We choose SMS for our protocol communicating between the payment server and the users (the sellers and the buyers), because it is widely available as already discussed and its communication cost is reasonable.

The remainder of this paper is organized as follows. Section 2 identifies the technical problems of current mobile payments, i.e. anonymity, security, and usability. Section 3 discusses the related protocol and Section 4 proposes a F2FMP protocol. Section 5 reports our experimental results and Section 6 evaluates our F2FMP protocol addressing the three problems identified in Section 2. Section 7 finally concludes our discussion.

2 Technical Problems with Mobile Payments

In this section, we investigate three important technical problems for the F2FMP, anonymity, security, and usability, and use these problems as criteria to evaluate the quality of the mobile payment technology:

Anonymity: Mobile payment is usually lack of anonymity. Cash payment is always anonymous, but an SMS-based mobile payment is inherently designed to disclose the identities of the seller and the buyer each other to confirm the transaction on the server. Once identity information becomes generally available due to careless management or malicious attacks, it may be misused for spamming or phishing. Therefore, the sellers or the buyers do not wish to disclose their identities.

Security: Mobile payment always carries security concerns, not only from its technology, but also from user experience. Cash can be counterfeited, is easily stolen, and is very hard to recover. On the other hand, mobile money cannot be stolen since its value is electrically exchanged. Even if the mobile phone is stolen, we already have several measures to disable the payments function, e.g. to prompt a password or remotely disable the function via Mobile Network Operator (MNO). Nevertheless, people still do not fully trust electronic forms of money, because of the security concerns. Security of the SMS-based mobile payments is widely addressed on communication channel, payment device, and the payment server, but we accept these concerns as givens, as discussed in Section 6. Other possible concerns are the mis-typing of the information to the mobile phone which will cause incorrect payment or mismatching of the seller-buyer pairs, and the falsification of the agreed price.

Usability: Mobile payment must be as simple as possible, since payments are basic and everyday actions. In some cases, cash is easy to pay. However, it often involves calculating and handling the changes, which bothers us. Or it sometimes takes time to find appropriate coins from the wallet. Although the SMS-based mobile payment released us from such burden of the changes, it typically requires a series of key inputs such as the seller's phone number, which is not ideal for the F2FMP in comparison to cash [15].

In addition, cost, availability, and portability are also important concerns. However, since they depend on the strategy of the MNOs, pervasiveness, and the performance of the mobile phones, not on the performance of the payment protocols, we do not evaluate them in this paper.

3 Related Protocol

In this Section, we discuss a typical SMS-based mobile payment protocol, M-PESA. This protocol is not intended for a F2FMP, but is the closest existing payment protocol that can be compared with our proposal, since it uses SMS and completes a transaction using only a pair of mobile phones.

A typical SMS-based payment, M-PESA, transfers a value from a buyer's (B) account to a seller's (S) account using following four steps (Fig. 1):

(I) The B and the S agree on a price (p) for the purchase.
(II) The B sends an SMS message containing the p and the S's identification, such as a phone number, to the payment server (V), which typically knows the telephone numbers of the senders of the messages.
(III) The V confirms that the payment transaction is requested by the B. This involves checking the identities of both the B and the S, and testing some other parameters.
(IV) The V sends a confirmation message, which includes the p and the identities of the two parties, to both the B and the S.

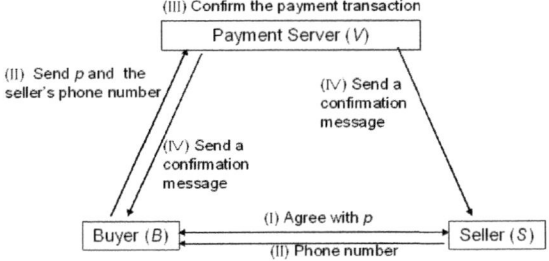

Fig. 1. Payment steps of M-PESA

4 Anonymous Face-to-Face Mobile Payment

In this Section, we propose a new protocol, an SMS-based anonymous F2FMP protocol, in which neither the seller nor the buyer needs to disclose their identities (Fig. 2). The main difference from the related protocol is that we use information uniquely associated with a transaction, such as the location and the time, and if necessary, the secrets only the pair of the seller and the buyer know, generated by our random number procedure, or the RN procedure hereafter, guarantee the anonymity and security. Also, our time shift procedure, or the TS procedure hereafter, improves the probability of correct pair-matching. Here are the steps:

(1) The B and the S agree on a price (p) for the purchase.
(2) The B and the S send messages containing the p to the V, respectively.
(3) The V collects all the messages from the sellers and the buyers received in the same time slot T_d, where d is a sequence number of each time slot as shown in Fig. 3 (a). The V first counts $N_s(d, c, p)$ and $N_b(d, c, p)$, the numbers of the messages which include only the price (p) received in the time slot T_d from the Cell ID c, from the sellers and the buyers, respectively. For pair matching, we introduce the TS procedure, which uses two buckets of time slots, D_1 and D_2, each shifted by $T/2$, as shown in Fig. 3, where T is the length of each time slot. We use these shifted time slots to avoid failing to find the pairs near the border of each time slot.

If there is only one seller-buyer pair with the same price, the same location, and the same time, or $N_s(d, c, p) = N_b(d, c, p) = 1$, the V determines that the seller and the buyer are paired, so processing goes to Step (8). If not, then the Ss and the Bs go through the RN procedure, as described in Steps (4) to (7). For example, in Fig. 3 (a), S_j and B_j corresponds to the seller and the buyer and they are pair if j's are identical. In this case, $N_s(d, c, p) = N_b(d, c, p) = 1$ is satisfied for the pair of S_2 and B_2 at T_2.

(4) If $N_s(d, c, p) = N_b(d, c, p) = 1$ is not satisfied but $N_s(d, c, p) + N_b(d, c, p) > 0$ in Step (3), the V assigns one unique random number to each of the seller and the buyer that already existed in T_{d-1}. Then the V sends the assigned random number to each of the seller and the buyer. For example, in Fig. 3 (a), S_1, S_3, and B_1 are in T_1 and all of them already existed in T_0. Therefore,

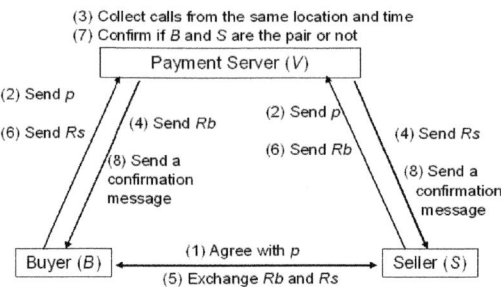

Fig. 2. Payment steps of the anonymous face-to-face mobile payment

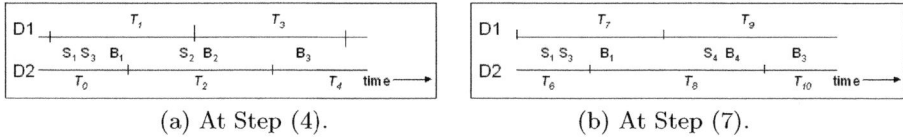

Fig. 3. Pair matching with the time shift procedure

random numbers are assigned to all of the three at T_1. As well, for B_3, a random number is assigned at T_4. S_2 and B_2 are already excluded in Step (3). None of the generated random number overlaps at least during $2T$, to avoid same random number being assigned to different buyer or seller.

Now random numbers R_b and R_s are assigned to the B and the S, respectively. The reason why the V needs different random numbers for both the B and the S, not only one random number for the S or the B, is discussed in the "Security" part in Section 6.

(5) The B and the S receive R_b and R_s, respectively, and exchange them.
(6) The exchanged random numbers are sent back to the V. This means the S and the B send R_b and R_s to the V, respectively.
(7) The V receives the messages with the random numbers. For each matching condition with the same time slot number d, the same Cell ID c, and the same price p, or (d, c, p) hereafter, the V first verifies the responses from the sellers and the buyers, respectively. If a same random number is included in two or more messages from the different buyers, the V excludes all of such buyers from the matching pairs. V also runs the same procedure for the sellers. The reason why the V excludes these is discussed in the "Security" part in Section 6.

After those are excluded, the V looks for a pair of response messages, $M_{bx} = (d, c, p, R_y)$ from the buyer B_x and $M_{sy} = (d, c, p, R_x)$ from the seller S_y, which satisfies $A(R_x) = B_x$ and $A(R_y) = S_y$, where $A(R)$ is the seller or the buyer that V assigns the random number R at Step (4). The V accepts such pairs as a real pair and continues with Step (8). For those sellers and buyers whose partner is not found, including those excluded earlier in this

Step, the V goes back to Step (4), by assigning new random numbers. At pair matching of this step, we also use two time slots as Step (4). Therefore, every seller or buyer has chances of pair-matching at two overlapping time slots. Fig. 3 (b) is an example of the time chart for a given (d, c, p). The B_2-S_2 pair matched at Step (3), and skips the RN procedure of Steps (4) to (7). In this case, B_1 and S_1 do not match at T_6 but match at T_7, and the V accepts them as a pair. As well, the B_4-S_4 pair matches at T_8, so the V accepts them as a pair. This pair also matches at T_9, but is already matched at T_8. B_3 and S_3 does not match in any time slot. Therefore, the transaction is unsuccessful and they must send the transaction request to the V again.

If the number of retries is greater than N_r, where N_r is maximum number of retries for some users, then the V terminates the transaction for such users. In such cases, the V should investigate the reasons, such as a malicious user among the sellers and the buyers involved in the matching attempts, and if malicious users are identified, then the V will exclude such users for the F2FMP from the next time.

(8) Now that a pair has been recognized, the V sends a confirmation message to each member of the pair. The confirmation message includes the values of d, c, p, and a unique random number that is exclusively common between the seller-buyer pair, so that the pairs can confirm that they are correctly matched.

The protocol above does not include the steps to match pairs in adjacent Cells. If the paired S and B are located at the border of a Cell ID, either or both of the seller and the buyer may fail to belong to the same Cell ID. However, the area covered by each Cell is normally duplicated to avoid the existence of the non-covered areas, so, even for such a case, they should belong to one or more same Cell IDs. If the V finds the same pairs with the same d and p at different Cell IDs, they should accept them only in one Cell ID.

5 Experimental Results

To evaluate the feasibility of the F2FMP, some concerns, e.g. some attacks for security, can be evaluated by an armchair theory as discussed in Section 6. However, it is not good enough for complete evaluation. In a sense, feasibility of the F2FMP should be evaluated both by the experimental results and the armchair theory.

In this Section, we report the results of two experiments which evaluate the anonymity, security, and usability, that cannot be evaluated just by an armchair theory: Experiment (a) evaluates of the probability of the pair-matching failure for anonymity and security, and Experiment (b) evaluates of the probability of skipping the RN procedure for usability.

We select Experiments (a) and (b) because the pair matching is the fundamental concern for our proposal, and because skipping the RN procedure most contributes to improve the usability, respectively.

Other concerns, such as some attacks for security, will be evaluated by discussion in Section 6.

For both of the two experiments, we simulate the case that both the Cell ID (c) and the agreed price (p) are the same for all the users. This can be possible in several cases, e.g. in the bargain sale with a same price, discount food shop at lunch time, in street stalls, especially in Asian countries, and so on. The following [1] to [4] show our assumptions:

[1] The occurrence of the payment transactions within a given time slot at a given location follows a Poisson distribution as

$$P(N = k) = e^{-\lambda}\lambda^k/k!, \qquad (1)$$

where $P(N = k)$ is the probability that the mobile payment transactions take place k times in a given unit time slot, e.g. one minute, and λ is a parameter of the expected number of transactions during the given time slot.

[2] The probability density function of the delay time t from a pair agreed on the price till V receives the random numbers from the B or the S, i.e. from Steps (1) to (3) follows normal distribution as

$$f(t - \Delta t) = \frac{2}{\sqrt{2\pi}\sigma} \exp(-\frac{(t - \Delta t)^2}{2\sigma^2}), t \geq \Delta t, \qquad (2)$$

where Δt is a minimum time delay, and σ is a parameter for the time delay.

[3] The delay from a seller-buyer pair exchange the random numbers till the random numbers arrive to the V, i.e. from Steps (5) to (7), follows Eq. (2) as well since both cases involves a single-way transaction from a seller or a buyer to the server.

[4] No congestion of SMS takes place.

We simulate both Experiments (a) and (b) by generating the mobile payment transactions with the probability in Eq. (1) with the delays in Eq. (2), each for the time period of $1000T$. Fig. 4 (a) shows the result of Experiment (a). In this experiment, we change T (x-axis) as the multiple of σ with $\lambda = 1$ and $\lambda = 5$, and calculate the probability that the pairs are not successfully matched (y-axis). We do not count the case of RN procedure being skipped. We also compare the results with and without the TS procedure (W/time shift and WO time shift in Fig. 4 (a)). We regard the mobile payment is successful if both the S and the B of a pair stay within a same time slot of T_d for some d without retries. Unsuccessful pair matching takes place when the difference of the delays of the seller and the buyer between Step (5) to (7) causes their messages arrive in different time slot. From this result, we find that the probability of matching failure is zero when $T \geq 4\sigma$ regardless of the value of λ. We also find that the TS procedure is effective, since we observe approximately 15% of matching failure without the TS procedure while 0% with the TS procedure at $\lambda = 1$.

Fig. 4 (b) shows the result of Experiment (b). In this experiment, we evaluate the probability to skip RN procedure (y-axis) by changing the value of λ as multiple of T (x-axis). To calculate the probability above, we count the number

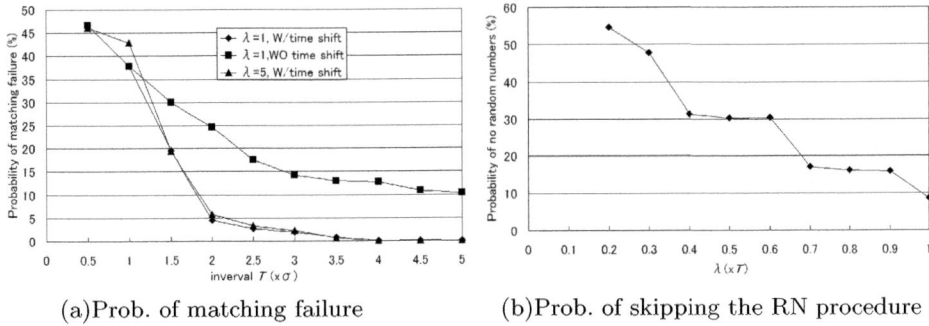

(a) Prob. of matching failure   (b) Prob. of skipping the RN procedure

Fig. 4. Experimental results

of the seller-buyer pairs which satisfy $N_s(d, c, p) = N_b(d, c, p) = 1$ with some d. $N_s(d, c, p) = N_b(d, c, p) = 1$ is not satisfied when the occurrences of the mobile payment transactions are frequent enough, or, even if transactions are not frequent, when the difference of the delays of the seller and the buyer between Step (1) to (3) is so large that the messages with the price do not arrive to the V in the same time slot. From this result, we find that when the occurrences of the mobile payment transactions are rare, e.g. $\lambda = 0.2T$, more than 50% of the pair matching can be done without RN procedure, but we need RN procedure with the probability of more than 90% when $\lambda \geq T$. Therefore, we can increase the probability of completing the payment transactions skipping the RN procedure by optimizing T according to the value of λ.

6 Discussion

In this section, we evaluate the proposed protocol based on the criteria discussed in Section 2, by comparing them with the related protocol:

Anonymity: For M-PESA, the sellers must disclose their phone numbers. In the proposed protocol, neither the seller nor the buyer needs to disclose the phone number, and the results from Experiment (a) prove that the pairs can be matched correctly without disclosing phone numbers by optimizing T.

At other times, people may be willing to disclose their identities to some extent. For example, the sellers may wish to disclose their shop name only, but not their phone numbers, to avoid some nuisance phone calls. Another case is that the sellers may wish to disclose their identities, perhaps by giving out receipts or business cards. The other case is that the buyers may wish to disclose some identities, such as in their favorite restaurants, shopping malls, or movie theaters, where they may have customer loyalty cards today and receive benefits as frequent customers. They will allow the sellers to analyze their purchase histories under certain conditions [16,17,18]. Therefore, we believe both the sellers and the buyers should be able to negotiate and control the level of anonymity depending on the payment situation. In that sense, there are typically three levels of anonymity appropriate to the payment types and amounts:

Case (1): Full exposure (e.g. M-PESA discussed in Section 3): The key information is disclosed, including the telephone number, name, and address.

Case (2): Partial anonymity: No information is disclosed but some ID other than the telephone numbers, e.g. shop names or nicknames.

Case (3): Full anonymity (e.g. our proposal discussed in Section 4): No information is disclosed.

We now describe the example of Case (2) since Case (2) is not discussed so far. An example is that in the F2FMP, the buyers do not get the sellers' phone number, but they get the possible list of the shops located near the buyers, which may be displayed on their mobile phones' display, then select the shop names to which they pay money from the list. One way to achieve Case (2) is to modify Step (4) and later of the proposed protocol to send such lists from the V to the B and the B sends back the seller's shop name to the V. In this case, Step (1) may be skipped.

If we support all of the three cases above, both the sellers and the buyers can select the anonymity level depending on their preference.

Many users may want to remain anonymous even from the payment server, the same way that cash payments can not even be traced by banks. Regarding this issue, we accept the anonymity, as well as security, of the payment server is given, as discussed in the next paragraph.

Security: First, we discuss the security of the communication channel and so on, then the comparison results between the related protocol and our proposal. We accept the security of the communication channel, the device hardware, and the servers are given. Regarding channel security, we can apply "Onion Routing" [19] that encrypts the messages including the IP headers. Regarding device security and server security, we can validate the software stack running on both servers and client devices by using the Trusted Platform Module (TPM)'s attestation feature, before the secure communication is established [20,21]. We can also apply Homomorphic Encryption [22] for server security.

For M-PESA, only the buyers input the agreed price. Therefore, if they falsify or mis-type the price, it would take time for the sellers to notice the injustice or the mistake, or they may not notice. In our protocol, since both the sellers and the buyers have to input the agreed prices, the matching will fail in such injustices or mistakes.

Regarding the risk of payments between a mismatched pair, we experimentally demonstrated how to minimize this risk without sacrificing the usability by studying the number of the payments and the communication delays. The nice property of the proposed protocol is that even if a matching fails, users can retry the F2FMP.

Other than the risk above, we studied the following two major possible attacks that could be directed against the sellers or the buyers.

The first attack by a seller would be to seek double payments from a buyer by claiming that the money had not yet been paid to the seller. For example, a malicious seller S prepares two mobile phones, M_1 and M_2. The S initiates the F2FMP as discussed in Section 4 using M_1, but at Step (8), the S could tell the buyer B that the payment has not been completed, by switching the phones to show the display of M_2. This attack would be possible when the server V generates only one random number, e.g. to the B, and the process requires only the S to return the random number sent to the B. However, in the proposed protocol, since each of the S and the B receives different random number and exchange them, V detects the injustice by the process of Step (7).

The second attack would involve stealing the random number from the buyer by some methods, impersonating the seller, and then trying to steal the buyer's payment. For example, when a legitimate seller-buyer pair S_1 and B_1 exchange their random numbers, and a hacker S_2 tries to impersonate S_1 to get the money from B_1 stealing the random number of B_1, and sends the random number to the server V. However, in this case, the V will not match the pair since the V will have received two messages with the same random number R_b from different sellers. The V will block such pairs and ask for a retry, as described in Step (7). If many of these retries occur, the V can contact the sellers that are frequently sending messages with duplicated R_b values and the V will exclude such malicious sellers from the next time.

Usability: For M-PESA, each buyer must input at least the agreed price and the seller's telephone number. In contrast, although each seller and buyer must input the agreed price for the proposed protocol, the buyer need not input any telephone number, which is usually around ten digits. Instead, the buyer may have to input random numbers.

However, in the experiment, we found that, by choosing the appropriate T depending on λ, we could minimize the necessity of inputting random numbers. Even if we have to input random numbers, the digits can be small. Random numbers are generated only for pairs whose Cell ID, time slot, and the agreed price are all the same. Also, for a random number, we can use alphabets of large and small capitals, and numeric numbers, which is 62 in all. Therefore, we need to input only $\log_{62}(2N-1)+1$ digits for N pairs, and even if $N = 100,000$, we need only three digits. In actual cases, such large number of pairs would be very rare. We may want random numbers sparse enough to avoid a random number assigned to another person which causes mismatching. In this case, we will need to add only one or two additional digits which will make the distribution of the random numbers 62 or 3844 times sparser. Therefore, it is expected that the digits of the random numbers we have to input would be at most between two to five.

Taking these discussions into consideration, we can improve not only anonymity, but security and usability compared with M-PESA.

So far, we discussed the protocol for the F2FMP that uses SMS. However, there are several alternative methods to improve the usability while guaranteeing the anonymity and security, if the mobile phones have more functionality. For

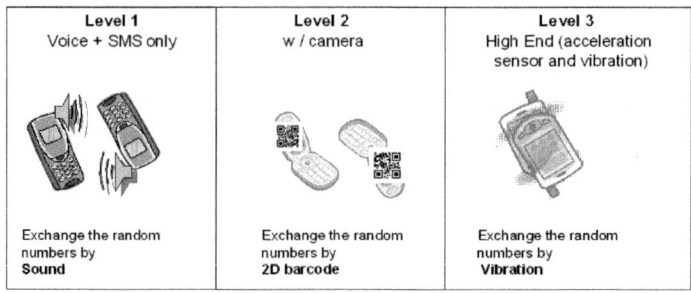

Fig. 5. Methods to exchange the random numbers with high end mobile phones

example, we can convert random numbers to sounds, barcodes, or vibrations, and exchange them, by capturing them with the mobile phones of the pair as shown in Fig. 5. To use barcodes or vibrations, the mobile phones of both users need to have cameras or acceleration sensors, as appropriate.

Instead of using random numbers, we can use the background noise, music, or image, to identify the pair. If the V can confirm that a pair of mobile phones are located close to each other, then the V can use that information, instead of or in addition to the Cell IDs, and match the pair for the transaction without using random numbers.

We may not have to input even p by taking photos of the price tag, if such an application is installed to the mobile phones.

7 Conclusion

In this paper, we introduced a F2FMP protocol in which the anonymity of both the buyers and the sellers is guaranteed. The technical contributions of the proposed protocol are the use of unique information, such as the same location and the time, and our TS and RN procedures for optimization. The random numbers used in our RN procedure and the time shift used in our TS procedure guarantee the anonymity and security of both the seller and the buyer. As well, the optimization to skip the RN procedure when possible improves the usability. We evaluated the proposed protocol from two points of view, pair matching failure and skipping the RN procedure, and showed that the proposed protocol is practical and offers advantages over the existing protocol. We also discussed methods for the F2FMP considering the capabilities of the mobile phones, showing that mobile phones with more features can improve the usability of the F2FMP. We believe that the proposed F2FMP protocol is attractive for both the sellers and the buyers for the protected anonymity, as well as easy and safe payment without carrying cash. We also believe our F2FMP protocol will further contribute to promote a cashless world, and a more effective digital economy.

References

1. Hammond, A., Kramer, W.J., Tran, J., Katz, R., Walker, C.: The Next 4 Billion: Market Size and Business Strategy at the Base of the Pyramid. World Resource Institute (2007)
2. United Nations Conference on Trade and Development, Information economy report 2007-2008 Science and technology for development: the new paradigm of ICT (2008)
3. Morawczynski, O., Miscione, G.: Examining Trust in Mobile Banking Transactions in Kenya: The Case of M-PESA. In: IFIP WG 9.4-University of Pretoria Joint Workshop, Pretoria, South Africa (2008)
4. Vaughan, P.: Providing the Unbanked with Access to Financial Services: The Case of M-PESA in Kenya, Presentation given during the Mobile Banking & Financial Services Africa Conference, Johannesburg, South Africa (2008)
5. Hughes, N., Lonie, S.: M-PESA: Mobile Money for the 'Unbanked': Turning Cellphones into 24-Hour Tellers in Kenya. MIT Press Journal, Innovations: Technology, Governance, Globalization 2(1-2), 63–81 (2007)
6. Mas, I., Morawczynski, O.: Designing Mobile Money Services Lessons from M-PESA. MIT Press Journal, Innovations: Technology, Governance, Globalization 4(2), 77–91 (2009)
7. Safaricom M-PESA, http://www.safaricom.co.ke/index.php?id=745
8. Voda M-Pesa, http://www.vodacom.co.tz/docs/docredir.asp?docid=3518
9. Zain Kenya Me2U, http://www.ke.zain.com/en/phone-services/me2u/index.html
10. Starcomms DashMe, http://www.starcomms.com/v_dashme.php
11. MTN MobileMoney Account, http://www.mtnbanking.co.za/
12. eTranzact, http://www.etranzact.com/Web/index.htm
13. GSMA, http://www.gsmworld.com/
14. Pay-Buy-Mobile Business Opportunity Analysis – Public White Paper Version 1.0, GSMA Association (2007)
15. Medhi, I., Gautama, S.N.N., Toyama, K.: A comparison of mobile money-transfer UIs for non-literate and semi-literate users. In: Proc. of ACM Conference on Computer Human Interaction (CHI), Boston, USA, pp. 1741–1750 (2009)
16. Karnouskos, S., Hondroudaki, A., Vilmos, A., Csik, B.: Security, Trust and Privacy in the SEcure MObile Payment Service. In: 3rd International Conference on Mobile Business, New York City, USA (2004)
17. Linck, K., Pousttchi, K., Wiedemann, D.G.: Security Issues in Mobile Payment from the Customer Viewpoint. In: Ljungberg, J. (Hrsg.) Proc. of the 14th European Conference on Information Systems, Gteborg, Sweden, pp. 1–11 (2006)
18. Lu, C.T., Liang, L.R.: Analysis of payment transaction security in mobile commerce. In: Proc. of the IEEE International Conference on Information Reuse and Integration, pp. 475–480 (2004)
19. Anderson, R.: Hiding Routing Information. In: Anderson, R. (ed.) IH 1996. LNCS, vol. 1174, pp. 137–150. Springer, Heidelberg (1996)
20. Trusted Computing Group, Trusted Platform Module, http://www.trustedcomputinggroup.org/developers/trusted_platform_module
21. Trusted Computing Group, Mobile Phone Work Group Mobile Trusted Module Specification, Version 1.0, http://www.trustedcomputinggroup.org/resources/mobile_phone_work_group_mobile_trusted_module_specification_version_10
22. Homomorphic Encryption, http://ja.wikipedia.org/wiki/

LINK: Location Verification through Immediate Neighbors Knowledge

Manoop Talasila, Reza Curtmola, and Cristian Borcea

Computer Science Department,
New Jersey Institute of Technology, Newark, NJ, USA
{mt57,crix}@njit.edu,borcea@cs.njit.edu

Abstract. In many location-based services, the user location is determined on the mobile device and then shared with the service. For this type of interaction, a major problem is how to prevent service abuse by malicious users who lie about their location. This paper proposes LINK (Location verification through Immediate Neighbors Knowledge), a location authentication protocol in which users help verify each other's location claims. This protocol is independent of the wireless network carrier, and thus works for any third-party service. For each user's location claim, a centralized Location Certification Authority (LCA) receives a number of verification messages from neighbors contacted by the claimer using short-range wireless networking such as Bluetooth. The LCA decides whether the claim is authentic or not based on spatio-temporal correlation between the users, trust scores associated with each user, and historical trends of the trust scores. LINK thwarts attacks from individual malicious claimers or malicious verifiers. Over time, it also detects attacks involving groups of colluding users.

Keywords: Secure location authentication, trust, smart phones.

1 Introduction

Recently, location-based services have started to be decoupled from the wireless network carriers, as illustrated by third-party services such as Loopt, Brightkite, and Google's Latitude. As such, the service providers must rely on the mobile devices to provide their location using GPS or other localization mechanisms. A major problem in this case is how to prevent service abuse by malicious users who tamper with the localization system on the mobile devices. For example, how can a store verify that only users in a 2-mile radius receive coupons? How can a cab company verify the location of a person who requested a cab? How can a news agency authenticate the claimed location of a geo-tagged photo uploaded by citizens located at an event of public interest?

Although a significant number of publications tackled the location authentication problem, all of them assumed support from the network infrastructure [1,2] or from a deployed localization infrastructure using distance-bounding techniques [3,4]. Typically, these solutions are based on signal measurements between the mobile devices and fixed beacons or base stations (e.g., cell towers,

WiFi access points) with known locations [5]. The problem tackled in this paper is different as we aim for a solution that works without any support from the network/localization infrastructure. Such a solution is important because wireless carriers may refuse to authenticate user location for third-party services due to legal and commercial reasons: they may not be allowed by laws to share any type of user location data, and and they may not want to help their competition in the location-based services area.

This paper proposes LINK (Location verification through Immediate Neighbors Knowledge), a secure location authentication protocol in which users help verify each other's location claims. LINK associates trust scores to users, and mobile neighbors with high trust scores play similar roles with the trusted beacons/base stations in existing solutions. The main idea is to leverage the neighborhood information available through short-range wireless technologies, such as Bluetooth which is available on most cell phones, to verify if a user is in the vicinity of other users with high trust scores.

LINK employs a Location Certification Authority (LCA) that interacts with the location-based services and with the mobile users over the Internet. Before submitting a location authentication request to the LCA, the claimers must broadcast a message to their neighbors using short-range wireless ad hoc communication. In response to this message, the neighbors send verification messages to the LCA over the Internet. The LCA decides the claim's authenticity based on spatio-temporal correlation between users and the trust score associated with each user. The protocol leverages the centralized nature of the LCA to compute the trust scores based on past interactions and historical score trends. While it works best in dense networks that provide enough neighbors, LINK was designed to be resilient to situations when users are alone.

Extensive simulation results and security analysis show that LINK can thwart attacks from individual malicious claimers or malicious verifiers. Over time, it can also detect more complex attacks involving groups of colluding users.

The rest of the paper is organized as follows. Section 2 defines the assumptions and the adversarial model. Section 3 describes the LINK protocol, and Section 4 analyzes its security. Section 5 presents the simulation results. The related work is discussed in Section 6, and the paper concludes in Section 7.

2 Preliminaries

This section defines the interacting entities in our environment, the assumptions we make about the system, and the adversarial model.

Interacting entities. The entities in the system are:

- *Claimer*: The mobile user who claims a certain location and subsequently has to prove the claim's authenticity.
- *Verifier*: A mobile user in the vicinity of the claimer (as defined by the transmission range of the wireless interface, which is Bluetooth in our implementation). This user receives a request from the claimer to certify the claimer's location and does so by sending a message to the LCA.

– *Location Certification Authority (LCA)*: A service provided in the Internet that can be contacted by location-based services to authenticate claimers' location. All mobile users who need to authenticate their location are registered with the LCA.
– *Location-based Service (LBS)*: The service that receives the location information from mobile users and provides responses as a function of this location.

System and Adversarial Model. We assume that each mobile device has means to determine its location. This location is considered to be approximate, within typical GPS or other localization systems limits. We assume the LCA is trusted and the communication between mobile users and LCA is secure. We also assume that each user has a pair of public/private keys and a digital certificate from a PKI. Similarly, we assume the LCA can retrieve and verify the certificate of any user. All communication happens over the Internet, except the short-range communication between claimers and verifiers.

We choose Bluetooth for short-range communication in LINK because of its pervasiveness in cell phones and its short transmission range (10m) which provides good accuracy for location verification. However, LINK can leverage WiFi during its initial deployment in order to increase the network density. This solution trades off location accuracy for number of verifiers.

LCA can be a bottleneck and single point of failure in the system. Currently, we do not address this issue, but standard distributed systems techniques can be used to improve the LCA's scalability and fault-tolerance. For example, an individual LCA server/cluster can be assigned to handle a specific geographic region, thus reducing the communication overhead significantly (i.e., communication between LCA servers is only required to access user's data when she travels away from the home region). Additionally, the geographic distribution of servers can improve response latency.

Any claimer or verifier may be malicious. When acting individually, malicious claimers may lie about their location. Malicious verifiers may refuse to cooperate when asked to certify the location of a claimer and may also lie about their own location in order to slander a legitimate claimer. Additionally, malicious users may perform stronger attacks by colluding with each other. A group of colluding malicious users may try to verify each other's false claims (we assume the attackers are able to communicate with each other using out-of-band channels).

We do not consider selfish attacks, in which users seek to reap the benefits of participating in the system without having to expend their own resources (e.g., battery). These attacks can be solved by leveraging the centralized nature of LCA, which can enforce a tit-for-tat mechanism (similar to those found in P2P protocols such as BitTorrent). For example, a user can be informed that she needs to perform a number of verifications for each submitted claim. Finally, we rely on that obtaining digital certificates is not cheap; this deters Sybil attacks [6].

3 Protocol Design

This section presents the basic LINK operation, describes the strategies used by LCA to decide whether to accept or reject a claim, and then details how trust scores and verification history are used to detect strong attacks from malicious users who change their behavior over time or collude with each other.

3.1 Basic Protocol Operation

All mobile users who want to use LINK must register with the LCA. During registration, the LCA generates a userID based on the user's digital certificate. At the same time, the LCA assigns an initial trust score for the user (which can be set to a default value or assigned based on other criteria). Trust scores are maintained and used by the LCA to decide the validity of location claims. A user's trust score is additively increased when her claim is successfully authenticated and multiplicatively decreased otherwise in order to discourage malicious behavior. This policy of updating the scores is demonstrated to work well for the studied attacks, as shown in section 5. The values of all trust score increments, decrements, and thresholds are presented in the same section. A similar trust score updating policy has been shown to be effective in P2P networks as well [7].

Figure 1 illustrates the basic LINK operation. In step 1, a user (the claimer) wants to use the LBS and submits her location. The LBS then asks the claimer to authenticate her location (step 2). In response, the claimer will send a signed message to LCA (step 3) containing *(userID, location, seq-no, serviceID)*. The sequence number is used to protect against replay attacks (to be discussed in Section 4). The LCA timestamps and stores each newly received claim.

The claimer then starts the verification process by broadcasting to its neighbors a location certification request over the short-range wireless interface (step 4). This message is signed and contains *(userID, seq-no)*, with the same sequence number as the claim in step 3. The neighbors who receive the message, acting as verifiers for the claimer, will send a signed certification reply message to LCA

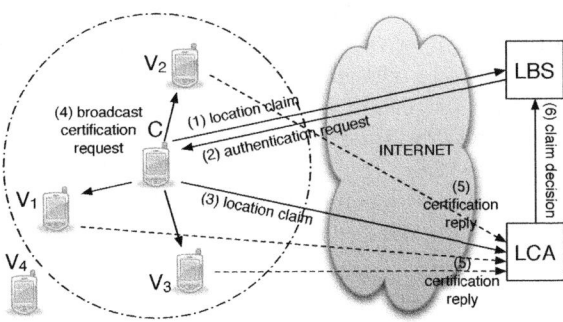

Fig. 1. Basic Protocol Operation (where C = claimer, V_i = verifiers, LBS = Location-Based Service, LCA = Location Certification Authority)

(step 5). This message includes *(userID, location, certify-request)*, where the userID and location are those of the verifier and certify-request is the certification request broadcasted by the claimer. The certification request is included to allow the LCA to match the claim and its certification messages. Additionally, it proves that indeed the certification reply is in response to the claimer's request.

The LCA waits for the certification reply messages for a short period of time and then starts the decision process (described next in section 3.2). Finally, the LCA informs the LBS about its decision (step 6), causing the LBS to provide or deny service to the claimer.

3.2 LCA Decision Process

In the following, we describe the LCA decision process. For the sake of clarity, this description skips most of the details regarding the use of historical data when making decisions, which are presented in Section 3.3.

Spatio-temporal correlation. The LCA checks the claimed location of the claimer with respect to the claimer's previously recorded claim. If it is not physically possible to move between these locations in the time period between the two claims, the new claim is rejected.

Contradictory verifications. If the claimer's location satisfies the spatio-temporal correlation, the LCA selects only the "good" verifiers who responded to the certification request. These verifiers must have trust scores above a certain threshold. We only use "good" verifiers because verifiers with low scores may be malicious. Nevertheless, the low score verifiers respond to certification requests in order to be allowed to submit their own certification claims (i.e., tit-for-tat mechanism) and, thus, potentially improve their trust scores.

After selecting the "good" verifiers, the LCA checks if they are colluding with the claimer to provide false verifications, and it rejects the claim if that is the case. This *collusion check* is described in detail in the next section. If the claimer and verifiers are not colluding, the LCA accepts or rejects the claim based on the difference between the sums of the trust scores of the two sets of verifiers, those who agree with the claimer and those who do not.

Low difference between the two sets of verifiers. If the difference between the trust score sums of two sets of verifiers is low, the LCA does not make a decision yet. It continues by checking the trust score trend of the claimer: if this trend is poor, with a pattern of frequent score increases and decreases, the claimer is deemed malicious and the request rejected. Otherwise, the LCA checks the score trends and potentially the location of the verifiers who disagree with the claimer. If these verifiers are deemed malicious, the claim is accepted. Otherwise, the claim is ignored, which forces the claimer to try another authentication later.

No verifiers. Finally, the LCA deals with the case when no "good" verifiers are found to certify the claim (this includes no verifiers at all). If the claimer's trust score trend is good and her trust score is higher than a certain threshold, the claim is accepted. In this situation, the claimer's trust score is decreased by a small value to protect against malicious claimers who do not broadcast a certification request to their neighbors when they make a claim. Over time, a user must submit claims that are verified by other users; otherwise, all her claims will be rejected.

3.3 Use of Historical Data in LCA Decision

The LCA maintains for each user the following historical data: (1) all values of the user's trust score collected over time, and (2) a list of all users who provided verifications for this user together with a verification count for each of them. These data are used to detect and prevent attacks from malicious users who change their behavior over time or who collude with each other.

Trust score trend verification. The goal of this verification is to analyze the historical trust values for a user and find malicious patterns. This happens typically when there are no good verifiers around a claimer or when the verifiers contradict each other with no clear majority saying to accept or reject the claim.

For example, a malicious user can submit a number of truthful claims to improve her trust score and then submit a malicious claim without broadcasting a certification request to her neighbors. Practically, the user claims to have no neighbors. This type of attack is impossible to detect without verifying the historical trust scores. To prevent such an attack, the LCA counts how many times has a user's trust score been decreased over time. If this number is larger than a certain percentage of the total number of claims issued by that user, the trend is considered malicious.

Colluding users verification. Groups of users may use out-of-band communication to coordinate attacks: For example, they can send location certifying messages to LCA on behalf of each other with agreed-upon locations. To mitigate such attacks, the LCA maintains an NxN matrix M that tracks users certifying each other's claims (N is the total number of users in the system). $M[i][c]$ counts how many times user i has acted as verifier for user c.

For each claim, the LCA uses weighted trust scores for verifiers. The weighted trust score of a verifier v is $W_v = T_v/\log_2(M[i][c])$, where T_v is the actual trust score of v. The more a user certifies another user's claims, the less its certifying information will contribute in the LCA decision. We choose a log function to induce a slower decrease of the trust score as the count increases. Nevertheless, a small group of colluding users can quickly end up with all their weighted scores falling below the threshold for "good" users, thus stopping the attack.

If the group of colluding users is larger, the weighted scores will be above this threshold for a longer time, improving the attack's effectiveness. To protect against this attack, LINK rejects a claim if the following conditions are satisfied for the claimer: (1) the number of claims verified by each potentially colluding user is greater than a significant fraction of the total number of claims issued by the claimer, and (2) the number of potentially colluding users who satisfy the first condition is greater than a significant fraction of the total number of verifiers for the claimer.

Eventually, repeated verifications from the same group of colluding verifiers will be ignored. However, it is possible that repeated verifications are from legitimate verifiers (e.g., close family or a few colleagues at work). If the number of repeated verifiers is small compared to the total number of verifiers for a given claimer, LINK will reset the weights of these verifiers to allow them to have a greater contribution in future verifications for the claimer. The detailed algorithm is presented in the companion technical report [8].

4 Security Analysis

The decision made by the LCA to accept or reject a claim relies on the trust scores of the users involved in this claim (i.e., claimer and verifiers). Thus, from a security perspective, the protocol's goal is to ensure that *over time* the trust score of malicious users will decrease, whereas the score of legitimate users will increase. LINK uses an additive increase and multiplicative decrease scheme to manage trust scores in order to discourage malicious behavior.

There are certain limits to the amount of adversarial presence that LINK can tolerate. For example, LINK cannot deal with an arbitrarily large number of malicious colluding verifiers supporting a malicious claimer because it becomes very difficult to identify the set of colluding users. Similarly, LINK cannot protect against users who accumulate high scores and very rarely issue false claims while pretending to have no neighbors (i.e., the user does not broadcast a certification request). An example of such situation is a "hit and run" attack, when the user does not return to the system after issuing a false claim. Thus, we do not focus on preventing such attacks. Instead, we focus on preventing users that *systematically* exhibit malicious behavior. Up to a certain amount of adversarial presence, our experimental evaluation in Section 5 shows that the protocol is able to decrease over time the scores of users that exhibit malicious behavior consistently and to increase the scores of legitimate users.

All certification requests and replies are digitally signed, thus the attacker cannot forge them, nor can she deny messages signed under her private key. Attackers may attempt simple attacks such as causing the LCA to use the wrong certification replies to verify a location claim. LINK prevents this attack by requiring verifiers to embed the certification request in the certification reply sent to the LCA. This also prevents attackers from arbitrarily creating certification replies that do not correspond to any certification request, as they will be discarded by the LCA.

Another class of attacks claims a location too far from the previously claimed location. In LINK, the LCA prevents these attacks by detecting it is not feasible to travel such a large distance in the amount of time between the claims.

Attackers may try to slander other nodes by intercepting their certification requests and then replaying them at a later time in a different location. However, the LCA is able to detect that it has already processed a certification request (extracted from a certification reply) because each such request contains a sequence number and the LCA maintains a record of the latest sequence number for each user.

We now consider individual malicious claimers that claim a false location. If the claimer follows the protocol and broadcasts the certification request, the

LCA will reject the claim because the claimer's neighbors provide the correct location and prevail over the claimer. However, the claimer may choose not to broadcast the certification request and only contact the LCA. If the attacker has a good trust score, she will get away with a few false claims. The impact of this attack is limited because the attacker trust score is decreased by a small decrement for each such claim, and she will soon end up with a low trust score; consequently, all future claims without verifiers will be rejected.

An individual malicious verifier may slander a legitimate user who claims a correct location. However, in general, the legitimate user has a higher trust score than the malicious user. Moreover, the other (if any) neighbors of the legitimate user will support the claim. The LCA will thus accept the claim.

A group of colluding attackers may try to verify each other's false locations using out-of-band channels to coordinate with each other. LINK deals with this attack by recording the history of verifiers for each claimer and gradually decreasing the contribution of verifiers that repeatedly certify for the same claimer (see Section 3.3). Even if this attack may be successful initially, repeated certifications from the same group of colluding verifiers will eventually be ignored.

Limitations and future work. The thresholds in the protocol are set based on our expectations of normal user behavior. However, they can be modified or even adapted dynamically in the future.

LINK was designed under the assumption that users are not alone very often when sending the location authentication requests. As such, it can lead to significant false positive rates for this type of scenario. Thus, LINK is best applicable to environments in which user density is relatively high.

A potential attack is when a group of colluding verifiers may try to slander a legitimate claimer. As long as at least one malicious verifier is near the legitimate claimer, it can use out-of-band communication to forward the claimer's certification requests and coordinate with the other malicious verifiers to slander the claimer. However, in order to target a specific claimer, the attackers would need to have a physical presence near the claimer. Since it is unlikely that the attackers would have a physical presence near an arbitrarily chosen claimer, we do not consider this attack in the paper.

We implicitly assume that all mobile devices have the same nominal wireless transmission range. One can imagine ways to break this assumption, such as using non-standard wireless interfaces that can listen or transmit at higher distances such as the BlueSniper rifle from DEFCON '04. In this way, a claimer may be able to convince verifiers that she is indeed nearby, while being significantly farther away. Such attacks can be prevented by using a "traditional" secure localization protocol that bounds the distance between a prover and a verifier based on the signal's time of flight [9].

Location privacy could be an issue for verifiers. Potential solutions may include rate limitations (e.g., number of verifications per hour or day), place limitations (e.g., do not participate in verifications in certain places), or even turning LINK off when not needed for claims. However, the tit-for-tat mechanism requires

Table 1. Simulation setup for the LINK protocol

Parameter	Value
Simulation area	100m x 120m
Number of nodes	200
% of malicious users	1, 2, 5, 10, 15
Colluding user group size	4, 6, 8, 10, 12
Bluetooth transmission range	10m
Simulation time	300min
User walking speed	1m/sec
Claim generation rate (uniform)	1/min, 1/2min, 1/4min, 1/8min
Trust score range	0.0 to 1.0
Initial user trust score	0.5
"Good" user trust score threshold	0.3
Low trust score difference threshold	0.2
Trust score increment	0.1
Trust score decrement - common case	0.5
Trust score decrement - no neighbors	0.1

the verifiers to submit verifications in order to be allowed to submit claims. To protect verifier privacy against other mobile users in proximity, the verification messages could be encrypted as well.

5 Performance Analysis

This section presents the evaluation of LINK using the ns-2 simulator. The two main goals of the evaluation are: (1) Measuring the false negative rate (i.e., percentage of accepted malicious claims) and false positive rate (i.e., percentage of denied truthful claims) under various scenarios, and (2) Verifying whether LINK's performance improves over time as expected.

5.1 Simulation Setup

The simulation setup parameters are presented in Table 1. The average number of neighbors per user considering these parameters is slightly higher than 5. Since we are interested to measure LINK's security performance, not its network overhead, we made the following simplifying changes in the simulations. Bluetooth is emulated by WiFi with a transmission range of 10m. This results in faster transmissions as it does not account for Bluetooth discovery and Piconet formation. However, the impact on security is minimal due to the low, walking speeds considered in these experiments. The second simplification is that the communication between the LCA and the users does not have any delay; the same applies for the out-of band communication between colluding users. Finally, a few packets can be lost due to wireless contention because we did not employ reliable communication in our simulation. However, given the low claim rate, their impact is minimal.

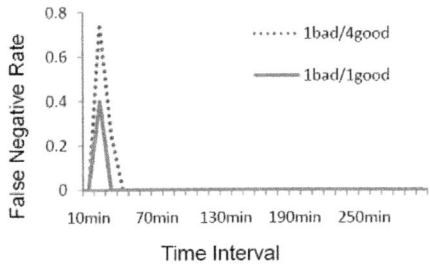
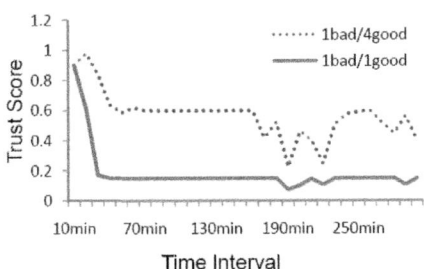

Fig. 2. False negative rate over time for individual malicious claimers with mixed behavior. The claim generation rate is 1 per minute, 15% of the users are malicious, and average speed is 1m/s.

Fig. 3. Trust score of malicious users with mixed behavior over time. The claim generation rate is 1 per minute, 15% of the users are malicious, and average speed is 1m/s.

5.2 Simulation Results

Always malicious individual claimers. In this set of experiments, a certain number of non-colluding malicious users sends only malicious claims; however, they verify correctly for other claims.

If malicious claimers broadcast certifying requests, the false negative rate is always observed to be 0. These claimers are punished and, because of low trust scores, they will not participate in future verifications. For higher numbers of malicious claimers, the observed false positive rate is very low (under 0.1%), but not 0. The reason is that a small number of good users remain without neighbors for several claims and, consequently, their trust score is decreased; similarly, their trust score trend may seem malicious. Thus, their truthful claims are rejected if they have no neighbors. The users can overcome this rare issue if they are made aware that the protocol works best when they have neighbors.

If malicious claimers do not broadcast certifying requests, a few of their claims are accepted initially because it appears that they have no neighbors. If a claimer continues to send this type of claim, her trust score falls below the "good" user threshold and all her future claims without verifiers are rejected. Thus, the false negative rate will become almost 0 over time. The false positive rate remains very low in this case.

Sometimes malicious individual claimers. In this set of experiments, a malicious user attempts to "game" the system by sending not only malicious claims but also truthful claims to improve her trust score. We have evaluated two scenarios: (1) Malicious users sending one truthful claim, followed by one false claim throughout the simulation, (2) Malicious users sending one false claim for every four truthful claims. For the first 10 minutes of the simulation, they send only truthful claims to increase their trust score. Furthermore, these users do not broadcast certifying requests to avoid being proved wrong by others.

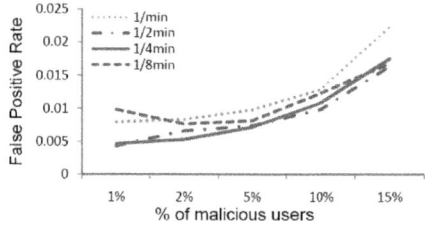

Fig. 4. False positive rate as a function of the percentage of malicious verifiers for different claim generation rates. The average speed is 1m/s.

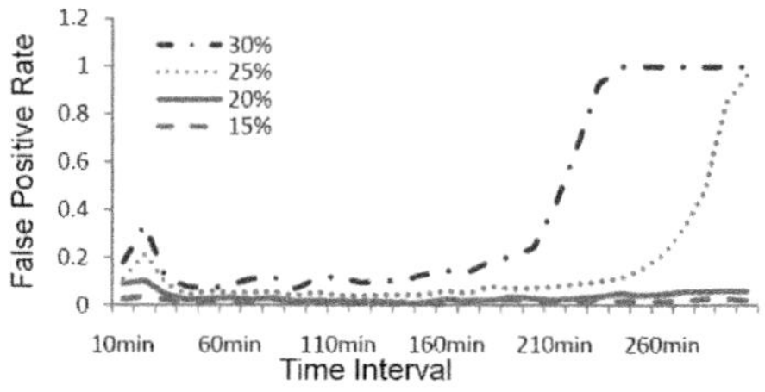

Fig. 5. False positive rate over time for different percentages of malicious verifiers. The claim generation rate is 1 per minute and the average speed is 1m/s.

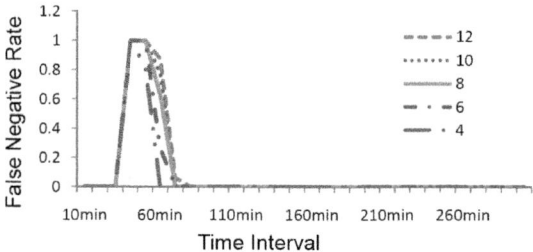

Fig. 6. False negative rate over time for colluding users. Each curve is for a different colluding group size. Only 50% of the colluding users participate in each verification, thus maximizing their chances to remain undetected.

Figure 2 shows that LINK quickly detects these malicious users. Initially, the false claims are accepted because the users claim to have no neighbors and have good trust scores. After a few such claims are accepted, LINK detects the attacks based on the analysis of the trust score trends and punishes the attackers.

Figure 3 illustrates how the average trust score of the malicious users varies over time. For the first type of malicious users, the multiplicative decrease followed by an additive increase cannot bring the score above the "good" user threshold; hence, their claims are rejected even without the trust score trend analysis. However, for the second type of malicious users, the average trust score is typically greater than the "good" user threshold. Nevertheless, they are detected based on the trust score trend analysis.

Always malicious individual verifiers. The goal of this set of experiments is to evaluate LINK's performance when individual malicious verifiers try to slander good claimers. In these experiments, there are only good claimers, but a certain percentage of users will always provide malicious verifications.

From Figure 4, we observe that LINK performs well even for a relatively high number of malicious verifiers, with a false positive rate of at most 2%. The 2% rate happens when a claimer has just one or two neighbors and those neighbors are malicious. However, a claimer can easily address this attack by re-sending a claim from a more populated area to increase the number of verifiers.

Of course, as the number of malicious verifiers increases, LINK can be defeated. Figure 5 shows that once the percentage of malicious users goes above 20%, the false positive rate increases dramatically. This is because the trust score of the slandered users decreases below the threshold and they cannot participate in verifications, which compounds the effect of slandering.

Colluding malicious claimers. This set of experiments evaluates the strongest attack against LINK. Groups of malicious users collude, using out-of-band communication, to verify for each other. Furthermore, colluding users can form arbitrary verification subgroups; in this way, their collusion is more difficult to detect. To achieve high trust score for the colluding users, we consider that they submit truthful claims for the first 30 minutes of the simulation. Then, they submit only malicious claims.

Figure 6 shows that LINK's dynamic mechanism for collusion detection works well for these group sizes (up to 6% of the total nodes collude with each other). After a short period of high false negative rates, the rates decrease sharply and subsequently no false claims are accepted.

6 Related Work

Location authentication for mobile users has been studied extensively so far. To the best of our knowledge, all existing solutions employ trusted network/localization infrastructure [3,4,10,11,12,13,14] to detect malicious users claiming false locations. Most of these solutions use distance bounding techniques, in which a beacon acting as verifier challenges the mobile device and measures the elapsed time until the receipt of its response.

None of these solutions, however, can be directly applicable to scenarios that involve interaction between mobile users and third-party services (i.e., services that do not have direct access to the network/localization infrastructure). The main novelty of LINK comes from employing mobile users (more exactly their mobile devices) to certify the location claimed by other users.

Similar to our work, SMILE [15] and Ensemble [16] use information collected by mobile devices (keys from nearby users or received signal strength – RSS – values) to provide mutual co-location verification for mobile users. However, they do not provide location verification. RSS signatures in conjunction with RSS fingerprinting could be used for location verification, but such solutions do not scale due to the very dense fingerprinting required to achieve good accuracy.

As it is based on trust scores, LINK shares a number of similarities with work on reputation systems for P2P and mobile ad hoc networks. For example, CONFIDANT is a protocol [17] that avoids node misbehavior by establishing trust relationships between nodes based on direct and indirect observations reported by other nodes. The CORE protocol [18] takes a similar approach and uses reputation to enforce node cooperation. In contrast with CONFIDANT, CORE requires reputation values received from indirect observations, thus preventing malicious nodes from wrongfully accusing legitimate nodes.

There are two main differences between this type of solution and LINK. First, LINK cannot monitor indirectly additional user actions (such as packet forwarding or file sharing) to assess the trust. Second, LINK employs the centralized LCA to have a global view of the the entire system. As such, it is able to detect malicious trust score trends and collusion attacks.

7 Conclusions

This paper presented LINK, a protocol for location authentication based on certification among mobile users. LINK can be successfully employed to provide location authentication for location-based services without requiring cooperation from the network/localization infrastructure. The simulation results have demonstrated that several types of attacks, including strong collusion-based attacks, can be quickly detected while maintaining a very low rate of false positives.

Acknowledgment. This research was supported by the National Science Foundation under Grant No. CNS 0831753. Any opinions, findings, and conclusions or recommendations expressed in this material are those of the authors and do not necessarily reflect the views of the National Science Foundation.

References

1. Kindberg, T., Zhang, L., Shankar, N.: Context authentication using constrained channels. In: Proc. of WMCSA 2002, pp. 14–21 (2002)
2. Wullems, C., Pozzobon, O., Kubik, K.: Trust your receiver? Enhancing location security. GPS World 1, 23–30 (2004)
3. Brands, S., Chaum, D.: Distance-bounding protocols. In: Helleseth, T. (ed.) EUROCRYPT 1993. LNCS, vol. 765, pp. 344–359. Springer, Heidelberg (1994)
4. Rasmussen, K., Čapkun, S.: Location privacy of distance bounding protocols. In: Proc. of the 15th ACM Conference on Computer and Communications Security (CCS 2008), pp. 149–160 (October 2008)
5. Vora, A., Nesterenko, M.: Secure location verification using radio broadcast. IEEE Trans. Dependable Secur. Comput. 3(4), 377–385 (2006)
6. Douceur, J.: The Sybil Attack. In: Proc. of IPTPS 2001, pp. 251–260 (2002)
7. Chu, X., Chen, X., Zhao, K., Liu, J.: Reputation and trust management in heterogeneous peer-to-peer networks. Springer Telecommunication Systems 44(3-4), 191–203 (2010)
8. Talasila, M., Curtmola, R., Borcea, C.: LINK: Location-verification through immediate neighbors knowledge. Department of Computer Science, NJIT, Tech. Rep. (2010)
9. Singelee, D., Preneel, B.: Location verification using secure distance bounding protocols. In: Proc. of the 2nd IEEE International Conference on Mobile Ad-hoc and Sensor Systems (MASS 2005), pp. 834–840 (November 2005)
10. Tippenhauer, N., Čapkun, S.: Id-based secure distance bounding and localization. In: Backes, M., Ning, P. (eds.) ESORICS 2009. LNCS, vol. 5789, pp. 621–636. Springer, Heidelberg (2009)

11. Chiang, J.T., Haas, J., Hu, Y.-C.: Secure and precise location verification using distance bounding and simultaneous multilateration. In: Proc. of the 2nd ACM Conference on Wireless Network Security (WiSec 2009), pp. 181–192 (2009)
12. Sastry, N., Shankar, U., Wagner, D.: Secure verification of location claims. In: Proc. of the 2nd ACM Workshop on Wireless Security (Wise 2003), pp. 1–10 (2003)
13. Chandran, N., Goyal, V., Moriarty, R., Ostrovsky, R.: Position based cryptography. In: Halevi, S. (ed.) CRYPTO 2009. LNCS, vol. 5677, pp. 391–407. Springer, Heidelberg (2009)
14. Shmatikov, V., Wang, M.-H.: Secure verification of location claims with simultaneous distance modification. In: Proc. of the 12th Annual Asian Computing Science Conference (Asian 2007), pp. 181–195 (December 2007)
15. Manweiler, J., Scudellari, R., Cox, L.: SMILE: Encounter-based trust for mobile social services. In: Proceedings of the 16th ACM Conference on Computer and Communications Security, pp. 246–255. ACM, New York (2009)
16. Kalamandeen, A., Scannell, A., de Lara, E., Sheth, A., LaMarca, A.: Ensemble: cooperative proximity-based authentication. In: Proc. of MobiSys 2010, pp. 331–344. ACM, New York (2010)
17. Buchegger, S., Boudec, J.-Y.L.: Performance Analysis of the CONFIDANT Protocol. In: Proc. of MobiHoc 2002, pp. 226–236 (2002)
18. Michiardi, P., Molva, R.: CORE: A collaborative reputation mechanism to enforce node cooperation in mobile ad hoc networks. In: Proc. of the IFIP TC6/TC11 6th Joint Working Conference on Communications and Multimedia Security, pp. 107–121 (2002)

Passport/Visa: Authentication and Authorisation Tokens for Ubiquitous Wireless Communications

Abdullah Almuhaideb, Phu Dung Le, and Bala Srinivasan

Faculty of Information Technology, Monash University, Melbourne, Australia
{Abdullah.Almuhaideb,Phu.Dung.Le,Srini}@monash.edu.au

Abstract. Ubiquitous connectivity faces interoperation issues between wireless network providers when authenticating visiting users. This challenge lies in the fact that a foreign network provider does not initially have the authentication credentials of the mobile users. The existing approaches are based on roaming agreement to exchange authentication information between the home network and a foreign network. This paper proposes Passport/Visa approach that consists of two tokens: Passport (authentication token) and Visa (authorisation token), to provide a flexible authentication method for foreign networks to authenticate mobile users. Our approach can be used when there is no roaming agreement between foreign networks and the mobile user's home network. The security analysis indicates that our protocol is resistant to well-known attacks, ant it efficiently ensures the security for both mobile users and network providers. The performance analysis also demonstrated that the proposed protocol will greatly enhance computation, and communication cost.

Keywords: authentication, ubiquitous mobile access, security protocols, roaming agreement, wireless roaming.

1 Introduction

The enhancement of mobile devices and wireless systems provide new opportunities for the next generation of mobile services, such as m-commerce, m-learning and m-government. This fact makes it desirable for m-internet users to be connected everywhere. It is estimated that half the world population pay to use mobile services [1]. When mobile users (MU) move from their home network (HN) domain to a foreign network (FN) domain, efficient cross-domain authentication and access control are necessary for multiple domains roaming [2-3]. There are security concerns from both the MUs and FNs perspectives, as they cannot establish a connection without being authenticated to each other. The traditional solution is to have a roaming agreement between the HN and a FN for verification. However if there is no roaming agreement, MUs cannot be authenticated and served by the FNs.

Problem Statement. A key challenge in such a ubiquitous heterogeneous network environment is authenticating unknown users by FN providers and preventing unauthorised access. This should take place when roaming to administrative domain without a pre-established roaming agreement with a MU's HN domain [4].

Our Approach and Contributions. This paper proposes a novel Passport/Visa authentication tokens as a practical solution to provide the MU with a flexible authentication and service access mechanism in a ubiquitous mobile access environment. One of the main features of the proposed scheme is the lack of having to authenticate a MU, via a trusted identity provider (IdP), every time the MU requests a service from the FN. In other words, it can eliminate re-authentication with the HN after the first successful authentication. Also, our scheme provides an efficient MU energy consumption, as the operation required by MU only involves symmetric cryptography. These features support both limited-resource (low-power) mobile devices and the low-bandwidth mobile communications. Security and performance analysis conducted to evaluate our proposed protocol.

Paper Organization. The rest of this paper is structured as follows. It starts with an overview of the ubiquitous mobile access model and the Passport/Visa approach, where Passport acquisition, Visa acquisition, mobile service provision, Passport and Visa revocation are illustrated (Section 2). This will be followed by a review and comparison of functionality of existing approaches to the problem (Section 3). We then demonstrate the security analysis (Section 4) and present the evaluation of performance in comparison to existing approaches (Section 5). Finally, our conclusion of this paper will be presented (Section 6).

2 The Proposed Solution

2.1 Ubiquitous Mobile Access Authentication Model

To achieve authentication for ubiquitous wireless access environments, there should be more flexible ways to establish trust without relying on roaming agreements. In The proposed model [5], the MUs are able to negotiate directly with potential FNs regarding quality of service, pricing and other billing related features in order to establish service agreement and get the authorization token. IdPs are required to verify the MU's identity and credentials, and IdP can provide this as a service to MUs. Identifying a MU is important for accounting and charging purposes by FN. MU is pre-registered with IdP to get identification token. The IdP role can be played by a trusted entity such as HN. To simplify the example, in this paper the HN will be considered as the IdP in this context. Also, FN providers are able to communicate directly with potential MUs and make trust decision whether or not to provide network service. For the FN provider to trust a MU, HN is used to verify the claimed identity of the MU. Also, Certificate Authority (CA) is engaged to establish a trust with both HN and FN. With the mutual trust, FN provider ensures that the service will get paid and MU ensures that the FN provider is a legitimate and trusted provider.

2.2 Passport/Visa Approach

This approach designed based on the above described model. It can be used when there is no roaming agreement between FNs and the MU's HN. It consists of two tokens: Passport and Visa. The "Passport" is an authentication token issued by the HN to the MU in order to identify and verify MU identity. The Passport in itself does not grant any access, but provides a unique binding between an identifier and the subject. The "Visa" is an authorisation token that granted to a MU via a FN. The Visa token can be used as an access control to ban individual users. In this paper, an improved version of our previous [6] Passport/Visa protocols is introduced. The followings are a set of protocols were developed to achieve the approach objective. Notations are clarified in the following table.

Table 1. Notations used in the protocols

Symbol	Description	Symbol	Description
MU	Mobile user	HN	Home network service provider.
id_A	Identity of an entity A	FN	Foreign network service provider.
CA	Certivicate Autheroity	SC	Smart card issued by HN for MU.
$Visa_B^A$	A visa that issued by A to B.	$Passport_B^A$	A passport that issued by A to B.
$Visa_{No}$	The visa number	$Pass_{No}$	The passport number
$PK_A(x)$	Encrypting a message X using the public key of A	$Sig_A(x)$	Signing a message X using the private key of A
$h(x)$	One-way has function	K_{A-B}	Symmetric Key shared between A and B
$Cert_A$	Certificate issued for A by the CA.	$valid_{A-B}$	Entity A has been validated by B.
T_A	Timestamp generated by an entity A	r_A	A random number generated by entity A
expiry	Passport or visa expiry date.	$VisaReq_{FN}$	Visa Request
RevOke	Revoke request	SerReq	Service Request
data	Consists of all other information such as type of Passport/Visa, type of MU, MU name, MU date of birth, date of issue, place of issue, issuer ID, and issuer name.In the Visa it may include number of access, duration of access, service type, service name, and times of access.		

2.2.1 Passport Acquisition Protocol

This protocol describes the MU registration process with HN (Passport issuer); by completing this protocol MU will receive a Passport. For any network service request from a FN, MU is required to have a Passport that registered with the HN. The registration with the HN takes place offline, and it occurs once. When completed, the HN issues a smart card (SC) to the MU. The SC information is encrypted with the MU's biometric (such as finger print). Every SC consists of three components:

$$SC = <K_{MU-HN}, Passport_{MU}^{HN}, Pass_{No}>$$

Every SC has a unique ID, which is combined with MU's biometric to generate a symmetric master Key. Key master is offline distributed, and stored on the Passport and the MU's SC. The HN's generate the Passport which is signed and encrypted with both the HN's Sig_{HN} and PK_{HN}, then stored in the SC. The HN's $Cert_{HN}$ is included for verification by FN and establishing trust with the HN using CA. The signature can be verified to ensure the integrity of the Passport. The Passport is given as:

$$Passport_{MU}^{HN} = \{Sig_{HN}(Pass_{No}, expiry, id_{HN}, PK_{HN}(id_{MU}, K_{MU-HN}, data)), Cert_{HN}\}$$

2.2.2 Visa and Service Acquisition Protocol

The MU will receive the required Visa from the FN after completing the identification and verification process with the HN successfully. When the MU has his/her Passport (authentication token) in hand, the authentication process can be started with the FN in order to obtain the required Visa. The protocol is demonstrated as follows (Fig. 1):

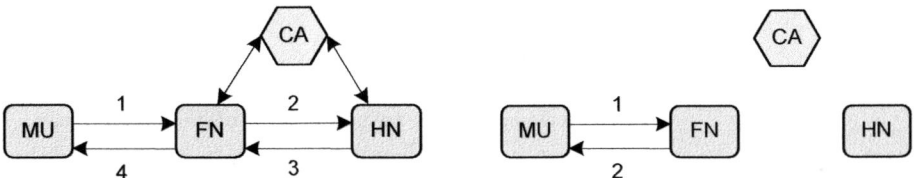

Fig. 1. Visa and service acquisition. **Fig. 2.** Mobile service provision.

Step 1: MU → FN: $VisaReq_{MU-FN}, Passport_{MU}^{HN}, \{id_{FN}, r_{MU}, T_{MU}\}_{SK_{MU-HN}}, T_{MU}, r''_{MU''}$

$$SK_{MU-HN} = h(K_{MU-HN}, id_{MU}, id_{FN}) \quad (1)$$

This protocol starts once the MU sends his/her Visa request, Passport, and $\{id_{FN}, r_{MU}, T_{MU}\}$ where they are encrypted by the session key SK_{MU-HN}. This key is generated using the formula (1) to establish a mutual authentication between the MU and the HN. Every time MU request service from different FN a new session key is generated, three factors are involved: K_{MU-HN}, id_{MU}, and id_{FN} are hashed using $h(x)$. The id_{FN} is used to enable the HN to verify it with the one in the FN certificate to make sure that it has not been modified by an attacker. The r_{MU} is used to authenticate the FN. Another MU's random number $r''_{MU''}$ is sent to the FN to be used as a factor in generating the initial key IK_{MU-FN} based on the formula (2).

Step 2: FN → HN: $Passport_{MU}^{HN}, \{id_{FN}, r_{MU}, T_{MU}\}_{SK_{MU-HN}}, Cert_{FN}, T_{FN}, r_{FN}$

Before processing the authentication with the HN, the FN checks T_{MU} and $Cert_{HN}$ whether it is valid or not, and if so, it forwards the Passport, $\{id_{FN}, r_{MU}, T_{MU}\}_{SK_{MU-HN}}$ with its $Cert_{FN}, T_{FN}$, r_{FN} to the HN as illustrated in Step 2. The $Cert_{FN}$ is sent to the HN for verification and establishing trust using the CA. The r_{FN} used to authenticate the HN. Both T_{MU} and T_{FN} are used to stop reply attacks.

Step 3: HN → FN: $PK_{FN}(Sig_{HN}(Pass_{No}, valid_{MU-HN}, r_{MU}, r_{FN}))$,

$$\{id_{FN}, valid_{FN-HN}, r_{FN}, r_{MU}, T_{HN}\}_{SK_{MU-HN}}$$

After receiving the message from the FN, the HN ensures if T_{MU} and T_{FN} are valid. If one of them is not, the HN replies with un-fresh session and terminates the request. Otherwise, the HN checks the validity of the $Cert_{FN}$ with the CA. If it was valid, the HN decrypts the Passport with its private key and then verifies the signature using the HN's public key. After the HN checks that the MU's Passport is genuine and valid, it gets the shared key (K_{MU-HN}) and its relevant information such as the date of expiry. The HN then generates the session key (SK_{MU-HN}) to decrypt the second part of the message $\{id_{FN}, r_{MU}, T_{MU}\}$. The HN compares the id_{FN} in this message with the one in the certificate to ensure the FN has not been changed. The HN then encrypts $\{id_{FN}, valid_{FN}, r_{FN}, r_{MU}, T_{HN}\}$ with the session key Sk_{MU-HN}.

Also, as the HN authenticate the MU, the HN then computes their digital signatures using its private key, then encrypt them ($Pass_{No}, valid_{MU}, r_{MU}, r_{FN}$) using the FN's public key. The HN then put both the FN and the MU authentication part in one message and sends it to the FN.

Step 4: FN → MU: $Visa_{MU}^{FN}, \{id_{FN}, valid_{FN-HN}, r_{FN}, r_{MU}, T_{HN}\}_{SK_{MU-HN}}$,

$$\{k_{MU-FN}, Visa_{No}\}_{Ik_{MU-FN}}, r''_{FN''}, \{Service\}_{SK_{MU-FN}}$$

Once the FN received the message from the HN, it decrypts its part using its private key and verifies it using the HN's public key. If the FN received the validity of the Passport and checks its random number, the Visa will be generated as follows:

$$Visa_{MU}^{FN} = PK_{FN}(Sig_{FN}(Pass_{No}, Visa_{No}, expiry, data, K_{MU-FN}))$$

The signature of the FN Sig_{FN} in the Visa is used to stop a forged Visa. The Visa is encrypted with the FN's public key, which means that only the FN can decrypt it. The FN stores the Visa information for future verifications. The field "valid" is set to FALSE once a Visa is revoked; otherwise it is set to TRUE.

The following is an example:

$$\{Pass_{No}; Visa_{No}; expiry; valid\}$$

Then the FN generates the initial key using formula (2). The initial key will be used once to distribute the master key K_{MU-FN} and $Visa_{No}$. Also, the FN forwards a new random number $r''_{FN''}$ to be used by the MU to generate the session key. The session key will be used to achieve mutual authentication between MU and FN and to deliver the services. This session key SK_{MU-FN} is generated using formula (3).

$$IK_{MU-FN} = h(Pass_{No}, id_{FN}, r_{MU}, r_{FN}, r''_{MU''}, r''_{FN''}) \qquad (2)$$

$$SK_{MU-FN} = h(K_{MU-FN}, Visa_{No}, Pass_{No}) \qquad (3)$$

After the MU receives the authorisation message from the HN through the FN $\{id_{FN}, valid_{FN}, r_{FN}, r_{MU}, T_{HN}\}_{SK_{MU-HN}}$, the MU decrypts it using the SK_{MU-HN}. The HN's T_{HN}, r_{MU}, and id_{FN} correctness will be checked. If they were incorrect, the Visa will be rejected, and if they were verified, the Visa will be kept for future service requests. The MU computes the Ik_{MU-FN} to get the shared master key K_{MU-FN} and $Visa_{No}$. Finally, The MU computes the SK_{MU-FN} to get the requested services.

2.2.3 Mobile Service Provision Protocol

This protocol (Fig. 2) illustrates how a MU can be granted further network services from a FN in secure manner. When the MU obtains a valid Visa, the MU will be eligible to request further network services from the FN based on the Visa condition.

Step1: MU → FN: $SerReq, Visa_{MU}^{FN}, \{r_{MU''}, Visa_{No}\}_{SK_{MU-FN}}$

To request an access to the FN services, the MU sends $SerReq$, the Visa, and both $r_{MU''}$ and $Visa_{No}$ encrypted by the first session key SK_{MU-FN} (formula 3).

Step2: FN → MU: $\{r_{FN''}, Pass_{No}\}_{SK'_{MU-FN}}, \{Service\}_{SK''_{MU-FN}}$

After the FN receives the service request, it decrypts the Visa with its private key to check its validity by its public key. If the Visa is considered as valid, the FN has to compute the SK_{MU-FN} to verify the $Visa_{No}$, and to get the new $r_{MU''}$. The $r_{MU''}$ will be used to generate the second session key SK'_{MU-FN} as follows (formula 4):

$$SK'_{MU-FN} = h(SK_{MU-FN}, K_{MU-FN}, r_{MU''}) \qquad (4)$$

The third session key will be used by the FN to encrypt its $r_{FN''}$ and $Pass_{No}$. Finally, the third session key will be generated SK''_{MU-FN} using formula (5).

$$SK''_{MU-FN} = h(SK'_{MU-FN}, SK_{MU-FN}, r_{FN''}) \qquad (5)$$

By having the third session key in hand both parties know that mutual authentication has been realized, and the service can be started. However, for the next access the MU is required to generate a new set of session keys.

2.2.4 Passport and Visa Revocation Protocol

This protocol will be used to stop requesting services with a stolen Passport or Visa. If a Passport or Visa is considered to be revoked (e.g., the mobile user's shared keys K_{MU-HN} or K_{MU-FN} expires, or the MU notices the FN revoking a Visa or the HN to revoke a Passport). The Passport revocation can be illustrated as:

$$\text{MU} \to \text{HN}: Passport_{MU}^{HN}, \{Pass_{No}, RevOke\}_{K_{MU-HN}}$$

The protocol starts when the MU sends the RevOke message to the corresponding HN. The HN decrypts the Passport with its private key and verifies the signature with its public key. The HN get the shared key from the Passport and decrypt the second part of the message. The HN checks if $Pass_{No}$ is already stored. If not, it means that there is no Passport issued with this Passport number. If it was stored, it stores the revoked Passport information and updates the status of the Passport as RevOke. The Visa revocation can be illustrated as:

$$\text{MU} \to \text{FN}: Visa_{MU}^{FN}, \{Pass_{No}, Visa_{No}, RevOke\}_{SK_{MU-FN}}$$

When FN receives a RevOke message from MU, the FN decrypts the Visa with its private key and verifies the signature with its public key. The FN gets the shared key from the Passport and generates the session key to decrypt the second part of the message. Then the FN decrypts the message with the session key SK_{MU-FN} (illustrated in (3)). The FN updates the status of the Visa as RevOke. Once a MU requests network services, the FN checks if the Visa was revoked. If it is revoked the service request will be rejected.

3 Related Works and Functionality Comparison

In this section we review a number of related works in the area of ubiquitous mobile access authentication. The review was based on following three key requirements. A flexible ubiquitous mobile access authentication solution should satisfy the following requirements: (A) Wireless Technology Independence: the proposed authentication solution should not be designed for a specific underlying wireless technology. It should be aimed to be designed at the network layer, or higher, of the OSI to avoid the differences in the link and physical layer. (B) Roaming Agreement-less: in the current solutions roaming agreement is used by cellular network to extend its services using other networks. However, it is not likely to set up formal roaming agreements with every possible provider by MU's HN [2, 7]. Therefore, the solution should not depend on roaming agreement between FN providers and the HN. (C) Home Network Independent: The solution should support direct negotiation to establish service agreement between the MU and any FNs, where the FN has full control over the authorisation process. As HN plays the role of an IdP, MUs can get the benefits of the HN partners and more. They could get more network service in areas not covered by the HN's partners with full freedom of choice. The below table summarise the comparison and indicates that our proposed approach can satisfy these requirements while the other related approaches cannot (Table.2).

Table 2. Functionality comparison between the existing approaches and our approach

Approach / Function	A	B	C
Proof-Token [4]	Yes	No	No
SSO architecture [8]	No (WLAN)	No	Yes
Mobile Bazaar [9]	No(Ad Hoc)	Yes	Yes
Homeless mechanism based on tickets [2]	Yes	No (Broker)	No
Sirbu et al.'s scheme (Kerberos-PK)[10]	Yes	No (Broker)	No
Lee et al. 's scheme (Ticket base) [11]	Yes	No (Broker)	No
Our Passport/Visa Approach	Yes	Yes	Yes

Tuladhar et al. [4] have proposed proof tokens authentication architecture and protocol. In their approach, they tried to solve two problems. The first problem is that the limited roaming agreement of the HN with FNs, and they proposed to allow MUs to access the partners of previously visited networks by that MU. The second problem is authentication delays, which they identified as a major cause for high latency. They propose the collaboration between adjacent networks. However, this approach still relies on roaming agreement for authentication, and does not support a direct negotiation and service agreement between the MU and FN.

Matsunaga et al. [8] have proposed a single sign-on (SSO) authentication architecture that confederates WLAN service providers through trusted IdPs. They argue that the dynamic selection of authentication method, and IdP will play a key role in confederating public wireless LAN service providers under different trust levels and with alternative authentication schemes. However, there are three limitations to this approach. The first limitation in this approach is the dependence on roaming agreement between network providers and IdPs, which may limit the MU roaming freedom. The second limitation is the dependency on a single wireless technology. Lastly, it is limited to web-based authentication using cookies [12].

Chakravorty et al. [9] proposed a mobile bazaar (MoB) , an open market architecture for collaborative wide-area wireless services by using reputation management. Their approach is based on short-term transient access network resource reselling by the network's subscribers to other users using an ad hoc network type solution. The limitation of this approach is the dependency on FN's users availability in trading and accessing the network.

The following four works are based on ticket model. Patel and Crowcroft [2] proposed a homeless mechanism based on the notion of tickets. Lei, Quintero and Pierre [13] presented a reusable tickets for accessing mobile services. In [10], Sirbu et al. proposed an extended Kerberos with PKC to improve the scalability and security. While, in [11], Lee et al. proposed a secure scheme for providing anonymous communications in wireless systems using ticket based authentication and payment protocol. The major disadvantage of this model is that a FN does not have a control over granting the authorisation token, as the tickets are approved by the ticket server (TS). The TS acts as a broker, where it requires FNs to have pre-established roaming agreements. The broker concept reduces the issue of one-to-one roaming agreement by having one-to-many service agreement. However, the broker approach will not work in case of there is no service level agreement between the TS and the potential FN. This solution does not support the open market environment as MUs depend on TS to access network providers. While in our solution, IdP does not require pre-established roaming agreement to authenticate their MU, and CA is used to establish trust between both FN and IdP.

4 Security Analysis

The SVO logic [14] has been use to prove the correctness of our protocol. The detailed proof is not included because of the pages limit. In this section, we analyse the security of the proposed protocol with respect to following security requirements:

Proposition 1. *The proposed scheme can prevent Passport/Visa forge.*

Proof. Since the Passport and the Visa contain the signature of the issuer, they cannot be generated by attackers in the name of the HN or FN. So it is impossible to fabricate or fake a Passport or a Visa as the issuer will check the integrity by verifying the signature.

Proposition 2. *The proposed scheme can provide mutual authentication.*

Proof. In the mobile service provision phase, the MU sends a message that consists of two parts: a Visa, and the encrypted new random number $r_{MU''}$. The FN decrypts the Visa with its public key and gets the shared key. Also as the FN signed the Visa, it can check the validation of the Visa. The FN uses the previous session key with $Pass_{No}$ and $Visa_{No}$ to generate the first session key which will be used to decrypt the second part of the message and get a new random number. The shared master key with the first session key, and $r_{MU''}$ will be used to generate the second session key. By decrypting the FN message, the MU can get the FN's random number. Now, both parties are able to generate the third session key and mutual authenticate each other.

Proposition 3. *Our protocol can resist replay and man-in-the-middle attacks.*

Proof. An attacker may sniff a valid Visa, however, the K_{MU-FN}, $Pass_{No}$, and $Visa_{No}$ cannot be obtained as they are encrypted in the Visa. The only party that can get the K_{MU-FN}, $Pass_{No}$ and $Visa_{No}$ from the Visa is the FN. In addition, timestamps are used in each communication between the three entities: MU, FN and HM to ensure the message has not been replayed.

Proposition 4. *The proposed scheme is safe against impersonation attacks.*

Proof. In our protocol, the stored information in SC (e.g. Passport) is encrypted with the MU fingerprint. Thus, when the SC has been stolen, it is infeasible for attackers to impersonate the MU to have an access.

Proposition 5. *The proposed scheme can withstand spoofing.*

Proof. Since a FN cannot get any information regarding to the MU unless the HN authenticates the FN, it is impossible for a malicious entity to masquerade as a legitimate FN to get the MU information. In other word, the MU can ensure that s/he is indeed communicating with a real service provider and not with a bogus entity.

Proposition 6. *The proposed scheme can provide key freshness.*

Proof. Only the MU and the FN know the shared master key K_{MU-FN}. In addition, it is not used to encrypt any message. Instead, a new session key is generated in every service request. This key is established by contributing the random numbers provided by both the MU and the FN. So the key freshness is guaranteed.

Proposition 7. *The proposed scheme can provide privacy and user anonymity.*

Proof. The MU's personal details are kept secretly with the HN. Therefore, when a MU wants to roam into a FN, s/he only needs to send his/her Passport without reveal any information related to his/her ID. Moreover, the HN only returns the $Pass_{NO}$ to the FN if the verification is true. This means that the FN has no idea about the ID of the owner of this Passport.

5 Performance Analysis

In this section, we will evaluate the proposed protocol in terms of computation, and communication cost, by comparing those of the existing schemes in [10] and [11]. Also, scalability analysis will be discussed to demonstrate the key management efficiency.

5.1 Computation Cost

In this subsection, the results of performance comparison of our scheme, the scheme of Lee et al. and the scheme of Sirbu et al. are shown in Figure 3 and Table 3. In the performance comparison, T_{sym} and T_{Asym} are used to denote the computational time of symmetric and asymmetric key cryptography, respectively.

In this performance analysis of the authorisation and service provision phase, the scheme of Sirbu et al. took $16T_{sym}+4T_{Asym}$, while Lee et al. and our schemes took $4T_{sym}+6T_{Asym}$ and $6T_{sym}+8T_{Asym}$, respectively. Obviously, the scheme of Sirbu et al. gains better performance as it requires less asymmetric encryptions/decryptions in this phase. Additionally, in the access service phase, it required $6T_{sym}+2T_{Asym}$ in our scheme, while the other two schemes require re-authentication and repeat the first phase. Our time calculations is based on [15], they indicated a symmetric encryption/decryption requires 0.87ms, and an asymmetric cryptography is approximately equal to 100 symmetric operations. Therefore, an asymmetric

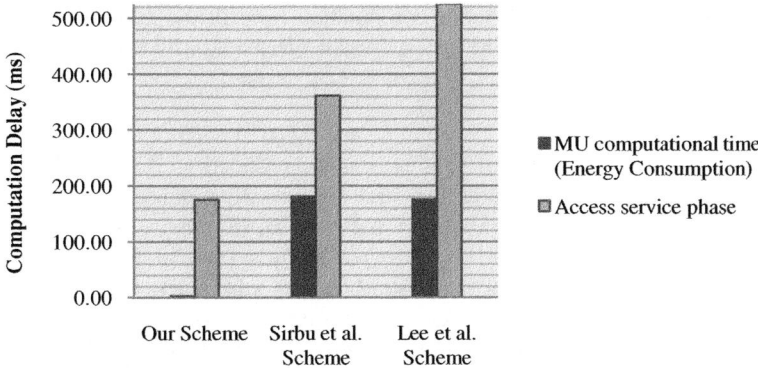

Fig. 3. Computation comparison among different protocols

encryption/decryption computation takes approximately 87ms. The computational costs of the one-way hash function (0.05ms) can be ignored since it is quite lighter, compared to asymmetric and symmetric operations.

Based on the above estimated times, the computational time for the access service phase were 179.22ms, 361.92ms, and 525.48ms in our scheme, Sirbu *et al.*, and Lee *et al.* schemes, respectively. Thus, our scheme is reduced to 49% and 34%, of access service phase computational cost of Sirbu *et al.*, and Lee *et al.* schemes, respectively. Therefore, the proposed scheme is highly efficient in the terms of service provision computational overheads, as shown in Figure 3. The other two schemes take more than the double of our access service phase computational time, as they requires MU to do both authorisation and access service phases (re-authentication) every time for service provision. For example, the scheme of Sirbu *et al.* is based on Kerberos which relies on timestamps for freshness indicator, therefore the time stamped ticket can only be valid for a single session [13].

Also, our scheme took around 2.61 ms ($3T_{sym}$ in both phases), while the scheme of Sirbu *et al.* required 180.96 ms ($8T_{sym}+2T_{Asym}$) and the scheme of Lee *et al.* took 175.74 ms ($2T_{sym} + 2T_{Asym}$). The MUs computational cost of our scheme is reduced to 2% of other schemes. In other words, our scheme outperforms the other two approaches in terms of the MUs computational cost, which affect the energy consumption of their limited power device. The proposed scheme is highly efficient in the terms of MUs computational overheads and energy consumption, shown in Figure 3, because of the elimination of asymmetric cryptosystems.

In terms of authorisation and service provision phase, our scheme is slower than the other two schemes by 2 s to 3 s. Since our scheme took 701.2 ms, while the scheme of Sirbu *et al.* required 361ms and the scheme of Lee *et al.* took 525.48ms. Most of our computation time is spent in the authorisation phase. However, compared with other two schemes, our scheme took less computation time in the access service phase, which will be performed more frequently than authorisation phase. For example, in our scheme MU performed authorisation phase (Visa acquisition) just once, then MU can access services any time based on the Visa expiration date. Table 3 summarizes the performance comparisons of our scheme with the schemes of Sirbu et al., and Lee et al. In summary, our scheme took more computation time in the authorisation phase, but achieves better performance in the access service phase and MU energy consumption.

5.2 Communication Cost

The communication cost in our scheme can be reduced to 33% of the schemes of Sirbu et al., and Lee et al. after the first authorisation phase. Since the proposed scheme can eliminate re-authentication with the HN compare to other schemes. In other words, FNs authenticate MUs with their HN just once in the authorisation phase (takes 4 round messages) to get the Visa, then they can access their services (takes 2 round message) multiple time, based on the Visa type, without the need for HN re-authentication. Most of the communication cost is in the authorisation phase; therefore eliminating the re-authentication will highly improve the performance.

Table 3. Efficiency comparisons between our scheme and other related schemes

Efficiency feature/Approach	Our Scheme	Sirbu et al.'s scheme (Kerberos-PK)[10]	Lee et al.'s scheme (Ticket base)[11]
Computation cost:			
Authorisation & service provision phase	$6T_{sym}+8T_{Asym}$ ≈701.2ms	$16T_{sym}+4T_{Asym}$ ≈361.92ms	$4T_{sym}+6T_{Asym}$ ≈525.48ms
Access service phase	$6T_{sym}+2T_{Asym}$≈179ms	≈$416T_{sym}$=361.92ms	≈$604T_{sym}$=525.48ms
MU computational time (Energy Consumption)	$3T_{sym}$=2.61ms	≈$208T_{sym}$=180.96ms	≈$202T_{sym}$=175.74ms
Communication cost:			
Eliminate HN re-authentication	Yes	No	No
Number of messages	4 then 2	5	6

In term of the total number of messages required in the full protocol, our scheme has better communication cost with 4 round messages, while the schemes of Lee et al. and Sirbu et al. required 5 and 6 round messages, respectively. Table 3 indicates that our proposed protocol can reduce computation and communication cost for the limited resource mobile device compares the other two schemes.

5.3 Scalability Analysis

The proposed scheme is scalable, as the increase of MUs subscription will not affect the HN and the FN storage space. Since both HN and FN do not store MUs shared master-key, which eliminate the maintenance of this key with every MU. Also, the large storage of these keys is eliminated. Moreover, this technique improves the scheme security as the compromise of the key storage in HN or FN will reveal all the symmetric keys to the attacker and will have to be revoked. This problem exist in the traditional Kerberos in the event of a KDC compromise [10]. Instead the master-keys are stored in both the Passport and the Visa to achieve an efficient key management. The Passport and Visa stored only in the MU's SC which provides tamper resistance.

6 Conclusion

This paper argued for the need of a flexible way to authenticate mobile users in ubiquitous wireless access environment. Thus, as a flexible and practical solution, we introduced the Passport/Visa approach as a roaming agreement-less based to enable MUs to authenticate themselves to FN providers via direct negotiation. Moreover, the FNs have full control over the authorisation process. In contrast to the existing approaches, we believe that our approach is more flexible and eliminates the need for roaming agreements. The security and performance analysis indicates that our protocol is secure and efficient to authenticate MUs and network service providers.

References

1. GSM Association: 20 Facts for 20 Years of Mobile Communications (2007), http://www.gsmtwenty.com/20facts.pdf (Date accessed: October 31, 2010)
2. Patel, B., Crowcroft, J.: Ticket based service access for the mobile user. In: MobiCom 1997: Proceedings of the 3rd Annual ACM/IEEE International Conference on Mobile Computing and Networking, pp. 223–233. ACM, NY (1997)
3. Wang, H., Zhang, Y., Cao, J., Varadharajan, V.: Achieving secure and flexible m-services through tickets. IEEE Transactions on Systems, Man, and Cybernetics 33, 697–708 (2003)
4. Tuladhar, S., Caicedo, C., Joshi, J.: Inter-Domain Authentication for Seamless Roaming in Heterogeneous Wireless Networks. In: Proceedings of the IEEE International Conference on Sensor Networks, Ubiquitous, and Trustworthy Computing (sutc 2008), pp. 249–255. IEEE Computer Society, Washington, DC (2008)
5. Almuhaideb, A., Alhabeeb, M., Le, P.D., Srinivasan, B.: Flexible Authentication Technique for Ubiquitous Wireless Communication using Passport and Visa Tokens. Journal of Telecommunications 1, 1–10 (2010)
6. Almuhaideb, A., Alharbi, T., Alhabeeb, M., Le, P.D., Srinivasan, B.: Toward a Ubiquitous Mobile Access Model: A roaming agreement-less approach. In: SNPD 2010: the 11th IEEE/ACIS International Conference on Software Engineering, Artificial Intelligence, Networking and Parallel/Distributed Computing, June 9-11. IEEE Computer Society, London (2010)
7. Shrestha, A., Choi, D., Kwon, G., Han, S.: Kerberos based authentication for inter-domain roaming in wireless heterogeneous network. Computers & Mathematics with Applications (2010)
8. Matsunaga, Y., Merino, A., Suzuki, T., Katz, R.: Secure authentication system for public WLAN roaming. In: Proceedings of the 1st ACM International Workshop on Wireless Mobile Applications and Services on WLAN Hotspots, pp. 113–121. ACM, New York (2003)
9. Chakravorty, R., Agarwal, S., Banerjee, S., Pratt, I.: MoB: a mobile bazaar for wide-area wireless services. In: MobiCom 2005: Proceedings of the 11th Annual International Conference on Mobile Computing and Networking, pp. 228–242. ACM, New York (2005)
10. Sirbu, M.A., Chuang, J.C.I.: Distributed authentication in Kerberos using public keycryptography. In: Proceedings Symposium on Network and Distributed System Security, pp. 134–141 (1997)

11. Lee, B., Kim, T., Kang, S.: Ticket based authentication and payment protocol for mobile telecommunications systems, pp. 218–221 (2001)
12. Shin, M., Ma, J., Arbaugh, W.: The Design of Efficient Internetwork Authentication for Ubiquitous Wireless Communications. Network 3, 1 (2004)
13. Lei, Y., Quintero, A., Pierre, S.: Mobile services access and payment through reusable tickets. Computer Communications (2008)
14. Syverson, P., Cervesato, I.: The logic of authentication protocols. In: Focardi, R., Gorrieri, R. (eds.) FOSAD 2000. LNCS, vol. 2171, pp. 63–137. Springer, Heidelberg (2001)
15. Chen, Y., Chuang, S., Yeh, L., Huang, J.: A practical authentication protocol with anonymity for wireless access networks. Wireless Communications and Mobile Computing (2010)

Virtualization for Load Balancing on IEEE 802.11 Networks

Tibério M. de Oliveira[1], Marcel W.R. da Silva[1],
Kleber V. Cardoso[2], and José Ferreira de Rezende[1]

[1] Universidade Federal do Rio de Janeiro - Rio de Janeiro, RJ, Brazil
{tiberio,marcel,rezende}@gta.ufrj.br
[2] Universidade Federal de Goiás - Goiânia, GO, Brazil
kleber@inf.ufg.br

Abstract. In IEEE 802.11 infrastructure networks composed by multiple APs, before a station can access the network it needs to make a decision about which AP to associate with. Usually, legacy 802.11 stations use no more than the signal strength of the frames received from each AP to support their decision. This can lead to an unbalanced distribution of stations among the APs, causing performance and unfairness problems. This work proposes a new approach that combines the number of associated stations and the current load of each AP plus the virtualization of client wireless interfaces. In this approach, stations frequently switch of association among APs and stay on each one of them for a time interval that is calculated based on the number of associated stations and the channel current load. Simulation results confirm the improvement obtained in the load balancing and fairness on network capacity allocation, while keeping the maximum network utilization.

Keywords: Wireless networks, scheduling, 802.11, association control.

1 Introduction

Nowadays, there is a large number of IEEE 802.11 access points (APs) available in both private and public access networks. Before a client station (STA) can have access to the data transmission service provided by such networks, it has to follow procedures of association and authentication with one of the APs in its transmission range. Initially, the STA detects APs in its vicinity by scanning wireless channels and collecting responses (probe responses and/or beacon frames) from them. Then the STA authenticates and associates with the AP from which it received frames with the highest RSSI (Received Signal Strength Indicator).

As presented in [1], this association metric does not ensure efficiency in the resources usage and may lead to a poor performance due to the unbalanced number of associated stations among the APs. Alternative approaches have been proposed [1,2,3], which perform a load balancing among APs by including load conditions in the frames in order to allow the STA to select the least loaded AP.

However, most of the proposals assume that the current load is given by the number of stations associated with the AP, not taking into account the amount of traffic generated by them.

Many 802.11 manufacturers also provide proprietary solutions for load balancing [16,17,18]. In general, there is not enough information about the algorithms applied in these solutions, so their performance can not be evaluated. Every proprietary solution is based on the AP hardware from that specific manufacturer which allows the use of extra and nonstandard features and protocols. This brings a well-known and undesirable drawback: cross hardware manufacturer incompatibility.

Our proposal takes into consideration the effective throughput achieved by the stations and uses wireless network interface virtualization [4] to perform load balancing. The virtualization scheme allows a single physical interface to offer simultaneous connectivity to more than one AP [5]. Only standards-compliant resources are employed in our solution. Through a simulation study, we show that this proposal outperforms the standard RSSI-based association control and other approach for load balancing known as DLBA (Dynamic Load Balancing Algorithm) [3].

The rest of this paper is organized as follows. In Section 2, we briefly present some important related works. Section 3 exposes some important concepts about load balancing and IEEE 802.11 network virtualization. Section 4 presents our proposal of load balancing through network virtualization. In Section 5, we show the results in comparison with some other approaches. Finally, in Section 6, we present our conclusion.

2 Related Work

Recently, virtualization has become an important tool in several areas, such as operating systems [6], faults detection and diagnosis [7]. In wireless networks, it has been applied in the handover process [8,9] and network/interfaces virtualization [10,11,4,5].

The IEEE 802.11 physical interface virtualization allows an STA to associate with multiple APs simultaneously. This capability can be used with the purpose of allowing the concurrent access to multiple networks or virtually increasing the connectivity in a unique infrastructure network [5]. In this last scenario, it is also required that STAs constantly change their association among the APs in order to announce their presence. Currently, the traditional association approach takes into account only the RSSI measured by STAs from multiple APs. The main drawback of this approach is that it can lead to an unbalanced distribution of STAs among the APs, which can drastically reduce network performance [3,1,9,12].

To circumvent this problem, some metrics for AP association that define a relation between RSSI and the amount of associated stations with an AP were proposed [3,1,2]. In this work, we propose a load balancing mechanism based on IEEE 802.11 physical interface virtualization, which uses a metric derived from the amount of associated STAs and channel load to determine how long an STA will stay associated with each AP. This proposal uses a virtualization scheme called Frequency Hopping (FH) and prioritizes the AP that has the lowest channel occupation.

3 Background

3.1 IEEE 802.11 Network Virtualization

The main virtualization focus is to allow a single substrate to execute multiple virtual experiments. The wireless medium can be virtualized by different techniques as presented in [4]. The most common are: FDMA, TDMA, Combined TDMA and FDMA, CDMA, SDMA, and Frequency Hoping (FH).

The FH scheme is a dynamic version of the Combined TDMA and FDMA scheme. It allows experiment partitioning by allocating a unique sequence of frequency and time slots for each virtual experiment. However, differently from the Combined TDMA and FDMA scheme, it allows that the same experiment uses different sequences. It is the most flexible and complete scheme to be used with IEEE 802.11 networks.

Our mechanism uses the FH scheme combined with an algorithm to choose the sequences, i.e. channels and time slot sizes to be used by each STA. The algorithm employs a metric based on media occupation and the amount of associated STAs with each AP. Our proposal also makes use of IEEE 802.11k and 802.11r standards in following manner. The proposed metric requires some data to be collected by the 802.11k, and the wireless interface virtualization employs 802.11r for fast handover.

3.2 IEEE 802.11k and IEEE 802.11r Standards

The IEEE 802.11k is an amendment to IEEE 802.11-2007 standard for radio resource management, which aims to increase the physical layer and medium access availability. For this purpose, it defines a sequence of measurements requests and reports about radio and network information that can be used by upper layers in different ways. It presents low overhead of control messages and low processing requirements, and provides enough accurate measurements for our proposal.

The radio measurements in wireless networks help applications to adapt automatically to the dynamic medium conditions, facilitating the management and maintenance of a WLAN. The standard specifies a generic pair of *Radio Measurement Request* and *Radio Measurement Report* frames, which can be specialized to acquire measurements such as channel load, STA statistics, number of neighbors, etc.

When an STA switches between APs in a handover process, it starts a reassociation with the new AP. During this process, the IEEE 802.11r implemented in the APs takes care of changing forwarding tables of L2 devices by issuing gratuitous ARP messages on the distribution system. It also allows the involved APs to exchange authentication information, which drastically reduces the time spent in the handover process.

4 Load Balancing by Using Virtual Interfaces

In general, a suitable load balancing provides an efficient use of resources, which in most cases improves fairness in their allocation. Thus, by properly distributing STAs across APs, the aggregate network throughput is increased and the fairness in the network capacity allocation is improved. The DLBA association control algorithm balances the number of STAs associated with each AP by using the RSSI. However, since STAs may present different traffic profiles, this is not enough to proper balance the network load, and hence, it is not suitable for providing throughput or delay fairness.

In this context, we propose a new load balancing mechanism based on the time slot scheduling of a frequency hopping virtualization technique. This mechanism is completely distributed and it is only needed in the STAs, therefore not requiring changes in the APs. Also, its implementation relies on standards (IEEE 802.11r and 802.11k).

In the proposed mechanism, the STAs manage associations with multiple APs using one virtual wireless interface for each AP. An STA stays associated with each AP for a time interval that is dynamically adjusted by the scheduling algorithm, which takes into account measurements of channel occupation and number of STAs associated with each AP. This time interval is always a fraction of a scheduling cycle and is called active time of a virtual interface.

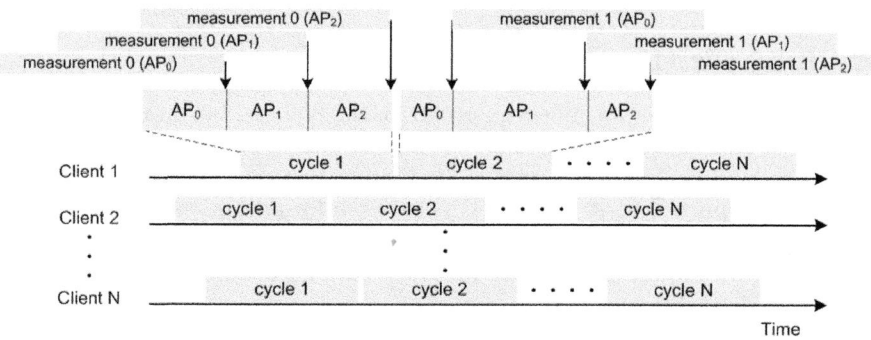

Fig. 1. Scheduling cycle and active times of virtual interfaces

Figure 1 illustrates the scheduling process in which an STA periodically chooses a different AP to associate with. When the APs in the STA's coverage area operate in different channels[1], the scheduling process becomes a client-based FH virtualization scheme.

---

[1] Our mechanism does not depend on an efficient channel allocation, but this is a common assumption in infrastructure networks with centralized control.

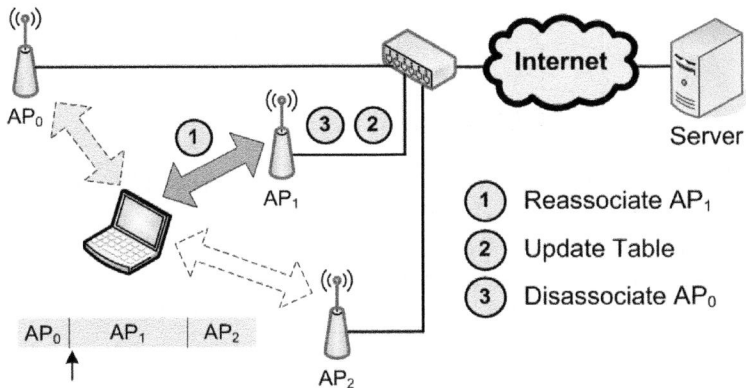

Fig. 2. Fast handover during virtual interface switching

Just before the end of an active time, the STA collects information from the current AP concerning channel occupation and number of associated STAs during the present active time. At the end of every scheduling cycle, the STA uses the collected measurements to compute fractions (or weights) for the calculation of the active time of each virtual interface in the next scheduling cycle. The measurements are performed and made available by the APs using IEEE 802.11k.

In Fig. 1, STA 1 has three virtual interfaces since it is in the coverage area of three APs[2]. To avoid synchronization of active times of different STAs on the same AP, every active time is randomly varied by 10% of its current computed value.

In order to keep the unawareness of the APs concerning the load balancing, our mechanism employs the IEEE 802.11r protocol to implement a fast handover process with authentication. It allows an uninterrupted forwarding of downlink traffic during virtual interface switching. As illustrated in Fig. 2, when an STA switches to a new AP (AP_1), it performs a fast authentication before reassociating, and then initiates an IEEE 802.1X authentication. After that, the IEEE 802.11r 4-way handshake is performed between the STA and the new AP. If this operation is successful, the new AP updates the L2 forwarding tables of network devices of the distribution system (DS). The whole process of migrating from one AP to another takes a non-negligible time. In IEEE 802.11r, this time is around 40 and 50 ms (excluding the scanning time) [13].

4.1 Active Time Computation Algorithm

The performance of the proposed load balancing mechanism is closely tied to the active time duration of each virtual interface. After each cycle, these durations are recomputed to be used in the next cycle. These updates are based on channel load and the number of associated STAs per AP. Initially, the mechanism computes an estimate of the idle time for each STA associated with $AP(i)$, as described by Equation 1:

---

[2] This number may vary from cycle to cycle.

$$\begin{cases} s \in AP(i), & \overline{I_{(i)}}(s) = \dfrac{1 - Ch_{Load}(i)}{NSt(i)}, \\ s \notin AP(i), & \overline{I_{(i)}}(s) = \dfrac{1 - Ch_{Load}(i)}{NSt(i) + 1} \end{cases} \quad (1)$$

where $\overline{I_{(i)}}(s)$ represents the average amount of time the $AP(i)$ has stayed idle per associated STA, $Ch_{Load}(i)$ is the last channel load reported by the $AP(i)$, and $NSt(i)$ is the number of STAs associated with $AP(i)$. When the STA s is not yet associated with $AP(i)$, the equation includes the STA s as if it was already associated with $AP(i)$ in order to artificially account for its contribution to the load.

The sum of all time slots is equal to the duration of a scheduling cycle, and the number of time slots is equal to the number of APs (N_{APs}). Time slot durations are determined by a weight that is assigned to each time slot. Hence, the weights ($W_{(i)}(s)$) computed by an STA s are normalized by the scheduling cycle duration to obtain a percentage of the cycle. This computation is described by Equation 2:

$$W_{(i)}(s) = \dfrac{\overline{I_{(i)}}(s)}{\sum_{k=1}^{N_{APs}} \overline{I_{(k)}}(s)} \quad . \quad (2)$$

Since the active time duration is adaptive, it allows our mechanism to dynamically track the network load. In other words, an STA allocates time slots proportionally to the amount of channel load and number of stations reported by APs. For example, an STA allocates larger time slots to APs that have a lower amount of load per associated STA. Therefore, the proposed virtualization mechanism provides a dynamic load balancing across the APs while performing a fair resource sharing among STAs.

To retain association with an AP, the amount of maintained state is low as it can be see in [5], which implements virtual interfaces at the driver level. Our proposal only adds the active time per AP to the state because this information summarizes the decision of the algorithm. Besides, the number of simultaneous association is not high, because infrastructured networks commonly have a deployment plan that tries to avoid too much superposition among APs coverage. This approach is taken because superposition in excess degrades performance and wastes resources.

5 Results

The ns-2 simulator, with a set of modifications, was used to evaluate the proposed load balancing mechanism. At the network and MAC layers, we added interface queues to handle the traffic of each virtual interface. Figure 3 depicts an example where STA 1 has three virtual interfaces. Each virtual interface is modeled as a transmission queue of a few packets capacity. In this example, the first virtual interface is at its active time, and is associated with AP 1 using channel 1. Modifications were also needed to provide 802.11k measurements

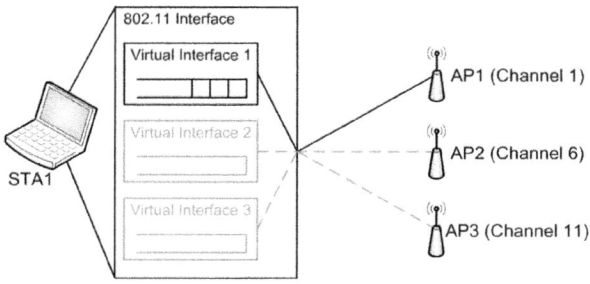

Fig. 3. Virtual interfaces implementation

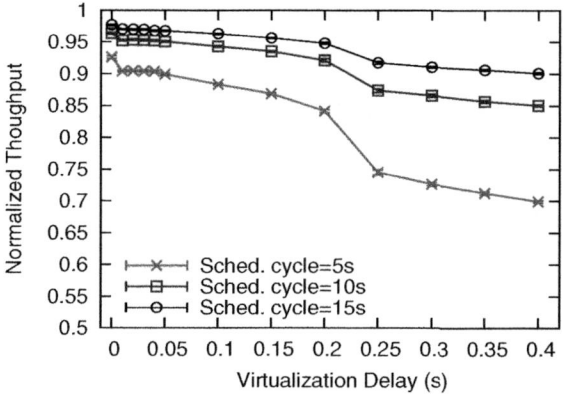

Fig. 4. Virtualization overhead

of the number of associated STAs, channel occupation and RSSI. Moreover, we have implemented the proposed load balancing mechanism and other association control mechanisms to make performance comparisons.

Before evaluating the proposed mechanism in commonly used IEEE 802.11 infrastructure scenarios, we have performed some experiments to assess the overhead imposed by the virtualization. The delay incurred when an STA virtualizes to a new AP, called here of virtualization delay, is due to channel switching, reassociation, authentication and L2 forwarding tables update. During this time, all in transit packets can be lost, degrading flows performance. To do this evaluation, we run simulations with only one virtualized STA in the range of three APs interconnected by a unique L2 switch. This STA performs an FTP download during all the experiment. Figure 4 shows the FTP flow average throughput normalized in respect to the throughput obtained by simulations when no virtualization is used. The normalized average throughput is plotted as a function of the virtualization delay for different scheduling cycle durations. According to [9], typical values for the virtualization delay are not larger than 40 ms. Results show that the throughput decreases by at most 10% in the range of 0 and 40 ms. The smaller is the duration of the schedule cycle, the more affected is the throughput since the STA virtualizes more often. When a channel scanning overhead, which has a typical value of 350 ms, is added to the virtualization delay, the performance degradation is severe. However, our mechanism does not require channel scanning since each STA already knows which APs it is virtualizing with.

To evaluate the performance of the proposed mechanism, the next simulations involve an infrastructure network composed by three APs disposed at the center of a square area, respecting a minimum distance of 120 meters among them. Each AP is configured to use one of the non-overlapping channels (1, 6 and 11).

Two classes of STAs were positioned in the communication range of these three APs: legacy and special STAs. Legacy STAs download Web traffic and use a traditional association method. Special STAs downloads FTP traffic during all simulation. In the simulations, special STAs use one of the following association control methods:

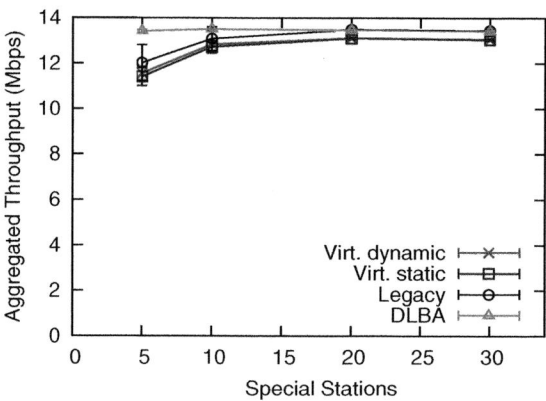
(a) Without legacy stations, only special stations

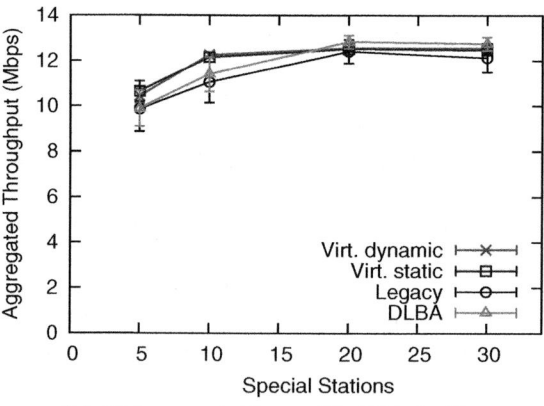
(b) 30 legacy stations with Web traffic

Fig. 5. Aggregated throughput of the FTP flows generated to special STAs as a function of the number of special STAs

- **Legacy**: Traditional association method, i.e. the same used by the legacy stations [14];
- **DLBA**: The DLBA association mechanism [3];
- **Static Virtualization**: The proposed association mechanism with static time slot duration;
- **Dynamic Virtualization**: The proposed association mechanism with dynamic time slot durations.

Each simulation run lasts for 200 s and 30 different STAs positioning scenarios were generated for each configuration set. The results presented are the mean of values for the 30 scenarios with confidence interval bounds at a confidence level of 95%.

Figure 5 presents the aggregate throughput of special STAs in two scenarios, without and with 30 legacy STAs. The aggregate throughput is given by the

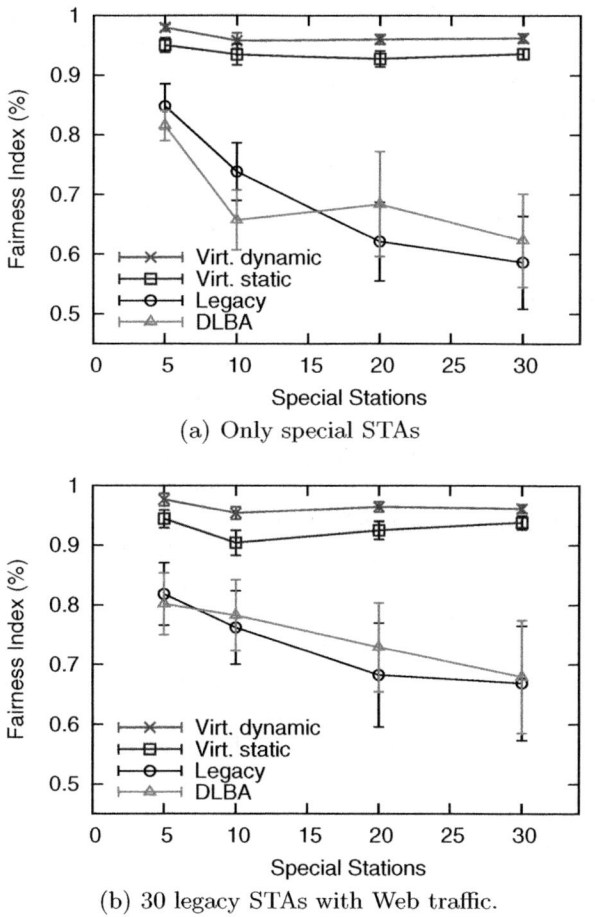

(a) Only special STAs

(b) 30 legacy STAs with Web traffic.

Fig. 6. Fairness index

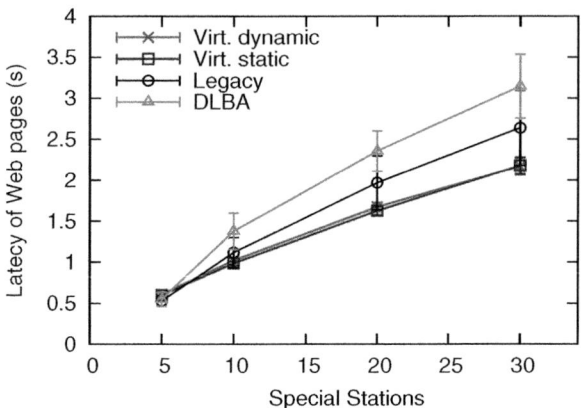

Fig. 7. Mean latency of web pages for legacy STAs

sum of the throughput of all FTP flows. Results show that with a small number of STAs the maximum network capacity is reached in all scenarios. This demonstrates that virtualized STAs are able to use all available capacity even subject to virtualization overhead.

Figure 6 shows the fairness index, as defined in [15], which was calculated over the throughput values obtained by the special STAs. It represents how fair the network capacity is distributed among the STAs. The closer this index is to 1, the fairer is the distribution. According to Fig. 6, the fairness index of virtualized stations is the best among all association algorithms. This demonstrates that our proposal provides a better load balancing, ensuring a fairer sharing of network resources. The difference between static and dynamic virtualized stations is related to the fact that the static algorithm keeps bad associations for a longer time: a large number of STAs may be associated with the same AP simultaneously. On the other hand, the dynamic algorithm is able to adapt to the time duration that the STA is associated with each AP according to the load measurements in each channel. This implies better fairness in resource sharing through load balancing in each AP. DLBA and Legacy association algorithms, which rely on RSSI measurements, present poor fairness and larger variance.

Other important performance metric is presented in Fig. 7. This figure shows the web page average latency obtained by legacy STAs when the amount of special STAs increases. This metric allows evaluating how friendly the association algorithm is to the legacy stations. Once again, DLBA and Legacy STAs present a lower performance with a significant increase in the average latency to visualize web pages. The virtualized methods have a low impact in the average latency of web pages as the amount of special STAs increases.

6 Conclusions

This work has presented a new mechanism that uses IEEE 802.11 network interfaces virtualization as a way to perform load balancing. Through simulations experiments, our proposal has been evaluated and compared with a traditional approach and another load balancing proposal that takes into account only the RSSI. The results show that the virtualization overhead is negligible and the full aggregate capacity of the network can be achieved. However, the most important result of the evaluation is that the proposed mechanism improves the fairness in the network capacity sharing. Considering the throughput experienced by client stations, our proposal provides to all clients equal average throughput. Additionally, our mechanism avoids performance degradation of legacy user communications.

The performance of the two presented virtualization techniques differs only in one aspect: the fairness index results. Slight differences in the fairness index [15] results represent large differences in the throughput distribution experienced by the clients. Thus, as the main objective of the proposed mechanism is to provide better load balancing to the clients, the dynamic virtualization technique is the best choice when compared to the other mechanisms evaluated. The improvements come at a low cost in complexity, since the measurements are simple and based on an IEEE standard. Besides, virtualization of network interface is a well-known mechanism that is becoming available natively in modern operating systems, including the ones that run on mobile devices.

As future works we intend to compare the performance of the dynamic virtualization to other mechanisms in the literature. We also intend to develop a prototype of the proposed virtualization mechanism using off-the-shelf 802.11 network devices supported by the open source MadWifi driver.

References

1. Bejerano, Y., Han, S.-J., Li, L.E.: Fairness and load balancing in wireless LANs using association control. In: MobiCom 2004: Proceedings of the 10th Annual International Conference on Mobile Computing and Networking, pp. 315–329. ACM, New York (2004)
2. Villegas, E.G., Ferre, R.V., Aspas, J.P.: Cooperative load balancing in IEEE 802.11 networks with cell breathing. In: IEEE Symposium on Computers and Communications, ISCC 2008, pp. 1133–1140 (July 2008)
3. Sheu, S.-T., Wu, C.-C.: Dynamic Load Balance Algorithm (DLBA) for IEEE 802.11 wireless LAN. Tamkang Journal of Science and Engineering 2, 45–52 (1999)
4. Paul, S., Seshan, S.: Technical document on wireless virtualization. GENI: Global Environment for Network Innovations, Tech. Rep. (September 2006)
5. Bahl, P., Bahl, P., Chandra, R.: MultiNet: Connection to multiple IEEE 802.11 networks using a single wireless card. Microsoft Research, Redmond, WA, Tech. Rep. (August 2003)
6. Barham, P., Dragovic, B., Fraser, K., Hand, S., Harris, T., Ho, A., Neugebauer, R., Pratt, I., Warfield, A.: Xen and the art of virtualization. In: SOSP 2003: Proceedings of the Nineteenth ACM Symposium on Operating Systems Principles, pp. 164–177. ACM, New York (2003)
7. Adya, A., Bahl, P., Chandra, R., Qiu, L.: Architecture and techniques for diagnosing faults in IEEE 802.11 infrastructure networks. In: MobiCom 2004: Proceedings of the 10th Annual International Conference on Mobile Computing and Networking, pp. 30–44. ACM, New York (2004)

8. Wang, W.-C., Hsu, C.-H., Chen, Y.-M., Chung, T.-Y.: SCTP-based handover for VoIP over IEEE 802.11 WLAN using device virtualization. In: The 9th International Conference on Advanced Communication Technology, vol. 2, pp. 1073–1076 (February 2007)
9. Ramine, I., Savage, S.: SyncScan - Practical fast handoff for 802.11 infrastructure network. In: INFOCOM 2005, March 13-17, vol. 1, pp. 675–684 (2005)
10. Smith, G., Chaturvedi, A., Mishra, A., Banerjee, S.: Wireless virtualization on commodity 802.11 hardware. In: WinTECH 2007: Proceedings of the Second ACM International Workshop on Wireless Network Testbeds, Experimental Evaluation and Characterization, pp. 75–82. ACM, Montreal (2007)
11. Mahindra, R., Bhanage, G.D., Hadjichristofi, G., Seskar, I., Raychaudhuri, D., Zhang, Y.Y.: Space versus time separation for wireless virtualization on an indoor grid. In: Next Generation Internet Networks, NGI 2008, pp. 215–222 (April 2008)
12. Athanasiou, G., Korakis, T., Ercetin, O., Tassiulas, L.: Dynamic cross-layer association in 802.11-based mesh networks. In: 26th IEEE International Conference on Computer Communications, INFOCOM 2007, pp. 2090–2098. IEEE, Los Alamitos (2007)
13. Bangolae, S., Bell, C., Qi, E.: Performance study of fast BSS transition using IEEE 802.11r. In: IWCMC 2006: Proceedings of the 2006 International Conference on Wireless Communications and Mobile Computing, pp. 737–742. ACM, New York (2006)
14. Part 11: Wireless LAN Medium Access Control (MAC) and Physical Layer (PHY) specifications, IEEE Std. 802.11 (August 1999)
15. Jain, R.K., Chiu, D.-M.W., Hawe, W.R.: A Quantitative Measure of Fairness and Discrimination for Resource Allocation in Shared Computer System. Digital Equipament Corporation, Maynard, MA, USA, DEC Research Report TR-301 (September 1984)
16. Cisco, Aggressive Load Balancing on Wireless LAN Controllers (WLCs), http://www.cisco.com/application/pdf/paws/107457/load_balancing_wlc.pdf (last access: July 25, 2010)
17. Aruba, Optimizing Aruba WLANs for Roaming Devices, http://www.arubanetworks.com/pdf/technology/DG_Roaming.pdf (last access: July 25, 2010)
18. Trapeze, Client Load Balancing, http://www.trapezenetworks.com/solutions/client_load_balancing/ (last access: July 25, 2010)

A Packet Error Recovery Scheme for Vertical Handovers Mobility Management Protocols

Pierre-Ugo Tournoux[1,2], Emmanuel Lochin[1,2],
Henrik Petander[3], and Jérôme Lacan[4]

[1] CNRS, LAAS, 7 avenue du colonel Roche, F-31077 Toulouse, France
[2] Université de Toulouse, UPS, INSA, INP, ISAE, LAAS, F-31077 Toulouse, France
[3] NICTA, Australian Technology Park, Eveleigh, NSW, Australia
[4] ISAE, 10 av. Edouard Belin - BP 54032 - 31055 Toulouse Cedex 4, France

Abstract. Mobile devices are connecting to the Internet through an increasingly heterogeneous network environment. This connectivity via multiple types of wireless networks allows the mobile devices to take advantage of the high speed and the low cost of wireless local area networks and the large coverage of wireless wide area networks. In this context, we propose a new handoff framework for switching seamlessly between the different network technologies by taking advantage of the temporary availability of both the old and the new network technology through the use of an "on the fly" erasure coding method. The goal is to demonstrate that our framework, based on a real implementation of such coding scheme, 1) allows the application to achieve higher goodput rate compared to existing bicasting proposals and other erasure coding schemes; 2) is easy to configure and as a result 3) is a perfect candidate to ensure the reliability of vertical handovers mobility management protocols. In this paper, we present the implementation of such framework and show that our proposal allows to maintain the TCP goodput (with a negligible transmission overhead) while providing in a timely manner a full reliability in challenged conditions.

1 Introduction

With the proliferation of new wireless access network technologies, mobile users can now access the Internet using multiple types of access network technologies. This heterogeneous network environment provides access through a varying range of network technologies. The characteristics of these access networks vary greatly; Wireless Local Area Networks (WLANs) provide high speed access with a network latency of tens of milliseconds, often at the price of fixed Internet access but with a very limited coverage. Wireless Wide Area Networks (WWANs) on the other hand provide wide coverage but have a significantly lower data rate, higher latencies up to several hundreds of milliseconds and a cost which may be several magnitudes larger than that of WLAN networks. For obvious cost and performance reasons, smartphone users frequently switch to WLAN when a hotspot is available although the cellular connection is almost always enabled. Therefore the ability to switch seamlessly between these different technologies allows a user to maximize his data rates and an operator to free resources in more expensive WWAN networks by maximizing the utilization of lower cost WLAN networks.

Seamless switching between heterogeneous access networks requires carefully managed vertical (inter-technology) handovers. Protocols, such as Mobile IP (see RFC 3344), can be used to ensure the handover does not break the on-going connections of a mobile node and that the mobile node remains reachable in spite of the handover. In a Mobile IP vertical handoff, on-going traffic is often disrupted due to protocol deficiencies [2]. Although, more advanced handoff protocols such the Safetynet architecture [13] (and Safetynet v.2 [12], where the use of FEC codes are suggested to mitigate the number of lost packets during the bicasting) can be used to reduce these packet losses, the challenging wireless link conditions triggering the handoff may cause unavoidable packet losses. This is especially the case for upward vertical handovers (i.e. handovers from WLAN to WWAN networks) which are typically performed only when the signal to noise ratio of the WLAN becomes too weak to offer a correct connectivity. This decrease of the signal strength may result in packet losses due to wireless errors or even a complete loss of connectivity with the Previous Access Router (PAR) during the time it takes to prepare the WWAN interface and link layer connection to the Next Access Router (NAR).

There are currently two main class of solutions that address the problem from a transport layer point of view. The first one aims to improve TCP tolerance to handover [7,3] while the second one uses multipath SCTP [14] version to benefit from this multiple connectivity capability [5,16]. In the present contribution, we show that the proposed coding scheme shares both advantages and allows the use of any kind of transport protocol without modification. In other words, this proposal is completely independent of the end-to-end transport solution deployed.

In this paper, we present the potential use of this "on the fly" coding scheme through an implementation called Tetrys [15], which can be applied to transport and layer-3 mobility management protocols to achieve a so called "soft handover" and significantly reduce the impact of the handover on the application traffic. A soft handover allows a mobile device to connect to multiple networks at the same time and receive coded streams of traffic from multiple routers or base stations at the same time and to combine those coded, partially redundant streams to a single, complete data stream. This allows the handover process to be very smooth since the ratio of data and the level of coding of the different streams can be dynamically adjusted to handle changing packet loss rates. So far soft handovers have been successfully used only in tightly controlled horizontal handovers in CDMA networks in which the traffic is synchronized between the different base stations and the mobile device. This paper explores whether a similar soft handover can be achieved in handovers between IP based WLAN and WWAN networks with the more challenging asymmetric and non-synchronized network conditions. The purpose of this study is not to propose yet another exhaustive mobility management architecture as this generic coding scheme could

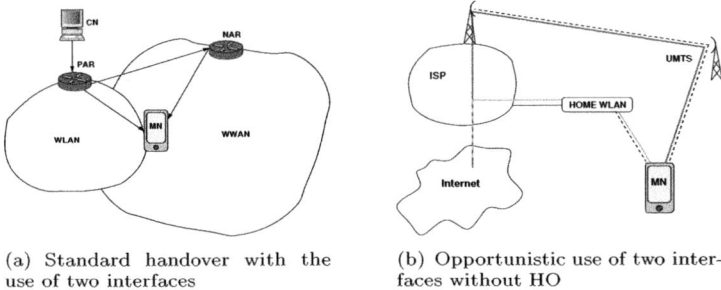

(a) Standard handover with the use of two interfaces

(b) Opportunistic use of two interfaces without HO

Fig. 1. Two illustrations of multipath

be used inside any mobility management protocol (such as Fast Handovers Mobile IPv6 for instance). We rather seek to demonstrate that our adaptive coding scheme can significantly reduce the impact of the challenged network conditions in a vertical handover by using a soft handover like approach. Thus, we evaluate our proposal for vertical soft handovers and the results obtained show that TCP remains close to its pre-handoff bitrate.

2 Background and Related Work

A soft vertical handover differs from a hard vertical handover as no disconnection occurs during the soft handover process. Although both are challenging, they impact the transport and application layers differently and also the methods used to minimize the handover effect at both layers are different.

The main problem in hard handovers, and also failed soft handovers, is the handover delay during which packets are lost. These lost packets cause a number of issues, discussed below. Solutions for reducing this delay, i.e. optimizations for hard handovers, consist of reducing the network detection period, the address configuration interval and the network registration time which reduce (see [2]). The solutions are also valid for reducing the chances of a soft handover turning into a hard handover.

The first and the most pressing issue that a transport protocol, such as TCP, must handle is the loss of its in-flight packets. This handover delay might trigger RTO in TCP and the resulting backoff procedure could lead to a connection stall. This problem can be mitigated by freezing the sending of TCP packets during the handover process [7]. Another challenge in hard and soft handovers is the possible differences in network characteristics between the new and the previous network in terms of propagation delay, bandwidth and packet loss rate (PLR). A plethora of work have been devoted to this problem and have mainly proposed solutions to adapt the TCP congestion windows length or to quickly update the TCP RTO value on the new link (involving active probing on the new link) in order to prevent RTO buffer overflow and/or RTO expiration [7,3].

As the multiple interfaces present on a mobile host can be enabled at the same time, many proposals provide soft handovers between heterogeneous networks.

This can be achieved with IP-level mobility solutions such as MIH, Multihoming MIP (see RFC 4908 [10]) at the cost of a devoted network infrastructure. This provides (see RFC 4980 [11]) improved reliability, load sharing between the different links and bandwidth aggregation.

Several propositions use multihoming to improve the quality of the communication by bicasting the flows via multiple available interfaces. When a handover becomes highly probable, packets are sent both from the PAR and the NAR. This allows a more robust service as it is known that the link signal can significantly decrease during the HO leading to high packet loss rate (PLR) as shown in Fig. 1(a). Instead of copying the same packets on both paths, the authors of [9] choose to send data packets on one path and redundant FEC packets on the other, thus reducing drastically the impact of losses on a video stream compared to the standard bicasting procedure. In [4], the authors obtain a similar result using staggered FEC.

Several modifications to the Stream Control Transmission Protocol (SCTP) have been proposed, enabling support for both real-time and non real-time traffic [5,16] taking benefit of the multihoming capability without the cost of a large network architecture deployment. For instance, the authors in [16] demonstrate that the use of SCTP allowed them to aggregate the bandwidth of different networks, thus providing a video with a better quality and a more robust service even in high mobility scenario.

To illustrate the generic character of our solution, another application where our proposal is of interest is illustrated in Fig. 1(b). In this figure, we represent a mobile node which has subscribed to both ADSL and 3G offer within the same operator and opportunistically uses its home wireless access (or any accessible local wireless spot) conjointly with a 3G access. The benefit of our proposal in this context will be highlighted in Section 4. Our scheme can be deployed both in a end-to-end fashion or from the edge router of the ISP to the terminal host while remaining transparent to the application layer. On the contrary, SCTP needs an end-to-end deployment and the use of specific applications build on top of SCTP socket.

Which Issues Have Still Not Been Addressed?

Adapting TCP to the context of handover would require changing every transport protocol stack already deployed to support these changes. In addition, defining the parameters of these TCP modifications (new values such as RTO and congestion window: *cwnd*, adapted to the new link) require a probing delay or a certain level of a-priori knowledge on the links characteristics which may not always be available. Finally they do not take advantage of the diversity of the links. The main drawback related to the solutions based on multipath SCTP is that they require modifying the interface between the application and the transport layer. Additionally, to date, the deployment of SCTP has been limited to Unix-like hosts. This requirement would make the deployment of these proposals harder. The bi-casting proposals, even when they involve FEC coding, result in halving the available bandwidth which might already be limited in the case of

WWAN interfaces. Further, with the exception of the bi-casting procedure, none of these proposals seem to perform well in challenged conditions with high PLR common before and after handover due to poor signal quality.

To the best of our knowledge, there currently exists no solution allowing to benefit from the robustness and bandwidth aggregation provided by the multihoming capability of new devices with an application on top of the standard TCP/IP protocol suite (i.e. without any modification of the suite).

All these facts motivate our proposal described in the following section 3.

3 Our Proposal

We present in this section the architecture and internal mechanisms that define our framework proposal.

3.1 Architecture, Coding and Handover

Our coding scheme, detailed in the next section, allows recovering in a timely manner from all the losses that occurred on a path, regardless of their distribution. The sole requirement is that in average the amount of redundant packet sent must be greater than the amount of the losses. The key idea is to use our proposal to enforce the part of a path that can be affected by PLR losses during a handover. The whole path can be protected by coding and decoding the packets at both ends of the connection if the Correspondent Node (CN) is aware of the different addresses of the Mobile Node (MN). Otherwise, as Fig. 2 suggests, in the case where multihoming is provided by an IP-level mobility solution, the coding/decoding can also be done between the Access Router (AR) and the MN. We propose to use our coding scheme at layer-3, thus hiding the losses to the transport layer. Fig. 2, describes a possible way to plug our coding scheme through the use of Divert Socket. In our case, we used the BSD implementation of divert socket which is also provided under GNU/Linux with `ipchains` API.

The sending of a data segment from the CN to the MN works as follows: as a first step, packets travels normally through the TCP/IP protocol stack as the CN is not involved in the coding process. Packets that reach the Access router cross

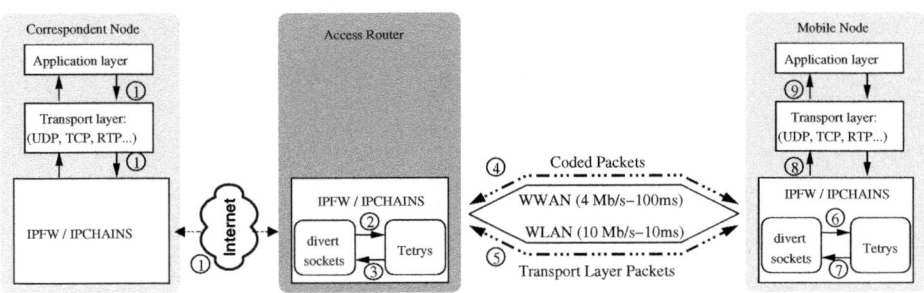

Fig. 2. How to plug our coding scheme in the network protocol stack

the IP forwarding rules (IPFW) which diverts (second step) the packets destined to the MN to the related Tetrys instance (there might be one instance by MN supported by the AR). Tetrys adds the packets to its encoding window and re-injects them (step 3 and 4) adding a packet sequence number (three bytes might be more than necessary) plus a bit to distinguish redundancy from source data packets inside the IP option field. Redundancy (i.e. coded) packets are injected with such an IP destination address that they go through the WWAN (step 4) or WLAN (step 5) links depending on whether it is an upward or downward vertical handover. The size of these coded packets is equal to the maximum size of the data packets currently in the Tetrys encoding window. Packets reach the MN through the different interfaces and are diverted to Tetrys (step 6) which decodes and rebuilds any lost packets. The whole packets received by the Tetrys encoder at step 2 are re-injected (step 7) without losses and ordered in-sequence. Finally at step 8, packets are transmitted to the transport and application layers in a transparent manner.

When our coding scheme is used to improve the link quality (with two interfaces) the source data packets are sent on the fastest (which is also the more lossy one in our experimental scenario) interface while the redundancy (coded) packets are sent through the WWAN.

During upward handover, the PLR is monitored and when it exceeds a given threshold (70% in the experiment), a coded version is sent over the WWAN for each data packet received from the source. When the WLAN is definitely out of range, packets are sent uncoded over the WWAN. During a downward handover, all the source data packets are sent coded over the WWAN and uncoded over the WLAN. When the WLAN PLR decreases below a threshold, the coded packets will be sent according to the redundancy ratio through the WWAN only until the PLR becomes negligible for TCP.

3.2 The Tetrys On-the-Fly Coding Scheme

The Tetrys sender uses an *elastic encoding window* (denoted W_{sender}) which includes all the source packets sent and not yet acknowledged. Let P_i be the source packet with sequence number i. Every k source packets, the sender sends a

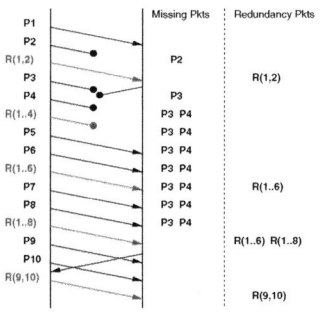

Fig. 3. Tetrys principle

(single) repair packet $R_{(i..j)}$, which is built as a linear combination (with random coefficients) of all the packets currently in W_{sender}. The receiver is expected to periodically acknowledge the received or decoded packets, and each time the sender receives an acknowledgment, the acknowledged packets are removed from W_{sender}. A receiver can decode lost packets as soon as the number of available repair packets is higher or equal to the number of lost packets (the lost packets are detected by the gaps they introduce in the sequence number of the received packets). Fig. 3 illustrates this principle. In the figure $k = 2$, which means that a repair packet is sent each time two source packets have been sent. The right side of this figure shows the list of packets that are lost and not yet rebuilt, as well as the repair packets kept by the receiver in order to recover them. During this data exchange, packet P_2 is lost. However, the repair packet $R_{(1,2)}$ successfully arrives and allows to rebuild P_2. The receiver sends an acknowledgement for packets P_1 and P_2, in order to inform the sender that it can compute the next repair packets from packet P_3. Unfortunately this acknowledgement is lost. However, this loss does not compromise the following transmissions and the sender simply continues to compute repair packets from P_1. After this, we see that P_3, P_4 and $R_{(1..4)}$ packets are also lost. These packets can be rebuilt using $R_{(1..6)}$ and $R_{(1..8)}$ since the number of repair packets becomes higher or equal to the number of lost packets.

The acknowledgement path is only used to optimize the encoding process and is not mandatory during a handover of a few seconds. However, these acknowledgements contain information about the PLR experienced by the MN that might be used to update or tune the redundancy ratio when the handover takes a longer time (we consider higher than 30sec). If the role of the coding scheme is to make the transport layer more robust, the redundancy ratio does not require to be frequently adjusted while coding for real-time application would need more accurate adaptation.

Unlike Tetrys, most of the forward error codes (FEC) used over packet erasure channels are block codes [8]. This means that at the encoder side, a set of repair packets (R) is built from a given set of source data (SD) packets and at the decoder side, these repair packets can only be used to recover SD packets from their corresponding set. If too many packets (among the SD and repair packets) are lost during the transmission, the recovery of the missing SD packets is then not possible.

As a result and compared to block codes:

- Tetrys is tolerant to any burst of source, repair or acknowledgement losses, as long as the amount of redundancy exceeds the PLR;
- the lost packets are recovered within a delay that does not depend on the RTT;
- the configuration is much easier and more robust to network variation than configuration for a block code. This is a key point in the context of handover;

These properties make Tetrys a perfect candidate to reduce packet loss and recovery delay during a handover process.

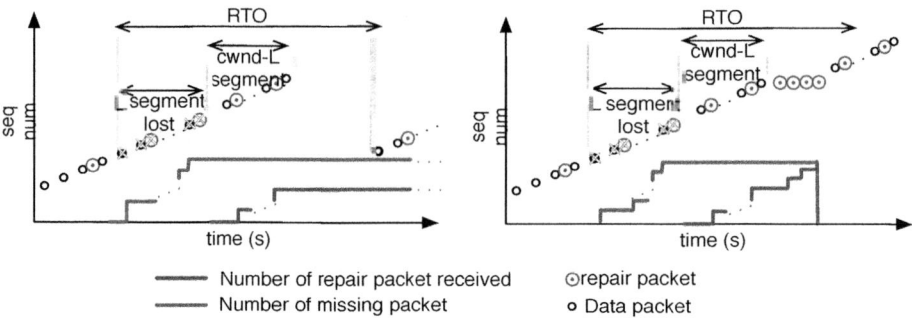

Fig. 4. Mechanism to prevent a blocking TCP window

3.3 Redundancy Emission/Allocation and Interaction with TCP

TCP is well-known for bad performance over lossy links as every lost packet is considered by TCP as an indication of congestion. A possible solution to mitigate this effect would be to use a FEC mechanism to correct losses due to error link. However in [1][1], the authors show that the joint use of end-to-end FEC with TCP does not solve the problem in case of significant PLR. This is due to the fact that TCP needs in-order delivery of data packets and is also strongly sensitive to RTT variations which trigger spurious timeouts resulting in a decreased throughput. A spurious timeout occurs when a non lost packet is retransmitted due to a sudden RTT increase (typically when the mobile node moves from a WLAN to WWAN) which implies an expiration of the retransmission timer set with a previous, and thus outdated, RTT value.

In a previous work [15], we have already shown that this code protects efficiently real-time traffic such as Voice over IP and video-conferencing over links with high PLR. Even if TCP behaves and performs better above Tetrys than above FEC as previously described, rebuilding a burst of L lost packets requires receiving at least L redundant packets. This means that if within a congestion window (and thus a RTT) there were more than $R \cdot cwnd$ lost packets, the decoder may not be able to rebuild the packets, freezing the connection until the emission of at least $\frac{L-R \cdot cwnd}{R}$ TCP retransmission. Needless to say, during our experiment we found that over Tetrys, TCP often entered into backoff mode and the connection stalled. To prevent such phenomenons, we propose to increase the redundancy based on a temporal frequency instead of a fixed ratio. The sole requirement to correct errors timely while minimising extra transmissions is thus to size the frequency so that $\frac{1}{RTT} > f_r \geqslant \frac{L-R \cdot cwnd}{4 \cdot RTT}$, with f_r the minimal frequency for the emission of redundancy packet (assuming the RTO is roughly four times the RTT). As explained in Fig. 4, the sending of 1 repair packet every k source data packets (left subfigure), L data losses would require $k * L$ more data packets to be sent by TCP. As TCP cannot send more than $cwnd - L$ data packets, if $k * L > cwnd - L$, RTOs may be triggered and the

---

[1] In their scheme, TCP is modified to ignore losses. In our case, we assume a complete separation between the coding layer and the transport protocols.

connection may stall. As the subfigure on the right suggests, this problem can be solved by sending repair packets at a minimal rate f_r when the TCP window is abnormally stalled (e.g. when the throughput drops below $k * f_r$).

4 Evaluation

Our testbed architecture is similar to the one presented in figure 2, except for the WLAN and WWAN links which are emulated with Netem [6] on the top of two Ethernet links. The CN and AR are connected with a 10Mbit/s Ethernet link with a negligible transmission delay. Default settings assume 10Mbit/s and an RTT of 10ms for the WLAN and 4Mbit/s with a RTT of 100ms for WWAN (this is realistic for UMTS HSDPA cellular networks). The default redundancy ratio is set to 20%.

4.1 Comparison with FEC over Lossy Links

Tab. 1 shows the throughput obtained by TCP on top of Tetrys or FEC when packets are sent over a single lossy link (over WLAN only). We can see that both Tetrys and FEC can handle a low loss rate efficiently maintaining a throughput

Table 1. Throughput/Std. dev. in Mb/s for $R = 0.2$, $B_D = 10Mbps$, $B_R = 4.0Mbps$ with data and repair packets sent on WLAN only

PLR	0.0	0.5	2	4	5
	8	10	12	16	20
TCP/Tetrys (4,5)	7.78/0.01	7.81/0.01	7.81/0.01	7.80/0.01	7.81/0.00
	7.82/0.01	7.81/0.01	7.81/0.01	7.18/0.02	4.6/0.26
TCP/FEC (4,5)	7.79/0.02	7.83/0.00	6.48/1.2	3.01/0.6	2.68/2.24
			Timeout		
TCP/FEC (8,10)	7.81/0.01	7.78/0.05	7.81/0.03	6.3/1.48	Timeout
			Timeout		
TCP/FEC (12,15)	7.82/0.02	7.82/0.01	7.79/0.03	7.54/0.10	4.06/5.05
			Timeout		
TCP/FEC (16,20)	7.81/0.01	7.81/0.02	7.82/0.01	7.825/0.01	Timeout
			Timeout		

Table 2. Throughput/Std. dev. in Mb/s for $R = 0.2$, $B_D = 10Mbps$, $B_R = 4.0Mbps$ with data sent on WLAN and repair sent on WWAN only

PLR	0.0	0.5	2	4	5
	8	10	12	16	20
TCP/Tetrys (4,5)	9.54/0.00	7.69/0.18	6.5/0.49	8.18/0.34	9.02/0.05
	8.55/0.7	8.66/0.3	8.45/0.57	7.10/0.31	5.12/0.2
TCP/FEC (4,5)	9.53/0.00	5.96/0.02	3.07/0.08	1.13/0.6	1.25/0.14
	0.17/0.05		Timeout		
TCP/FEC (8,10)	9.54/0.00	7.19/0.08	4.72/0.32	2.71/0.46	1.85/0.40
			Timeout		
TCP/FEC (12,15)	9.55/0.00	7.7/0.11	4.79/1.36	3.73/1.13	Timeout
			Timeout		
TCP/FEC (16,20)	9.53/0.00	7.37/0.65	5.83/0.68	3.73/1.13	0.84/1.18
			Timeout		

of 8Mbit/s for the TCP flow. Similarly to previous work on TCP/FEC [1], we observe that with a significant loss rate, the TCP/FEC throughput decreases and the connection often stalls. In contrast to this, TCP over Tetrys is not severely impacted by these loss rates, and in fact the TCP throughput starts to decrease only after $PLR = 14\%$. We have to remark that the code rate is fixed during the experiments. As a matter of fact, there would not a decrease of the TCP throughput by dynamically adjusting the code rate as a function of the PLR.

Tab. 2 shows the throughput obtained by TCP on the top of Tetrys or FEC when data packets are sent over the (10Mbit/s, 10ms) WLAN link lossy link and the coded packets over the (4Mbit/s, 100ms) WWAN link.

We can notice that the results of TCP/FEC are even worse than in the one link only experiment (Tab. 1). We can make the same observation for TCP/Tetrys under small PLR (for 0.5% or 2%). This is explained by the delay asymmetry between the two links and the TCP *cwnd* which fits the bandwidth delay product (BDP) corresponding to the WLAN link. When losses occur, their reconstruction requires to wait for the coded packet that arrives 90ms later. During this time, no packet reaches the TCP receiver and thus there is no acknowledgement sent to the TCP sender that would slide the congestion window. The RTT perceived by TCP increases with the PLR and the *cwnd* also increases until it reaches the BDP corresponding to the slowest link. This explains the poor performance[2] of TCP/Tetrys for small PLR and the improvement observed when the PLR is higher. These two facts: 1) the link delay asymmetry, and 2) the few losses not recovered by FEC, impact the TCP throughput over FEC more significantly.

In spite of our testbed not enabling bandwidth aggregation (as in SCTP), these results show that in contrast to previous coding proposals, Tetrys allows transport protocols such as TCP to remain efficient in spite of the deteriorated link conditions during and around a handover.

4.2 Illustration of the Mechanism during a Handover Scenario

Fig. 5 shows the results for various WWAN bandwidths (300kbit/s, 1Mbit/s and 4Mbit/s) It takes 0.5 second to the WLAN link to change from "up" state to "down" state and the same for the opposite transition. We can see that even if these various parameters have an impact on the TCP throughput, they do not significantly impact the amount of time required by TCP to reach its average throughput with Tetrys. Similar results hold for different values of the WWAN RTT.

Fig. 6 shows the impact of the redundancy ratio ($R = 1/5, R = 1/6, R = 1/8, R = 1/9$) on TCP during a handover. In this case, it takes 10 seconds for the WLAN PLR to switch from 0 to 100% and inversely. The three sub-figures show different runs of the experiment. We can see that the configuration of the Tetrys redundancy ratio does not require to be timely adjusted as there is no impact on the throughput achieved by TCP.

---

[2] This can be solved by artificially delaying the packet at the speed of the slowest interface.

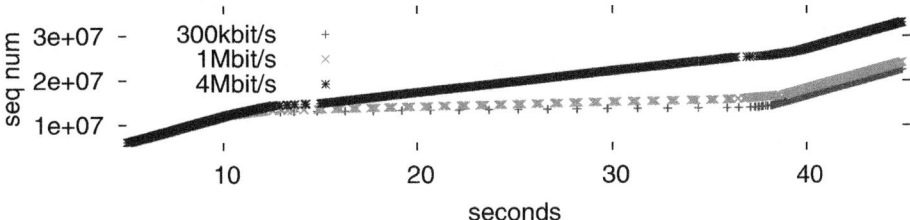

Fig. 5. Handover scenario with various WWAN bandwidth and (10Mb/s, 10ms) WLAN

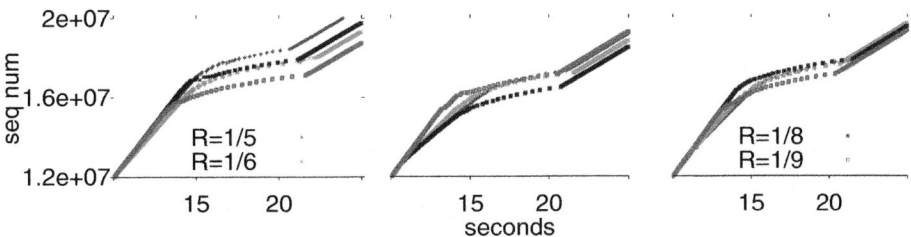

Fig. 6. Handover scenario with various redundancy ratio with (10Mb/s, 10ms) WLAN and (4Mb/s, 100ms) WWAN

Compared to block codes (characterized by a specific FEC coding configuration) where we would have to dynamically reconfigure the redundancy parameters $((k, n))$ as a function of the size of the loss burst, Tetrys is resistant to any kind of loss burst patterns and does not need to be dynamically adjusted (as already highlighted Section 3.2 illustrated and Tab. 1, 2). Furthermore, although the increase of the redundancy parameters allows to correct larger burst of losses, they trigger TCP timeout as the decoding process can be longer than the RTO value.

This last result illustrates that compared to block codes (such as a specific FEC coding configuration), Tetrys is resistant to any kind of loss burst patterns.

5 Conclusion

In this paper, we evaluate the benefits of using an "on the fly" coding scheme to reduce packet losses during a soft vertical handover due to low signal quality. The experimental evaluation suggests that the use of this type of coding scheme may be an interesting complementary strategy to vertical handover management protocols due to its fast configurability and in the context of multipath communications. Our experiments clearly show that this coding scheme allows to maintain the TCP throughput during a handover by taking advantage of the multiple wireless interfaces present in today's smartphones. Particularly, results show that it significantly improves the quality of TCP flows in terms of delivery ratio. As a next step, we are planning to integrate the implementation of this error recovery algorithm called Tetrys as a part of the SafetyNet architecture and evaluate the performance empirically using our SafetyNet implementation.

References

1. Anker, T., Cohen, R., Dolev, D.: Transport layer end-to-end error correcting. Tech. Rep. School of Engineering and Computer Science, Hebrew University (2004)
2. Chakravorty, R., Vidales, P., Subramanian, K., Pratt, I., Crowcroft, J.: Performance issues with vertical handovers - experiences from GPRS cellular and WLAN hot-spots integration. In: IEEE PERCOM (2004)
3. Daniel, L., Kojo, M.: The performance of multiple TCP flows with vertical handoff. In: ACM MobiWAC (2009)
4. Liu, H., et al.: A staggered FEC system for seamless handoff in WLANs: Implementation experience and experimental study. In: IEEE ISM (2007)
5. Fiore, M., Casetti, C.: An adaptive transport protocol for balanced multihoming of real-time traffic. In: IEEE GLOBECOM (2005)
6. Hemminger, S.: Network emulation with netem. In: Australia's National Linux Conference (LCA), Canberra, Australia (2005)
7. Kim, S.-E., Copeland, J.A.: TCP for seamless vertical handoff in hybrid mobile data networks. In: IEEE Globecom (2003)
8. Lin, S., Costello, D.: Error Control Coding: Fundamentals and Applications. Prentice-Hall, Englewood Cliffs (1983)
9. Matsuoka, H., Yoshimura, T., Ohya, T.: A robust method for soft IP handover. IEEE Internet Computing 7, 18–24 (2003)
10. Nagami, K., et al.: Multi-homing for small scale fixed network using mobile IP and NEMO. RFC 4908 (2007)
11. Ng, C., et al.: Analysis of multihoming in network mobility support. RFC 4980 (October 2007)
12. Petander, H., Lochin, E.: Safetynet version 2, a packet error recovery architecture for vertical handoffs. In: ICST MONAMI, Santander, Spain (2010)
13. Petander, H., Perera, E., Seneviratne, A.: Multicasting with selective delivery: a safetynet for vertical handoffs. Wirel. Pers. Commun. 43(3), 945–958 (2007)
14. Stewart, R., et al.: Stream Control Transmission Protocol. RFC 4960 (2007)
15. Tournoux, P.U., Bouabdallah, A., Lacan, J., Lochin, E.: On-the-fly coding for real-time applications. In: ACM Multimedia (2009)
16. Wang, J., Zhu, X.: Latent handover: A flow-oriented progressive handover mechanism. Computer Communications 31(10), 2319–2340 (2008)

A Quantitative Comparison of Communication Paradigms for MANETs

Justin Collins and Rajive Bagrodia

University of California, Los Angeles
Los Angeles, CA
{collins,rajive}@cs.ucla.edu

Abstract. Mobile ad hoc networks (MANET) present a challenging area for application development. The combination of mobile nodes and wireless communication can create highly dynamic networks with frequent disconnections and unpredictable availability. Several language paradigms have been applied to MANETs, but there has been no quantitative comparison of alternative approaches. This paper presents the first quantitative evaluation of three common communication paradigms (publish/subscribe, RPC, and tuple spaces) compared within realistic MANET environments using real applications. We investigate the application-level performance of the paradigms and present a summary of their relative strengths and weaknesses. We also demonstrate the impact of wireless and mobility on application-level metrics, the most dramatic being delivery rates dropping to nearly 25% and round trip times increasing up to 2000% in a mobile scenario.

1 Introduction

In mobile ad hoc networks (MANET), high nodal mobility causes frequent topology and route changes, making it difficult to maintain connections between nodes. Many routes in the network span multiple wireless hops which may experience dramatic and unexpected fluctuations in quality. The combination of mobility and wireless communication creates highly dynamic network topologies in which frequent, possibly permanent disconnections are commonplace. The dynamics of the network and the wireless channel requires changes to the networking stack and alternative solutions at the application level.

Middleware, frameworks, libraries, and languages have been proposed for meeting the challenges of developing mobile and ubiquitous applications. Quantitatively comparing these projects is difficult, since they are implemented in different languages, with different feature sets, and with varying levels of completeness. In this paper, we examine the fundamental differences in communication paradigms commonly used in projects, rather than the projects themselves. We accomplish this by evaluating representative implementations of each paradigm.

We also use real applications utilizing each of the paradigms, allowing us to investigate the impact of the paradigms on application-level metrics. Of prime importance to MANET applications is the performance of the communication model when the wireless channel and mobility are introduced. Therefore, we have evaluated the communication paradigms using a high-fidelity emulation of the network stack and detailed models of the wireless channel.

Since these communication models are common across multiple projects and are likely to be used in future projects, the results of this study have wide applicability. While previous work has qualitatively compared a subset of these paradigms or reported experimental results at the individual project level, we have investigated the performance characteristics of the underlying communication models themselves.

In this paper, we present the first quantitative comparison of three communication paradigms - publish/subscribe [1], remote procedure calls [2], and tuple spaces [3] - using canonical implementations within real applications. Our results show wide variation in paradigm performance within the same scenario. Publish/subscribe provides fast, cheap message delivery, with a message overhead of 357 bytes and median round trip times of $<400ms$ even with mobility. RPC supplies good delivery ratios when a reply is expected, achieving a 94% delivery ratio with mobility, while publish/subscribe and tuple spaces dropped to 75% and 72%. In the whiteboard application, however, tuple spaces were able to deliver all messages in 5 out of 6 scenarios, including a scenario in which RPC only delivered 25%.

2 Evaluation Architecture

The architecture used for this comparison has three layers: a network emulator, which provides a scalable and realistic MANET environment; the communication components, which implement the paradigms; and the applications which utilize the paradigms. The application is built on top of the communication component, which communicates over a regular wired network. The traffic from the network is routed through the network emulator, which provides a high fidelity simulation of the wireless network, intermediate nodes, and mobility. This allows applications to be written independently of whether the network is real or emulated.

Network Emulation. Because application-level performance is affected by variations in the wireless channel, mobility, and disconnection patterns, it is essential to have an accurate representation of the network stack and the wireless channel [4,5]. EXata provides a high fidelity emulation of the entire network stack, using real MAC and routing protocols, as well as detailed simulation of the effects of the wireless channel and mobility, such as fading, shadowing, and path loss [6]. The emulator allowed us to run actual applications, rather than models, in a realistic representation of the MANET environment while retaining precise control of the variables in each experiment. This ensures our results are a fair comparison and not influenced by transient environmental effects.

Communication Components. Each of the three communication paradigms are implemented in Java on top of Apache MINA[1], a high-performance net-

---
[1] Multipurpose Infrastructure for Network Applications: http://mina.apache.org/

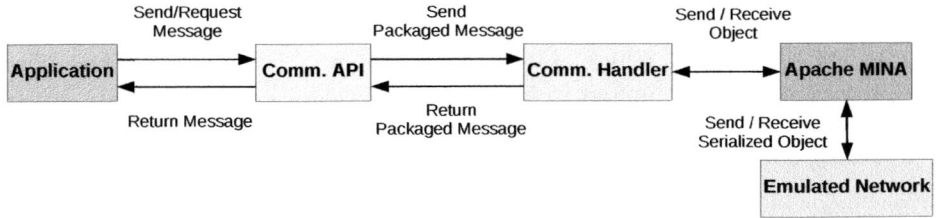

Fig. 1. Communication Components

working library. A summary of the interaction between the application, the communication components, and the networking library is illustrated in Figure 1. The libraries are intended to be functionally equivalent implementations of each paradigm to keep the comparison as fair as possible.

For publish/subscribe, we implemented a simple topic-based system. Publications to topics are broadcast to all subscribers available at time of publication. Our RPC implementation uses a reflection-based mechanism for invoking methods on remote objects, which are addressed by class. Parameters are copied to the remote machine and there is always a return value. The tuple space implementation we used is largely modeled on LIME [7] and uses the same local tuple space library. All three implementations support unicast, multicast, synchronous, and asynchronous communication.

Applications. The first application used for these experiments is a simple client-server application which can send messages between hosts. This provides a baseline for the performance results and allows us to easily test performance with varying message sizes and frequency. The second application is a shared whiteboard. Collaborative applications are often cited as use cases for MANETs and the shared whiteboard is a common example [8–11]. This provides a non-trivial, realistic test case for each of the three communication paradigms.

3 Experimental Results

In the following sections, we present measurements of message delay and message delivery reliability for unicast and group communication, as well as for a non-trivial whiteboard application. We also examine the message overhead and the influence of routing algorithms. These experiments demonstrate the impact of the wireless network and mobility at the application level.

We compared application-level metrics using unicast and group communication in three network scenarios which are used throughout the experiments: a single hop, static network; a multi-hop, static network; and a fully mobile network. Each node in the emulated network is equipped with an 802.11b wireless interface. The two-ray model is used for path loss. Based on preliminary results, we used DSR [12] as the routing protocol for the static scenarios and AODV [13] for the mobile scenario.

The mobile scenario uses random waypoint mobility with a pause time of 30s and maximum speed of 1 meter/second, representing pedestrians carrying handheld devices. The nodes move within a $1500m$ x $1500m$ indoor space where transmission range is limited to 50m. To avoid network segmentation, the scenario ensures there are always possible routes between any two nodes by having four fixed nodes. The mobility pattern in each experiment is identical.

3.1 Unicast Communication

Message Overhead

Application Overhead. The first step of our experimental evaluation of these three paradigms is discovering the basic cost of communication. Table 1 provides an overview of the sequence of messages involved when using each of the communication paradigms in the simple case of a single sender and a single receiver sending a 1KB payload. The total size includes the 1KB payload. Publish/subscribe requires only two messages to be sent: one to subscribe to a topic and one to publish. Since publish/subscribe only needs to add a string indicating a topic, there is very little overhead added to the original message.

RPC first sends out a query to find the desired remote object. Once found, it sends a second message to invoke the method and transfer any arguments. The final message in the sequence is the return value from the method, which is dependent on the size of the return value.

Tuple spaces require the same number of messages as RPC, but the overhead is 2.3 times higher. Except for the search reply messages, all messages include a tuple object, making them larger than the simple messages exchanged in RPC.

Table 1. Message Sequence Overview

Paradigm	Sender	Receiver	Size (bytes)	Overhead (bytes)
Publish/Subscribe				
		Subscribe	175	
	Publish		1182	
		Total	1357	357
RPC				
	Search		146	
		Search Reply	187	
	Invoke		1238	
		Return Value	152	
		Total	1571	571
Tuple Space				
		Search	608	
	Search Reply		133	
		Tuple Request	588	
	Tuple Reply		1586	
		Total	2915	1915

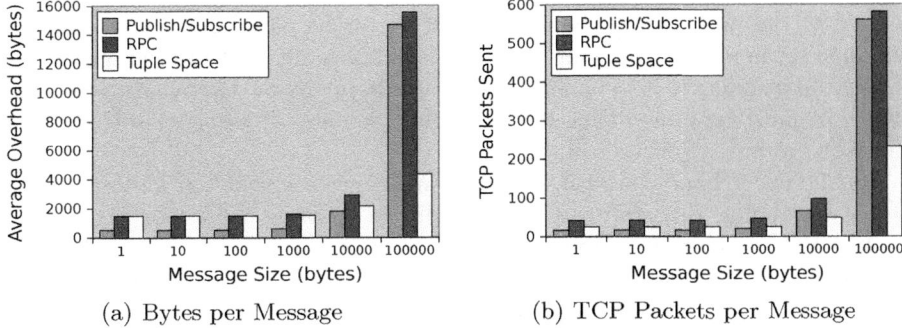

(a) Bytes per Message (b) TCP Packets per Message

Fig. 2. Average Message Overhead

Network Overhead. While Table 1 indicated the overhead added at the application layer, Figure 2(a) shows the average amount of TCP traffic which is sent over the network for a single message, calculated as *bytes sent - message size*. These results use the single hop static scenario and are averaged from 50 messages.

The results are fairly constant until the packet size is exceeded. There is some increase at 10KB, and a dramatic increase at 100KB. Figure 2(b) shows the same data in terms of TCP packets and indicates the cause of the sharp increase in traffic at 100KB is the result of packet fragmentation.

Despite having large message sizes, tuple spaces have much lower overhead in terms of TCP traffic. This difference arises from a side issue related to TCP send window sizes. For tuple spaces, where the receiver initiates the connection, the TCP send window size grows to accommodate larger packet sizes. With RPC and publish/subscribe, the send window size remains constant, causing the large messages to be split into many more packets.

Message Reliability. How reliably a communication paradigm handles message delivery has a direct impact on the application layer. The more reliable the communication paradigm, the less responsible the application is for handling lost messages. We measured reliability in terms of message delivery. In the single hop scenario, all paradigms achieved 100% delivery and figures 3(b) and 3(b) indicate nearly perfect message delivery for all the paradigms in the unicast scenario. Publish/subscribe performed the worst and still only lost 4 messages.

Message Delay. Message delay is another important application-level metric, as it determines how quickly information is transferred and the freshness of the application's information. Figures 4(a), 4(c), and 4(e) show delay in terms of round trip times for each paradigm in a single hop scenario. The majority of the messages in each paradigm are under the $200ms$ mark, with just a few wayward messages taking longer. Even for tuple spaces, 80% of the messages take less than $400ms$ to complete their round trip. However, some messages take much longer, up to $8s$. For tuple spaces, this is partially due to the complexity and overhead of the messages required to perform the round trip message delivery.

However, the time delay for tuple spaces in the single hop scenario is also related to the pull (rather than push) nature of the paradigm. A tuple is timestamped when it is output, but the tuple is not actually sent to the receiver until the receiver requests it. The same situation happens on the return trip, when the tuple must be pulled back to the original sender. Any delays in this process cause the round trip time to increase.

On the other hand, publish/subscribe messages are sent out almost immediately after being timestamped. Nearly all the delay is caused by the network itself. RPC has more potential for delays since it must find the remote method before invoking it. However, the return message can reuse the existing TCP connection, which appears to provide an advantage over tuple spaces.

The mobile scenario introduces even greater delays. Routes are changing frequently and may be several hops long. While the publish/subscribe and RPC results are clustered around $100\,ms$ and remain under $500\,ms$, the tuple space values are considerably higher with a median at $256\,ms$ and a high of nearly $20\,s$. This is again due to the pull nature of tuple spaces and the overhead seen in Section 3.1.

3.2 Group Communication

Group or multicast communication is a useful but more complex part of MANETs, where information and resources are often disseminated in a peer-to-peer manner. Group communication differs significantly from unicast communication. Given the mobile characteristics and decentralized nature of MANETs, a group's membership may be in constant flux, so it is unlikely a sender has perfect knowledge of the members of the group. The time difference between replies from members of the group may vary greatly, and the initiating node cannot know how many replies to expect.

We have investigated how well each paradigm handled group communication by again evaluating message delay and message delivery reliability, but with multiple receivers.

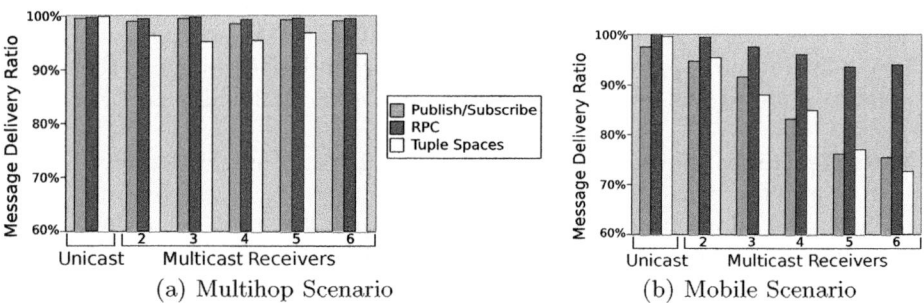

(a) Multihop Scenario   (b) Mobile Scenario

Fig. 3. Message Reliability

Message Reliability. With this application, message reliability refers to messages which make the circuit from the sender to the receiver and back to the sender. This is useful, for example, in situations where a sink node aggregates information from other nodes.

Figures 3(a) and 3(b) show the percentage of messages successfully completing the round trip. The single hop scenario is not shown, as all paradigms achieved $> 99\%$ reliability in that scenario. In the multi-hop scenario, there are more losses even without mobility, but there is no significant trend as the number of receivers increases.

RPC has a slight advantage with this metric, as it will wait until at least one receiver is available. Publish/subscribe and tuple space will send out messages whether or not any receivers are available at the time. However, none of the communication paradigms will retry a message which is lost in transit. A message lost anywhere in the circuit causes the entire attempt to be reported as a failure.

This contributes to tuple spaces showing the lowest delivery ratio (93%) in the multi-hop scenario and a low delivery ratio (72.6%) in the mobile scenario. While tuple spaces can easily handle the delivery of the outgoing tuple, it is more difficult to guarantee the return of the reply tuple. If a node is not available to receive the request broadcast for a reply tuple, then the reply will never be sent even if the original outgoing tuple is received.

Message Delay. We again consider round trip time for each of the paradigms, but this time with an increasing number of receivers. Figures 4(a) - 4(i) show the results for each paradigm and scenario.

For the single hop and multi-hop scenarios, where there is no mobility, the majority of the round trip times are fairly fast. The bottom 75% of the messages have very similar results, while the top 25% varies much more. This indicates that an application can expect most messages to be delivered quickly or not at all, but about a quarter of the messages may arrive up to minutes later.

The median delay does increase as receivers are added, especially in the mobile scenario. In the static scenario, the median delay publish/subscribe increased $121\,ms$ from two receivers to six receivers. RPC increased $147\,ms$, and tuple spaces increased $140\,ms$. For the mobile scenario, the median times for publish/subscribe increased $255\,ms$, RPC increased $237\,ms$, and tuple spaces increased by $2035\,ms$. The maximum delay values varied much less predictably. For tuple spaces, the static scenarios have unusually long delays with two and three receivers. In the static scenarios, the first three receivers are located in close proximity. One node would dominate the channel for several seconds before relinquishing it. Once again, this shows how influential the wireless channel is on the performance and behavior of applications in MANETs.

The median and maximum tuple space results are much longer than the other two paradigms. The median delay for tuple spaces ranges from twice as much as publish/subscribe in the single hop scenario up to 6 times as high in the mobile scenario. For publish/subscribe and RPC, the majority of delays can only be caused by the network, since they do not attempt to retransmit messages. Tuple spaces, on the other hand, can have very large delays due to the paradigm itself.

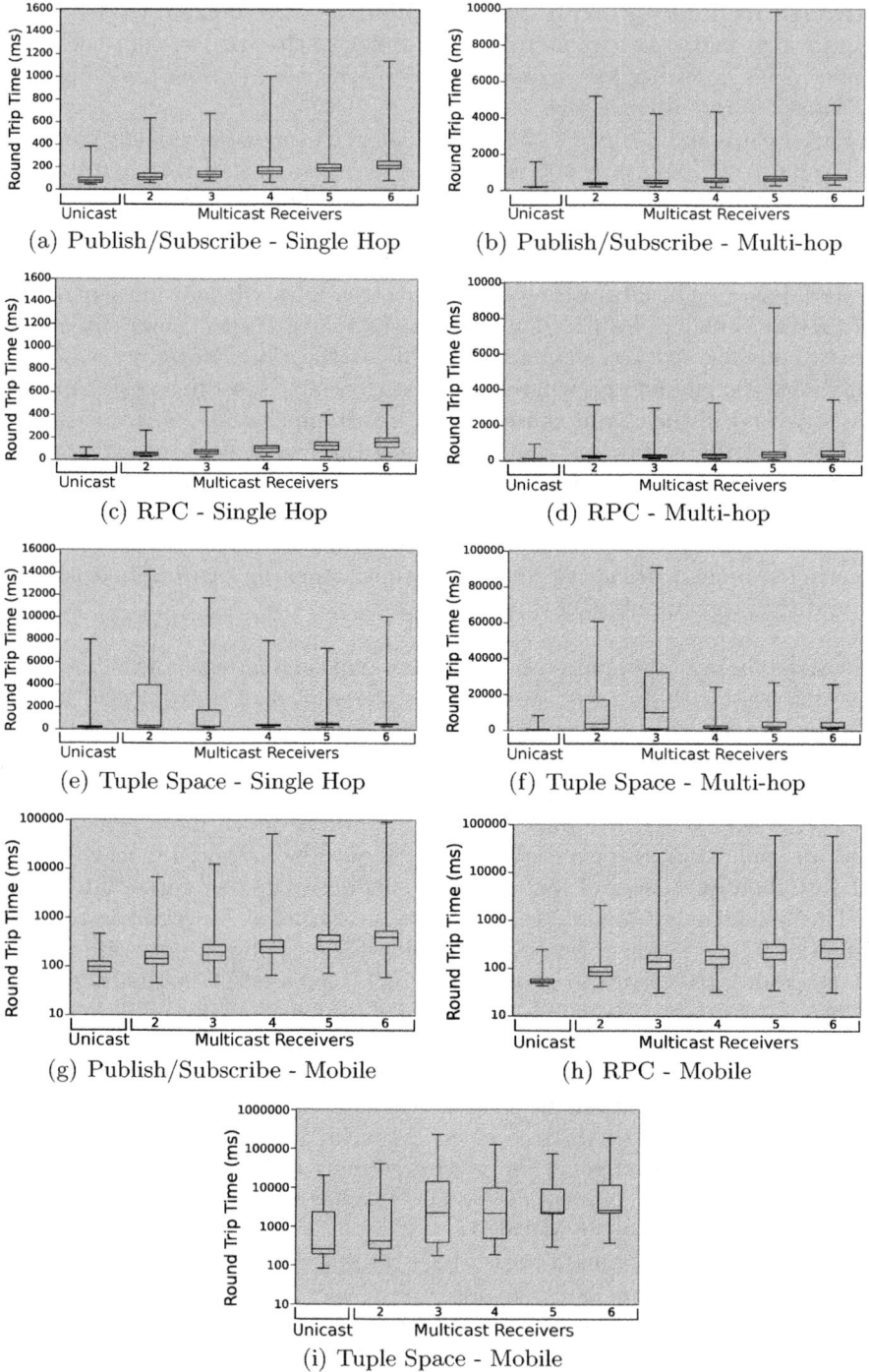

Fig. 4. Round Trip Times

If a receiver is "behind" it may spend time receiving older tuples before the newest tuple is requested. This causes the round trip times to increase while only improving one-way message delivery.

3.3 Shared Whiteboard Application

When testing the whiteboard application, we considered the metrics which a user might care about at the application level: how reliably and quickly users receive updates. In the results below, a single user is updating the whiteboard and the updates are propagated to 6 receivers. We used traffic traces from Coccinella[2] to ensure our implementation accurately represented a typical whiteboard application. For these experiments, 250 whiteboard update messages of varying sizes were sent out over a 10 minute period at varying intervals.

Furthermore, we tested the whiteboard application under the two different routing protocols we have been using, AODV and DSR. This is not meant to be an exhaustive comparison of the routing protocols themselves, but is intended to show how the choice in routing protocols might affect the performance of the communication models in a nontrivial application.

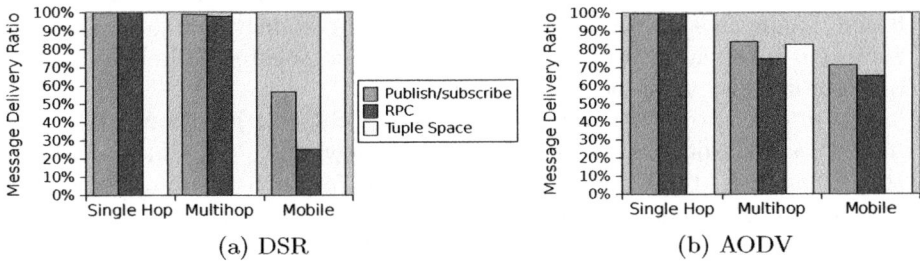

Fig. 5. Whiteboard Message Delivery

Message Reliability. Unlike the previous results, these represent one-way communication from the whiteboard user to the receivers. Message reliability determines how accurately the receivers' views reflect the state of the shared whiteboard.

Figures 5(a) and 5(b) show the percentage of whiteboard messages delivered for DSR and AODV, respectively. As before, the results are nearly 100% for all paradigms and both protocols in the single hop network. AODV performs poorly on the multi-hop scenario, while DSR achieves nearly 100% delivery for all paradigms. On the other hand, DSR performs much worse in the mobile scenario, with the delivery ratio for RPC only reaching 25%.

The reliability of tuple spaces is considerably better in these experiments than in the round trip scenario, with 100% delivery in all but the AODV multi-hop scenario. The difference between these results and Section 3.2 is the lack of a return message. Each receiver is responsible for requesting the whiteboard

---

[2] http://thecoccinella.org/

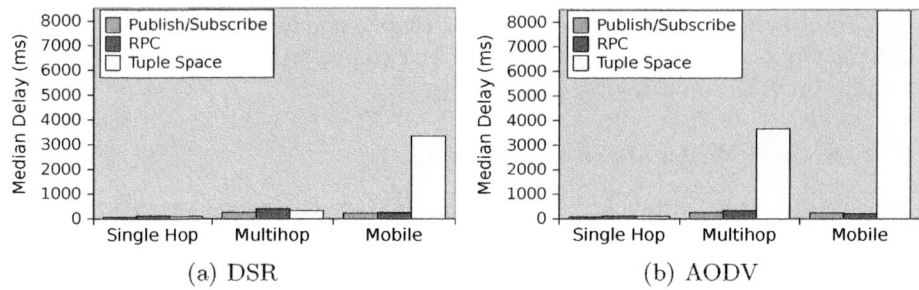

Fig. 6. Whiteboard Message Delay

updates, so the blocking request will be retried until the tuples are received. The only exception is the multi-hop scenario with AODV, in which all three paradigms perform much worse. Since all three paradigms are affected equally, these results must be directly due to the behavior of AODV in this scenario. Investigation of this phenomenon is outside the scope of this paper.

Message Delay. Message delay is measured as the time from when a whiteboard update is sent by the application until it is delivered to the receiver's whiteboard. Update delays are very noticeable in a shared whiteboard application, so the delay time should be minimized.

Figure 6(a) shows the results when using DSR and Figure 6(b) shows the AODV results. Not unexpectedly, tuple spaces have the highest median latencies of $8,486ms$ with AODV and $3,357ms$ with DSR. For publish/subscribe and RPC, the median delay remained under $400ms$.

With DSR, tuple spaces report a nearly 100% delivery ratio in every scenario, yet the delay times are $<400ms$ in the static scenarios. In contrast, AODV causes long delays for tuple spaces in both the multi-hop and mobile scenarios. Since tuple spaces will repeatedly attempt to deliver messages, retries are expected to contribute to the majority of the delays. This is supported by the long delay times experienced by tuple spaces with AODV in the multi-hop scenario. However, in the mobile scenario tuple spaces achieve 100% delivery with AODV and DSR, but the median delay with DSR is less than half as with AODV.

From the mobile reliability results, we can infer that DSR does not maintain viable routes, because the results of publish/subscribe and RPC are poor. However, the delay results suggest DSR is faster than AODV at finding new routes when they become available.

4 Related Work

There are many projects using publish/subscribe, RPC, and tuple spaces specifically in MANETs. M2MI (Many-to-Many Invocation) [14] adapts RPC to a MANET context. LIME (Linda in a Mobile Environment) [7] is a tuple space implementation intended for mobile devices. STEAM (Scalable Timed Events and Mobility) [15] is an example of an event-based middleware which uses publish/subscribe for communication.

Middleware for MANETs is surveyed in [16] and [17], but no quantitative results are presented. Projects using different communication paradigms were compared in [18]. [19] implements tuple spaces in terms of a modified publish/subscribe, but does not provide quantitative results.

5 Conclusions

Publish/subscribe, remote procedure calls, and tuple spaces are three communication paradigms which have been applied to MANETs. They have been used as the basis for many projects and applications intended to operate in MANETs. This paper presented a quantitative comparison of these three paradigms in three different network scenarios, with a focus on application level metrics. The results show the relative strengths and weaknesses in each of the three paradigms, as well as how they varied within the same scenario.

Publish/subscribe and RPC provide fast delivery of messages (best times were $<100ms$), but provide little message reliability (as low as 25% delivery ratio for RPC). Tuple spaces, on the other hand, pay a speed penalty (median round-trip times are 2-6 times slower than publish/subscribe), but provide better reliability, since messages will persist until explicitly removed from the tuple space. When used to implement a whiteboard application, tuple spaces achieved a 100% delivery ratio in all but one scenario.

The wireless channel itself can cause unexpected delays in message delivery. In the single hop scenario, publish/subscribe had a maximum delay of $1.6s$ and tuple spaces had one message require $14s$ to deliver. Introducing multi-hop routes without mobility caused median delay times to double for publish/subscribe and increase by a factor of 10 for tuple spaces. Mobility and multi-hop wireless routes, both defining characteristics of MANETs, strongly influenced the application-level performance and reliability of these paradigms.

Our results provide essential quantitative data for deciding which communication model should be used for new projects. While the paradigms presented here are essentially interchangeable in terms of functionality, their performance varies widely according to traffic and wireless conditions. Since we have tested canonical implementations of each paradigm, these results are applicable to basic versions of the paradigms in general and can be used to inform future work in application development for MANETs.

References

1. Eugster, P.T., et al.: The many faces of publish/subscribe. ACM Comput. Surv. 35(2), 114–131 (2003)
2. Birrell, A.D., Nelson, B.J.: Implementing remote procedure calls. ACM Trans. Comput. Syst. 2(1), 39–59 (1984)

---

[3] http://www.isi.edu/nsnam/ns/

3. Gelernter, D., Carriero, N.: Coordination languages and their significance. Commun. ACM 35(2), 97–107 (1992)
4. Takai, M., Martin, J., Bagrodia, R.: Effects of wireless physical layer modeling in mobile ad hoc networks. In: MobiHoc 2001: Proc. of the 2nd ACM Intl. Symp. on Mobile Ad Hoc Networking & Computing (2001)
5. Varshney, M., Bagrodia, R.: Detailed models for sensor network simulations and their impact on network performance. In: MSWiM 2004: Proc. of 7th ACM Intl. Symp. on Modeling, Analysis and Simulation of Wireless and Mobile Systems (2004)
6. Scalable Networks. Exata: An exact digital network replica for testing, training and operations of network-centric systems. Technical brief (2008)
7. Murphy, A.L., et al.: Lime: A coordination middleware supporting mobility of hosts and agents. ACM Trans. on Software Engin. and Methodology (July 2006)
8. Lien, Y.-N., et al.: A manet based emergency communication and information system for catastrophic natural disasters. In: ICDCSW 2009: Proc. of the 29th IEEE Intl. Conf. on Distributed Computing Systems Workshops, pp. 412–417 (2009)
9. Badache, N.: A distributed mutual exclusion algorithm over multi-routing protocol for mobile ad hoc networks. IJPEDS 23(3), 197–218 (2008)
10. Leggio, S., et al.: Session initiation protocol deployment in ad-hoc networks: a decentralized approach. In: 2nd Intl. Workshop on Wireless Ad-hoc Networks, IWWAN (2005)
11. Sung, M.Y., Lee, J.H.: Desirable mobile networking method for formulating an efficient mobile conferencing application. In: Yang, L.T., Guo, M., Gao, G.R., Jha, N.K. (eds.) EUC 2004. LNCS, vol. 3207, pp. 377–386. Springer, Heidelberg (2004)
12. Johnson, D.B., Maltz, D.A.: Dynamic source routing in ad hoc wireless networks. In: Mobile Computing, pp. 153–181. Springer, US (1996)
13. Perkins, C.E., Royer, E.M.: Ad-hoc on-demand distance vector routing. In: WM-CSA 1999: Proc. of the 2nd IEEE Workshop on Mobile Computer Systems and Applications, p. 90 (1999)
14. Kaminsky, A., Bischof, H.-P.: Many-to-many invocation: a new object oriented paradigm for ad hoc collaborative systems. In: OOPSLA 2002: 17th Conf. on Object-Oriented Programming, Systems, Langs., and Apps. (2002)
15. Meier, R., Cahill, V.: Steam: Event-based middleware for wireless ad hoc network. In: ICDCSW 2002: Proc. of the 22nd Intern. Conf. on Distributed Computing Systems, pp. 639–644 (2002)
16. Hadim, S., et al.: Trends in middleware for mobile ad hoc networks. Journal of Communication 1(4), 11–21 (2006)
17. Paroux, G., et al.: A survey of middleware for mobile ad hoc networks. Technical report, Ecole Nationale Supérieure des Télécommunications (January 2007)
18. Collins, J., Bagrodia, R.: Programming in mobile ad hoc networks. In: WICON 2008: Proc. of the 4th Annual Intl. Conf. on Wireless Internet (2008)
19. Ceriotti, M., et al.: Data sharing vs. message passing: synergy or incompatibility?: an implementation-driven case study. In: SAC 2008: Proc. of the ACM Symp. on Applied Computing, pp. 100–107 (2008)

Multi-modeling and Co-simulation-Based Mobile Ubiquitous Protocols and Services Development and Assessment

Tom Leclerc, Julien Siebert, Vincent Chevrier,
Laurent Ciarletta, and Olivier Festor

MADYNES & MAIA - INRIA Lorraine, Nancy Université, France
`first_name.name@loria.fr`

Abstract. Mobile and Ubiquitous Computing is about interconnected computing resources embedded in our daily lives and providing contextual services to users. The real influence between user behavior and ubiquitous communication protocols performance and operation needs to be taken into account at the protocol design stage. Therefore, we provide a generic multi-modeling approach that allows us to couple a user behavior model with a network model. To allow both assessment and benchmarking of ubiquitous solutions, we define formal reference scenarios based on the selection of a set of environmental conditions (contexts). We illustrate the use of the framework through its application to the study of mutual influences of mobility models and ad hoc network protocols.

1 Introduction

1.1 Context and Motivation

Ubiquitous or Pervasive Computing is about interconnected often dynamic and mobile computing resources embedded in our daily lives and providing services to users in a changing context and environment.

Several scientific domains (network, AI, physics, sociology, ...) are involved in the field of Ubiquitous Computing together with their own vocabulary, habits, needs and culture. To deal with the interacting complex models of ubiquitous computing, no single universal model exists. Experience in ubiquitous systems demonstrates that advanced research in such a complex topic cannot be pursued by only broadening an initial domain with unavoidably partial knowledge from others. A typical example is the design of mobile services where the user carries devices that contribute to the delivery of data to other users. In this case, the behavior of the users like (e.g. their mobility patterns) in a crowd highly impact the overall operation of the service and need thus to be considered early in the service design.

We therefore propose a methodology and a novel distributed framework to design, implement and assess "mobiquitous" communication related technologies. Our solution is build on two key elements : model interaction and multi-simulation engine.

First, our approach enables the combined use of reference models and simulators coming from different specific domains (figure 1). Through a simple interface implemented for each simulator, the presented framework eases interaction among both models and simulators. This significantly improves the initial design of the EXiST (EXperImental Simulation Tool [4]) co-simulator by both providing decentralization support and a better formalization.

1.2 Case Study

In the domain of dynamic networks addressed here, wireless technologies, ad hoc or mesh routing protocols, or ubiquitous services are often studied (designed, experimented, assessed) using network simulators. Indeed, real world experimentations with a representing set of devices is excessively time and money consuming, especially in the case of ad hoc networks or large scale peer-to-peer environments. It is even scientifically of little relevance since reproducing a scenario / an experiment is not possible due to the ever changing experimental conditions. Therefore, a lot of models and simulators have been developed in the field of ubiquitous computing over time [10,9]. They aim at simulating the network layers in more or less details and indeed most of them are not designed for doing more, like advanced node dynamics for example, or users goals. In fact, in ubiquitous computing, one key element of the equation is "the human" and more specifically his behavior.

As a case study, in this paper we focus on mobility in MANETs (Mobile Ad-hoc NETworks), as an example to demonstrate our approach. MANETs are wirelessly connected devices connecting spontaneously without any preexisting infrastructure. In MANET simulation, nodes move according to a mobility model. Most mobility models are computed by merely considering the user as a random walker without goal or decision process, and without any knowledge of how the network actually behaves. Unfortunately this is what is generally considered sufficient to give the system its "dynamic" characteristic, and therefore used to prove the validity and demonstrate the performance of protocols which later fail when deployed in the real world. Our approach circumvents this limitation.

As a proof of concept, we combine two existing simulators: a mobility simulator (based upon a multiagent model) and a network simulator. By doing this, we combine sociological research achieved in urban simulation community with network research. Our experiments in Section 5) show the possibilities that the framework offers and also the importance of the mutual influences between the network and user behavior. We believe that the originality of our approach is to allow to close the loop between the users behaviors and their mobile ubiquitous environment.

The remainder of this paper is organized as follows. Section 2 motivates the usage of multi-modeling and co-simulation. In Section 3 we present our conceptual framework and a prototype implementation. We focus in Section 4 on mobility modeling and present the multiagent paradigm applied to the modeling of users behaviour. Experiments and results are described in Section 5. Section 6 summarizes the contribution, ongoing and future work.

2 Multi-modeling and Co-simulation Motivation

Our approach is built on the use of multi-modeling and co-simulation in order to take into account both users behaviors and network performance within an integrated study. The framework offers a way for protocol and service designers to get a "bigger picture" early in the design phase.

As stated in the introduction, we argue that the study and the design of mobile ubiquitous applications cannot be achieved efficiently by taking into account only one point of view. By point of view, we mean the physical medium aspects, the network aspects (protocols, services, messages, topology ...), the users behavior aspect (mobility, sharing resources ...), etc. (figure 1). Depending on the study and questions asked, omitting some of these points of view may lead to non-significant simulation results. For example the authors of [2] show the impact of taking into account different physical medium models for the wireless communication. Moreover as many models and simulators already exists and have been validated reusing them is the best approach.

We propose to use multi-modeling and co-simulation in order to represent all the different aspects or point of view needed for the simulation to be more significant. We rely upon a meta-model and a framework called AA4MM which allows us to couple different existing models and simulators in order to build a more complex and more accurate simulation. These simulations are used, on one hand, to evaluate protocols and services against different usage scenarios and, on the other hand, to design new protocols and services by taking into account some global usage scenarios (the user behavior and the environment parameters). By co-simulation or multi-simulation, we mean the ability to combine multiple simulators and/or real implementations (prototype, software and/or hardware) at the same time.

The main advantage of this approach is to achieve a good separation of concerns. Computer network scientist and designers only focus on the network aspect (protocol and services definitions, network parameters: radius, bandwidth, latencies ...) and cognitive and human scientists focus on the user behavior modeling (mobility, user needs, (ir)rationality). The whole simulation efficiency is our main limitation. Reusing existing simulators - that may not have been designed for distributed simulations - may be less efficient than a single multi-model implemented in a natively distributed simulator. However we consider that the advantages brought by the separation of concerns are conceptually more important and that the simulation efficiency is a technical question that can be targeted later.

3 The AA4MM Meta-model and Platform

We develop a multi-modeling platform called AA4MM (Agent and Artefact for Multiple Models [12]). Its main goal is re-usability and interoperability of different simulators like HLA (High Level Architecture), EXiST or MSI (the Multi Simulation Interface) but its software architecture is completely decentralized

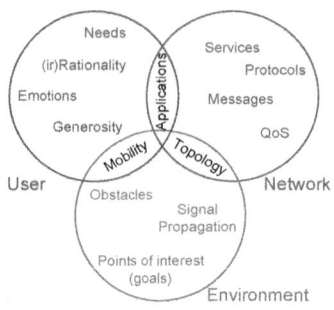 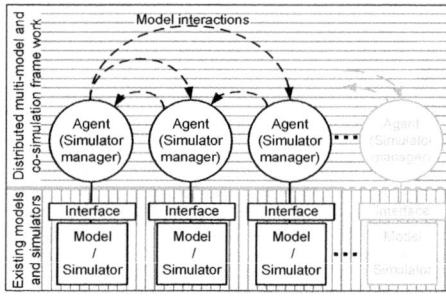

Fig. 1. Abstraction levels **Fig. 2.** Framework model interactions

and based upon multiagent paradigm. This paradigm allows us to take into account solutions developed in complex systems modeling and multiagent community such as automated parametrization or simulation control [5,3]. In this paper we only highlight how existing simulators interacts withing the framework.

Figure 2 describes the composition of the AA4MM framework. Each simulator is controlled by a simulator manager (formally an agent) which is an autonomous entity. All these manager agents cooperate in order to run the whole simulation and to take care about the interaction problematics. To make different simulators interact in the AA4MM platform, the following steps are required:

- define a simulator interface (one for each simulator): implement 6 basic functions as described in table 1 directly from the source code or from the api, or more laborious by extracting an api if only binaries available; this is the only modification to make.
- create the specific AA4MM entities:
  - for each simulator create an entity (called an agent) in order to manage the simulator (input/output data flows, model execution and simulation time management).
  - for each link between the simulators create an entity in charge of the data flow exchange (called an artifact).

It is relatively simple to add a model and its simulator within the interfacing (Table 1). The interactions problematic (e.g. simulators synchronization and model compatibilities) are managed by the AA4MM platform itself by creating entities external to the original tools : the agents and the artifacts. This presents some advantages: it is easy to reuse existing models and simulators without knowing anything of distributed simulations. The few modifications brought to the models and simulators allow us to use them either as a standalone application or inside the multi-simulation.

Artifacts created can be seen as a distributed tuple space where agents exchange simulation data timestamped with a specific validity interval [13]. The whole platform is sustained by a distributed simulation algorithm (a Chandy-Misra-Bryant algorithm variant) that allows the agents to manage the whole

Table 1. Interface to define for a simulator to work within the AA4MM platform

	Fonctions	Description	Example
1	Initialization	Initialize, passing parameters	
2	Model execution	Execute 1 step of the model	Execute 1 simulation step, 1 simulation event or a given time interval
3	Get simulation time	Obtain current simulation time	
4	Data input	Provide data to the simulator	Input information from another simulator
5	Data output	Retrieve data from the simulator	Output information going to another simulator or for logging purposes
6	(Optional) Finalize simulation	Finalize simulation after last step	Retrieve logs from / execute logging scripts

simulation process in a distributed way, whatever the execution policy of their simulators (discrete event, step by step, continuous time).

Our framework uses a series of XML configuration files that allow for the simple description and tweaking of the different simulators involved and of the global simulation.

4 Case Study: MANETs and Users Behaviors

4.1 Mobility Modeling: A Quick Survey

There are many ways to model the different types of mobility. Classical mobility models are well documented and can be classified, as surveyed in [1], in 4 categories: random models (e.g. Random Waypoint), models with temporal (e.g. Gauss-Markov), spatial (e.g. Reference Point Group) or geographical (e.g. obstacle mobility) dependencies.

There is no formal model combining some of those classical models. And most critically, none ever considered any feedback from the network to the user behavior (e.g. impact of perceived QoS). Our work allows both by proposing to use the agent paradigm as a unique tool for modeling the largest and various sort of mobility.

4.2 The Multiagent Paradigm

The multiagent paradigm is a way to model sets of autonomous interacting entities within an environment. It is a well known paradigm used in human sciences, ecology or in robotics. It describes the systems into, at least, these different components: agents, environment, interactions. The agents are autonomous and proactive entities, situated in an environment. They only have a partial (local) view of it and decide which action to take dealing with their own perceptions and reasoning.

MABS (Multiagent Based Simulation) offers us the right level of description when we want to model users' behavior, goals and actions. Instead of using a global equation to model users' trajectories, we can, via the agent based model, re-create the way users move. It means that we can directly model behaviors such as "if an obstacle is present in front of you, then avoid it" or "reach a goal, stay nearby during five minutes and then go".

More generally, with this approach, we can model more complex behaviors such as willingness to use and share a service depending on the bandwidth consumed or the generosity of a user ; or the reaction to unpredictable events.

Mobility has already been studied and modeled via the multiagent paradigm. Here an agent can describe human, animal or robots. In [11], Craig Reynolds worked on bird flocks modeling where each agent tries to stay inside the flocks only by computing a small set of forces (Boids). Individual-based pedestrian modeling is also used in urban simulations [7,14]. This paradigm is also used to model crowd scenes in movies (as battlefields in Lord of the Ring) and implemented in video animation software such as MASSIVE[1].

4.3 User Model Description

Our agent based mobility model is inspired from urban research and pedestrian modeling [5,7], but can also model classic mobility behaviors (e.g. random waypoint). Each mobile node (a user) is represented by an agent (named a_i). The agent behavior can be seen as a combination of simple behaviors resulting in a complex one. For example random waypoint is implemented as the simple following rule: Each every time period each agent changes its direction. More complex behaviors such as obstacle avoidance, flocking or goal attraction are modeled as a function, a sum of forces, resulting in a node movement. Each force/behavior describes an interaction of the agent with its environment and the other agents. The agent has a limited perception (figure 3): these interactions are effective only on the neighborhood of the agent. In our case, the movements of the agents are computed by applying laws of mechanics: namely point kinematics. These models are easily extensible, easy to implement and can express a large set of behaviors by weighting each force. The examples below (figures 4,5 and 6) depict force oriented behaviors.

Fig. 3. Perceptions of an agent (a user)

Fig. 4. Repulsive force for obstacle avoidance model

---

[1] http://www.massivesoftware.com

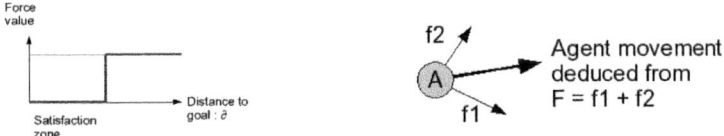

Fig. 5. Attractive force to the goal   **Fig. 6.** Movement computation

4.4 Modeling Network Aware Users

Integrating network aspects into the agent decision process is achieved easily and straightforward. Indeed, once the agent perceived the network information (e.g. connectivity presence/loss, quality of services) a simple rule defines its reaction. For example in Section 5 we describe users that slow their speed or stop moving when they perceive good connectivity.

4.5 Synthesis

This model respects the constraints cited in [1]: temporal dependency, spatial dependency and geographical dependency. Describing sophisticated movements is straightforward: for example from our two simple movements we have nodes avoiding obstacles and following multiple succeeding goals. Moreover, we can easily model mobility of groups of people just by adding a force that attracts agents that go in the same direction (as shown in Section 5).

We've develop a set of mobility models, from simple random waypoint or restricted random waypoint to advanced particles engine, flocking or explorer behavior, that are fully parametrized. Using MABS to simulate basic behaviors such as random waypoints seems probably overkill at first. However, since this modeling approach is individual-based, we can easily tune each behavior and describe heterogeneous ones. Indeed, the highest level of granularity can be reached by implementing a different model of behavior per agent. Thus, we can describe, for example, different kind and mixes of populations. Finally, with our approach, a user can dynamically switch from one behavior to another.

5 Experiments and Results

As a proof of concept of our vision and framework we coupled a users behaviors simulator that we developed (MASDYNE: MultiAgent Simulator of DYnamic Network usErs) with a MANET simulator (JANE: The Java Ad hoc Network Emulator [6]).

The goals of the following experiments are on the hand, to show the simplicity of a realistic usage scenario design and implementation, and on the other hand, to show the effects of having interactions between the user behavior model and the network model.

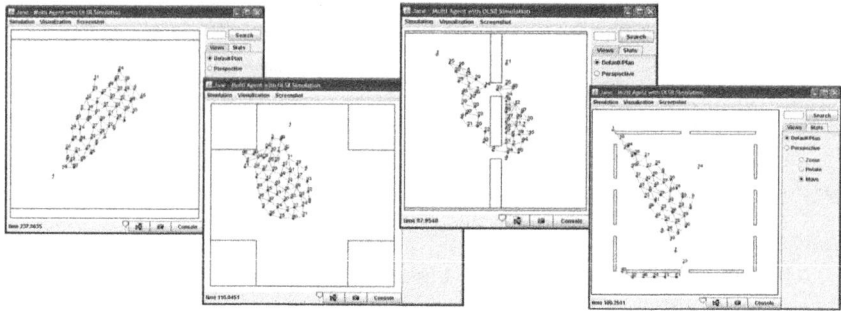

Fig. 7. Museum visit example in multiple environments

5.1 Building Realistic Usage Scenarios

The goal of the first experiment is to obtain a mobility scenario (figure 7): A group of students visiting a museum. This scenario fits with our goal to test and deploy in the future an ad hoc network within a museum (ANR SARAH project).

This mobility model is based upon force oriented behaviors: the user behavior, the interaction between the users and the environment are represented by simple forces. We use, for this scenario, four simple force oriented behaviors: *goal force, avoid walls, repulsive force, attraction force*. This provides us the following scenario: Agent 1, a tour guide, follows goals unknown to the students. The other agents, the students, follow agent 1. The combination of these simple behaviors, done by summing the forces, results in a complex and more realistic behavior.

- All Agents have a Goal: e.g. the students follow Agent 1, if visible.
- All Agents avoid walls: Repulsive force from the walls.
- Agents have a repulsive force from each other (comfort zone).
- Agents are attracted by other agents that go in the same direction.

Figure 7 show this usage scenario in different environments (e.g. corridor, crossroad, doors, museum). Parameterization of the model is done according to [8]. We observe that in different environments the student group is clearly following the Tour guide even when walls are involved.

5.2 Network and User Behavior Mutual Influences

Even the simplest network protocol such as a broadcast obviously would perform very well if the students follow seriously the rules. However, what happens if not all the students follow the rules and react for example to network events (e.g. network connectivity). The AA4MM framework allows us to respond to this kind of questions. To show the effects of these mutual influence, the second experiment is a scenario with network feedback (figure 8). The aim of this experiment is not so much about being realistic but more about showing the possibilities offered by our approach.

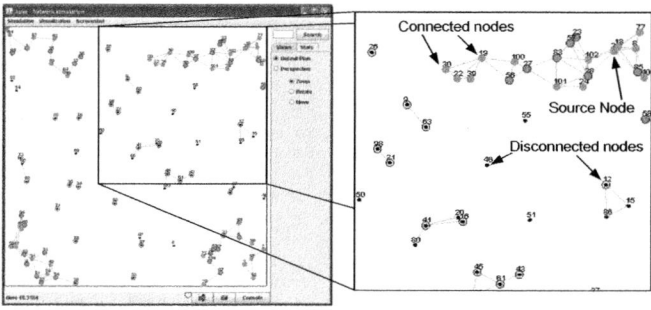

Fig. 8. 4 source nodes: green nodes, 100 moving nodes: black when disconnected, orange when connected to a source

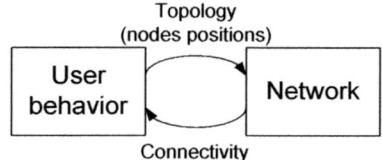

Fig. 9. Interactions between MASDYNE and JANE

Experimental protocol: The usage scenario is the following: 4 source nodes (access points) are placed in every corner of a place. 100 nodes/users want to connect to a source. To keep it simple to explain and to avoid errors or bias induced by the algorithm of a protocol, we used a basic flooding algorithm that simply rebroadcasts every non seen message. Messages already seen are silently discarded.

The user behavior is the following. At beginning the user moves randomly. He is aware of its connection status: connected or not connected. Then we propose 3 basic behaviors: continue to move randomly (user is not aware of network feedback or doesn't care), slow down speed (user continues walking but only slowly), stop (user sits down to enjoy the connection). Figure 9 depicts the interactions between both simulators.

We measured the evolution of the percentage of connected nodes. For each experiment we use 2 behaviors. We vary the percentage of agents having each behavior. For example in figure 10 the curve marked 40% means that 40% of the agents implement the stop behavior while the remaining 60% implement the random behavior. Each experiment was done using 50 distribution seeds.

Results: We observe that the more the users slow down or stop the better is the connectivity rate. The stopping nodes create a sort of backbone for the other nodes while the backbone created by slowing nodes is only temporary until nodes move out of range.

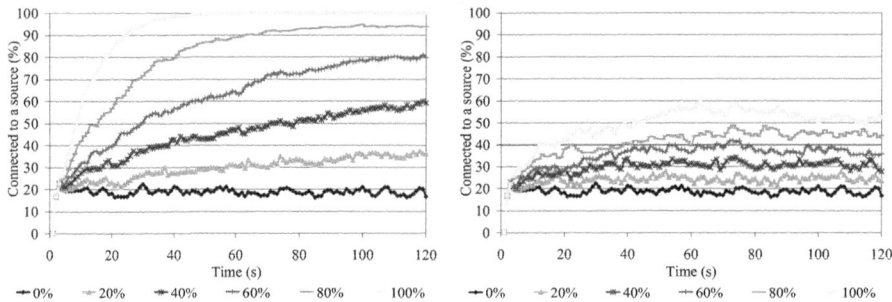

Fig. 10. Percentage of nodes stopping when connected, remaining nodes move randomly

Fig. 11. Percentage of nodes that divide their speed by 6 when connected, remaining nodes move randomly

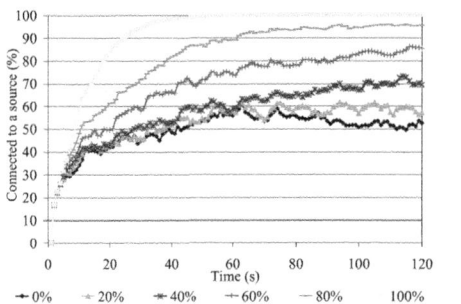

Fig. 12. Percentage of nodes stopping when connected, remaining nodes divide their speed by 6 when connected

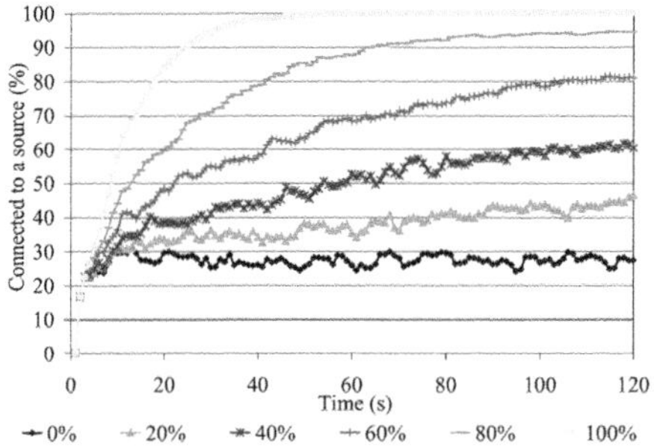

Fig. 13. Percentage of nodes sStopping when connected, remaining nodes divide their speed by 3 when connected

In figure 10 we increased the percentage of nodes stopping while the remaining ones continue moving randomly. With 100% stopping nodes, after 60 seconds all nodes reached an access point. With already 60% of nodes stopping, 80% of the nodes are connected. In figure 11, the nodes divide their speed by 6 when connected. Again we increased the percentage of nodes slowing while remaining nodes move randomly. Performances compared to stopping nodes are worse. In figure 12 and 13 every node reacts on connection. We varied the percentage of stopping nodes while the remaining nodes divide their speed. We observe that the results significantly differ when using random waypoint model or more complex usage scenario.

5.3 Synthesis

This experimental work show that our approach has: 1) the ability to take mutual influences of users behaviors and network performances into account;

2) the ability to design usage scenarios with heterogeneous users behaviors;
3) the ability to benchmark a network protocol against a wide range of usage scenarios.

In order to consider this work from a higher standpoint, we don't assume that the users behaviors will always be predictable. But, with this approach, we are able to predict that if only a percentage of the users behave like we predict, then the network performances will be better or worse.

6 Conclusion

When using classical modeling approaches, it is not straightforward to take into account the users behaviours and their interactions with the network performances. We presented the conceptual framework and a prototype implementation. With our multi-modeling approach, different existing models can be easily coupled in a loose and generic way. We focus in this paper on mobility modeling which is a key point in evaluating wireless technologies and services, and described a couple of experiments.

We presented the multiagent paradigm applied to the modeling of users behaviour. We argued and shown that while it still provides the usual mobility models, it is very simple to design, fine-tune, redesign those models or even design completetly new ones. Multiagent allows the description of heterogeneous behaviours. The new mobility models can take into account networks or more generally environment inputs, basically having a closed-loop system where something closer to the "human behavior and real-life" is considered.

Our approach offers a basis for valid comparison of wireless technologies and services but it can be extended to any dynamic environment, such as P2P networks for example. It is very well-suited for every situation where there are interactions between the users, the networking and the physical environments.

Our experiments demonstrate that closing the loop leads to new ways of evaluating technologies. Even a basic protocol, such as our flooding example, can have strong performances if the users follow their directives. Some can be difficult to enforce (complete stop when detecting a connection), but other could be reasonable in a real world (slowing down).

In the short term, we plan to show on the opposite the disruptive effect of non conforming behavior, and to extend our experiments to more advanced protocols and scenarios. In parallel, we continue to work on the theoretic and practical aspects of the AA4MM simulation framework. Our platform will be extended by defining and implementing more standard and novel mobility models (node/users behaviors), and reference environments.

In the longer term, the framework will be enriched with a set of mobility models, a set of environments models. We also plan to have a real setup (a typical existing room or building or city modeled in 3D from real data for example). These sets can serve as references that could be used to assess the performances and applicability of a solution, and validate it in certain contexts. This could be a good basis to provide the ubiquitous computing community with a benchmarking evaluation toolkit.

References

1. Bai, F., Helmy, A.: A Survey of Mobility Models. In: Wireless Adhoc Networks. University of Southern California, U.S.A. (2004)
2. Ben Hamida, E., Chelius, G., Gorce, J.-M.: Impact of the Physical Layer Modeling on the Accuracy and Scalability of Wireless Network Simulation. Simulation 85, 574–588 (2009)
3. Bonneaud, S., Redou, P., Chevaillier, P.: Pattern oriented agent-based multi-modeling of exploited ecosystems. In: 6th EUROSIM Congress on Modelling and Simulation, September 9-13 (2007)
4. Ciarletta, L., Iordanov, V., Dima, A.: Using Intelligent Agents to assess Pervasive Computing Technologies. In: IAWTIC 2001, Las Vegas, USA, p. 10 (2001)
5. Gaud, N., Galland, S., Gechter, F., Hilaire, V., Koukam, A.: Holonic Multilevel Simulation of Complex Systems. Application to real-time pedestrian simulation in virtual urban environment. Simulation Modelling Practice And Theory (SIMPAT) 16(10), 1659–1676 (2008)
6. Gorgen, D., Frey, H., Hiedels, C.: Jane - the java ad hoc network development environment. In: ANSS 2007, pp. 163–176. IEEE Computer Society, Los Alamitos (2007)
7. Helbing, D., Buzna, L., Johansson, A., Werner, T.: Self-organized pedestrian crowd dynamics: Experiments, simulations, and design solutions. Transportation Science 39(1), 1–24 (2005)
8. Helbing, D., Molnar, P., Farkas, I.J., Bolay, K.: Self-organizing pedestrian movement. Environment and Planning B: Planning and Design (2001)
9. Kurkowski, S., Camp, T., Colagrosso, M.: Manet simulation studies: The incredibles. ACM SIGMOBILE Mobile Computing and Communications Review 9, 50–61 (2005)
10. Naicken, S., Basu, A., Livingston, B., Rodhetbhai, S.: A survey of peer-to-peer network simulators. In: Proceedings of The Seventh Annual Postgraduate Symposium, Liverpool, UK (2006)
11. Reynolds, C.W.: Flocks, herds, and schools: A distributed behavioral model. Computer Graphics 21, 25–34 (1987)
12. Siebert, J., Ciarletta, L., Chevrier, V.: Agents and Artefacts for Multiple Models coordination. Objective and decentralized coordination of simulators. In: 25th Symposium on Applied Computing, SAC 2010, Lausanne Suisse. ACM, New York (2010)
13. Siebert, J., Rehm, J., Chevrier, V., Ciarletta, L., Mery, D.: Aa4mm coordination model: event-b specification, rr-7081. Technical report, INRIA (2009)
14. Teknomo, K., Takeyama, Y., Inamura, H.: Review on microscopic pedestrian simulation model. In: Proceedings Japan Society of Civil Engineering Conference (2000)

TERMOS: A Formal Language for Scenarios in Mobile Computing Systems

Hélène Waeselynck[1,2], Zoltán Micskei[3], Nicolas Rivière[1,2],
Áron Hamvas[3], and Irina Nitu[1,2]

[1] CNRS; LAAS
7 av. Colonel Roche
F-31077 Toulouse, France
[2] Université de Toulouse
UPS, INSA, INP, ISAE ; LAAS
F-31077 Toulouse, France
[3] Budapest University of
Technology and Economics
Muegyetem rkp. 3
1111 Budapest, Hungary
{waeselyn,nriviere}@laas.fr, micskeiz@mit.bme.hu

Abstract. This paper presents TERMOS, a UML-based formal language for specifying scenarios in mobile computing systems. TERMOS scenarios are used for the verification of test traces. They capture key properties to be checked on the traces, considering both the spatial configuration of nodes and their communication. We give an overview of the TERMOS design and semantics. As part of the semantics, we present the principle of an algorithm that computes the orders of events from a scenario. Two proof-of-concept prototypes have been developed to study the realization of the algorithm.

Keywords: Mobile computing systems, UML Sequence Diagrams, formal semantics, testing.

1 Introduction

Graphical scenario languages (e.g., Message Sequence Charts [1], UML Sequence Diagrams [2]) allow the visual representation of interactions in distributed systems. Typical use cases, forbidden behaviors, test cases and many more aspects can be depicted. The popularity of graphical scenarios is due to their user-friendly syntax, which facilitates communication while opening the door for formal treatments (this however requires that the used notation has a precise semantics). We investigate one of such formal treatments, namely the automated analysis of test traces. In our work, a scenario captures a key property to be checked on the traces. Scenario-based verification is not a novel approach, the originality here is that we apply it to a specific class of distributed systems: mobile computing systems.

Mobile computing systems involve devices (handset, PDA, laptop, intelligent car) that move within some physical areas, while being connected to networks by means of wireless links (Blue-tooth, IEEE 802.11, GPRS). Such systems differ from "traditional" distributed systems in many aspects: frequent connections and disconnections of mobile nodes, communication with unknown partners in a local vicinity, context awareness. These novelties require us to revisit the notion of scenario for mobile settings.

Preliminary work led us to conclude that some extensions are necessary [3]. One of these extensions is to consider the spatial relations between nodes as a first class concept. We proposed to use labeled graphs to depict the spatial configurations of a scenario. We then noticed that, due to this extension, the checking of test traces against scenarios has to combine event order analysis and graph matching.

Since the preliminary workshop paper, the graph matching part has been addressed in [4]. This paper now aims at giving an overview of the whole approach, and at presenting the remaining event order analysis. This analysis considers both communication and configuration change events.

The contributions are the following:

- We show how the general extensions proposed in [3] can be introduced into one of the existing scenario languages, UML Sequence Diagrams (Section 3).
- We then define a specialization of Sequence Diagrams, called TERMOS (Test Requirement language for Mobile Setting). TERMOS is given a formal semantics suitable for the checking of traces, at the price of some syntactic constraints and some interpretation choices of UML constructs (Section 4).
- We explain the principle of the algorithm that computes orders of events from a TERMOS scenario according to the chosen semantics (Section 5.1, see also our technical report [5] for a detailed presentation of the algorithm).
- We present two prototypes we developed as a proof-of-concept of the algorithm (Section 5.2).

2 Related Work

Previous work has investigated how to incorporate mobility into UML scenarios [6, 7, 8]. However, the focus was more on logical mobility (mobile computation) than physical mobility (mobile computing). It induces a view of mobility that consists of entering and exiting administrative domains, the domains being hierarchically organized. This view is adequate to express the migration of agents, but physical mobility requires further investigation, e.g., to account for dynamic ad-hoc networking. Also, there is not always a formal semantics attached to the proposed notations.

Having a formal semantics is crucial for our objective of analyzing traces. We had a thorough look at existing semantics for UML 2.0 Sequence Diagrams [9]. We also looked at other scenario languages distinguishing potential and mandatory behavior, which makes them well suited for the expression of properties [10, 11]. The most influential work for the TERMOS semantics was work on Live Sequence Charts (LSC) [10], and more specifically Kloses's version of the semantics [12], as well as work adapting LSC concepts into UML Sequence Diagrams [13, 14].

3 Graphical Interaction Scenarios for Mobile Computing Systems

Graphical scenario languages are used to represent interactions in distributed systems. Lifelines are drawn for the individual participants in the interaction, and the partial orders of their communication events are shown. To represent complex orderings, the languages offer operators like choice, iteration, parallelism and sequencing. As mentioned in Section 2, some languages also distinguish potential and mandatory behavior. A point-to-point view of communication is usually adopted, with one sender and one receiver for a given message. The underlying connection topology is not the focus, it is not supposed to change during the shown interaction.

Such characteristics are not sufficient for mobile computing systems, where the movement of nodes inherently yields an unstable topology. Links with other mobile and infrastructure nodes may be established or destroyed depending on the locations. Moreover, nodes may dynamically appear and disappear as devices are switched on and off, run out of power or go to standby. Interaction scenarios should thus explicitly account for the spatial configuration of nodes and how it evolves.

In addition to usual point-to-point communication, a natural communication for mobile computing systems is local broadcast. In this class of communication, a node broadcasts a message in its neighboring environment. Whoever is at transmission range may listen to, and react to, the message. For example, local broadcast is used as a basic step for the discovery layer in mobile-based applications (group discovery for group membership services, route discovery in routing protocols, etc.).

We propose that existing scenario languages be extended to accommodate local broadcast and make the spatial configuration of nodes a first class concept. More precisely, we proposed in [3] the following three extensions:

1. introduction of a spatial view,
2. consideration for spatial configuration change events,
3. representation of broadcast communication events.

The extensions can be introduced into the various existing languages. We now elaborate a solution for UML SD.

3.1 Introduction of a Spatial View

A scenario in a mobile setting contains two connected views: the event view (as classically done), and a spatial view describing the topological configurations of nodes. Fig. 1 illustrates these two views in UML SD. The represented scenario comes from a case study we performed [15], a Group Membership Protocol (GMP) for ad hoc networks. In this GMP [16], groups split and merge according to the location of mobile nodes. The protocol uses the concept of safe distance to determine which nodes should form a group.

Conceptually, the spatial view consists of a set of labelled graphs, corresponding to the various configurations that occur in the scenario. We depict them using UML object diagrams (Fig. 1a). The shown scenario involves two successive spatial configurations (C1 and C2). Nodes are depicted by instances. They are identified by

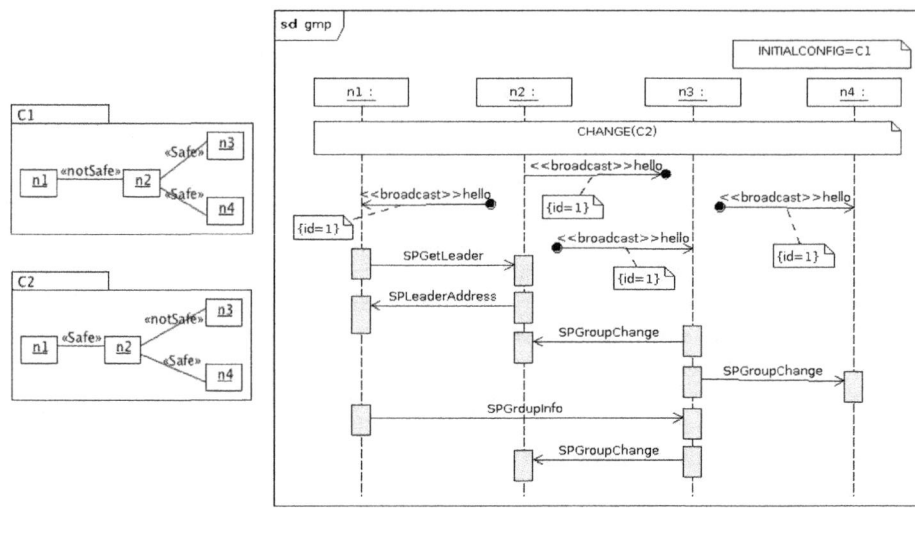

(a) Spatial view (b) Event view

Fig. 1. A concurrent split and merge scenario for the GMP

an id (e.g., *n1*, *n2*, ...) and could have contextual attributes (not shown here). Edge labels characterize the connection of nodes. They are represented as stereotypes. In the GMP example, labels indicate whether the nodes are at a safe distance *(«Safe»)*, or whether they are at communication range but not at a safe distance *(«notSafe»)*.

3.2 Consideration for Spatial Configuration Change Events

To connect the spatial and event views, configuration changes are introduced as global events in the event view. The event view in Fig. 1b has an initial configuration (shown in the *INITIALCONFIG* comment box). Configuration changes are then represented by global events of the form *CHANGE(name_of_new_config)*. In the GMP scenario, the two successive configurations correspond to node *n2* getting close to *n1*, while getting away from *n3*.

In other scenarios, configuration changes may involve the dynamic creation and shutdown of nodes. As there is no convenient way to describe such a dynamic structure in sequence diagrams, we adopt the convention that every node mentioned in at least one configuration has a lifeline in the event view. If a node is not active at some point of the scenario, then it is not supposed to participate in any communication interaction. The event view makes it explicit which communication event occurs in which spatial configuration. Checks can then be provided to warn the scenario specifier whenever communication is not compatible with the spatial view. In Section 5, we will see a tool that includes such checks.

Finally, note that the representation of configuration changes as global events is an abstraction for the fact that the topology is a global system property. A configuration change event does not result from an active synchronization of nodes, it just happens in their physical world. In the GMP example, the change to C2 remains unnoticed until *n2* broadcasts its new location by a *hello* message.

3.3 Representation of Broadcast Communication Events

To represent the local broadcast, we use the concepts of lost and found messages and a special stereotype. Lost messages are messages with no explicit receiver. Similarly, found messages do not have an explicit sender. In order to distinguish broadcasts from "usual" lost/found messages, we assign them the «broadcast» stereotype. A broadcast involves one send event followed by one or several receive events. A tagged value is attached to the corresponding lost/found messages, so that each receive event of the diagram can be paired to the send event that caused it.

To conclude, we believe that the three proposed extensions address needs that are recurring when modeling scenarios in mobile settings. They allow us to depict non trivial scenarios, like the concurrent group merge and split of Fig. 1. Test experiments have shown that this interaction yields a GMP failure [15]. The spatial and event views are complementary to describe this fail scenario, which combines a specific evolution of configuration and a specific ordering of messages after the configuration change.

4 Analyzing Test Traces wrt Requirements and Test Purposes

Graphical scenarios may be used at various phases of a development process, ranging from the elicitation of requirements to late testing phases. Our work focuses on the use of scenarios to analyze execution traces of mobile computing systems. We assume that the traces are collected on a test platform offering log facilities. The logged trace contains both communication data and contextual data (e.g., location data) from which the system spatial configurations can be retrieved. We then want to check whether the test trace exhibits some behavior patterns described by scenarios.

We consider three classes of scenarios exemplified by Fig. 2. *Positive requirements* capture key invariant properties of the form: whenever *A* happens in the trace, then *B* always follows. *Negative requirements* describe forbidden behaviors that should never occur in the trace. *Test purposes* describe behaviors to be covered by testing, that is, we would like these behaviors to occur at least once in the trace. The TERMOS language (Test Requirement language for Mobile Setting) is a UML-based notation we developed to capture the three classes of scenarios. It incorporates the extensions previously proposed to account for mobile settings. The notation has a formal semantics, so as to allow the automated analysis of test traces. We introduce here the general principle of this analysis.

A TERMOS scenario ends with an *assert* fragment. Everything before the *assert* represents a potential behavior, while the content of the *assert* is mandatory. The content of the *assert* characterizes the class of the scenario. A negative requirement contains just a false invariant. Since *FALSE* never holds, this means that the behavior before the *assert* should never happen. A test purpose contains just a true invariant. It holds whenever the *assert* is reached. A positive requirement contains a fragment different from a trivial true or false invariant.

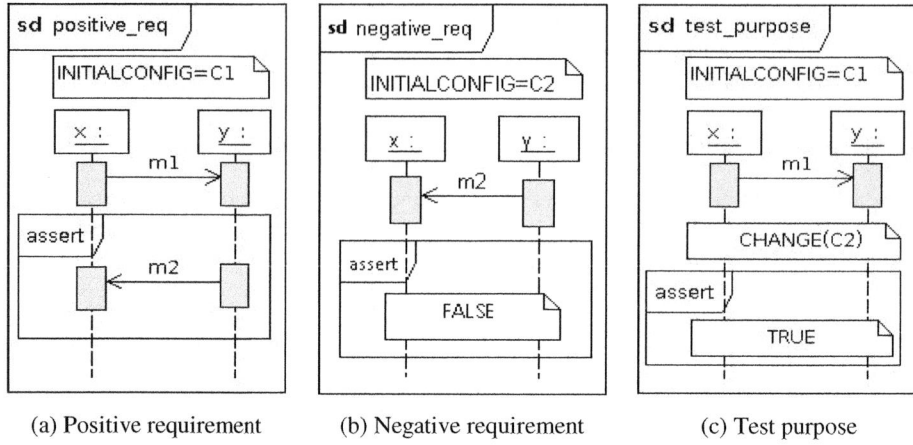

Fig. 2. Requirement and test purpose scenarios (event views)

We interpret TERMOS scenarios as generic behavior patterns that may be matched by various subsets of the system during the test run. In Fig.2, the node ids x and y are *symbolic* node ids. For example, the positive requirement (Fig.2a) is interpreted as:

"Whenever two nodes exhibit spatial configuration C1, and the node matching x sends message *m1* to the node matching y, then the node matching y must answer with message *m2*."

At some point of a test run, we may have two simultaneous instances of *C1*, one with system nodes n_1 and n_2 matching x and y, and one with n_1 and n_3. At some later point, system node n_1 may play the role of y in yet another instance of *C1*.

Given a scenario, the analysis of a test trace thus involves two steps:

1. Determine which physical nodes of the trace exhibit the (sequence of) configuration(s) of the scenario, and when they do so.
2. Analyze the order of events in the identified configurations.

Assuming that system configuration graphs can be built from the contextual test data, step 1 amounts to a graph matching problem. We explained in [4] how subgraph isomorphism can be used to search for all instances of the scenario configurations in a trace. We developed a tool, *GraphSeq*, and performed experiments using randomly generated graphs, contextual data from a mobility simulator, and test traces from the GMP case study [4].

In this paper, we focus on the second step of the trace analysis. The principle is to process the event view of a scenario to build a symbolic automaton, having variables that depend on the spatial configurations. The automaton can then be instantiated according to *GraphSeq* outputs, and used to analyze the order of communication and configuration change events for all found spatial matches.

Before describing the building of the automaton, we discuss some design choices made for the TERMOS language.

5 Design Choices for the TERMOS Event View

When developing a new language both its syntax and semantics have to suit the high-level goal of the language. The primary purpose of TERMOS scenarios is to describe various verification artifacts, and to use these scenarios for checking execution traces. Hence the language should be designed to make it possible to determine whether or not a test trace fulfils a target property by (i) limiting non-determinism, and (ii) providing an operational semantics that can be implemented in a toolset.

As TERMOS is based on UML 2 Sequence Diagrams, as a first step the capabilities of Sequence Diagrams were investigated. We surveyed the existing formal semantics proposed for Sequence Diagrams [9]. It turned out that many semantic choices exist even for simple diagrams and elements, e.g., whether a diagram is a complete or a partial interaction or how a trace is represented. If more complex language elements are allowed which can express alternatives or negation, it becomes harder and harder to decide whether a given execution trace conforms to a scenario. Therefore we selected those options for TERMOS from the choices collected in [9] which make checking of traces possible, and added further syntactic restrictions. This section highlights these design decisions (for a complete list see [5]).

- *Interpretation of a basic Interaction:* In TERMOS the execution traces can be categorized as valid, invalid or inconclusive (as opposed to some approaches where only two categories are sufficient). In a requirement or test purpose scenario only the relevant part of the behaviour is depicted (subset of nodes, subset of messages). Hence a TERMOS scenario describes only a partial interaction: a prefix or suffix is allowed in the execution traces and extra messages can interleave.
- *Introducing CombinedFragments:* Alternative fragments, parallel compositions or even negation can be expressed with CombinedFragments. However, in UML there is no synchronization mechanism amongst lifelines when entering or exiting fragments. This could present several challenges when verifying traces (e.g., there is no common point to evaluate guards or the scope of the operator is unclear). For this reason, in TERMOS, we interpret the entering and exiting of a CombinedFragment as a synchronization point for the participating lifelines.
- *Computing partial orders:* The orderings between the elements of the diagrams are encoded in a state-based formalism. A crucial point is how to handle the alternative fragments in the diagrams. In TERMOS to make verification possible there is a global time point when all the participating lifelines evaluate the guards and choose one alternative (this will be represented by a common transition in the formal semantics). Moreover, only a deterministic form of guarded choice is allowed (similar to an if-then-else construct). Finally, variables in guards and state invariants can only refer to message parameters previously sent or received and to node attributes in the current configuration. This guarantees that unrepresented nodes and messages cannot change the valuation of a predicate.
- *Interpretation of conformance-related operators:* The conformance-related operators (*assert, neg, consider, ignore*) modify the categorization of a trace as valid, invalid or inconclusive. Their usage is heavily restricted in order to make the checking of a trace feasible. Negation can be used only for the whole interaction as a global false predicate. The diagram can have only one *assert* box at the end of the

diagram, which should cover all lifelines. *Consider* and *ignore* change the alphabet of messages that are allowed to interleave with the ones depicted in the diagram. In TERMOS, the default interpretation is that any extra message may interleave, hence the *ignore* operator is not needed. We use *consider* to reduce the set of valid traces, i.e. to indicate that some of the extra messages are not allowed. Only one level of nesting is allowed for conformance-related operators (e.g., *assert* into a toplevel *consider*). Furthermore the type of nesting is also restricted (e.g., no double assertion).

We implemented a prototype tool that checks conformance to the syntactic restrictions that TERMOS puts on UML interaction models. The tool is an Eclipse plug-in and can be used to check diagrams imported from a UML editor. In addition to syntactic checks, it performs some semantic verification: for each message, it checks whether the communication between the sender and the receiver can take place in the current spatial configuration (nodes are active and connected by an edge).

6 Semantics of the TERMOS Event View

We now provide a high-level description of the semantics of the TERMOS event view. A detailed, technical description is to be found in [5]. Here, we illustrate the principle by the scenario shown in Fig. 3. The interaction takes place in a hypothetical application where infrastructure nodes with fixed locations periodically broadcast information. Any mobile node that approaches an infrastructure node may ask for additional details, in which case the details must be delivered to it.

For space constraints, the illustrative scenario is kept simple. It contains no alternative or parallel fragments. Obviously, our algorithm for building the automaton accommodates all such constructs from the TERMOS syntax (see [5]).

6.1 Construction of the Automaton

The construction of the symbolic automaton corresponding to the TERMOS event view involves two steps.

Pre-processing: First, the diagram is parsed to extract its atomic elements and the ordering relations between them. For example, the atoms appearing on the infrastructureNode lifeline are: the lifeline head, the configuration change, the sending of information, the receiving of getDetail, the entering of *assert*, the sending of details, the exiting of *assert*, the lifeline end. Atoms are grouped into classes capturing simultaneity. For example, all lifelines enter the *assert* at the same time. Precedence and conflict relations between classes are then computed. For example, the sending of getDetail precedes its receiving, and both precede the entering of assert. Conflict relations would concern atoms located in different operands of an *alt* CombinedFragment, this is not illustrated by Fig.3.

Unwinding: In the second step, the symbolic automaton is built by gradually unwinding the classes of atoms, until all of them have been processed. A state of the automaton is a global state of the scenario, capturing the progress of all lifelines. The algorithm starts in an initial state with the lifelines heads unwound. Then, it uses the precedence and conflict relations to search for the enabled classes of atoms and to compute the successor states. Each transition is labeled according to the currently unwound class. A label can be an event expression, consuming a trace event that matches it, or a predicate to be evaluated without consuming an event. Both kinds of labels may involve variables, and event consumption may trigger update actions. If the trace analysis reaches a state where no transition can be fired, the automaton exits and returns a verdict that depends on the category of the state.

We illustrate these notions informally on the example.

The automaton of Fig. 4 has three categories of states. Double circle nodes represent trivial accept states: they are used to categorize traces that do not exhibit the potential behavior before the *assert*. Single circle nodes are the reject states (trace is invalid) and triple circle nodes the stringent accept states (trace successfully reaches the end of the *assert*).

In Fig. 4, the states are labelled by an id and the set of variables that are currently valuated. For example, in initial state *0*, the only valuated variables are the ones from the current C1 configuration: we know the identity of the nodes playing the roles of infrastructureNode and mobileNode.

Transitions coming out of states *0* and *1* have labels illustrating event expressions. Transition from *0* to *1* consumes a configuration change to C2. The self-loop on *0* consumes any event that does not match a configuration change (i.e., it consumes any communication event). Transition from *1* to *2* consumes the sending of *information*. The event expression is a triplet, where *!information* denotes the sending of the message, *insfrastructureNode* is the node performing this event, and *$1* is a symbolic message id. Since *$1* is a free variable in state *1*, it can be matched by any id generated by the instrumentation functions of the test platform. Transition to state *2* updates this variable, hence *$1* is no longer free when waiting for the receive event.

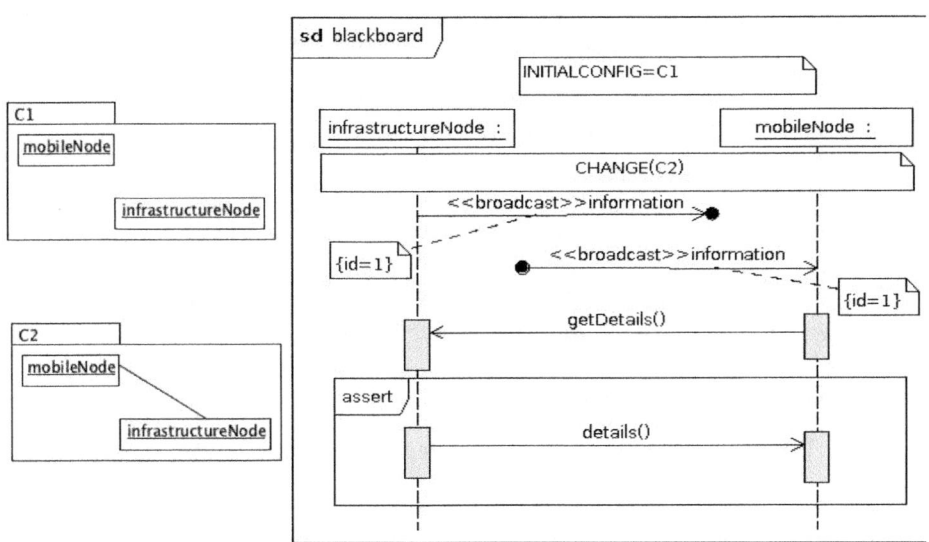

Fig. 3. A TERMOS scenario

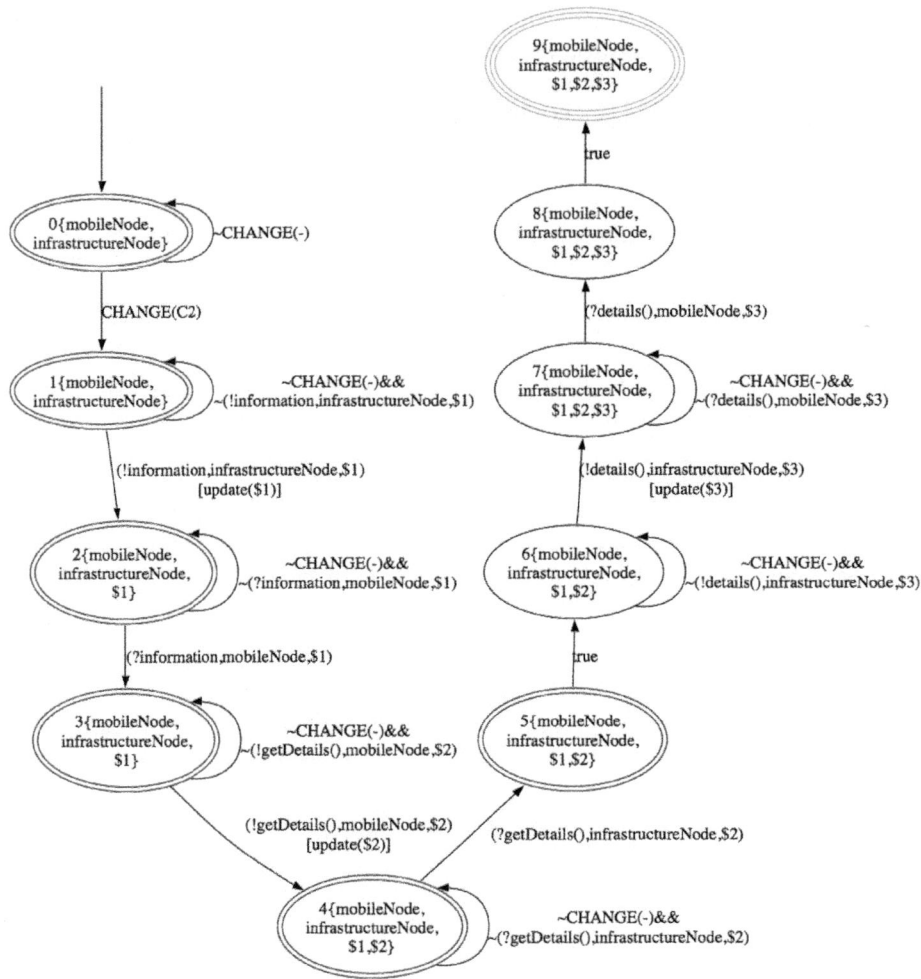

Fig. 4. Symbolic automaton from the scenario in Fig.3

Transitions that are not labelled by event expressions do not consume trace events. Examples are 5->6 and 8->9, corresponding to the entering and exiting of *assert*.

6.2 Proof-of-Concept Prototypes

We developed two prototypes, demonstrating the building of the automaton from different standpoints.

The first prototype, developed at LAAS, focused on the study of the algorithm itself. The aim was to convince us that it captures the intended meaning of diagrams. Accordingly, the tool provides a graphical visualization of the generated automaton structure (Fig. 4 has been produced by this tool, which uses the Graphviz open source package). The graphical visualization allowed us to manually check the result of the algorithm for a sample of diagrams, illustrating the various TERMOS constructs. Note that the prototype extracts the atoms from raw diagrams exported in the SVG (Scalar Vector Graphics) format. This allowed us to get quick feedback on the algorithm, as we did not bother specializing UML editors and parsers to a TERMOS profile. The solution is however not satisfactory for the long run.

The second prototype, developed at Budapest, aimed to demonstrate the integration of TERMOS into UML support technology. The tool was developed as an Eclipse plug-in, as the Eclipse platform has extensive built-in support to manipulate UML models. A TERMOS UML profile was developed to tag TERMOS scenarios and configurations with special stereotypes. The plug-in loads an Eclipse UML2 compliant model, and operates on the UML elements directly. It first searches for the elements with the TERMOS stereotypes, and can perform the syntactic and semantic checks described in Section 5. As the next step it generates the automaton for a TERMOS interaction, and stores it as an XML file. Finally, a test trace specified in a simple textual format can be checked against the automaton, and the trace is categorized as valid, invalid or inconclusive. Note, that the tool is not connected to *GraphSeq*, i.e. it requires that nodes in the traces should be mapped to nodes defined in the scenario manually.

7 Conclusion

This paper presented an approach that uses graphical scenarios, describing key interactions in mobile computing systems, for test and verification activities. The formal language TERMOS, based on UML Sequence Diagrams, is at the core of the approach. For space constraints, the presentation of TERMOS remained high-level. Its aim was to convey the main concepts underlying the language:

- TERMOS incorporates three elements we found useful to model scenarios in mobile settings. It has a spatial view, allowing us to abstract movement and node creation/shutdown by a sequence of labeled graphs. Its event view may contain configuration change events. Communication events include local broadcast.
- TERMOS is used to specify properties for subsets of nodes exhibiting predefined patterns of spatial configurations. The properties concern the partial orders of their communication and configuration change events. They come in various forms: positive requirements, negative requirements and test purposes.
- The TERMOS semantics combines graph matching and an operational state-based semantics for UML Sequence Diagram constructs.

The aim of TERMOS is the automated checking of test traces. We do not have yet a complete tool chain for this analysis, but the major pieces are there. The graph matching tool, developed at LAAS, has been presented in [4]. The production of the automaton was presented in this paper. Future work will build on the existing tools to integrate the two parts of the semantics. We will retain the principle of using UML support technology, demonstrated by the Budapest contribution.

Acknowledgement. This work was partially supported by the ReSIST network of Excellence (IST 026764) and the HIDENETS project (IST 26979) funded by the European Union under the Information Society Sixth Framework Program.

References

1. ITU-T: Recommendation Z.120: Message Sequence Chart (MSC) (2004)
2. Object Management Group: UML 2.2 Superstructure Specification, formal/09-02-02 (2009), http://www.omg.org/docs/formal/09-02-02.pdf
3. Nguyen, M.D., Waeselynck, H., Rivière, N.: Testing Mobile Computing Applications: Toward a Scenario Language and Tools. In: Int. Workshop on Dynamic Analysis (WODA 2008), pp. 29–35 (2008)
4. Nguyen, M.D., Waeselynck, H., Rivière, N.: GraphSeq: a Graph Matching Tool for the Extraction of Mobility Patterns. In: 3rd IEEE Int. Conf. on Software Testing, Verification and Validation (ICST 2010). IEEE CS Press, Paris (2010)
5. Waeselynck, H., et al.: Refined Design and Testing Framework, Methodology and Application Results, Hidenets D5.3 (2008),
http://www.hidenets.aau.dk/Public+Deliverables
6. Baumeister, H., Koch, N., Kosiuczenko, P., Stevens, P., Wirsing, M.: UML for Global Computing. In: Priami, C. (ed.) GC 2003. LNCS, vol. 2874, pp. 1–24. Springer, Heidelberg (2003)
7. Grassi, V., Mirandola, R., Sabetta, A.: A UML Profile to Model Mobile Systems. In: Baar, T., Strohmeier, A., Moreira, A., Mellor, S.J. (eds.) UML 2004. LNCS, vol. 3273, pp. 128–142. Springer, Heidelberg (2004)
8. Kusek, M., Jezic, G.: Extending UML Sequence Diagrams to Model Agent Mobility. In: Padgham, L., Zambonelli, F. (eds.) AOSE VII / AOSE 2006. LNCS, vol. 4405, pp. 51–63. Springer, Heidelberg (2007)
9. Micskei, Z., Waeselynck, H.: The Many Meanings of UML 2 Sequence Diagrams: a Survey, Software and Systems Modeling, Online first (2010), doi:10.1007/s10270-010-0157-9
10. Damm, W., Harel, D.: LSCs: Breathing Life into Message Sequence Charts. Form. Methods Syst. Des. 19(1), 45–80 (2001), doi:10.1023/A:1011227529550
11. Sengupta, B., Cleaveland, R.: Triggered Message Sequence Charts. IEEE Trans. on Software Engineering 32(8), 587–607 (2006)
12. Klose, J.: Live Sequence Charts: a Graphical Formalism for the Specification of Communication Behavior. PhD thesis, C. v. O. Universitat Oldenburg (2003)
13. Küster-Filipe, J.: Modelling Concurrent Interactions. Theoretical Computer Science 351(2), 203–220 (2006)
14. Harel, D., Maoz, S.: Assert and negate revisited: Modal semantics for UML sequence diagrams. Software and Systems Modeling 7(2), 237–253 (2008)
15. Waeselynck, H., Micskei, Z., Nguyen, M.D., Rivière, N.: Mobile Systems from a Validation Perspective: a Case Study. In: 6th Int. Symp. on Parallel and Distributed Computing (ISPDC 2007), July 5-8. IEEE Press, Hagenberg (2007)
16. Huang, Q., Julien, C., Roman, G.: Relying on Safe Distance to Achieve Strong Partitionable Group Membership in Ad Hoc Networks. IEEE Trans. on Mobile Computing 3(2), 192–204 (2004)

Enforcing Security Policies in Mobile Devices Using Multiple Personas

Akhilesh Gupta[1,3], Anupam Joshi[1,2], and Gopal Pingali[1]

[1] IBM Research - India,
Plot 4, Block C, Vasant Kunj Institutional Area,
New Delhi 110070, India
akhilesh.iitdelhi@gmail.com, anupam.joshi@in.ibm.com, gpingali@us.ibm.com
[2] CSEE Department,
University of Maryland, Baltimore County
[3] Department of Computer Science,
Stanford University

Abstract. Cell phones are becoming increasingly more sophisticated, and such "Smart" phones are a growing front end to access the web and internet applications . They are often used in a multiple modes – for instance for both personal and business purposes. Enterprises that allow employees to use the phones in this dual mode need to protect the information and applications on such devices and control their behavior. This paper describes an approach that integrates declarative policies, context and OS level device control to enforce security by creating multiple personas for the device. We describe the approach, and present a proof of concept implementation on Android.

1 Introduction

Mid to high end smartphones have become commonplace in corporate settings. The same smartphone is used in a corporate as well as personal setting, has both types of applications, and stores both personal and corporate data. This can lead to significant security concerns. Sensitive data can be compromised by making the device behave inappropriately using one of the installed applications, or by hacking into the device using its connectivity to public networks. Data can even be compromised by the device being lost, or someone stealing the removable storage media or the device itself. While we illustrate this problem using a corporate / personal dual persona situation, more generally we might want the device to have multiple personas. For instance, one when it is connected to a trusted network, another when it is in a particular location (office vs home), and yet another when we are using a particular application. We address this problem by forcing the device to isolate its functionality and behave differently when being used in different contexts using a middleware layer.

We argue that a set of applications, data, and device capabilities are relevant to a context, which we call a *device persona*. Each such persona should have its own sandbox to run, and should be isolated from other personas. What is permitted within this sandbox needs to be dictated by the corresponding policy. We propose a security mechanism in which the device functions in distinct modes, and specifically show an implementation for separating personal use from enterprise use.

For example, the enterprise may want to specify in a policy that device capabilities like 802.11, Bluetooth, and GPRS are unrestricted while in the personal mode. However, if the user wishes to read their official email, it requires the device to switch to the enterprise mode. In this mode, the policy directs the enforcement mechanisms on the device to disable the Cellular radio and Bluetooth, and limit access via 802.11 interfaces to the enterprise VPN. The device platform is modified to actually enforce the issued policies. The enterprise VPN server is entrusted with the task to verify the login credentials of the user requesting to switch to the enterprise mode and to verify the authenticity of the device before granting network resources to the device. The device authentication is carried out by ensuring that the device is actually running the customized build of the platform distributed by the enterprise. We describe a proof of concept implementation carried out by customizing the Android Donut Mobile platform. Note that our approach builds on top of existing mechanisms such as link encryption and user authentication – it does not supplant them. Moreover, it also does not obviate the need for malware detection and remediation systems.

2 Related Work

Most of the current systems for device level security work on a per application basis, typically at install. For instance in Android[1], each application typically gets its own uid. All permissions it needs are predeclared (and signed) by the developer. At install time, these permissions are either granted to the uid or not based, amongst others, on interactions with the user. This is very coarse grained control. Ideally, the decision on whether or not to grant a request to access device features from an application should be based on the context of the device and the user. For instance, an enterprise may have a policy that cell phones with cameras are not allowed to take pictures inside the office, or that they must be in vibrate mode when in a meeting room. These are all instances of the context of the device, as captured in the policy of the space it happens to be in. Similar context dependant policies arise from the user's perspective. A user might want to make his Bluetooth device discoverable when at home or office, but not otherwise. We posit that a context dependent, policy driven security mechanism is best suited for device capability security.

Access control is a very well studied topic in literature, with a variety of models that have been developed over four decades of work. We focus here on two works of immediate relevance. The first, by Jansen et al[2] suggests that devices such as PDAs be provided enterprise security policies (in XML) on a smart card. These policies would describe access/noaccess decisions for specific device features. The PDA would run a trusted kernel which would enforce these policies.

---

[1] http://www.android.com/

The focus there was to provide policies, not base them on changing context. We extended this approach in our prior work[5] and showed that a pointer to such a policy could be sent over the 802.11 beacons, so that a "smart space" could point a device to its acceptable use policy. Our proposed approach builds upon some of these ideas. Susilo [6] identified the risks and threats of handheld devices connected to the internet. Since the mobile device is not subject to physical security as are the fixed computers in the wired networks, it is susceptible to attacks and hence can potentially host malicious code while it is in some untrusted network and try to propagate it, once back in the home network. Another scenario involves a temporary user granted access to the network injecting malicious code into the network.

Our system uses a declarative policy-based approach, where the rules of behaviour, or the boundaries of the sandbox, of entities in a variety of environments are described in a machine-understandable specification language. Policy driven systems can even be engineered so as to be extremely lightweight on resource constrained devices[5]. Semantic web languages such as RDF and OWL prove a natural choice for such policy languages.

Policy driven security has been looked at by several academic research groups over the last decade. Rei [3] is an example of a *declarative policy language* that uses Semantic Web technologies to describe policies as constraints over allowable and obligated actions on resources in the environment. Rei is, of course, just one of the recent efforts to develop declarative policy languages. Most are not, unlike Rei, motivated by the security and privacy issues in open systems such as ubiquitous computing. These include industry standards such as XACML [4], but also academic efforts such as Ponder [1].

3 Design Approach

Android is a free, open source, and fully customizable mobile platform by Google and the Open Handset Alliance. Before we discuss our implementation in the next section, there is a present a quick overview of three relevant components of the Android platform viz. *services, intent broadcasts and security*. A more general discussion of Android is beyond the scope of this paper. However, note that security is enforced at the process level through standard Linux facilities, such as user and group IDs that are assigned to applications. The permissions required by an application are declared statically in that application, so they can be known up-front at install time and will not change after that.

A *service* runs in the background for an indefinite period of time. Most core system functions are carried out by services. A *broadcast receiver* receives and reacts to broadcast announcements. Many broadcasts originate in system code to notify the rest of the system about a state change. Services and broadcast receivers are activated by asynchronous messages called *intents*. These *intents* name the action being announced making the concerned service handle the state as required.

There are several possible approaches to provide each persona with its own sandbox. The most obvious approach is to build in separation at the application level using the access control mechanisms provided by the OS. However, these are generally limited in most mobile platforms. More importantly, they can be easily overriden by the user. Another possibility is to introduce modifications at the kernel level. This approach is based on the idea of Jansen et al.[2]. However, this approach adds an extra layer to check to every access request thus raising the overhead significantly.

For Android, we consider an alternate approach that creates each sandbox as a "mode" similar to built in modes like the "Airplane mode". While such modes merely switch off certain functionality by default, they can be extended to perform the functions we need. We add such modes to the stock Android distrubution (1.6) and build a customized image from source to achieve our objectives. We utilized the Android system services to disable interfaces at the framework level. This also allows us to work with the Android security mechanisms and disable user controls in device settings preventing manipulation of the functioning in the modes, and not allowing functionality to be added to the sandbox by the user. Moreover, this approach makes it easier to work with Android networking for authentication procedures and the Volume daemon for working with accessibility of file-systems, both of which are useful for providing the sandbox. To keep the initial implementation simple, we start with the policy expressed in XML, and directly describe the device constraints.

4 Implementation

For our implementation on the Android platform, certain setup functions are performed during initialization of the device. An application has been created for the purpose of accepting the user's enterprise login credentials. The user settings interface presents a button to issue the command for switching back and forth between the two modes.

During the boot process, the device reads the enterprise policy from the policy file (included as part of the build) and configures enterprise mode accordingly. The volume daemon detects the external storage card and stores its configuration and mount point. The device boots up in the personal mode. On receiving the boot completed *intent*, the mount *service* on the device unmounts the external storage reserved for the enterprise mode.

The user opens the application for entering enterprise login credentials. When the user issues the command for switching the mode, the credentials are sent to the enterprise server for validation. In our prototype, we use IBM's internal single sign on system to validate the user.

An *intent* is generated to indicate the switch to the new mode and is broadcast to inform the rest of the device. The *services* corresponding to each of the radio interfaces execute the handler for this *intent*. The handler decides if the radio interface is sensitive to the enterprise mode and enables or disables the corresponding radio based on the policy. The status icons on the user interface are updated to present the current status of various wireless interfaces on the device. The user interface buttons corresponding to each of these radios also handle this *intent* and enable or disable their interface presence. In our example policy demonstrated in the screenshots, all network connections are disabled when in enterprise mode, and only the VPN connection to the enterprise gateway is permitted. This will prevent data leaks on any network path.

The mount *service* invokes the volume daemon to mount or unmount the external SD storage. We mount it for enterprise mode, and unmount it for regular usage. For added security, the filesystem on the SD card could be encrypted. The key, instead of being stored locally on the device, would be provided by the enterprise gateway via the encrypted VPN channel upon successful authentication. Note that we chose to keep the on device flash filesystem encryption free to comply with FCC mandates on smartphones that they be allowed to boot into a mode allowing calls to 911. An Android application can store data on the device memory itself which is private to that application ensured through linux file-system permissions. However, any data stored on the external storage is not secured. Thus, in our implementation, the external storage is reserved for enterprise data and can be accessed only in the enterprise mode. The external storage in Android is handled by the Volume daemon. The external storage is unmounted as soon as the user switches to the personal mode. We show in Figure 1, the disabling of the radios/network and mounting of the external SD card in response to a successful authentication into the enterprise mode.

The user can go back to the personal mode by unselecting the enterprise mode. In our present implementation, no authentication is done for this reverse switch, although it could be easily added if desired.

Fig. 1. Authentication with enterprise server successful. Radios mentioned in enterprise policy get disabled. SD card mounted.

An Android source patch has been created for ease of application of the modification made to the platform to the original source code. This modified source can be built for an actual android device to generate a system image. This system image can then be signed by the enterprise for installing onto real devices. The customized Android build can be flashed to devices and then distributed to employees of the organization. New policy files and build patches can be supplied by the enterprise network to the devices wirelessly using the Android OTA (over-the-air) update mechanism. We note that arbitrary personas can be created using our system that are specific to the particular needs of an enterprise or organization. Each persona can specify an arbitrary combination of functionality that is available.

5 Conclusion and Future Work

Devices such as smartphones are increasingly being used in both personal and professional/enterprise context. This creates serious security concerns about loss of data from the compromise or loss of the phone. Devices can also be attacked using their network connections. We have proposed a solution to this problem by having the device take on multiple personas, each corresponding to a sandbox where certain applications and device functionalities are allowed. What functionalities are allowed are based on the needs of the mode (e.g. personal, enterprise, travel hotspot, etc.) and are specified by a declarative policy.

References

1. Damianou, N., Dulay, N., Lupu, E., Sloman, M.: The ponder policy specification language. In: Sloman, M., Lobo, J., Lupu, E.C. (eds.) POLICY 2001. LNCS, vol. 1995, pp. 18–37. Springer, Heidelberg (2001)
2. Jansen, W.A., Karygiannis, T., Gavrila, S., Korolev, V.: Assigning and Enforcing Security Policies on Handheld Devices. In: Proceedings of the Canadian Information Technology Security Symposium (May 2002)
3. Kagal, L., Finin, T., Joshi, A.: A Policy Language for A Pervasive Computing Environment. In: Proceedings of the IEEE 4th International Workshop on Policies for Distributed Systems and Networks (June 2003)
4. Moses, T., et al.: eXtensible Access Control Markup Language (XACML) Version 2.0. OASIS Standard, 200502 (2005)
5. Patwardhan, A., Korolev, V., Kagal, L., Joshi, A.: Enforcing Policies in Pervasive Environments. In: International Conference on Mobile and Ubiquitous Systems: Networking and Services. IEEE, Cambridge (2004)
6. Susilo, W.: Securing Handheld Devices. In: 10th IEEE International Conference on Networks (August 2002)

Scalable and Efficient Pattern Recognition Classifier for WSN

Nomica Imran and Asad I. Khan

School of information Technology,
Monash University, Clayton, Victoria, Australia
{nomicac,asad.khan}@infotech.monash.edu.au

Abstract. We present a light-weight event classification scheme, called Identifier based Graph Neuron (IGN). This scheme is based on highly distributed associative memory. The local state of an event is recognize through *locally* assigned identifiers. These nodes run an iterative algorithm to coordinate with other nodes to reach a consensus about the global state of the event. The proposed approach not only conserves the power resources of sensor nodes but is also effectively scalable to large scale WSNs.

1 System Model

In this section, we model the wireless sensor network. Let there are N sensor nodes in the wireless sensor network. Let $\mathcal{E} = \{e_1, e_2, \ldots, e_E\}$ be a non-empty finite set of such data elements sensors sense from their surroundings. Input data is in the form of *patterns*. A pattern over $\mathcal{E}$ is a finite sequence of elements from $\mathcal{E}$. The length of a pattern is the number of sensors in the system. We define $\mathcal{P}$ as a set of all possible patterns of length L over $\mathcal{E}$. We model GN as a structured overlay $\mathcal{G} = \{(\mathcal{E} \times \mathcal{L})\}$ where $\mathcal{L} = \{1, 2, \ldots, L\}$, where $n(e_i, j)$ is a node in $\mathcal{G}$ at i-th row and j-th column. GN can be visualized as a two dimensional array of L rows and E columns. Total number of nodes in the $\mathcal{G}$ are $E \times L$. We refer all the nodes in the $(j-1)$ column as the left neighbors of any node $n(*, j)$ in j-th column. Similarly, all the nodes in the $(j+1)$ column are called as the right neighbors of $n(*, j)$. In a GN-based classifier[Raja,2008],[Nasution,2008], each node $n(e_i, j)$ is programmed to respond to only a specific element e_i at a particular position j in a pattern. That is, node $n(e_i, j)$ can only process all those patterns in $\mathcal{P}$ such that e_i is at the j-th position in that pattern. Each node maintains an *active/inactive* state flag to identify whether it is processing the incoming pattern or not. Initially all nodes are in inactive state. Upon arrival of a pattern, if a node finds its programmed element e_i at the given position in the pattern, it switches its state to active otherwise it remains inactive. Only active nodes participate in our algorithm and inactive nodes remain idle. At any time, there can be exactly L active nodes in a GN. Hence, there are exactly one active left-neighbor and exactly one active right-neighbor of a node $n(e_i, j)$ where $j \neq 0, l$. Whereas terminal nodes $n(e_i, 0)$ and $n(e_i, L)$ has only one active left and right neighbor respectively.

2 Proposed Protocol

On arrival of an input pattern $\mathcal{P}$, each active node $n(e_i, j)$ store e_i in its j_{th} position. Each node $n(e_i, j)$ sends its matched element e_i to its active neighbors $(j+1)$ and $(j-1)$. The GNs at the edges will send there matched elements to there penultimate neigbours only. Upon receipt of the message, the active neighboring nodes update there bais array. Each active node $n(e_i, j)$ will assign a local state L_s to the received $(e_i,)$ value. The generated local state L_s will be $Recall$ if the the added value is already present in the bais array of the active node and it will be a store if in-case its new value. An $<ID>$ will be generated against each added value. The rules for assigning ids are as under:

- Rule 1. $Store_{(Si)} > Recall_{(Ri)}$: If node $n(e_i, j)$ self local state L_s is $Recall$ but it receives a $Store$ command from any of its neighbors, $(j+1)$ or $(j-1)$, it will update its own state from $Recall_{(Ri)}$ to $Store_{(Si)}$.
- Rule 2. All New Elements: If any of the elements presented to $\mathcal{G}$ is not existing in the bais array of any of the active nodes $n(e_i, j)$ suggests that its a new pattern. Each active node $n(e_i, j)$ will create a new <ID> by incrementing the already stored maximum <ID> in there bias array by 1.
- Rule 3. All Recalls with Same ID: If e_i presented to $\mathcal{G}$ is the recall of previously stored pattern with same <ID>, means that its a repetitive pattern. The same <ID> will be allocated to this pattern.
- Rule 4. All Recalls with Different IDs: If all the e_i of the pattern $\mathcal{P}$ presented to $\mathcal{G}$ are the recall of already stored patterns with different <IDs> indicates that it's a new pattern. Each active node $n(e_i, j)$ will find out the $max(ID)$ in there bias array and will increment it by 1.
- Rule 5. Mix of Store and Recall: If the end decision is to $Store$ due to mix of $Store$ and $Recall$, each active node $n(e_i, j)$ will again find out the $max(ID)$ in there bais array and will increament it by 1.
- Transition Rule 1 If the active node $n(e_i, j)$ has received a greater value from its left neighbor $(j+1)$, it will upgrade its local state and transfer this updated value to its right $(j-1)$ neighbor only.
- Transition Rule 2 In case if the received value from both the neighbors $(j+1)$ and $(j-1)$ are smaller than the local value, node $n(e_i, j)$ will upgrade its value and transfer this new value to both of its neighbors. Once the pattern has been stored, a signal is sent out within the network informing all the IGN nodes that the pattern has been stored. This is called the pattern resolution phase.

3 Conclusion and Future Work

In this paper we have proposed an in-network based pattern matching approach for providing scalable and energy efficient pattern recognition within a sensor network. Our proposed scheme is base station independent. The proposed scheme is also independent of preprocessing steps such as patterns segmentation or training for its processing. Through parallel processing, the scalability issues in WSN are catered well.

Pervasive Integrity Checking with Coupled Objects

Paul Couderc, Michel Banâtre, and Fabien Allard

INRIA Rennes, Campus Universitaire de Beaulieu, 35042 Rennes Cecex France
{paul.couderc,banatre,fallard}@inria.fr
http://www.inria.fr/recherche/equipes/aces.fr.html

Abstract. Integrity checking is important in many activities such as. While the computing and telecommunication worlds commonly use digital integrity checking, many activities from the real world do not beneficiate from automatic mechanisms for ensuring integrity. We propose the concept of coupled objects where groups of physical objects are tagged with RFID chips enabling pervasive and autonomous integrity checking.

Keywords: Pervasive computing, security, integrity, RFID.

Checking for integrity of a set of objects is often needed in various activities, both in the real world and in the information society. The basic principle is to verify that a set of objects, parts, components, people remain same along some activity or process, or remains consistent against a given property (such as a part count).

In the real world, it is a common step in logistic: objects to be transported are usually checked by the sender (for their conformance to the recipient expectation), and at arrival by the recipient. When a school get a group of children to a museum, people responsible for the children will regularly check that no one is missing. Yet another common example is to check for our personal belongings when leaving a place, to avoid lost. While important, these verification are tedious, vulnerable to human errors, and often forgotten.

Because of these vulnerabilities, problems arise: E-commerce clients sometimes receive incomplete packages, valuable and important objects (notebook computers, passports etc.) get lost in airports, planes, trains, hotels, etc. with sometimes dramatic consequences.

While there are very few automatic solutions to improve the situation in the real world, integrity checking in the computing world is a basic and widely used mechanism: magnetic and optical storage devices, network communications are all using checksums and error checking code to detect information corruption, to name a few. The emergence of Ubiquitous computing and the rapid penetration of RFID devices enables similar integrity checking solutions to work for physical objects. However, RFID raises serious concerns regarding privacy. In a world where many personal objects are electronically identified, the activities of individuals could be traceable in a similar, though much more comprehensive, way to googling someone today on the Internet.

An important cause of this issue is that RFID systems are usually based on the concept of global identification, associated directory services or tracking databases. However, technically RFID are just small memory devices that can be addressed by near-field communication, and although identification has been the main application target, these devices provide support for alternative mechanisms.

We developped such an alternative approach for pervasive integrity checking, called *coupled objects*, which does not rely on global identifification. We briefly present the principle on the simple application that addresses a common problem when travelling: it is unfortunately easy to forget something. For example, security procedures in airports require that your personal effects are checked separately from you by X-rays. Forgetting one of your items, or mistakenly exchanging a similar item with someone else occurs frequently. Solutions have been proposed for this problem, based on active tags attached to the items that are monitored by an owner tag. This is impractical for several reasons: active tags are expensive, they require batteries (and hence regular maintenance), radio emissions may be restricted by regulations (on planes for example), and temporarily separating an item from its owner would require the alarm to be disabled.

1 Autonomous Integrity Checking for a Set of Physical Objects

Coupled objects enables another solution using RFID tags attached to the items. It is possible to write in the memory of the tags the data required to check the integrity of the group of items. We compute a signature from the identifiers of all the items using a hashing function. An important aspect is that the identifiers associated with each item can be regenerated regularly (eg for each trip): they are only used for a locally computed integrity check, not for identification. The values could be written in the tags at, for example, the airport check-in, the train station, or even when leaving home. Then, at relevant points after the area in which people are vulnerable to item loss or exchange, we deploy checking gates (such as the exit of the security check in airports, or the exit gate of a plane or a train). These gates would ensure the integrity of groups of items crossing them, warning people in the case of a missing item or the presence of someone elses item.

This solution is a distributed system where only local properties are checked in order to ensure a global goal. In fact, it uses a principle similar to the transmission of a file in independent fragments over a packet network, where integrity is verified by checking sequence number coherency or checksums, except that here the data are carried by physical fragments. Such a solution is interesting because while providing a security service, it avoids the privacy concerns of many other RFID approaches. Specifically, tracking of individuals is not easy, since the tags content may change often for the same person and same set of objects. Further, the system is not based on identification, ensuring greater privacy.

Another interesting aspect is that checkpoints and association points are autonomous and only carry local processing. The system is therefore not dependent on a remote information system. This has important benefits in terms of extensibility, reliability and deployment costs.

2 Perspectives and Conclusion

We presented the principles of an approach enabling the design of pervasive integrity checking solution for many applications. The strong points of this solution are its independence of any remote information system support or network support, and user's privacy respect as it is anonymous and does not relies on global identifiers. It differs from many RFID systems where the concept of identification is central and related to database supported information systems. The approach also differs from some personal security systems based on PAN and active monitoring [3]. Coupled objects is in the line of the idea of RFID used to store in a distributed way group information over a set of physical artifacts, due to Bohn and Mattern [1], and *SmartBox* [2], where abstractions are proposed to determine common high level properties (such as completeness) of groups of physical artifacts using RFID infrastructures.

The approach we presented has benefits in various potential application where privacy and/or autonomy is a concern. Yet there are still challenges to overcome. In some applications where many tags have to be read at once on mobile objects, the performance of current hardware with respect to inventory read reliability and speed can be an issue. Another issue is security: the tags used in any RFID security solution should resist to tag cloning attacks or tag destruction attempt. This typically involve tag level cryptography logic, which require more execution cycles, more power, and more time to be read. Other perspectives are other application scenarios, we are in particular examining green-IT solutions for waste recycling using coupled objects to ensure the quality of the returned materials. Finally, we are also developing the mapping of complex data structures such as tree on a set of memory-limited tags considered as fragments.

References

1. Bohn, J., Mattern, F.: Super-distributed rfid tag infrastructures. In: Markopoulos, P., Eggen, B., Aarts, E., Crowley, J.L. (eds.) EUSAI 2004. LNCS, vol. 3295, pp. 1–12. Springer, Heidelberg (2004)
2. Floerkemeier, C., Lampe, M., Schoch, T.: The smart box concept for ubiquitous computing environments. In: Proceedings of sOc 2003 (Smart Objects Conference), Grenoble, pp. 118–121 (May 2003)
3. Kraemer, R.: The bluetooth briefcase: Intelligent luggage for increased security (2004), http://www-rnks.informatik.tu-cottbus.de/content/unrestricted/teachings/2004/

Service Discovery for Service-Oriented Content Adaptation

Mohd Farhan Md Fudzee, Jemal Abawajy, and Mustafa Mat Deris[*]

School of Information Technology,
Deakin University, Geelong, Australia

Abstract. Service-Oriented Content Adaptation (SOCA) has emerged as a potential solution to the content-device mismatch problem. One of the key problems with the SOCA scheme is that a content adaptation task can potentially be performed by multiple services. In this paper, we propose an approach to the service discovery problem for SOCA and it is demonstrated to perform well.

1 Background and Related Work

Content adaptation is a multi-step process involving a number of services each performing a specific adaptation task. The service discovery problem of interest can be formulated as follow:

Let $T = \{t_1, t_2, \ldots, t_n\}$ and $S = \{s_1, s_2, \ldots, s_m\}$ be a set of adaptation tasks and a set of available services respectively such that $n \ll m$. Let Q represent the quality of service criteria (e.g., time, rating, reputation, cost) of each available service $s_i \in S$. Given S, T and Q, the central problem is how to discover and select a set of composite services that are capable of performing series of tasks.

Existing QoS-based service discovery systems for web service [1] and for pervasive computing require that both the services and the client QoS must be known a priori. However, this may not be practical as most end-users may find it difficult to indicate all their QoS precisely. Moreover, there may not be services that match users' indicated QoS thus, resulting faulty QoS matchmaking. The proposed approach addresses these problems.

2 Proposed Solution and Evaluation

Fig.1(a) shows a high level service discovery architecture for the SOCA platform. The provider layer has the platform providing the services. A service provider advertises its services in one or more service registries. Access to a service is made possible through its service handle in a reference form. The discovery layer deals with those aspects that describe how service discovery is operated to facilitate incoming client requests. Fig.1(b) outlines the algorithm for the service discovery. It is composed of three inter-related steps: adaptation function matching, candidates' discovery and composite service selection. For details of the proposed discovery architecture and algorithm, please refer to [2].

---

[*] Faculty of Comp.Sc.& Info.Tech., Tun Hussein Onn University of Malaysia.

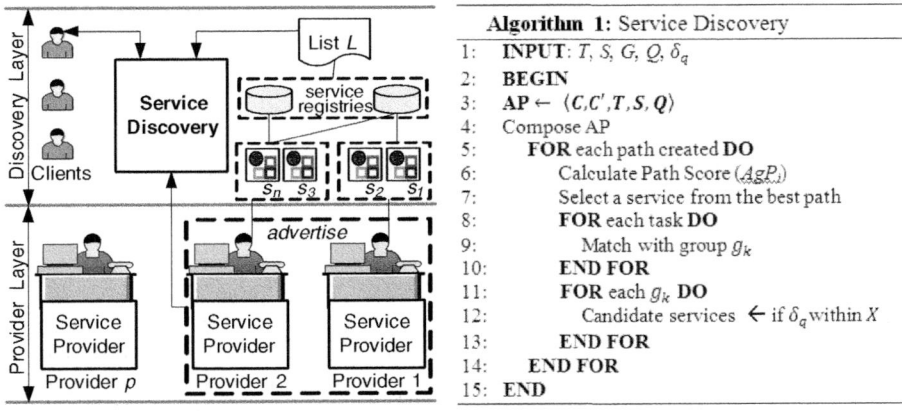

Fig. 1. (a) Service Discovery Architecture and (b) Service Discovery Algorithm

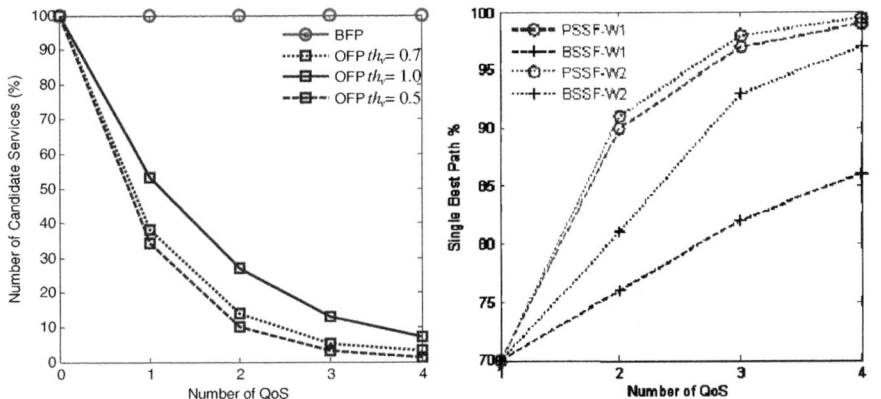

Fig. 2. (a) Number of Candidates and (b) Single Best Path, versus QoS Variations

Fig.2(a) demonstrates that our optimize filter (*OFP*) variations reduce the number of candidates compared to the baseline filter (*BFP*). The suitable *OFP* threshold th_v for providing sufficient candidates is around 0.8 to 1.0, especially when more QoS is taken into account. As shown in Fig.2(b), our objective function (PSSF) generated higher percentages of single path generation for both W1 (equal weight) and W2 (unequal weight) compared to the baseline function (BSSF). For a detailed explanation on experimental setup and result, please refer to [2].

References

1. Song, X., Dou, W.: A Workflow Framework for Intelligent Service Composition. Future Generation Computer Systems (2010), doi:10.1016/j.future.2010.06.008
2. Fudzee, M.F.M., Abawajy, J.H.: On the Design and Evaluation of a QoS-based Service Discovery Approach. TRC10-4. School of IT, Deakin University Australia (2010)

A Hybrid Mutual Authentication Protocol for RFID

Harinda Fernando and Jemal Abawajy

School of Information Technology, Deakin University, Geelong, Australia
{hsf,jemal.abawajy}@deakin.edu.au

Abstract. Out of the large number of RFID security protocols proposed in recent years none have proven to be truly secure and the creation of a truly secure security protocol for systems employing low cost RFID tags remains an open problem. In this paper we develop and present a RFID security protocol which not only allows mutual authentication and secure transmission of data between the reader and tag but is also secure against a number of common attacks.

1 Introduction and Related Work

RFID is a tagging technology that allows an object, place, or person to be automatically identified from a distance without visual or physical contact. Recently a number of RFID security protocols employing cryptographic hashes, public key encryption and PRNG on tag side have been proposed [1]. But these protocols proved to be too resource intensive to me implemented on low cost tags. Therefore protocols using ultra-light-weight cryptography was developed. Some of the more notable protocols in this area are the UMAP family and Gossamer [2]. Unfortunately these protocols were compromised soon after their publication. In this paper we present a mutual authentication protocol for networked RFID system. By using a hybrid of simple one way hash functions, reader side PRNG and bitwise operations we have developed a provably secure protocol whose resource requirements are compatible with low cost RFID tags.

2 Proposed Protocol

Both the tag and the backend system store the EPC and an associated IDS, K1 and K2 for each tag. On coming into contact with the radio wave field of a RFID reader the tag responds with its current IDS. If the received IDS is not recognized the protocol is terminated.

Once the reader has received the IDS it selects the matching EPC, K1 and K2 and generates the PRNG R. It then calculates M1 = H(EPC+K1) and M2 = (K1+R) and M1∥M2 is transmitted to the tag. On receiving M1∥M2 the tag uses the K1 and EPC it has on its memory to calculate C1= H(EPC+K1) and compares it to the received M1. If the two values match the reader is authenticated by the tag. If the values don't match the protocol terminates. Then the tag uses the K1 saved on its memory retrieves R from M2 and calculates M3 = H(EPC+K2+R) and transmits it to the reader. On receiving M3 the reader uses the EPC and K2 it received from the backend database to calculate C2 = H(EPC+K2+R) and compares it to M3. If they match the reader authenticates the tag else the protocol terminates.

Then the reader generates PRNG R2 and calculates M4 = (K1+R2) and M5 = H(EPC+R2) and broadcasts M4||M5. On receiving M4||M5 the tag retrieves R2 and calculates C2= H(EPC+R2). If C2 == M5 then the tag accepts the R2 else it discards and requests a new R2. Once a secure R2 is received the tag starts its key updating. To do this it updates its IDS. K1 and K2 as IDS = IDS + R2, K1 = (K1Right||K2Left) + R2 and K2 = (K2Right||K1left) + R2. Simultaneously the reader does the same calculations and transmits the IDSnew, K1new and K2new along with the EPC to the back end database which updates those values for that EPC.

3 Protocol Evaluation

Our protocol does not contain any of the common weaknesses from other protocols:

- Most protocols use multiple weakly encrypted messages containing the secret keys during authentication which can be leveraged by the attacker using crypto attacks. Our protocol transmits only one message which holds the secret values.
- Most RFID protocols broadcast the EPC of the tag allowing for data leakage attacks. Our protocol only broadcasts the EPC in one way hashed form.
- Some of the protocols employ bitwise AND and OR which have poor statistical properties. Our protocol only employs the XOR function which does not have a biased output eliminating this weakness.

The security analysis of the protocol further showed that the hybrid protocol successfully implements the core security concepts of mutual authentication, transmission confidentiality and integrity, anonymity and availability. It is also secure against a large number of common attacks including main-in-the middle attacks, eavesdropping attacks, replay attacks, tag cloning and spoofing, reader impersonation and de-synch attacks.

We also compared the performance of our protocol against a number of protocols. The analysis showed that our protocol compared favorable to both the ultra-light-weight protocols and the traditional protocols on the metrics of storage and required and overhead bandwidth. Under the metric of number of operations it was definitely less resource intensive than traditional protocols. While it required a hash function on the tag it also required significantly less boolean operations to be carried out when compared to the only currently unbroken ultra-light-weight protocol: Gossamer [2]

References

1. Lim, J., Oh, H., Kim, S.: A New Hash-Based RFID Mutual Authentication Protocol Providing Enhanced User Privacy Protection. In: Chen, L., Mu, Y., Susilo, W. (eds.) ISPEC 2008. LNCS, vol. 4991, pp. 278–289. Springer, Heidelberg (2008)
2. Peris-Lopez, P., Hernandez-Castro, J.C., Tapiador, J.M.E., Ribagorda, A.: Advances in Ultralightweight Cryptography for Low-cost RFID Tags: Gossamer Protocol. In: Chung, K.-I., Sohn, K., Yung, M. (eds.) WISA 2008. LNCS, vol. 5379, pp. 56–68. Springer, Heidelberg (2009)

E2E Mobility Management in Ubiquitous and Ambient Networks Context

Rachad Nassar and Noëmie Simoni

TELECOM ParisTech, LTCI, UMR 5141 CNRS
46, rue Barrault F 75634 Paris Cedex 13, France
{rachad.nassar,noemie.simoni}@telecom-paristech.fr

Abstract. With the rapid evolution of the Next Generation Networks (NGN) concept in the communication industry, mobility requirements have become major challenges that raise the complexity of resource management. In order to maintain continuous end-users sessions, an End-to-End (E2E) mobility management solution must be applied. For this purpose, we introduce the Community of Interest (CoI) concept, also called Virtual Community (VXC) concept, at each NGN layer (service, network, equipment, user). In this paper, we propose to gather elements into VXCs according to two main interests: first, "Ubiquity" which refers to elements that have equivalent QoS and functionality, and second, "Location" which refers to elements that have the same mapping address. These gatherings guarantee the user-centric mobile approach by taking into consideration end-users needs, their demanded QoS and real-time changes in their ambient environments. At he end of this paper, a feasibility study that is based on JXTA (Juxtapose) and EJB (Enterprise Java Bean) technologies is provided.

Keywords: E2E Mobility Management, Community of Interest, Virtual Community, Ubiquity, Location, Context-awareness.

1 Introduction

Nowadays, with the emergence of successive network generations and technologies (2G, 3G, 4G, etc.) and the convergence towards IP-based core networks, the Next Generation Networks (NGN) concept has seen a rapid evolution. Thus, new complex challenges arise. The ubiquity, fix/mobile convergence, and ambient Internet aspects are some of the problems that need to be answered. Such an evolution also induces more complex environments and contexts for end-users, especially by stimulating the growth of their nomadic behavior. Normally, end-users claim the access to any service without any temporal, geographical, technical or economical barriers. They aim to establish a dynamic session that best suits their preferences and their QoS requirements. According to this user-centric context, end-users want to preserve their session while continuously executing their services whatever the access-network, the core-network, equipment and service providers are. All the previously mentioned issues impose the need of an End-to-End (E2E) mobility management that overcomes the mobility challenges on the four NGN visibility levels:

- **Terminal Mobility:** represents the ability of a terminal to switch between access networks while preserving the execution of the same subscribed services list.
- **Network Mobility:** represents the ability to move a gateway router without interrupting the opened session.
- **User Mobility:** represents the ability of an end-user to switch between terminals during the access to his/her services.
- **Service Mobility:** represents the ability to execute the subscribed services even if the end-user is connected to foreign service providers.

Our main objective is to guarantee this E2E mobility management while taking into consideration the end-users preferences and QoS requirements, without neglecting the real-time changes in their ambient environments. In order to ensure an efficient E2E management, mobility solutions should simplify the heterogeneous end-users contexts through the regrouping and partitioning of their ambient resources. Therefore, the way of gathering ambient resources becomes the first challenge to overcome. Once the gathering is insured, managing the resources becomes another challenge. Many questions need to be answered: Based on which criteria this gathering should be performed? Can it satisfy the user-centric flexible context? How to manage the gathered resources so that a seamless and dynamic E2E mobility management is guaranteed?

The remainder of this paper is organized as follows. Section 2 discusses existing mobility solutions. In Section 3, we detail our E2E mobility management solution that answers the aforementioned questions. It is based on gathering NGN resources into virtual communities according to a novel concept, namely the Community of Interest (CoI) concept. Section 4 shows the feasibility of the propositions by providing implementations using JXTA (Juxtapose) and EJBs (Enterprise Java Beans) technologies. Finally, conclusion and future perspectives are presented in Section 5.

2 Overview on Existing Mobility Solutions

In the NGN context, mobility management has become one of the main issues of modern communication. In the last decade, several studies evolved to overcome this challenge. In this section, we provide an overview of existing mobility solutions.

- **Handover:** It is a technology for managing terminal mobility at the transmission layer. It maintains the continuity of communication, either when users change the connectivity type (vertical Handover), or when they switch among attachment points in the cellular system (horizontal Handover). For this purpose, several Handover technologies were proposed [1], such as Soft Handover, Softer Handover, and Hard Handover.
- **Mobile IP:** It is introduced at the network layer in order to allow end-users to move through various access networks with their applications in progress, and while maintaining a permanent IP address [2]. For each visited network, the end-user acquires a temporary IP address called Care of Address (CoA). The latter is different from the permanently obtained address called Home Address (HoA).

- **IMS/SIP:** IP Multimedia Subsystem (IMS) is an architectural platform used for the delivery of IP multimedia in mobile networks. The signaling protocol used by IMS to communicate with the application layer is the Session Initialization Protocol (SIP) [3]. This protocol mainly considers session mobility in the case where end-users change their terminals.
- **HIP:** Host Identity Protocol (HIP) separates the two IP address roles: localization and identification [4]. This separation is guaranteed by introducing a new layer between the network and application layers of the TCP/IP model, as well as a new identifier called Host Identity. Hence, user displacement becomes transparent for applications since they only see the end-user's permanent Host Identity instead of his/her variable IP Address.

Having briefly mentioned existing mobility solutions, we note that each solution is limited to its functional layer and none of them can answer all the aforementioned NGN mobility types. Therefore, a new solution with a global mobility vision is needed. In this paper, we propose an innovative mobility management approach that guarantees a continuous E2E mobile session for each end-user. We should mention that the originality of our solution is in its architectural model that is based on Peer-to-Peer (P2P) auto-managed components. Effectively, on the performance level, our novel gathering model is not comparable to other mobility solutions since it manages a larger scope of the problem. It supports all layers and mobility types in order to reach a continuous E2E session.

3 Proposition

Mobility is a main challenging issue for NGN and a main cause of instability in any NGN environment. To overcome this problem and to guarantee continuous end-users sessions, we propose in this paper a dynamic seamless E2E mobility management solution which is based on regrouping elements into Virtual Communities (VXC, X: Service, Connectivity, Equipment or User). This novel concept is also called the Community of Interest (CoI) concept. Each CoI defines a group of elements that share a common interest. Community members are auto-managed and can exchange information with each others in pursuit of shared goals. The CoI creation and management concepts are applied at four layers (user, equipment, network and service) and can consequently handle the four aforementioned mobility types (user mobility, terminal mobility, network mobility and service mobility). In order to take into consideration the user-centric approach, the sessions E2E QoS and the real-time changes in end-users ambient environments, we propose in the next sub-sections to gather elements into communities according to two main interests: "Ubiquity" (Section 3.1) and "Location" (Section 3.2). In the last sub-section (Section 3.3), we combine these two community types in order to create ubiquity and location based VXCs. Therefore, throughout end-users movements, we anticipate their demands by creating the communities that represent their ambient environments and that respond to their preferences.

3.1 Ubiquity-Based Virtual Community: VXCU

One of the main aspects of our user-centric approach is to guarantee all services anytime, anyhow and anywhere. To reach this goal, ubiquitous elements should be deployed on each NGN layer. For this purpose, we propose to apply "Ubiquity" as the first "Interest" and to gather ubiquitous elements into Ubiquity-based Virtual Communities (VXCUs, U: Ubiquity). In our context, "Ubiquity" represents elements having equivalent QoS and the same functionality. The VXCU concept is applied on the service, network, equipment and user layers in order to guarantee an E2E mobility management. Normally, in a mobile context, some components in end-user's session may not continue to fulfill their SLA or the end-user's QoS requirements. To solve these mobility problems, we dynamically replace the current component by a ubiquitous counterpart that belongs to the same VXCU. Therefore, we seamlessly guarantee a continuous end-user's session while maintaining the end-user's E2E demanded QoS.

Actually, the VXCU model is based on a Peer-to-Peer (P2P) self community management process. Each component is auto-managed and acts as a peer. In the following, we divide the communities creation and management into two phases, namely the deployment and exploitation phases.

In the deployment phase, each component notifies other peers about his functionality and his current QoS. In the first case, the peers that support the same functionality and guarantee an equivalent QoS reply to the notification. These peers are members of a specific VXCU. Consequently, the component joins this VXCU and puts the Community ID into its real-time profile. The latter is the subject of another working group. It instantly describes each NGN component. It is a real-time representation of this component. In the second case, the component did not receive any reply to his notification. Thus, it considers that there is no existing community that answers its QoS and its functionality. Hence, it creates a new community that adopts the component's own characteristics (QoS and functionality). In addition, the newly created VXCU contains a new Community ID that is put into the component's real-time profile.

After explaining what happens in the deployment phase and how we can create and update these ubiquity based VXCs (VXCUs), we need to manage these communities in the exploitation phase. In this phase, all NGN elements confront instant QoS variation. In order to dynamically manage the communities that contain these elements, we develop a QoS-Agent in each NGN component. This QoS-Agent permits the auto-management of its corresponding component. In this auto-management process, each element compares its current QoS with the QoS range of its corresponding community. If the current QoS is into the community's QoS range, then the QoS-Agent sends a notification called "IN Contract" in order to inform the other community members that the community contract is still respected. If not, the QoS-Agent sends a notification called "OUT Contact". In this case, the element exits from the community and tries to find another VXCU with a new Community ID.

It should be mentioned that all these communities (VXCUs) and real-time profiles are stored in a novel knowledge base called Infoware [5]. It is the subject of another working group in our UBIS ("User-centric": uBiquity and Integration of Services) project. This Infoware manages efficiently and dynamically the decisional and reactive information. It is well structured and acts as a real-time informational inference. In this sub-section, we explained how the E2E end-user's session uses ubiquitous VXCU members in order to be easily adapted to the changes caused by mobility. In the next sub-section, we apply the CoI concept to overcome the spatial mobility challenge by using the end-user's location as an "Interest".

3.2 Location-Based Virtual Community: VXCL

Another main aspect of our user-centric approach is the end-user's context awareness [6]. The context is any relevant information that can be provided to characterize the end-user's ambient environment. End-users are context aware when they are capable to adapt their ambient sessions according to the received ambient information. As opposed to fixed distributed systems, mobile systems applications are executed in an extremely dynamic context. Specifically, "Location" is one of the main aspects that cause great modifications in end-users context. For this purpose, context aware end-users should receive information according to their locations. Many concepts are treating location based information. Ambient Networks (ANs) [7] is one of these concepts. It tries to incorporate location based information into service provisioning, network composition, and service adaptation. Exploiting this information provides end-users with added-value services and an enhanced communication experience. Actually, ANs intervene by adapting network interconnections in order to provide a wide range of services. Consequently, we notice that ANs main contribution is on the network and service layers. In this paper, we enhance the ANs concept, enrich end-users experiences, and anticipate end-users demands by providing information about all end-users E2E location based ambient resources (services, networks, equipment). To reach this goal, we apply the CoI concept while using the end-user's "Location" as "Interest" for creating and managing Location-based Virtual Communities (VXCLs, L: Location). The latter are location aware communities that gather E2E resources that have the same location as the end-user. For each end-user's location, three VXCLs are created (VSCL, VCCL and VECL).

In the first step, we introduce the four basic services that are used in the creation and management of these VXCLs:

- **Location Basic Service (LBS):** This component determines an element position (Mapping Address: e.g. 11/300 George St, Sydney NSW 2000, Australia) while having its Element ID as an input.
- **Discovery Basic Service (DBS):** This component launches a search in order to discover new elements that meet some demanded criteria (indicated at the input).
- **Presence Basic Service (PBS):** This component filters an obtained list according to the resources states. It selects among a list of Element IDs the ones corresponding to "Availabe" (Accessible by the end-user), "Activable" (activated by the provider but not yet used) and "Activated" (Activable and used) elements.
- **Sorting By Type Basic Service:** This component sorts by type (service, network, equipment) a list of Element IDs received at the input.

In our context, "Location" is represented by a Mapping Address that appears in the end-user's real-time profile. Therefore, Mapping Address might change when end-users

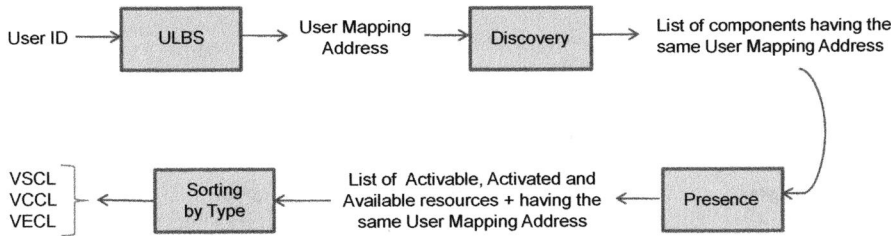

Fig. 1. VXCLs creation workflow

change their terminals (user mobility) or change their location (terminal mobility). In order to easily adapt the E2E session to this spatial mobility, VXCLs are dynamically and seamlessly created and managed according to the hereafter workflow (Fig. 1) that is composed out of six steps:

- **Step 1:** When an end-user changes his/her Mapping Address during an activated session, he/she directly notifies the Infoware that contains the real-time profile in order to update its Mapping Address field.
- **Step 2:** The Location Basic Service localizes the end-user by using the end-user's ID as input. ULBS gives the new Mapping Address at its output.
- **Step 3:** The Discovery Basic Service finds all ambient resources (services, networks, equipment) that have the same Mapping Address as the end-user. For this purpose, DBS uses the end-user's Mapping Address given at its input.
- **Step 4:** The Presence Basic Service is applied on the discovered resources list. It chooses among these resources the "Available", "Activable", and "Activated" ones.
- **Step 5:** The Sorting By Type (Service, Network, Equipment) basic service classifies the obtained result into three lists (VSCL, VCCL and VECL).
- **Step 6:** If the end-user prefers to receive the VXCL lists that correspond to his/her location, he/she should activate a service called the "VXCL Service". The latter permits to launch the workflow and then to send the obtained results to the end-user.

In this sub-section, we introduced the location based virtual community approach that permits to support the ANs concept and to enhance the context awareness for end-users. It should be mentioned that these VXCLs are not stored for a long time because they expire once the end-user changes his/her location. After explaining the VXCU and VXCL approaches, we study in the next sub-section the possibility of their combination.

3.3 CoIs Combination: VXCL&U

In order to manage mobility challenges, we introduce the CoI concept with two main "Interests". Each "Interest" permits the creation and management of a type of VXCs. However, the combination of these interests is considered as an obvious inference that enhances the E2E mobility management process and the end-user's context awareness. In fact, it is possible to have several elements that belong to the same VXCU and VXCL. Consequently, they form what we call a Location and Ubiquity-based Virtual Community (VXCL&U).

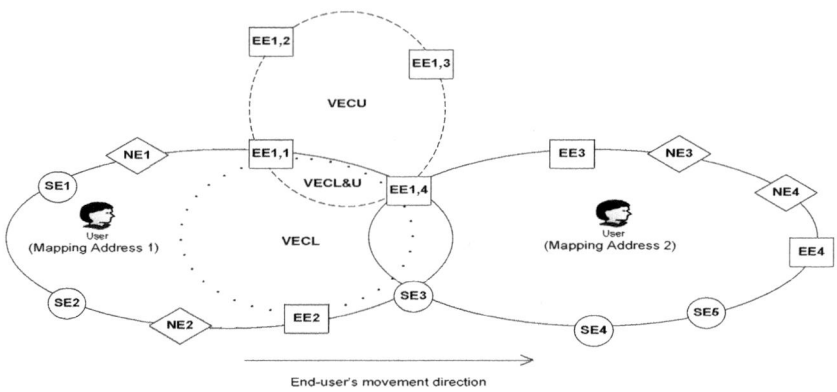

Fig. 2. VXCL&U: VXCU and VXCL combination

An example of VECL&U creation is given in figure 2. In fact, for each Mapping Address, the end-user gets a VECL. Among the VECL members, *EE1,1* and *EE1,4* are QoS and functionally equivalent (belong to the same VECU). Thus, they form a new VECL&U. If the end-user faces a QoS degradation in one of these equipment, he/she can switch to the second equipment in order to maintain his/her continuous session.

4 Feasibility

To validate the feasibility of our CoI proposition, we implement the VXCU approach by using the P2P network proposed by JXTA (Section 4.1), and we implement the VXCL approach by using EJBs (Section 4.2).

```
<?xml version="1.0" ?>
<!DOCTYPE jxta:PD (View Source for full doctype...)>
<jxta:PD type="jxta:PD" xmlns:jxta="http://jxta.org">
    <PID>urn:jxta:uuid-59616261646162614A787461503250339 62AABD49FFD4CDC9F54E187EABE4CE303</PID>
    <Name>PeerES</Name>
    <ElementID>ElementID1</ElementID>
    <Type>Service</Type>
    <Funct>Funct1</Funct>
    <Desc>A specification</Desc>
    <SURI />
    <Vers>1.0</Vers>
    <QOS>QOS1</QOS>
    <SipAddr>127.0.0.1:5061</SipAddr>
</jxta:PD>
```
(a)

```
<?xml version="1.0" ?>
<!DOCTYPE jxta:PGD (View Source for full doctype...)>
<jxta:PGD type="jxta:PGD" xmlns:jxta="http://jxta.org">
    <GID>urn:jxta:uuid-8AAB12F4997C46EBBB6D9F36FAE17AE202</GID>
    <CommunityID>CommunityID2</CommunityID>
    <Funct>Funct2</Funct>
    <type>VSCU</type>
    <SURI />
    <Vers>1.0</Vers>
    <QOS>QOS2</QOS>
</jxta:PGD>
```
(b)

Fig. 3. Advertisements: (a) PeerAdvertisement, (b) PeerGroupAdvertisement

```
myDiscoveryService.addDiscoveryListener((DiscoveryListener) this);
myDiscoveryService.getRemoteAdvertisements(null,DiscoveryService.GROUP,"Funct",Funct,10);
```

Fig. 4. JXTA remote discovery methods

4.1 JXTA for Feasible VXCU

The JXTA platform [8] is based on a P2P network and a set of open-source protocols developed by SUN. JXTA abstracts the complexity of the underlying layers (network, transport) and permits the conception of auto-managed and stateless components.

In a P2P manner, these components (service, network, equipment) called Edge Peers, communicate, collaborate and share their resources. For this purpose, they publish advertisements (XML models) that contain their own descriptions. In this paper, we describe two types of advertisement: the first one is the PeerAdvertisement (Fig. 3(a)) which describes each peer. It contains the peer's name, ID, type, functionality, QoS, etc. The second one is the PeerGroupAdvertisement (Fig. 3(b)) which describes a peer group. The latter represents our VXCU. It contains the VXCU's communityID, type, functionality, QoS, etc. In the following, we limit the feasibility part to the service layer. Thus, each peer refers to a Service Element (SE).

In the deployment phase, each peer discovers its ubiquitous peers (with equivalent QoS and functionality) and creates or subscribes to a community called VSCU. For this purpose, we conceive and integrate a JXTA-Agent into each peer. Its role is to instantiate an object (peer) in the JXTA platform by using the *newNetPeerGroup()* method. The latter permits the peer creation and its access to all basic services that are proposed by the JXTA default group (NetPeerGroup). When we are deploying a peer, the latter publishes a PeerAdvertisement. Our JXTA-Agent launches the *searchCommunity()* method in order to discover the VSCU that contains its ubiquitous peers. First, the JXTA-Agent starts the search in its NetPeerGroup. If the SE does not locally find its desired VSCU, it spreads the search into the whole JXTA network by using a Rendez-vous Peer. The SE uses a DiscoveryListener to keep listening to the replies during a fixed timeout. According to figure 4, we use the *getRemoteAdvertisements()* method while having the "SE functionality" as the discovery criteria. Thus, a list of PeerGroupAdvertisements is obtained. On this list, we apply a newly developed method in order to filter the results according to the QoS criteria. Therefore, we can consider that the discovery is launched according to the QoS and functionality criteria. Hence, two cases appear:

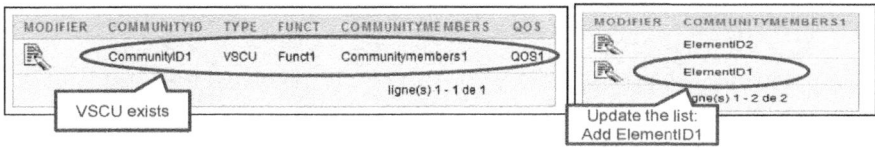

Fig. 5. First case: VSCU update

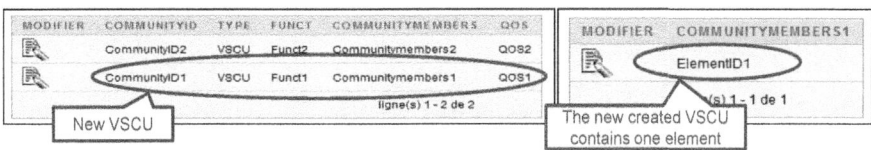

Fig. 6. Second case: VSCU creation

In the first case, the SE receives a local or a remote PeerGroupAdvertisement describing the VSCU that suits its QoS and functionality. Consequently, the SE joins the VSCU by using the *joinToGroup()* method. In order to conserve all the communities and elements information, we create three tables (CommunityProfile, CommunityMembers and RealtimeProfile) in the database. They are managed by the JXTA-Agent by using a newly developed class called *CommunityManagement*. In this first case (Fig. 5), the JXTA-Agent of the SE launches an *update()* method to add the SE's Element ID (*ElementID1*) to the found VSCU CommunityMembers table (*CommunityMembers1*). It also launches another *update()* to add the found VSCU Community ID (*CommunityID1*) to the SE RealtimeProfile table.

In the second case, the SE did not receive a PeerGroupAdvertisement before the timeout, thus it did not find any VSCU that suits its QoS and functionality. Consequently, the SE creates a new VSCU with a new Community ID. The JXTA-Agent publishes the PeerGroupAdvertisement that describes this new VSCU. Then, it uses the *CommunityManagement* methods to update the database tables (Fig. 6). First, it adds the new community into the CommunityProfile. Second, it creates a new table (*CommunityMembers1*) that only contains one SE (*ElementID1*). Third, it updates the SE RealtimeProfile table by adding the new VSCU Community ID (*CommunityID1*).

After explaining the deployment's feasibility part, we pass to the exploitation phase. In this part, the QoS-Agent launches "IN/OUT Contract" messages. In the case of the "OUT Contract", the JXTA-Agent uses the *leaveFromGroup()* method of the *CommunityManagement* class in order to delete the SE's Element ID from the CommunityMembers table, and to delete the SE's Community ID from the RealtimeProfile table.

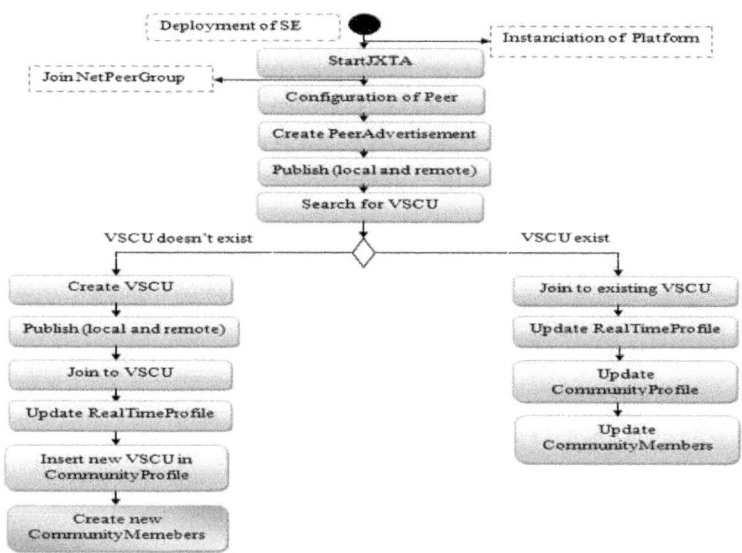

Fig. 7. Logic process of JXTA-Agents

The figure 7 represents the aforementioned logic process of our JXTA-Agent. The same process can be done on the network and equipment layers.

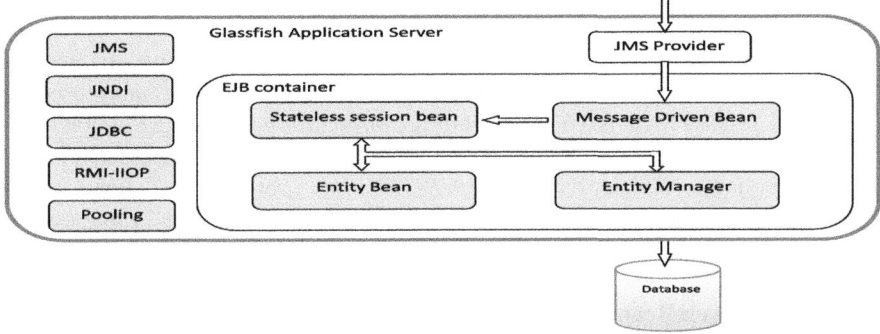

Fig. 8. EJB architecture in the Glassfish application server

4.2 EJBs for Feasible VXCL

The VXCLs creation process is based on a workflow that relates four basic services (Location, Discovery, Presence and Sorting By Type). In order to validate our proposition, we have developed the aforementioned basic services as independent EJBs [9]. The latter permit the creation of different autonomous and loosely coupled components. Besides, EJB provides transparency for distributed transactions and helps the creation of portable and scalable solutions. EJBs are deployed on Glassfish v3 application server [10]. Glassfish v3 is JEE6 certified and consequently supports different APIs such as JMS, JNDI, JDBC and RMI-IIOP. Actually, each EJB service component has the following structure (Fig. 8):

MODIFIER	ELEMENT_ID	TYPE	STATUS	ADDRESS	COMMUNITYID
🔍	ElementID1	service	activated	address1	CommunityID1
🔍	ElementID4	user	accessible	address1	-
🔍	ElementID6	netw	available	address1	CommunityID4
🔍	ElementID7	equip	available	address4	CommunityID3
🔍	ElementID3	service	activatable	address2	CommunityID2
🔍	ElementID5	equip	activated	address1	CommunityID5
🔍	elementID9	service	non activated	address1	-
🔍	ElementID2	service	activated	address1	CommunityID1
🔍	ElementID8	user	accessible	address3	-
🔍	ElementID10	service	activatable	address1	communityID6

ligne(s) 1 - 10 de 10

Fig. 9. The "Database" table

- **Entity Bean:** it is a lightweight persistence domain object that represents one or more tables in a relational database. In our project, the information is stocked in an Infoware. The latter is not used in our feasibility study; instead we assembled all the needed information in one Oracle table that we called "Database". In figure 9, we present a database sample containing different elements with different Types, Statuses, Mapping Addresses and Community IDs.

- **Entity Manager:** it is used to create and remove persistent entity instances, to find entities by their primary key, and to query over entities.
- **Stateless Session Bean:** All beans hold *Conversations* with clients at some level. A *Conversation* is a number of method calls between the client and the bean. A *Stateless Session Bean* is a bean that holds *Conversations* that span a single method call. It is stateless because it does not require state to be maintained across multiple clients requests. Hence, each *Stateless Session Bean* serves many clients and permits to easily pool and reuse components. We must note that these *Stateless Session Beans* are used to implement our stateless components but they interfere in statefull sessions.
- **Message Driven Bean:** it allows the asynchronous messages processing.

```
INFO: MESSAGE BEAN: Message received:
INFO: ElementID1, ElementID6, ElementID5, ElementID2, ElementID10
INFO:
 The VSCL list is :[ElementID1, ElementID2, ElementID10]
 The VECL list is[ElementID5]
 The VCCL list is :[ElementID6]
```

Fig. 10. The VXCL created lists

In this feasibility part, we consider a moving user. His/her address changes from "address 0" to "address 1". Thus, DBS finds all the resources that have an identical Mapping Address. In consequence, PBS filters the DBS result and provides a list that only contains "Available", "Activable" and "Activated" components. Finally, the "Sorting By Type" component sorts the obtained list according to elements types (service, network, equipment). Three lists that represent the different VXCLs are obtained at the output (Fig. 10).

5 Conclusion and Perspectives

Mobility is seen as a main challenge for NGN. We should meet all mobility needs in order to maintain a continuous end-user's session. For this purpose, an E2E mobility management solution must be performed. In this paper, we have introduced the Community of Interest (CoI) concept, also called the Virtual Community (VXC) concept, and we apply it on all NGN layers (service, network, equipment, user). This novel concept consists of decreasing the NGN complexity by gathering elements according to specific interests. The latter have been particularly chosen to suit E2E mobility management needs. Therefore, we applied "Ubiquity" as the first interest. Henceforth, for each element degradation, the latter is dynamically and seamlessly replaced by an ubiquitous element (having the same functionality and an equivalent QoS) that belongs to the same Ubiquity-based Virtual Community (VXCU). Furthermore, in order to overcome the spatial mobility challenge, we have increased end-users context-awareness, by applying "Location" as the second interest. In fact, all elements that have the same end-user's Mapping Address are gathered into Location-based Virtual Communities

(VXCL). Consequently, these created VXCLs guarantee a continuous end-user session by anticipating the end-user's demands. Four basic services (Location, Discovery, Presence, Sorting By Type) are conceived and interconnected in order to create and manage these VXCLs. By combining the VXCU and VXCL communities, the end-user could easily choose an ubiquitous element among the VXCL members in order to replace any damaged element in its session. In the feasibility part, we guaranteed the VXCU logic process by using the P2P JXTA platform, and the VXCL logic process by implementing the basic services into EJBs.

In our future work, we will treat the temporal mobility that is based on end-user's activities. For this purpose, an activity-based virtual community approach seems to be a good proposition. Furthermore, other user interests might be used to create and manage communities and to decrease the complexity of mobile contexts.

Acknowledgement. The authors would like to thank Amen LABIADH and Salah BEN SALAH for their precious contribution and valuable work that benefited greatly this paper. The authors would also like to thank all the participants in the UBIS project, financed by the French ANR VERSO 2008 in which is situated our work.

References

1. 3GPP: Manifestations of Handover and SRNS Relocation. 3GPP TS 25.832, V4.0.0 (2001)
2. Xie, J., Howitt, I., Shibeika, I.: IEEE 802.11-based mobile ip fast handoff latency analysis. In: The IEEE International Conference on Communications, pp. 6055–6060. IEEE Press, Glasgow (2007)
3. Schulzrinne, H., Widlund, E.: Application-layer mobility using sip. ACM SIGMOBILE Mobile Computing and Communications Review 4(3), 47–57 (2000)
4. Al-Shraideh, F.: Host identity protocol. In: The Fifth International Conference on Networking and the International Conference on Systems, pp. 203–203. IEEE Press, Port Louis (2006)
5. Simoni, N., Yin, C., Du Chene, G.: An intelligent user centric middleware for ngn: Infosphere and ambientgrid. In: The Third International Conference on Communication System Software and Middleware, Bangalore, India, pp. 599–606 (2008)
6. Dey, A.K.: Understanding and Using Context. Personal and Ubiquitous Computing 5(1), 4–7 (2001)
7. Giaffreda, R., et al.: Context-aware communication in ambient networks. In: The 11th Wireless World Research Forum, Oslo, Norway (2004)
8. SUN Microsystems, https://jxta.dev.java.net/
9. Sringanesh, R.P., Brose, G., Silverman, M.: Mastering Entreprise JavaBeans 3.0. Wiley Publishing Inc., Chichester (2006)
10. SUN Microsystems, https://glassfish.dev.java.net/

Energy-Aware Cooperative Download Method among Bluetooth-Ready Mobile Phone Users

Yu Takamatsu, Weihua Sun, Yukiko Yamauchi, Keiichi Yasumoto, and Minoru Ito

Nara Institute of Science and Technology, 8916-5 Takayama, Ikoma, Nara 630-0192, Japan
{yu-ta,sunweihua,y-yamauchi,yasumoto,ito}@is.naist.jp

Abstract. In this paper, we propose a cooperative download method to save cellular network bandwidth when mobile phone users download large-size files. Our method allows users who want to acquire the same file to exchange part of the file called *chunks* when coming closer, while suppressing the power consumption by Bluetooth as much as possible. Moreover, our method lets each user actively use cellular network so as to acquire the whole file by the specified deadline.

Keywords: Cooperative Download, Energy-Awareness, Bluetooth, DTN.

1 Introduction

Recently, mobile phone users who download large-size contents from the cellular network are increasing as smartphone users increase and video sharing services such as YouTube become popular. However, if many users download large-size contents at the same time, quality of other communication services such as e-mail and web may be significantly deteriorated. In this paper, we propose a cooperative download method by which users exchange fragments of a content called *chunks* through Bluetooth with neighboring users while moving.

As a cooperative download method to reduce the load of network, BitTorrent has been used. However, BitTorrent is designed for the wired network and we cannot expect the good performance in mobile network. There are some cooperative download methods for mobile users. In [1], Conti, et al. studied how the MANET factors affect performance and overhead when implementing Gnutella in MANET and applied the cross-layer optimization to it. In [2], Rajagopalan, et al. designed and implemented the architecture of BitTorrent for MANET. In [3], McNamara et al. proposed a method where a user terminal identifies a specific terminal that is likely to have useful files while using public transportation. In [4], a cooperative download method for Bluetooth-ready terminals was proposed. However, these existing methods neither guarantee the acquisition of the whole file by the deadline nor consider efficiency in acquiring large-size files. As a cooperative download method that guarantees the acquisition of the whole file by the specified deadline, Hanano, et al. proposed a method utilizing both cellular and WiFi communications [5]. However, this method always turns on WiFi device and consumes a lot of power for WiFi communication.

We think that cooperative download methods for mobile phones should consider the following factors: (i) reducing cellular network load, (ii) suppressing power consumption, and (iii) acquiring contents by the specified deadline.

Our method computes a schedule of communication actions for each user in order to reduce the cellular network load and to make users receive as many chunks as possible. Based on the schedule, user terminals (called *node*, hereafter) download some chunks and exchange them with encountered nodes. Our basic idea is that each user node retrieves a *contact table* specifying the probability and time to encounter other nodes from a server and determines a schedule of communication actions to exchange chunks based on it. Moreover, to reduce power consumed by Bluetooth device, we control each node's Bluetooth state and turn off the device during unnecessary time.

2 Problem Definition

In this section, we define the problem of the cooperative download utilizing both cellular and Bluetooth communication.

Communication Model. Each user terminal[1] u can communicate with other terminals which are in its communication range through Bluetooth. We do not consider the constraint of the number of sessions in the common radio range.

System Model. Each file that users want to acquire is called a *content*. Each content consists of multiple chunks of the same size. The deadline of each chunk is equal to the deadline of each content desired by user. There is a server in the Internet that stores the set of contents. Each terminal can download any chunk of any content stored in the server through cellular network regardless of present location and time.

Mobility Model. The field where users move is represented by a weighted graph $G = (V, E, w)$. The set of vertices V consists of intersections and spots (stations and malls, etc.) where users enter and leave spots but just pass through intersections. The set of edges E is the set of roads on which users move. The weight of each link $(v, v') \in E$ is the distance between vertices v and v' and denoted by $L(v, v')$. Each user appears at a spot (called *departure spot*) and moves to a destination spot selected by the user. A path (*route*) from the departure spot to the destination is decided when the user departs. However, a terminal does not know which route its user has selected. Each user moves on the beeline between the vertices and does not return to the edge the user has passed.

Terminals. Each terminal can utilize cellular and Bluetooth communications at the same time. We assume that the version of Bluetooth is 2.0+EDR. A terminal has information about the field G and knows which edge and location its user is currently moving by its built-in GPS device.

Server. A server has information about the field G. The server does not know the moving route of each user, but it statistically knows the probability that a user at an intersection v moves to an adjacent intersection v', denoted by $MP(v, v')$. In addition, the server knows the speed of each user.

Problem Definition. We consider the following constraints. First, each user must download every chunk through the cellular network or receive it through Bluetooth.

---

[1] Hereafter, we use the term *user*, *terminal*, and *node*, interchangeably.

Second, every chunk must be obtained by the deadline. Third, a terminal can send a chunk only to the terminals in its Bluetooth transmission range. Fourth, to send a chunk, a terminal must receive the chunk or download it in advance. Our target problem is to derive the set of actions (send or receive a chunk via Bluetooth or download a chunk via cellular network) of mobile terminals that satisfies the above constraints and minimizes the cellular network usage.

3 Cooperative Download Method Utilizing Bluetooth and Cellular

In our method, in order to allow each node to know the probability and the time of encountering other nodes, the server calculates a *contact table* for the node. Hereafter, we call the node under consideration (i.e., the owner of the contact table) *owner node*, and other nodes that will encounter the owner node *peer nodes*. Each entry of a contact table consists of the ID of a peer node, the meeting probability, the time when the owner node and the peer node will be in the communication range, and the set of chunks the peer node retains.

Our method consists of two phases, *contact table construction phase* and *action phase*. These two phases are executed repeatedly to update the contact table whenever a user registers its present location and retaining chunks with the server.

3.1 Contact Table Construction Phase

Contact table construction phase is executed by the following procedure.

(1) Information Registration to the Server. When a node u leaves intersection v_u at time T_u, u records v_u and T_u. Then, as soon as u recognizes the next intersection to reach[2], say v'_u, u informs the server of the following information: ID of u, v_u, T_u, v'_u, retaining chunks, and desired chunks. Whenever u passes through an intersection, u registers this information with the server. The size of this information is at most several KBs and overhead to register the information is sufficiently small.

(2) Refining Encounter Candidate Node. To save cellular network usage, the server should be able to reduce the size of a contact table sent to each node.

A contact table should include the information that can identify the chunks with smaller opportunities to obtain them from peer nodes than other chunks. We call such a chunk with low acquisition opportunity *rare chunk*. In our method, the server selects entries of a contact table according to the following two conditions. First condition is whether the peer node u' of the entry will encounter the owner node u before u arrives at the next intersection. Second condition is whether the peer node u' retains some chunks that u wants to obtain. The server calculates and writes in the contact table the meeting time and probability of only the peer nodes that satisfy the above two conditions.

(3) Calculation of Meeting Time and Meeting Probability. The server calculates the time and probability that the owner node meets with the peer nodes. Meeting time

---

[2] We assume that each node can identify the next intersection from the position measured by GPS some time after passing the previous intersection.

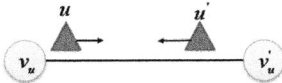

(a) two nodes are approaching each other on the same edge

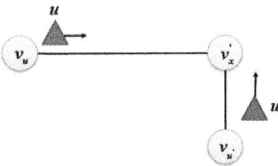

(b) two nodes are moving toward the same intersection

Table 1. Parameter for Simulation

Number of chunks per content	100
Chunk size	150KB
Cellular network bandwidth per node	1.2Mbps
Bluetooth bandwidth per node	700Kbps
Bluetooth radio radius	10m
Number of desired contents per node	2 (Zipf distribution)
Node speed	1.0m/s
Number of chunks initially retained	0–50
Time length to deadline	8 min.
Simulation time	60 min.
Threshold α	0.25
Threshold β	100%
Number of contents	10, 20, 40, 100
Number of nodes	500

Fig. 1. Two Cases that Two Nodes May Encounter

is calculated from length of road and speed of each node which the server knows in advance and the information which each node has registered with the server. There are the following two cases that two nodes u and u' encounter: **(case 1)** u and u' are moving on the same edge as shown in Fig. 1 (a); and **(case 2)** u and u' are moving on different edges but toward the same intersection as shown in Fig. 1 (b). Let v_u and v'_u denote the intersection that node u left and the next intersection toward which u is moving, respectively. If node u encounters node u', the sum of moving distance of u and u' will be equal to $L(v_u, v'_u)$ in (case 1) or the sum of $L(v_u, v'_u)$ and $L(v_{u'}, v'_{u'})$ in (case 2). Taking this into consideration, the server calculates the meeting time of node u with node u', denoted by $CT_u(u')$, by formula (1). Here, a is the speed of node, and T_u is the time when node u left the last intersection.

$$CT_u(u') = \begin{cases} \frac{1}{2}(\frac{L(v_u,v'_u)}{a} + T_u + T_{u'}), & \text{case 1} \\ \frac{1}{2}(\frac{L(v_u,v'_u)+L(v_{u'},v'_{u'})}{a} + T_i + T_h), & \text{case 2} \end{cases} \quad (1)$$

Meeting probability of node u with node u', denoted by $P_u(u')$, is 1 in (case 1) or $MP(v'_u, v_u)$ in (case 2).

The server constructs a contact table by calculating the meeting time and meeting probability of node u with possible peer nodes and sends the contact table to u.

It is possible to reduce cellular network load by including in the contact table only entries whose meeting probability is higher than a predefined threshold α.

3.2 Action Phase

Selection of Acquired Chunks. To exchange chunks efficiently among users, it is desirable that each node acquires rare chunks prior to others. The opportunity that node u receives chunk ch from node u' ($u \neq u'$), denoted by $w_u(u', ch)$, is calculated by formula (2). Here, $D_{u'}$ is the set of chunks which node u' retains.

$$w_u(u', ch) = \begin{cases} P_u(u'), & \text{if } ch \in D_{u'} \\ 0, & \text{otherwise} \end{cases} \quad (2)$$

From the information of retaining chunks of all nodes registered in a contact table, the overall opportunity that node u receives chunk ch, denoted by $Wh_u(ch)$, is calculated by formula (3). Here, N_u is the set of peer nodes registered in node u's contact table and M_u is the set of chunks that node u wants to obtain.

$$Wh_u(ch) = \begin{cases} \sum_{u' \in N_u} w_u(u', ch), & \text{if } ch \in M_u \\ \infty, & \text{otherwise} \end{cases} \quad (3)$$

Since a chunk with small $Wh_u(ch)$ has small opportunity to be received, node u acquires such chunks prior to others.

Control of Bluetooth State. To establish Bluetooth connection between nodes, it is necessary to take synchronization between a terminal in inquiry state called *inquiry node* and a terminal in inquiry scan state called *inquiry scan node*. Total energy consumption for inquiry scan state is lower than inquiry state [4]. In our method, Bluetooth device is turned on only when the node encounters other nodes according to the meeting time in its contact table. In addition, each node searches with inquiry state peer nodes whose meeting probability is higher than a specified threshold β (e.g., 80%). For other peer nodes, each node waits inquiry packets with inquiry scan state.

Download Chunks through Cellular Network. There will be some chunks that cannot be received through Bluetooth by deadline. To meet the deadline, each node downloads chunks through the cellular network so as to keep the ratio of retaining chunks to be proportional to the ratio of the elapsed time to the time to deadline, as in [5].

4 Experimental Evaluation

In this section, we evaluate the proposed method with our own simulator that reproduces the Bluetooth session establishment behavior, inquiry and inquiry scan.

4.1 Simulation Settings

Parameters used in each experiment are shown in Table 1. The target field, moving routes of nodes, movement probabilities of nodes at intersections are shown in Fig. 2. The field is a $500m \times 500m$ square. We set four spots A–D on the center of some roads where nodes appear and disappear. Each node moves between spots. When a node finishes movement, it disappears from the field and a new node appears. Bluetooth session establishment time was set to 2 sec. according to [4].

We compared the performance of the proposed method with the following methods.

Always Activated Method (AAM). Nodes always turn on Bluetooth device. Bluetooth state is set to inquiry and inquiry scan alternately. When downloading and exchanging chunks, the node randomly selects some of the desired chunks.

Contact Oracle. Each node exactly knows when it meets each peer node by oracle. The node turns on Bluetooth device and starts communication only when peer nodes with desired chunks are in the Bluetooth communication range. The ways of selecting desired chunks are the same as the AAM.

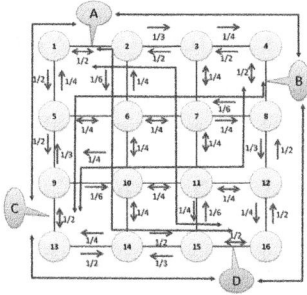 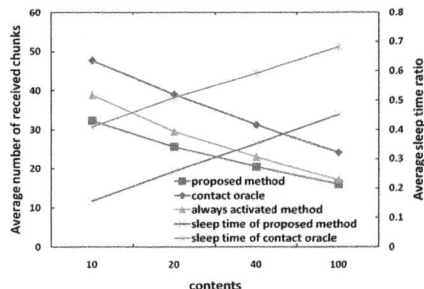

Fig. 2. Field, Moving Route and Moving Probability

Fig. 3. Number of Received Chunks and Total Sleep Time vs. Number of Contents

In order to evaluate the reduction of the cellular network usage, we measured the average number of chunks that were received from peer nodes during simulation time. In addition, we measured the average ratio of the time during which node turned off the Bluetooth device.

4.2 Experimental Result

The experimental result is shown in Fig. 3. The number of received chunks and total sleep time ratio in our method are 70% and 50 – 70% of contact oracle, respectively. Our method achieved 15 – 45% total sleep time ratio, while the number of received chunks in our method gets closer to AMM as the number of contents increases.

5 Conclusion

In this paper, we proposed a cooperative download method, aiming to reduce usage of cellular network and save power consumption. In our method, user terminals utilize a contact table calculated by the server to make it turn on Bluetooth device only when they can exchange chunks. As part of future work, we will implement the proposed method on smartphones and evaluate the power consumption with real devices.

References

1. Conti, M., Gregori, E., Turi, G.: A cross-layer optimization of gnutella for mobile ad hoc networks. In: Proc. of ACM MobiHoc 2005, pp. 343–354 (2005)
2. Rajagopalan, S., Shen, C.-C.: A Cross-layer Decentralized BitTorrent for Mobile Ad hoc Networks. In: Proc. of Mobiquitous 2006, CD-ROM (2006)
3. McNamara, L., Mascolo, C., Capra, L.: Media Sharing based on Colocation Prediction in Urban Transport. In: Proc. of ACM MobiCom 2008, pp. 58–69 (2008)
4. Lee, U., Jung, S., Chang, A., Cho, D.-K., Gerla, M.: P2P Content Distribution to Mobile Bluetooth Users. IEEE Trans. on Vehicular Technology 59(1), 344–355 (2010)
5. Hanano, H., Murata, Y., Shibata, N., Yasumoto, K., Ito, M.: Video Ads Dissemination through WiFi-Cellular Hybrid Networks. In: Proc. of IEEE PerCom 2009, pp. 322–327 (2009)

Measuring Quality of Experience in Pervasive Systems Using Probabilistic Context-Aware Approach

Karan Mitra[1,2], Arkady Zaslavsky[1,2], and Christer Åhlund[2]

[1] DSSE, Monash University,
900, Dandenong Road, Caulfield East, Victoria, Australia, 3145
karan.mitra@monash.edu
[2] Luleå University of Technology,
SE-97187, Luleå, Sweden

Abstract. In this paper, we pioneer a context-aware approach for quality of experience (QoE) modeling, reasoning and inferencing in mobile and pervasive computing environments. The proposed model is based upon Context Spaces Theory (CST) and influence diagrams (IDs) to handle uncertain and hidden complex inter-dependencies between user-perceived and network level QoS and to calculate overall QoE of the users. This helps in user-related media, network and device adaptation, creating user-level SLAs and minimizing network churn. We perform experimentation to validate the proposed approach and the results verify its modeling and inferencing capabilities.

Keywords: Context-awareness, influence diagrams, quality of experience (QoE), quality of service (QoS).

1 Introduction

In mobile and pervasive computing environments, understanding user-perceived QoE is an important and a challenging task. This can be attributed to the fact that QoE about a particular technology, network service or application depends on user expectations, as well as his/her cognitive, psychological and behavioral aspects. There are several stakeholders who are interested in understanding what users think and perceive about the services being provided to them in terms of new products and applications. For example, telecommunication companies want to understand how to minimize network churn by providing better service to the users. From the state-of-the-art we gather that a unifying framework to model, reason and infer QoE is missing. Also, techniques that can simultaneously handle user-centric context, subjective and objective assessment metrics are required. Thus, in this paper we pioneer an approach to integrate context-awareness and decision-theoretic reasoning to model, reason and infer QoE in an efficient manner.

2 Context-Aware QoE Modeling, Reasoning and Inferencing

At the lowest level, context information ($a_i^t \in A_i^t$) such as location ($a_{location}^t$) and bandwidth ($a_{bandwidth}^t$) collected from the network, user, device and the surrounding environment. This context information is then modeled using the CST [1] and IDs [2]. At the intermediate level, we define causal mappings of context attributes with the QoE classes represented as context states (S_i^t) such as user satisfaction ($S_{usersatisfaction}^t$) and technology acceptance ($S_{technologyacceptance}^t$) which are hidden. Once the QoE states are inferred probabilistically, they are then assigned utilities. U($S_{usersatisfaction}^t$) and U($S_{technologyacceptance}^t$) as in e.q 1. Where, h_n is the hypothesis and e is the evidence variable. $P(\bullet)$ represents the belief of the agent in a hypothesis and $U(\bullet)$ encodes the preference on the numerical scale. We consider a Likert-like scale of 1 to 5. 1 being "Poor" and 5 means "Excellent". In order to decide the context state (S_i^t), the agent chooses the decision alternative which gives the maximum expected utility (MEU) as:

$$MEU(S_i^t) = max_{s_i^t \in S_i^t} \sum_n P(h_n|e) U(s_i^t, h_n) \qquad (1)$$

At the top-most level, these context states are then fused together to determine the overall situation of QoE (R_{QoE}) of the user which is a global utility as in e.q. 2. We consider a global utility which comprises of several QoE related classes such as cognitive (user satisfaction in terms of the MOS) and behavioral (technology acceptance). Each context state can contribute to the global utility differently. Thus, we assign weights to these states which sum to 1. It can be written as MEU(S_i^t, S_{i+1}^t):

$$MEU(S_i^t, S_{i+1}^t) = \max_{s_i^t \in S_i^t, s_{i+1}^t \in S_{i+1}^t} \sum_{h_i \in H_i, h_{i+1} \in H_{i+1}} P(h_i, h_{i+1})$$
$$(U(s_i^t, h_i) + U(s_{i+1}^t, h_{i+1})) \qquad (2)$$

Based on the calculated R_{QoE}, we can determine whether the overall QoE is "Poor", "Fair", "Good", "Very good" or "Excellent". For results evaluation we developed a prototype using GeNIe/SMILE 2.0 platform. We consider several cases like a user using a voice application at different locations such as home and office. At each location the social context is different and thus the QoE inferred also varies where the QoS may or may not vary much. The proposed model correctly inferred the QoE in terms of user satisfaction and technology acceptance and in-turn, the overall QoE is correctly calculated as the MEU.

References

1. Padovitz, A., Loke, S., Zaslavsky, A., Burg, B., Bartolini, C.: An approach to data fusion for context awareness. In: Dey, A.K., Kokinov, B., Leake, D.B., Turner, R. (eds.) CONTEXT 2005. LNCS (LNAI), vol. 3554, pp. 353–367. Springer, Heidelberg (2005)
2. Russell, S., Norvig, P.: Artificial intelligence: A modern approach (2003)

Task-Oriented Systems for Interaction with Ubiquitous Computing Environments

Chuong C. Vo, Torab Torabi, and Seng Wai Loke

Department of Computer Science, La Trobe University, VIC 3086, Australia
{c.vo,t.torabi,s.loke}@latrobe.edu.au

Abstract. We present the design and implementation of a task-oriented system, which dynamically binds user tasks to available devices and services within a ubiquitous environment. The system provides users with a task-based interface for interacting with the environment. This interface is re-oriented around user tasks, rather than particular functions of individual devices or services. The system has two main features: (1) context-aware task recommendation; and (2) task guidance, execution, and management. The fundamental input into the system is task descriptions, which are XML specifications that describe how tasks are executed. The abstraction of these descriptions allows tasks to be executed in diversely ubiquitous computing environments.

Keywords: Task-Based Interface, Smart Space, Ubiquitous Computing.

1 Introduction

A ubiquitous computing environment consists of a rich set of casually accessible and often invisible devices and software services. They are expected to seamlessly integrate and cooperate in support of human tasks. However, it is often difficult for users to use such an environment to accomplish their tasks. According to a recent study [1], half of all reportedly malfunctioning devices returned to stores are in full working order; customers were simply unable to work out how to operate them. In another user study [2], some participants saw technologies as something excessive, sometimes useless and invasive while others (particularly mature adults with non-technical backgrounds) only used them to perform basic tasks. A study on usability of smart conference rooms [3] also found that the users tended to rely on experts (wizards) for technical operations. This phenomenon is often referred to as the *usability crisis* in digital products including ubiquitous environments. There are at least two factors that lead to this crisis: *overload of functions* and *complexity of user interfaces* [4].

As our environment is increasingly embedded with computers and devices, together with the devices we carry with us, the range of available tasks has burgeoned. This overwhelms users, hindering them from recognising and using the technological features which are available to them. Since devices and services are distributed and sometimes invisibly vanish into the environment's physical infrastructure, users may not be aware of many of them [5]. Moreover, to accomplish a task, a user must determine how to split it into sub-tasks for each device or service and then find the particular functions of each device or service to accomplish the sub-tasks [6]. The explosion of functions makes this work difficult for the user, especially in unfamiliar environments.

The complexity of the interfaces of many devices is essentially due to three reasons. First, since devices have no knowledge of what task a user intends to accomplish, they often provide the user with the full range of functions, often in the form of buttons and menus. The user needs to switch between the devices' interfaces and to appropriately combine buttons and menus to accomplish the task. This requires the user to understand the task decomposition and the devices' capabilities. Second, since there is little or no goal-action consistency [7] between the interfaces, the user must learn to operate them separately (i.e., there is no "transfer of training"). Third, it is easy to add features to devices, hence resulting in complexity that exceeds the capacity of user interfaces [4]. Therefore, controls become overloaded, leading to heavily-moded interfaces, push-and-hold buttons, and long and deep menus [8].

Our solution to these problems is based on the concept of *task-driven computing* [9,10,11]. We have developed a task-oriented system called TASKOS, that links user tasks and available functions of a ubiquitous environment and that shields users from variations in devices' and services' availability. TASKOS provides users with a task-based user interface, where interfaces are organised according to users' tasks. In particular, TASKOS has the following features:

- *Context-aware task recommendation:* recommends tasks for users based on their context and available devices and services in the current environment.
- *Proactive task guidance:* provides users with instructions to complete a task.
- *Task execution management:* manages task execution, multi-tasking, resource conflicts, and adaptation to variations in availability of devices and services.

The remainder of this paper is organised as follows. Section 2 presents an overview of how TASKOS works. Section 3 presents a generic conceptual architecture for TASKOS. Section 4 presents an implementation and a performance evaluation of TASKOS. Section 5 presents related work. The conclusion and future work are given in Section 6.

2 Envisioned Operation of TaskOS

Developers create or modify a task in a *task description*, an XML specification describing how the task is executed. They register the description with a *task repository*, where it becomes available as a recommendation to users according to their context. Fig. 1 shows a description of the *"checkout a book"* task.

Automatically or when asked by a user (let's call him Bob), TASKOS recommends him for relevant tasks possible in the current environment based on his context. He can also issue queries to search for an intended task. Once Bob selects a task to perform, TASKOS executes its description and guides him to

```
<taskModel about="models:BookCheckOut"                    <task id="SwipeBook">
  xmlns="http://ce.org/cea-2018"                            <dc:title>swipe the book over the barcode reader</dc:title>
  xmlns:dc="http://purl.org/dc/elements/1.1">               <script>Packages.TaskOS.callService("scanbookid");</script>
                                                          </task>
  <task id="Borrow">
    <dc:title>check out a book</dc:title>                 <task id="GetBook">
    <subtasks id="borrowing">                               <output name="book" type="String"/>
      <step name="go" task="GoToCheckOutDesk"/>             <script>
      <step name="identifybook" task="IdentifyBook"/>         $this.book = Packages.Event.getValue();
      <step name="identifyuser" task="IdentifyUser"/>         Packages.TaskOS.print("The book ID is " + $this.book);
      <step name="checkout" task="CheckOut"/>               </script>
      <binding slot="$checkout.book" value="$identifybook.book"/>   </task>
      <binding slot="$checkout.user" value="$identifyuser.user"/>
    </subtasks>                                           <task id="SwipeIdCard">
  </task>                                                   <dc:title>swipe your ID card over the card reader
  <task id="GoToCheckOutDesk">                              </dc:title>
    <dc:title>go to the checkout desk</dc:title>            <script>
  </task>                                                     Packages.TaskOS.callService("scancardid");
                                                            </script>
  <task id="IdentifyBook">                                </task>
    <dc:title>identify the book</dc:title>
    <output name="book" type="String"/>                   <task id="GetUser">
    <subtasks id="IdentifyingBook">                         <output name="user" type="String"/>
      <step name="scanbook" task="SwipeBook"/>              <script>
      <step name="getbook" task="GetBook"/>                   $this.user = Packages.Event.getValue();
      <binding slot="$scanbook.external" value="true"/>       Packages.TaskOS.print("Your ID is " + $this.user);
      <binding slot="$this.book" value="$getbook.book"/>    </script>
    </subtasks>                                           </task>
  </task>
                                                          <task id="CheckOut">
  <task id="IdentifyUser">                                  <input name="book" type="String"/>
    <dc:title>identify the borrower</dc:title>              <input name="user" type="String"/>
    <output name="user" type="String"/>                     <script>
    <subtasks id="IdentifyingUser">                           Packages.TaskOS.callService("checkoutbook " +
      <step name="scanidcard" task="SwipeIdCard"/>              $this.book + " " + $this.user);
      <step name="getuser" task="GetUser"/>                   Packages.TaskOS.print("The book " + $this.book +
      <binding slot="$scanidcard.external" value="true"/>       " is checked out by " + $this.user + ".");
      <binding slot="$this.user" value="$getuser.user"/>    </script>
    </subtasks>                                           </task>
  </task>                                               </taskModel>
```

Fig. 1. Description for the *"checkout a book"* task written in CE-TASK [12]

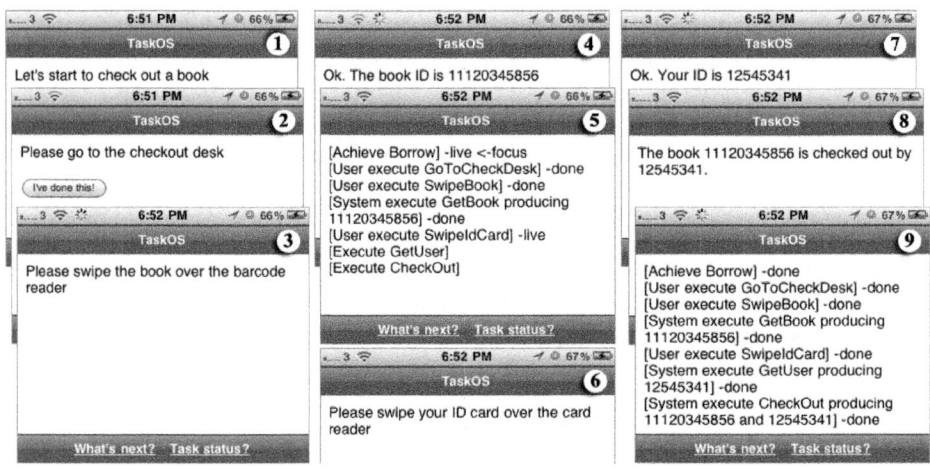

Fig. 2. Task-based interface for executing the *"checkout a book"* task

complete it via a task-based user interface. Fig. 2 shows a number of screens TASKOS presents to him for completing the *"checkout a book"* task.

Bob can perform multi-tasks concurrently. When he comes back to an unfinished task, TASKOS presents its current step to him, so he can continue it.

He can suspend the task interface on one interaction device and resume it on another device. He can query the status of a task to see what steps have or haven't been done. For example, screens 5 and 9 in Fig. 2 show two statuses of a task. He can also request TASKOS to cancel, undo, or redo an ongoing task.

3 Generic Conceptual Architecture

Fig. 3 illustrates the architecture for TASKOS. The *Service & Device Manager* (SDM) manages devices and services. The *Context Information Manager* (CIM) manages context information and answers context queries to other components. This paper focuses on the *Task Execution Engine* (TEE), *Context-Aware Task Recommender* (TASKREC), *Task-Based User Interface* (TASKUI) components and *task descriptions*.

TASKREC suggests to users possible and relevant tasks based on their context and environment capability. TASKUI is an interface for users to interact with TASKOS. TEE loads, verifies, and executes task descriptions. While executing a task, if TEE needs input from users or wants to present information or instructions to them, it sends an interface request to TASKUI. How TASKUI presents this request depends on the current interface modality; it may be visual, spoken, tactile, or multimodal. If the users perform manual operations on devices, TASKUI will simply present them with necessary instructions.

A task description specifies the steps involved in completing a task. It includes *user steps* performed by humans and *system steps* performed by machines. A *primitive step* is a manual operation on a device or an abstract call to a service. To execute a call, TEE passes it to SDM for reasoning about an available service or device's function. Hence, developers can abstractly specify calls in a task description; these calls can be satisfied in diverse environments where the availability of devices and services is often unforeseen by the developers.

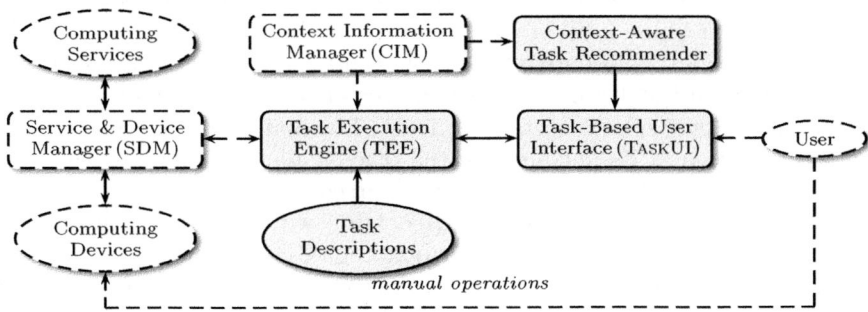

Fig. 3. Generic conceptual architecture for TASKOS. The components with grey-fill and solid border are our focus in this paper.

4 Implementation

This section presents how we describe tasks, how we implement TASKREC that recommends tasks based on user context, and how we implement TEE and TASKUI that executes tasks and provides task guidance.

4.1 Task Description

To describe tasks, we use CE-TASK [12]. CE-TASK allows us to describe high-level, multi-device tasks abstractly from networking technologies and interaction modality. Since it is a mark-up language, CE-TASK descriptions are human-readable and machine-interpretable. Its key expressive features include tasks, input and output parameters, preconditions and postconditions, grounding, task decomposition, temporal order, data flow, and applicability conditions. Fig. 1 shows a CE-TASK description for the task *"checkout a book"*.

4.2 Task Recommendation

To recommend tasks for users, we implement a *place-based* and *pointing-based* task recommender called TASKREC. We model a place as a hierarchy of sub-places. The lowest level in the hierarchy includes devices' zones. We associate a task with one or more places. TASKREC keeps inquiring CIM, which monitors users' locations, to obtain a user's current place. Then it recommends to the user tasks which are associated with that place. In our experiment, it took no longer than *two seconds* to generate task recommendations for a user once s/he changed the place. This latency includes the network latency for sending location information to CIM, the time for updating location at CIM, the one-second interval for each place enquiry at TASKREC, and the time for generating recommendations and sending them to the user device. In the following, we present how to estimate users' current place and pointing directions.

To identify outdoor places, we use the global positioning system. We represent a location of an outdoor place by two attributes: a *geo-coordinate* (latitude and longitude) representing the place's centre and a *radius* representing the place's area. So, if the distance between a user's and a place's geo-coordinates is less than its radius, s/he is seen to be located in that place, hence s/he is also seen to be located in its ancestors in the place hierarchy. The user can switch between these places to obtain different task recommendations.

To estimate outdoor pointing direction, we use compass heading. For each place (called i) in the user's surrounding, CIM computes an angle (β_i) formed by three points: North Pole (P_0), the user's current geo-coordinate (P_u), and the place's geo-coordinate (P_i) using Formula 1. It compares β_i with the current compass heading (α) to get a difference (δ_i). The smaller δ_i the more likely the user is pointing at the place i. Therefore, if the smallest difference δ_j is less than an angle threshold, CIM infers that the user points at the place j.

$$\beta_i = \arccos\left(\frac{\overrightarrow{P_uP_0} \cdot \overrightarrow{P_uP_i}}{|\overrightarrow{P_uP_0}| \times |\overrightarrow{P_uP_i}|}\right)$$

$$= \arccos\left(\frac{(x_0 - x_u)(x_i - x_u) + (y_0 - y_u)(y_i - y_u)}{\sqrt{(x_0 - x_u)^2 + (y_0 - y_u)^2} \times \sqrt{(x_i - x_u)^2 + (y_i - y_u)^2}}\right), \quad (1)$$

where (x_u, y_u), (x_i, y_i), and (x_0, y_0) represent the latitude and longitude of P_u, P_i, and P_0 respectively. In our implementation, we selected P_0 at $(90, y_u)$.

To identify indoor places, we use the Bluetooth and RFID technologies. Each user registers with TASKOS for one or more Bluetooth addresses and each place is associated with a Bluetooth scanner. Once the scanner discovers a Bluetooth device, it sends to CIM the discovered Bluetooth address, allowing CIM to infer the user's current place. However, if a user is in an overlapped coverage of two or more scanners, the user's place changes as frequently as one of these scanners detects him/her; causing the task list to be unstable. To address this problem, CIM uses a time threshold where if a user keeps changing places within this threshold, CIM will use the nearest ancestor of these places in the place hierarchy as the current place of the user. To avoid these overlapping cases, we use short-range RFID technology. Specifically, a place is associated with a 10cm-range RFID reader. A user swipes a passive RFID tag over the reader to update her/his current place.

To estimate indoor pointing direction, we use the Cricket system [13]. Cricket consists of beacons and listeners. It uses Time-Difference-Of-Arrival between radio frequency and ultrasound to estimate a distance between a beacon and a listener. A beacon is worn by a user. Each device (e.g., a television and a printer) is attached with a listener. The listener keeps scanning for beacons within its range. Once it discovers a beacon, it sends CIM a four-tuple of ⟨*device' id, discovered beacon' id, distance, time-stamp*⟩. CIM continually compares fresh distances (within the last *two-seconds* in our experiment) between the same beacon and the listeners (which report these distances to CIM) to infer the device currently nearest to the user.

4.3 Task Execution and Task-Based User Interface

We implement TEE based on a reference implementation of a task engine in [4]. TEE can listen to CIM for external events and react accordingly. For example, the "*checkout a book*" task has three user sub-tasks: "*go to the checkout desk*", "*scan the book' id*", and "*scan the ID-card*". For the first sub-task, we assume that the system cannot recognise the user is already at the checkout desk, so it needs her/him to confirm once s/he has done this. So, it asks TASKUI to present her/him a button "I've done this" (see Step 2 in Fig. 2). For the second and third sub-tasks, CIM notifies TEE once the user has done each of them. There is no button "I've done this" on TASKUI as shown in Steps 3 and 6 in Fig. 2.

By introducing SDM in TASKOS, a service call in a task description is abstractly defined as `TaskOS.callService('<service name and arguments>')`. TASKOS will pass the call to SDM for reasoning about a specific service.

We implement TASKUI as a web-page using Ajax technologies [14]. An advantage of using a web-based user interface is the portability of the interface from one computer to another without the need to install the system. Ajax allows us to hide communication between TASKUI and other components of TASKOS from the user, and to update TASKUI autonomously.

5 Related Work

InterPlay [15] allows users to express their tasks via a pseudo-English interface, and then the system automatically accomplish them. But the users must learn how to express their tasks properly. Similarly, a task execution engine [16] can discover and bind the best available resources for executing a task. However, these systems do not recommend to users possible tasks or provide users with task guidance.

Huddle [6] automatically generates interfaces for tasks involving multiple connected devices, yet it only supports multimedia tasks that rely on data flows.

A Task Computing Environment (TCE) [10] assists users in completing tasks. It represents a task as a composition of web services. Like workflows, since web services and compositions thereof are autonomous, TCE only supports autonomous, one-step, or batch tasks[1], such as *"exchange e-business card"*. In particular, the task of exchanging e-business cards is a batch execution of two services: one returns someone's contact, the other adds this contact into someone else's contact list. TCE does not support multi-step, interactive tasks such as *'borrow a book'*. To complete these kinds of tasks, users must interact with several devices and/or with an interactive user interface to provide inputs (e.g., user identity and book call number) and to obtain instructions for completing tasks, e.g., where the book is located and how to get there.

The Aura system [17] supports the migration of tasks between environments. Since Aura's task description language is designed only for capturing the status of tasks, it cannot describe envisioned tasks, which will be executed in the future.

6 Conclusion

This paper has presented the concept, design, and implementation of a task-oriented system called TASKOS for ubiquitous computing environments. TASKOS provides users with a task-based user interface called TASKUI. The system is able to recommend users with possible tasks based on users' location, surrounding devices, and users' pointing gestures. It can execute task descriptions and guide users through the task executions.

The future work includes developing a graphical tool for authoring task descriptions, implementing other features of TASKOS such as undo, redo, task-recording, why menus, adaptation to variations in resource availability, developing algorithms and techniques for efficient and effective storing, retrieving task descriptions, extending CE-TASK to support recognising possible tasks in an environment, and mining task instructions on the internet.

---

[1] A batch task requires its inputs to be given before being executed.

Acknowledgment. The research was carried out whilst the first author was supported by La Trobe University Postgraduate Research and Tuition Fee Remission Scholarships.

References

1. Den Ouden, E.: Developments of a Design Analysis Model for Consumer Complaints: revealing a new class of quality failures. PhD thesis, Technische Universiteit Eindhoven (2006)
2. Brugnoli, M.C., Hamard, J., Rukzio, E.: User expectations for simple mobile ubiquitous computing environments. In: Proceedings of the Second IEEE International Workshop on Mobile Commerce and Services, pp. 2–10 (2005)
3. Golovchinsky, G., Qvarfordt, P., van Melle, B., Carter, S., Dunnigan, T.: DICE: Designing conference rooms for usability. In: Proceedings of the 27th International Conference on Human Factors in Computing Systems, pp. 1015–1024 (2009)
4. Rich, C.: Building task-based user interfaces with ANSI/CEA-2018. Computer 42(8), 20–27 (2009)
5. Shafer, S.A.N., Brumitt, B., Cadiz, J.J.: Interaction issues in context-aware intelligent environments. Hum.-Comput. Interact. 16(2), 363–378 (2001)
6. Nichols, J., Rothrock, B., Chau, D.H., Myers, B.A.: Huddle: Automatically generating interfaces for systems of multiple connected appliances. In: Proceedings of the Symposium on User Interface Software and Technology, pp. 279–288 (2006)
7. Monk, A.: Noddy's guide to consistency. Interfaces 45, 4–7 (2000)
8. Lieberman, H., Espinosa, J.: A goal-oriented interface to consumer electronics using planning and commonsense reasoning. Know.-Based Syst. 20(6), 592–606 (2007)
9. Wang, Z., Garlan, D.: Task-driven computing. Technical report, School of Computer Science, Carnegie Mellon University (2000)
10. Masuoka, R., Parsia, B., Labrou, Y.: Task computing—the semantic web meets pervasive computing. In: Fensel, D., Sycara, K., Mylopoulos, J. (eds.) ISWC 2003. LNCS, vol. 2870, pp. 866–881. Springer, Heidelberg (2003)
11. Loke, S.W.: Building taskable spaces over ubiquitous services. IEEE Pervasive Computing 8(4), 72–78 (2009)
12. Consumer Electronics Assoc. Task model description (CE Task 1.0), ANSI/CEA-2018 (March 2008)
13. Priyantha, N.B., Chakraborty, A., Balakrishnan, H.: The Cricket location-support system. In: Proceedings of International Conference on Mobile Computing and Networking, pp. 32–43 (2000)
14. Garrett, J.: Ajax: A new approach to web applications. Online (February 2005)
15. Messer, A., Kunjithapatham, A., Sheshagiri, M., Song, H., Kumar, P., Nguyen, P., Yi, K.H.: InterPlay: A middleware for seamless device integration and task orchestration in a networked home. In: Proceedings of International Conference on Pervasive Computing and Communications, pp. 298–307 (2006)
16. Ranganathan, A.: A Task Execution Framework for Autonomic Ubiquitous Computing. PhD thesis, University of Illinois at Urbana-Champaign (2005)
17. Garlan, D., Siewiorek, D.P., Smailagic, A., Steenkiste, P.: Project Aura: Toward distraction-free pervasive computing. IEEE Pervasive Computing 1(2), 22–31 (2002)

Context Data Management for Mobile Spaces

Penghe Chen[1], Shubhabrata Sen[2], Hung Keng Pung[2],
Wenwei Xue[3], and Wai Choong Wong[2]

[1] NUS Graduate School for Integrative Sciences and Engineering, NUS, Singapore
[2] National University of Singapore, Singapore
[3] Nokia Research Center, Beijing, China
{g0901858,g0701139,dcsphk,elewwcl}@nus.edu.sg,
wayne.xue@nokia.com

Abstract. Context-aware applications can monitor and detect the surrounding Context-aware applications can monitor and detect the surrounding situation changes of context entities, and adapt to these situations automatically. As context-aware applications can reside in mobile entities such as cars and mobile phone, an efficient context data management mechanism is required for mobile context-aware applications. In this paper, we propose a new mechanism to model and represent mobile entities. In addition, we also propose new system services to support mobile applications so that disruptions caused by mobility can be minimized.

Keywords: Context-aware, ubiquitous computing, context data management, mobile space, mobile application, availability, callback, mobility.

1 Introduction

The recent advances in mobile computing and wireless communication technology have resulted in the proliferation of context-aware applications on mobile devices. In order to realize the context-awareness, more and more context sources will be utilized. Consequently, an effective context-data management is quite necessary for context-aware applications inside mobile devices whereby mobility is one of the most significant problems that should be solved by the context-data management. Figure 1 illustrates how mobility can affect the on-going context-aware applications. Our paper is just trying to solve this mobility problem by proposing the concept of "Mobile Space" and designing new system services based on our previous system – Coalition [1].

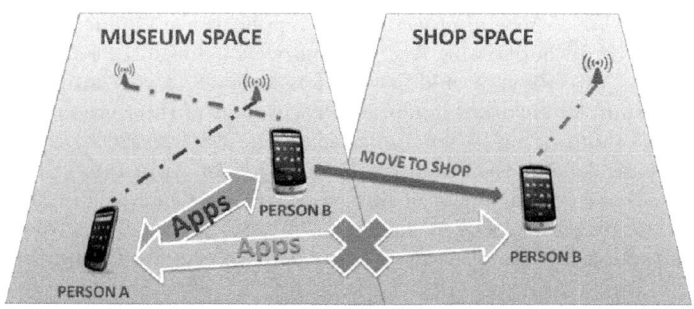

Fig. 1. Application failure due to mobility

2 Mobile Space and Data Management

For any operating environment whose entities including Physical Space Gateway (PSG) [1] *are moving together, we call this environment a Mobile Space, the corresponding* Context Domain as *a Mobile Context Domain, and the PSG is denoted as an M-PSG.*

To be consistent with the schema concept [1] of Coalition, three special attributes are defined for "Mobile Space". *location* -- closely related with mobility is an important context element which can be used to identify the surrounding situations of a mobile space. *visitingSpace* -- the physical space in the mobile space is immersed which supplies different context applications to its inside entities and varies when mobile space moves. *powerPSG* -- the current power level of an M-PSG and can be used as a situation context to assess the state of an M-PSG.

The current context data management of Coalition specifies common functions for context spaces, such as how data is collected, processed, managed and provisioned. In order to better serve "Mobile Spaces", we propose and design one new system service: Availability Updating Service to handle the availability, which defines as the reachability through communication network, of mobile spaces for context acquisition via network. Each mobile space achieves one unique ID during registration with Coalition that can be used to update the new availability information with Coalition when availability information varies. In addition, in order to make this availability change transparent to applications, we propose and design another system service: Application Callback Service to manage the callbacks, which defined as notifications of availability information changes, of mobile spaces on behalf of context-aware applications running on mobile devices. Applications running on MPSGs can issue callbacks to Coalition when failures happen due to availability information of service provider PSGs vary, and receive notifications when new availability information updated. As a result, application failures caused by mobility are solved.

3 Conclusion

In this paper, we propose the concepts of "Mobile Space" to model mobile physical spaces and extend the context data management. In addition, we propose two new system services: Availability Updating Service and Application Callback Service to help "Mobile Spaces" handle mobility problem and relief its influence to context-aware applications. Our future work will be to derive the detail reasons of application disruption on behalf of applications.

Acknowledgments. This work is partially support by project grant NRF2007IDM-IDM002-069 on "Life Spaces" from the IDM Project Office, Media Development Authority of Singapore.

References

1. Pung, H.K., et al.: Context-aware middleware for pervasive elderly homecare. IEEE Journal on Selected Areas in Communications 27, 510 (2009)

Efficient Intrusion Detection for Mobile Devices Using Spatio-temporal Mobility Patterns

Sausan Yazji[1], Robert P. Dick[2], Peter Scheuermann[1], and Goce Trajcevski[1]

[1] EECS Dept., Northwestern University, Evanston, IL. 60208
[2] EECS Dept., University of Michigan, Ann Arbor, MI. 48109

Introduction

Mobile phones are ubiquitous and are used for email, text messages, navigation, education, and as a pyment tool (e.g., Mobile Money – extensively used in China and Japan [1]). Consequently, mobile devices carry a lot of personal data and, if stolen, that data can be more important than the loss of the device.

Most of the works on mobile devices security have focused on physical aspects and/or access control, which do not protect the private data on a stolen device that is in the post authentication state. However, some existing works, e.g. Laptop Cop [2] aim to protect data on stolen devices by remotely and manually deleting it, which requires user intervention. It may take hours before the user notices the loss of his device.

The main goal of this work is to efficiently detect a theft of a mobile device based on the intruder's anomalous behavior. In a previous study [3], we used network access and file system activities to build a behavioral model and were able to detect attacks on portable devices within 5 minutes with 90% accuracy. In this study, we use spatio-temporal information and trajectory analysis for modelling user behavior and anomaly detection.

While some works [4,5,6] have proposed mobility-based intrusion detection, to the best of our knowledge, this is the first mobile intrusion detection solution that is based on spatio-temporal information and trajectory analysis enabling a detection of an attack in 15 minutes with 81% accuracy. The simple data structure used to represent user model (2- and 3-dimension matrix), allows efficient lookup-based attack detection.

System Architecture

Our main objectives are to: (1) develop an efficient algorithm to derive a user model based on spatio-temporal information and trajectory analysis; (2) determine the accuracy of distinguishing individual users based on their motions patterns, and (3) provide high detection accuracy with the smallest possible delay at low energy cost. Our methodology is based on the following observations:

- most mobile systems have GPS receivers and can gather location traces.
- individuals tend to have small set of locations that they visit every day [7].
- individuals tend to take the same path when moving between the same locations [7].

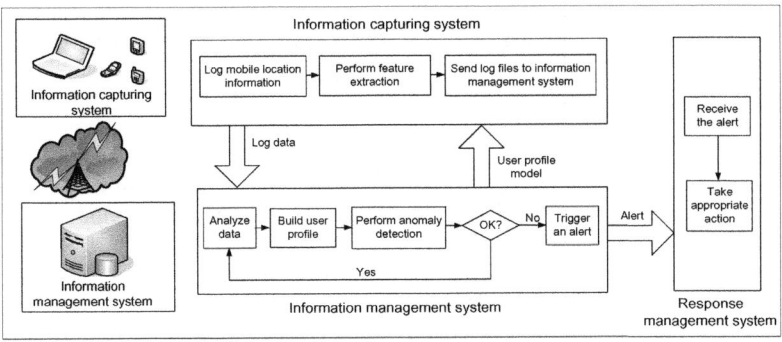

Fig. 1. System architecture

To achieve our objectives, we developed a system to automatically generate mobility models and detect behavioral anomalies. Figure 1 illustrates the system architecture which consists of: (ICS) – the *information capturing system*, residing on the mobile device, with a custom-developed application to track the device location; register it continuously in a new log every T minutes. It also contains the feature extraction module. (IMS) – the *information management system*, which collects the log-files from the ICS and resides on a computer with higher performance and much looser power consumption constraints than the mobile device. It is responsible for building mobility models and performing anomaly detection. Upon building the user model, the IMS sends it to the mobile device allowing it to detect attacks in the absence of wireless connection at some power consumption penalty. (RMS) – the *response management system* resides on both the mobile device and the remote server that hosts the IMS. Upon receiving an alert, the IMS identifies the appropriate action to be taken to protect data on the mobile device. These actions could be a notification to the device owner, locking device, or deleting data automatically.

References

1. Chen, L.-D.: A Model of Consumer Acceptance of Mobile Payment. J. IJMC 6(1), 32–52 (2008)
2. Laptop COP Software, http://www.laptopcopsoftware.com/index.html
3. Yazji, S., Chen, X., Dick, R.P., Scheuermann, P.: Implicit User Re-authentication for Mobile Devices. In: Zhang, D., Portmann, M., Tan, A.-H., Indulska, J. (eds.) UIC 2009. LNCS, vol. 5585, pp. 325–339. Springer, Heidelberg (2009)
4. Sun, B., Yu, F., Wu, K., Xiao, Y., Leung, V.: Enhancing Security Using Mobility-Based Anomaly Detection in Cellular Mobile Networks. IEEE Trans. Vehicular Technology (2007)
5. Hall, J., Barbeau, M., Kranakis, E.: Anomaly-Based Intrusion Detection Using Mobility Profiles of Public Transportation Users. In: Proc. WiMob (2005)
6. Yan, G., Eidenbenz, S., Sun, B.: Mobi-Watchdog: You Can Steal, But You Can't Run! In: Proc. WiSec (2009)
7. Gonzalez, M.C., Hidalgo, C.A., Barabasi, A.L.: Understanding Individual Human Mobility Patterns. J. Nature 453 (2008)

A Study on Security Management Architecture for Personal Networks

Takashi Matsunaka[1], Takayuki Warabino[1], Yoji Kishi[1],
Takeshi Umezawa[2], Kiyohide Nakauchi[2], and Masugi Inoue[2]

[1] KDDI R&D Laboratories, Inc.,
Ohara 2-1-15, Fujimino, Saitama, Japan
{ta-matsunaka,warabino,kishi}@kddilabs.jp
[2] The National Institute of Information and Communications Technology,
Nukui-Kitamachi 4-2-1, Koganei, Tokyo, Japan
{umezawa,nakauchi,inoue}@nict.go.jp

Abstract. The authors have studied the security management architecture for Personal Networks (PN). The main feature of the proposed architecture is to exploit a trusted cellular system, namely, an IMS (IP Multimedia Subsystem), to provide security functions for PNs over open networks like the Internet. They also proposed two security functions to solve security issues for a PN, PE ID and Key Assignment (PIKA) function and a PN Key Sharing (PNKS) function. The PIKA function assigns an ID and key to non IMS–compliant devices (Peer Equipment: PE) to authenticate the user of the PE with the assistance of the IMS–compliant terminal (User Equipment: UE). The PNKS function makes it possible for PEs to share a common cipher key (PN Key) in a PN, which is used to protect a PN against eavesdropping on application data.

Keywords: Personal Network, Mobile Computing, Key Sharing.

The authors proposed the security management architecture for Personal Networks (PN) [1] (Fig. 1 shows an example of PN concept). The architecture has two features, (1) to exploit a trusted cellular network system, the IMS (IP Multimedia Subsystem) [2], to provide security functions for PNs over open networks like the Internet, and (2) to reduce the management cost on the centralized server in order to deal with multitudes and a wide variety of information devices (e.g. a computer, an information appliance, a sensor device and a specific-use device such as a portable music player). Fig. 1 shows an overview of the proposed architecture. There is a PN Server (PN–S) to provide PN establishment service. The PN–S is accommodated in the IMS as an IMS Application Server and also is connected to the Internet. The PN–S plays the role of mediator for the security functions and interworking with the IMS. Users have an IMS–compliant terminal (User Equipment (UE)), e.g. a cellular-phone, and non–IMS compliant devices (Peer Equipment (PE)). UEs are connected to the IMS, and PEs are connected to the Internet over several access systems. PEs in a PN directly exchange application data with each other without the intermediation of the PN–S.

To realize the secure establishment of PNs, two security issues need to be considered. (1) Accommodation of PEs in a PN securely, and (2) Secure communication in a PN. To solve the issues, the authors proposed two security functions, a PE ID and Key Assignment (PIKA) function, and a PN Key Sharing

(PNKS) function. The PIKA function provides an identity and a key (a PE ID and a PE Key) to a PE so that PN–S can authenticate the user of the PE. In the proposed approach, a PE ID includes information to derive the PN Key so that the PN–S can generate the PE Key corresponding to the PE on demand. This approach alleviates key management cost in PN–S since the PN–S does not need to keep all PE Keys. The PNKS function enables PEs to share a common key (PN Key) in a PN, which prevents outsiders of PN from eavesdropping application data in the PN. For secure PN Key delivery, the function employs Broadcast Encryption (BE) [3], which has a revocation method to prevent digital contents being viewed by nonsubscribers without changing all the key information users own. In addition, from the viewpoint of scalability, the function adapts the hierarchical membership management to reduce the computational load of device management imposed on the PN–S. A PN convener's UE only manages the PN participant's UE membership of a PN, and a PN participant's UE only the PE membership. In Fig. 1, Alice's UE (UE_{PN}) recognizes Bob's UE (UE_b) is a member of her PN (PN_y), and UE_b recognizes PE_4 is a member of PN_y. The overview of the PN Key delivery flow is as follows, (1) a PN convener's UE sends the PN Key to each PN participant's UE via the IMS, (2) a PN participant's UE encrypts the Key with BE so that only PEs in the PN can decrypt it, and sends it to the PN–S, (3) a PE receives the encrypted Key from the PN–S, and decrypts it.

The authors also implement the proposed architecture and perform a qualitative analysis. Through the analysis, security and scalability of the proposed architecture are validated. It is notable that the PNKS function reduces the key management cost of the PN–S from $O(N_{all})$ to $O(N_U)$, where N_U be the number of participants in a PN, N_{all} be the number of all PEs which could potentially join a PN, compared with the centralized approach where only the PN–S manages all PEs' security information.

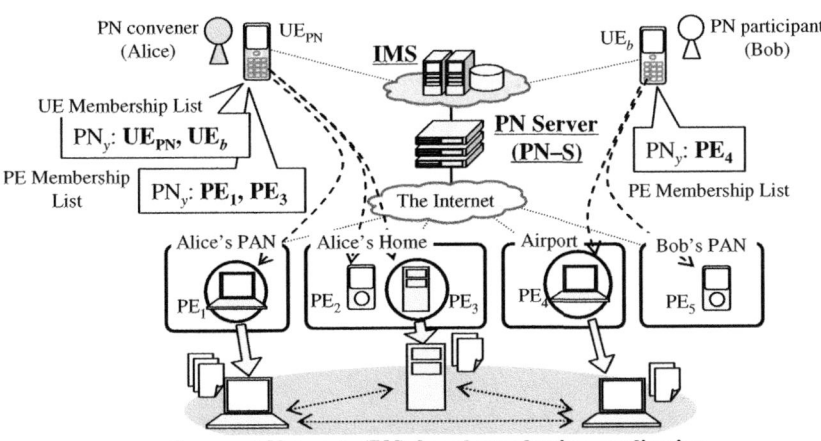

Fig. 1. Overview of Personal Networks and Hierarchical Membership Management

This is the case where Alice (PN convener) wants to provide a photo sharing application using her PC at home (PE_3). She wants to look at her private photos on her laptop PC (PE_1), and Bob (PN participant) also wants to look at them on a laptop PC at the airport lounge (PE_4). Alice constructs a PN, which consists of PE_1, PE_3 used by Alice and PE_4 used by Bob.

References

1. IP Multimedia Subsystem (IMS) Stage 2 (Release 8), 3GPP Technical Specification 23.228
2. My personal Adaptive Global NET,
   http://www.telecom.ece.ntua.gr/magnet/index.html
3. Naor, D., Naor, M., Lotspiech, J.: Revocation and Tracing Schemes for Stateless Receivers. In: Kilian, J. (ed.) CRYPTO 2001. LNCS, vol. 2139, pp. 41–62. Springer, Heidelberg (2001)

Preliminary Results in Virtual Testing for Smart Buildings*

Julien Bruneau[1], Charles Consel[1], Marcia O'Malley[2],
Walid Taha[3,2], and Wail Masry Hannourah[4]

[1] INRIA, Bordeaux, France
{julien.bruneau,charles.consel}@inria.fr
[2] Rice University, Houston, Texas, USA
{omalleym,taha}@rice.edu
[3] Halmstad University, Sweden
Walid.Taha@hh.se
[4] HVAC Consultant, Cairo, Egypt
wail_hannourah@yahoo.com

Abstract. Smart buildings promise to revolutionize the way we live. Applications ranging from climate control to fire management can have significant impact on the quality and cost of these services. However, a smart building and any technology with direct effect on the safety of its occupants must undergo extensive testing. Virtual testing by means of computer simulation can significantly reduce the cost of testing and, as a result, accelerate the development of novel applications. Unfortunately, building physically-accurate simulation codes can be labor intensive.

To address this problem, we propose a framework for rapid, physically-accurate virtual testing of smart building systems. The proposed framework supports analytical modeling and simulation of both a discrete distributed system as well as the physical environment that hosts it.

Keywords: Virtual testing, Smart buildings, HVAC, Mixed (virtual/ real) world infrastructures, Energy efficiency.

1 Introduction

Buildings are designed to achieve a wide range of goals. For example, ensuring occupant safety and health requires specialized appliances such as fire alarm systems and proper ventilation. These concerns are very real: Each year, hundreds of people die from carbon monoxide poisoning due to poor ventilation design. Similarly, ensuring occupant comfort (which is generally essential for ensuring their productivity) also requires specialized appliances. The stringent uptime requirements placed on building appliances also give rise to concerns about energy, cost, and environmental impact. Traditional technologies such as brick and mortar, standard heating and airconditioning units, have all been carefully scrutinized with respect to their impact on all of these goals, and have passed the test of extended timelines.

---

* Funded by National Science Foundation (NSF) Embedded and Hybrid Systems (EHS), award number 0720857.

Smart Buildings. Recently, there has been an increasing interest in smart building technologies. Such technologies bear a promise to revolutionize the way we live. Applications ranging from climate control to fire alarm systems, to lighting management, to security systems can be found in smart buildings.

Intrinsic to the idea of smart buildings is the introduction of higher levels of active control into the traditional components of a building. Higher levels of active control are achieved by the use of more sophisticated control algorithms, more extensive sensing of the physical environment, more actuation of various physical subsystems, and more communication between different components of the system. However, because of the direct impact that buildings have on humans and on their resources, smart building technologies must be held to stringent standards for correctness, availability, and a wide range of other software quality considerations.

Virtual Testing. Testing is a crucial tool for eliminating poor designs, and developing a degree of confidence in promising designs. Testing is equally important for traditional physical design and design involving active control. But testing smart building technologies in physical buildings can be slow and prohibitively expensive. Addressing these two problems can accelerate the rate of innovation and deployment of successful smart building applications. In particular, using computer simulations to carry out some part of the testing virtually can help achieve this goal.

2 Our Approach

The goal of our approach is to address three technical challenges that must be overcome in order to enable effective virtual testing of smart buildings. The first is to accurately capture the distributed and networked nature of the active devices in the system. The second is to accurately capture the physical properties of the building. The third is to automatically map such models directly to executable simulation codes.

Existing approaches only cope at most with one of the three challenges raised by the virtual testing of smart buildings. For instance, MATLAB/Simulink [1] allows to simulate active devices but does not attempt to use analytically sound models of the physical environment surrounding such devices. Numerous existing approaches allow to model and execute this physical environment [2,3]. However, they do not allow the modeling of active devices.

By addressing these three challenges, our approach makes smart building systems easier to write, more reliable and safer:

- The active components of smart homes are modeled explicitly in DiaSpec [4], a domain-specific language with specialized support for the modeling of active systems.
- The complete models, containing both physical models and active system models, are mapped to executable codes. This is achieved by combining (1) the simulation capability of Acumen [2], a physical environment modeling tool, and (2) the DiaSim simulator [5].
- Virtual experiments using different building models and control strategies can be analyzed.

References

1. Simulink, http://www.mathworks.com/products/simulink/
2. Zhu, Y., Westbrook, E., Inoue, J., Chapoutot, A., Salama, C., Peralta, M., Martin, T., Taha, W., O'Malley, M., Cartwright, R., Ames, A., Bhattacharya, R.: Mathematical Equations as Executable Models of Mechnical Systems. In: ICCPS 2010 (2010)
3. Tiller, M.: Introduction to Physical Modeling With Modelica. Kluwer Academic Publishers (2001)
4. Cassou, D., Bertran, B., Loriant, N., Consel, C.: A generative programming approach to developing pervasive computing systems. In: GPCE 2009 (October 2009)
5. Bruneau, J., Jouve, W., Consel, C.: Diasim, a parameterized simulator for pervasive computing applications. In: Proceedings of Mobiquitous 2009 (2009)

A Model-Based Approach for Building Ubiquitous Applications Based on Wireless Sensor Network

Taniro Rodrigues[1], Priscilla Dantas[1], Flávia C. Delicato[1], Paulo F. Pires[1],
Claudio Miceli[2], Luci Pirmez[2], Ge Huang[3], and Albert Y. Zomaya[3]

[1] Federal University of Rio Grande do Norte
[2] Federal University of Rio de Janeiro
[3] The University of Sydney
{tanirocr,pridnt,fdelicato,paulo.f.pires,
cmicelifarias,luci.pirmez,plutohg}@gmail.com,
albert.zomaya@sydney.edu.au

Abstract. This paper presents a MDA approach to build WSN applications that allows domain experts to contribute in the developing of applications without knowledge on WSN platforms while allowing network experts to program nodes meeting application needs without needing specific knowledge on the application domain.

1 Introduction

There are several platforms that support the development and implementation of wireless sensor networks (WSNs) applications, each one with its own requirements and execution environments. Most of WSNs are designed for a target platform, operating system, and addressing the requirements of a single application. Application developers need to know several network specificities and built programs either by using the low-level abstractions provided by the sensor OS or directly over the hardware. The high coupling between the application logic and the sensor platform along with the lack of a methodology to support the development lifecycle of WSN applications result in projects with platform dependent code that are hard to maintain, modify and reuse. Therefore, building WSN applications imposes several challenges to the developer, requiring on one hand, specific knowledge of platforms and the need to deal with low level abstractions, and on the other hand, specific knowledge about the application domain. It is not usual that a same developer has both these expertise.

We argue that a promising solution to facilitate developing WNS applications and to promote a clear separation between the specification of requirements at the application level and the specification of such requirements in a given sensor platform is to adopt the Model-Driven Development approach, more specifically the Model-Driven Architecture (MDA) [3]. MDA is a development methodology proposed by OMG which defines a software development process based on successive model refinement, where more abstract models are (automatically) transformed to the more concrete models until it reaches the source code, considered as the most concrete representation of a system. In the MDA approach the system development is divided into three levels of abstraction: CIM, PIM and PSM, where the developed models pass from a higher abstraction level to a lower abstraction level through a set of transformations. MDA aims to provide a larger reuse of software artifacts independently on the target platform which the system will run in. With an MDA approach, WSN systems can be broken down into levels of abstraction dependent or not on the sensor platform, and the design of each level is upon the responsibility of their respective expert. This paper presents a MDA infrastructure and associated process to build WSN applications. The knowledge of the application domain is represented at the PIM level using a Domain Specific Language (DSL). The knowledge representing sensor platforms is specified at the PSM level. Therefore, the proposed MDA infrastructure encompasses different PSM meta-models, one for each WSN platform.

2 A MDA Based Solution to Build WSN Ubiquitous Applications

WSN applications are built according to two viewpoints: one is the application domain expert's (biologists, engineers, etc.) and the second is network expert view. Building WSN application using our MDA process promotes the division of responsibilities among developers of these different standpoints, allowing them to use their specific knowledge and unburdening them from the need to deal with requirements that do not belong to their expertise field.

The first activity in the proposed process "Requirements Analysis", is performed by both the experts, where they get all information needed to build the application. The software artifacts produced as outcome of this activity (UML diagrams as Use Cases, textual documents, etc) represent the system requirements and compose the CIM, which will be used in further phases for both the developers. The requirements document includes functional requirements (related to the application logic) and non-functional requirements (related to the configuration of the WSN platform). We do not address the built of CIM. CIM is used by the domain expert in the activity "Model application with DSL", to specify the PIM model. This model is based on the previously developed DSL meta-model. CIM is also used by the network expert in the activity "Choosing Platform", where he/she will evaluate the available platforms and choose the one that best meets the elicited requirements. Following, the activity "Apply transformation M2M" is performed by the MDA infrastructure. Such activity takes as input the PIM model, with its associated meta-model, and the PSM meta-model of the WSN platform chosen by the network expert, to generate as output a PSM instance that represents the realization of the application in this specific platform. Such PSM is refined by the network expert to augment the model with information referring to network related specificities of the target platform. Finally, the activity "M2T Transformation" is accomplished, taking as inputs: the PSM model refined by the network expert and the chosen platform code templates and generating as output the application source code to be deployed in the sensor nodes. The generated code is then refined by both the developers to add improvements as application specific functions or protocol parameters that are not automatically generated by the process.

To specify the PIM metamodel of our infrastructure we chose the DSL described in [2]. Such DSL includes structural and behavioral characteristics of the application. We specified PSMs for two sensor platforms: TinyOS [5] and SunSPOT/J2ME [4], upon an extensive review of works about WSN development using such platforms (see project site[1] for details). The PSMs include a set of implementation characteristics that the DSL does not encompass (for instance, event or command creation in the TinyOS), since these low-level features are often out of scope for the application domain expert. The designed meta-model defines the basic feature of any application implemented in nesC [1] or JME [4] to

---

[1] http://labdist.dimap.ufrn.br/twiki/bin/view/LabDistProjects/WSN

run in TinyOS 2.x or SunSPOT/J2ME platforms, respectively. For each sensor platform added in the MDA infra-structure, the respective M2M and M2T transformations were defined. Two M2M transformations (DSL to tinyOS and DSL to SUNSpot) and two M2T transformations (tinyOS to nesC and SUNSpot to java) are currently available in our MDA infra-structure. Code templates for nesC and J2ME languages are available in the web page[1] of this work. It is worth noting that M2M and M2T transformations are defined only once, being reused in several WSN systems that share the same target platform (i.e., the same PSM).

References

1. Gray, D., Levis, P., Behren, R.: The nesC Language: A Holistic Approach to Networked Embedded Systems. ACM SIGPLAN, 1–11 (2003)
2. Losilla, F., Vicente-Chicote, C., Álvarez, B., Iborra, A., Sánchez, P.: Wireless Sensor Network Application Development: An Architecture-Centric MDE Approach. In: Oquendo, F. (ed.) ECSA 2007. LNCS, vol. 4758, pp. 179–194. Springer, Heidelberg (2007)
3. Model Driven Architecture, http://www.omg.org/mda/
4. Sun SPOT World, http://www.sunspotworld.com
5. TinyOS Community Forum, http://www.TinyOS.net

Probabilistic Distance Estimation in Wireless Sensor Networks

Ge Huang[1], Flávia C. Delicato[2], Paulo F. Pires[2], and Albert Y. Zomaya[1]

[1] Centre for Distributed and High Performance Computing
School of Information Technologies
The University of Sydney
NSW 2006, Australia
[2] DIMAp, Federal University of Rio Grande do Norte, Brazil

Abstract. Since all anchor-based range-free localization algorithms require estimating the distance from an unknown node to an anchor node, such estimation is crucial for localizing nodes in environments as wireless sensor networks. We propose a new algorithm, named EDPM (Estimating Distance using a Probability Model), to estimate the distance from an unknown node to an anchor node. Simulation results show that EDPM reaches a slightly higher accuracy for distance estimation than the traditional algorithms for regularly shaped networks, but reveals significantly higher accuracy for irregularly shaped networks.

Keywords: Probability Model, Estimating Distance, Wireless Sensor Networks.

1 Introduction

As an essential aspect in wireless sensor networks (WSNs), localization has attracted much research attention over the years. Proposed WSN localization algorithms fall into two main categories: range-based and range-free. Range-free approach does not rely on characteristics of the wireless signal, and is commonly employed in large scale networks, where energy efficiency is a crucial issue.

Being an important class of range-free localization algorithms, DV-Hop algorithms have the advantages of being simple but providing reasonably high accuracy. However, one of the major shortcomings of DV-hop algorithms is that they are not suitable for irregularly shaped networks, which largely affects its potential applicability in a wide range of WSNs domains. The reason lies in the following. Like all range-free localization algorithms, DV-hop algorithms face the problem of estimating the distance from an unknown node to an anchor node. For any given anchor node Q and unknown node P, in order to achieve a high accuracy in distance estimation, DV-Hop algorithms not only require that there is no large obstacle between P and Q, but also require that there are no large obstacles between all other participating anchor nodes(at least two) and Q.

Our proposed EDPM algorithm only requires that there are no large obstacles between P and Q to have a good estimation of the distance from P to Q. Hence, compared to DV-Hop algorithms our algorithm works more effectively in irregularly shaped networks, where presence of large obstacles is more likely.

2 Related Work

In all the DV-Hop algorithms, each anchor node A needs to calculate the average hop distance δ_A. DV-Hop algorithms that use formula (2.1) to calculate δ_A are categorized as unbiased DV-Hop algorithms such as in [1, 4, 5], while others that use the formula (2.2) are categorized as least mean square DV-Hop algorithms such as in [3].

$$\delta_A = \frac{\sum_Q \|A-Q\|_2}{\sum_Q h(A,Q)} \qquad (2.1)$$

$$\delta_A = \frac{\sum_Q h(A,Q)\|A-Q\|_2}{\sum_Q h^2(A,Q)} \qquad (2.2)$$

Besides, DV-Hop algorithms that require all anchor nodes participate in average hop distance calculation are designated as nonselective DV-Hop algorithms such as in [1, 3, 5], while others that select certain anchor nodes to participate are designated as selective DV-Hop algorithms such as in [4, 6].

Most DV-Hop algorithms are unbiased and nonselective with their distance estimation method named as UNDE (Unbiased Nonselective Distance Estimation). UNDE can achieve reasonable accuracy in isotropic and regularly shaped density WSNs, but has relatively large error in randomly distributed or irregularly shaped networks. Ji and Liu [3] proposed a nonselective least mean square DV-Hop algorithm with its distance estimation method named as BNDE (Biased Nonselective Distance Estimation). BNDE can reach reasonably high accuracy in regularly shaped WSNs, but it generates relatively large error in irregularly shaped WSNs. Authors in [4] proposed a so-called "convex hull test method", to select anchor nodes to participate in the average hop distance calculation. We name the distance estimation algorithm used in [4] as CHTDE (Convex Hull Test Distance Estimation). In C-shaped and O-shaped WSNs, CHTDE reaches better accuracy, while in star-shaped WSNs, the error is quite large. Moreover, CHTDE has a high computational complexity, and does not work as effectively as other DV-hop algorithms in large scale regularly shaped networks.

3 Estimating Distance Using Probability Model (EDPM)

In the formulating of our solution, we assume that every node in the WSN has the same communication radius r. For any nodes P and Q and considering the positive real number a, we use $R(P,Q,a)$ to represent the rectangle with line segment $\overline{PQ}$ as its middle line, and width a. For any two nodes P and Q, we call P and Q line-of-sight connected, if there exists a hop path from P to Q in $R(P,Q,r/2)$. Note that, if there is a large propagation obstruction between an anchor node Q and an unknown node P, by intuition we can tell that we cannot reach ideal accuracy for distance estimation no matter how efficient the estimation method is. Thus, we only estimate the distance between such a pair of an unknown node and an anchor node that are line-of-sight connected. For lack of space, the proof of the following theorem is omitted.

Theorem 3.1. Suppose the anchor node Q and the unknown node P are line-of-light connected, and the node density within $R(P,Q,r/2)$ is ρ. Let $\delta = 2/\sqrt{\pi\rho}$ and n be the smallest integer not less than $\|Q-P\|_2/\delta$. Suppose $\mu=r/\delta$ is an integer. Let $Q_1,\ldots,Q_n$ be n geometric points on the ray $\overrightarrow{QP}$ such that the distance from Q_m to Q is $(2m-1)\delta/2$. Let C_m be close disk centered at Q_m and with radius $\delta/2$ (according to the definition of δ, there must exist nodes inside). Let $M=\cup C_m$. Tag Q as 0, the nodes in M and within the communication range of Q as 1, and the untagged nodes in M and within the communication range of some nodes tagged by 1 as 2. Repeat this process until P is tagged. Suppose P is tagged as h. Let k(0)=0. For every 1<s<h, let k(s)=max{m I there exists node tagged as s in C_m}. Then,

(1) For every 1<s<h, there exists $P_s \in C_{k(s)}$ such that $Q=P_0, \ldots, P_h=P$ is the smallest hop count path from Q to P.
(2) For every $1 \leq s < h$, let $\omega_s = \delta[k(s)-k(s-1)]$, then $\omega_s = r$ or $r-\delta$.
(3) The probability that ω_1 has value r is 1; if the value of ω_s ($1 \leq s < h$) is r, then the probabilities that ω_{s+1} takes value r and r−δ are both likely to be 1/2; if the value of ω_s ($1 \leq s < h$) is r−δ, then the probability that ω_{s+1} has value r is quite high, and that of value r−δ is very low.
(4) Let $\zeta = \delta[n-k(h-1)]$, then ζ has a distribution which is close to normal distribution.

In the remainder of this section we suppose that Q is an anchor node and P is an unknown node which is line-of-light connected with Q, and $Q=P_0, \ldots, P_h=P$ is the smallest hop count path from Q to P. Let $AN(Q,P) = \left(\sum_{i=0}^{h-1} N(P_i)\right)/h$, where $N(P_i)$ is the number of the neighbours of P_i ($0 \leq i < h$). It is obvious that the node density within $R(P,Q,r/2)$ ρ is approximately $(AN(Q,P)+1)/\pi r^2$. Let $\delta = 2/\sqrt{\pi\rho}$. Suppose τ is a random variable with (0-1) distribution such that the probability that it is 1 is 1/2. Suppose ζ is a random variable that follows a normal distribution with mean $r/2$, and standard deviation $(r-2\delta)/\sqrt{70}$. By theorem 3.1, we propose the following algorithm.

Algorithm EDPM (Estimating Distance Using Probability Model Method)

1: $flag \leftarrow true, sum \leftarrow 0$.
2: for i=1 to h-1 step 1 do
3:    if flag==true then
4:       $sum \leftarrow sum + r$;
5:       $flag \leftarrow \tau$;
6:       if flag==true then
7:          $sum \leftarrow sum - \delta/3$;
8:       else
9:          $sum \leftarrow sum + \delta/3$;
10:      end if
11:   else
12:       $sum \leftarrow sum + (r-\delta)$;
13:       $flag \leftarrow true$;
14:   end if
15: end for
16: $sum \leftarrow sum + \zeta$

4 Simulation Experiments

To analyse the performance of EDPM experimentally, we use C++ language to develop a simulator, which can generate various scale and distribution of networks. In experiments 1 to 4 we simulate respectively one type of regularly shaped networks (a rectangular network) and three types of irregularly shaped networks: O-shaped, C-shaped, and star-shaped networks, and compare the performance of EDPM under such network topologies with the algorithms UNDE, BNDE and CHTDE. In every experiment, we randomly select one anchor node and ten unknown nodes. Results of experiments 1 and 2 are shown in Figure 1 (a) and (b), and of experiments 3 and 4 are shown in Figure 2 (a) and (b). Overall, the experiments demonstrated that our EDPM algorithm works effectively under all topologies. Moreover, the more complex the topology is the more our algorithm outperforms other algorithms. Results also confirm the dependence of DV-hop algorithms on the uniformity and regularity of the network.

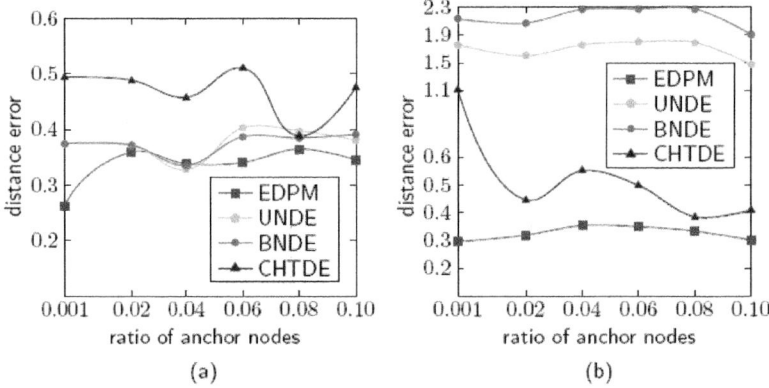

Fig. 1. Distance estimation accuracy in (a) rectangular (b) O-shaped network

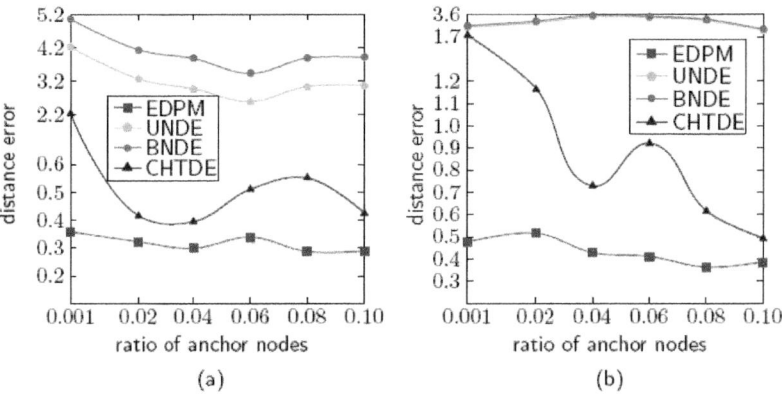

Fig. 2. Distance estimation accuracy in (a) C-shaped (b) star-shaped network

From experiments 1-4 we can tell that when the number of anchor nodes is no greater than one thousandth of the total number of nodes, DV-hop algorithms do not perform well, which is the same conclusion as [2].

Since UNDE requires all anchor nodes to participate in the calculation of δ_A, the more irregular the topology is, the worse UNDE performs. A similar behavior occurs for BNDE. CHTDE, which only selects a portion of anchor nodes to participate in the calculation of δ_A, has better performance than UNDE and BNDE under irregular shaped topology, but performs worse than both UNDE and BNDE in regularly shaped networks because of its selection strategy. Both UNDE and BNDE that have poor performance in irregularly shaped networks can reach similar accuracy in regularly shaped networks.

5 Conclusion

Our experiments show that the proposed distance estimation method can reach much higher accuracy compared to popular distance estimation methods based on the average hop-distance UNDE, BNDE and CHTDE. We demonstrated that even if exists only one anchor node in the network, it won't affect EDPM accuracy, while distance estimation methods based on the average hop-distance almost have no practical use when only few anchor nodes exist. Moreover, EDPM has higher energy efficiency, since it can save the communication cost that anchor nodes use to broadcast the computed average hop distance to the entire network, while such cost is necessary for all DV-Hop algorithms.

Acknowledgements. This work is partially supported by CNPq and CAPES, through processes 201090/2009-0, 477229/2009-3, 480359/2009-1 and 4073-09-06.

References

[1] Niculescu, D., Nath, B.: DV Based Positioning in Ad hoc Networks. Journal of Telecommunication Systems 22(1-4), 267–280 (2003)
[2] He, T., et al.: Range-free localization schemes in large scale sensor networks. In: 9th Annual International Conference on Mobile Computing and Networking, USA (2003)
[3] Ji, W., Liu, Z.: Study on the Application of DV-Hop Localization Algorithms to Random Sensor Networks. Journal of Electronics & Information Technology 30(4) (April 2008)
[4] Niu, Y., Zhang, S., Xu, X., Huo, H., Gao, S.: An Enhanced DV-hop Localization Algorithm for Irregularly Shaped Sensor Networks. In: Zhang, H., Olariu, S., Cao, J., Johnson, D.B. (eds.) MSN 2007. LNCS, vol. 4864, pp. 694–704. Springer, Heidelberg (2007)
[5] Guo, Y., et al.: Hop-based Refinement Algorithm for Localization in Wireless Sensor Networks. Computer Engineering 35(3), 145–147 (2009)
[6] Cheng, K.-Y., et al.: Improving aps with anchor selection in anisotropic sensor networks. In: International Conference on Networking and Services, pp. 49–49 (2005)

Context Aware Framework

Sridevi S., Sayantani Bhattacharya, and Pitchiah R.

Centre for Development of Advanced Computing,
Chennai, India
{sridevis,sayantanib,rpitchiah}@cdac.in

Abstract. In a dynamic environment, one of the challenges in building context aware applications is the communication paradigm between context providers and end applications. Synchronous communication may not be suitable since all the entities may not be live all the time. In this paper we present the architecture of Context Aware Framework developed using publish / subscribe paradigm to provide asynchronous communication between sensor layer which provides the context information and application layer which consumes the context in order to act according to the context. Sensors keep publishing the data in the XML format using Publish-API and Applications subscribe to the interested context using Subscribe-API. The framework also maintains a knowledge base to store context information in OWL (Web Ontology Language) format and does context interpretation, context modeling, rule based reasoning and event notification. It helps in rapid development of context aware applications across various domains.

Keywords: context awareness, publish / subscribe, ontology based middleware, context modeling and reasoning.

1 Introduction

Three important aspects of context are: where you are, who you are with, and what resources are nearby. Context encompasses more than just the user's location, because other things of interest are also mobile and changing [1]. According to Anind K Dey [2], Context is any information that can be used to characterize the situation of an entity. Context Aware Computing is a key aspect of the future computing environment, which aims to provide relevant services and information to users, based on their situational conditions [3]. Building and deploying context aware systems in open, dynamic environments raises a new set of research challenges. We have developed a framework addressing various challenges of context aware systems to reduce the difficulty and cost of building context aware applications. The rest of the paper is organized in the following manner: - Section 2 elaborates the architecture and sub systems of Context Aware Framework, Section 3 describes about implementation of publish / subscribe paradigm in the framework, Section 4 explains about the framework being used in a medical application, Section 5 shows the result of performance test of the Context Aware Framework, Section 6 gives an insight about related works and finally Section 7 concludes with our future work plan.

2 Architecture and Sub Systems

Fig. 1. shows the overall architecture of Context Aware Framework. It has three sub systems: - Context Management, Device Management and Service Management. Context Management sub system interacts with the Service Management sub system to know the available services and with the Device Management sub system to deliver the required services to the applications based on context. In our model, contexts are described using ontology written in OWL. Modeling context using an ontology based approach allows us to describe contexts semantically in a way which is independent of programming language, operating system or middleware.

Device Management manages the metadata information about the devices (sensor, actuators or any other devices) used in the applications. Sensor has to be registered in to the framework using a Context Builder Tool to provide the metadata information of the data provided by the sensor. Once the sensor is registered, a XML template is auto generated and provided to the sensor layer to publish data. Any new device can be dynamically added to the system. It collects information like devices used type of data the devices transmit, how the data relate to the entity present in the environment, possible commands to control the devices, number of rooms in the environment, the devices available in each room etc. For example Room A may have one temperature sensor and one Air Conditioner (AC). If the temperature sensed by sensor in Room A is very high and AC is off then our framework will send an event notification to the application layer to switch on the AC in Room A. It also provides an interface to create rules for the Environment. Some rules may be generic and can be applied to any environment (e.g. temperature is high then switch on AC). Some rules may be specific depending on application (e.g. upload presentation materials to a computer in

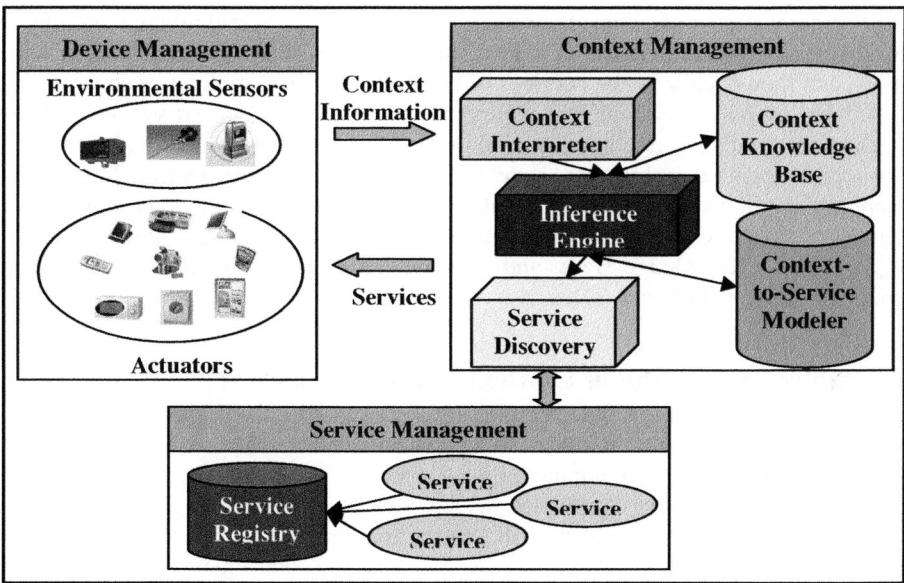

Fig. 1. Architecture and sub systems

a conference room if the speaker enters the room). Rules can be added dynamically and classified as low level (if-then rules) and high level context rules. High level rules are constructed by combining various low level rules.

Service Management maintains a registry for various services available in the framework. Any application can query the context knowledge base and know about the status of any existing context entity. Framework exposes the interfaces in various forms such as Java APIs, Web Services and OSGi services. At the context server end, one can see the context changes of various entities using Entity Watcher Graphical Tool. Framework also contains in-built services for sending SMS, sending E-mail, for knowing the current weather condition, for receiving weather forecast for a location and for getting news updates.

Context Management manages context parameters of various entities such as people, places, smart artifacts / objects etc. Various modules of the context management are the Context Interpreter, Knowledge Base (Rule-base), Inference Engine, Service Discovery and Context-to-Service modeler. The Context Interpreter gathers contextual information about a certain entity from Context Providers (sensors or external sources), aggregates gathered context, converts the gathered context into OWL format and sends to Inference Engine for further processing. Context Knowledge Base is a persistent store which manages the context information in the form of an OWL file and also stores the information about the context provider as meta-data in the XML form. It keeps the context history, since it is essential to allow context inference based on past occurrences. Rule based inference engine is developed using Jena framework along with SPARQL queries. Whenever any context information is given from the Context Interpreter, Inference Engine analyzes the context and decides on the action to be taken using context-to-service modeler. Rule matching is an important task of Inference Engine. Whenever a context of an entity E is changing, it has to fetch all the low level and high level rules defined for the context parameters of E. Once the rules are fetched, it has to find all the matching low level rules. With the matched low level rules, it has to identify matching high level rules, and then decide on the services to be provided to the application layer. Service Discovery module communicates with various service providers or service registries. It checks periodically about the availability of the existing services and also checks for any new services that may have been added. Context-to-Service Modeler is a mapping module in the Context Builder Tool where the context is mapped with relevant services. Context-to-Service modeler finds the list of all context information from Context Knowledge Base, and gets available services in the environment by communicating with Service Discovery module.

3 Publish Subscribe in Context Aware Framework

As distributed systems on wide area networks grow, the demands of flexible, efficient, and dynamic communication mechanisms increase. The publish/subscribe communication paradigm provides a many-to-many data dissemination [4]. In systems based on the publish/subscribe interaction paradigm, subscribers register their interests in an event or pattern of events, and are subsequently asynchronously notified of events generated by publishers [5]. In our framework, Publish / Subscribe paradigm has been developed using RMI Callback and interfaces are exposed for publish, subscribe, unsubscribe operations. Whenever a publisher publishes context information, Context Interpreter parses the published data, searches for a Context Entity E associated and updates the entity E if it exists or creates a new Context Entity E. Then it interprets the context based on the data gathered from various other sensors. If there is any change in the Entity due to the newly published data, then Context Manager generates a Context Event for Entity E and notifies all subscribers of E. Sensors act as publishers and publishes information like health parameters, location of a person, temperature of a room etc. Applications like Medicare, Intelligent Home and Discussion Room subscribe for information required for them. Subscription Manager manages the subscription using a Hash table <K, Vector<S>> where K is a context item and S is the reference of a subscriber. Subscriptions are added and removed by calling subscribe(c, k) and unsubscribe(c, k) operations where c is the connection to context server and k is the interested context item. Whenever there is a change in a context item K, Subscription Manager forwards the event to all the set (Vector<S>) of subscribers of K using the reference S.

4 Usage of Framework in Healthcare Application

Context aware computing is a research field which often refers to healthcare as an interesting and potential area of application [6]. We have developed a Context Aware Health Monitoring System [7] which shows the usage of our framework in health domain. In our experiment, we used Bluetooth enabled wireless medical devices: - Pulse Oximeter, Heart Monitor and Blood Pressure Monitor. Patients visit the Primary Health Care Centres and a nurse records the patient's health parameters using these devices. These health parameters are collected by a mobile phone and transmitted to our framework. Health Service Provider is a service running in the application layer which subscribes for any change in the health parameters. Health parameters are parsed by the Context Interpreter and the Inference engine compares the parsed health data against the Rule-base to find if any abnormalities are present. If any abnormality is detected, an event will be triggered and notified to the application layer requesting to send SMS / Email to the domain experts depending on the state of the abnormalities.

5 Performance Test on Context Aware Framework

Performance Test has been conducted to test the scalability of the framework and Fig. 2 shows the results of the test. Context Server has been tested on a computer with Intel Pentium 4 CPU @ 3.00GHz, 1.48GB RAM. Sensor Layer and Application Layer are simulated, tested on a computer with Intel Core2Duo E6750 @ 2.64GHz processor, 1GB RAM. Totally 6 sensors were simulated to publish data for various number of context entities at 10 minute intervals in a local area network with 100mbps speed. Time delay in event notification has been recorded after loading 100, 200, 400, 600, 800, 1000 context entities in the context server. Results depict that even though the server is loaded with more number of entities, time delay is only in the order of a few milliseconds.

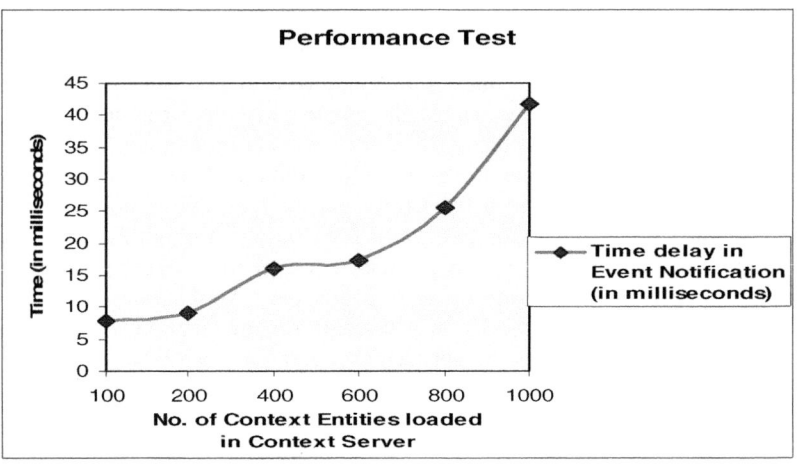

Fig. 2. Performance Test Results

6 Related Work

Java Context-Awareness Framework – JCAF, [8] is a Java-based context-awareness infrastructure and programming API for creating context-aware computer applications. Service-oriented Context-aware Framework [9] generalizes the location-based services to context-based services in mobile or desktop environments. On Context-Aware Publish-Subscribe [10] paper proposes a context-aware extension to publish subscribe model of communication. A Rule Based Publish/Subscribe Context Dissemination Middleware [11] provides the base functionalities for mobile devices to publish their context information, and in addition subscribe to the context information published by their mobile peers. Context Toolkit [12] aids in development and deployment of context aware services. Context Broker Architecture (CoBrA) [3] describes about the architecture for smart spaces. Service Oriented Context aware Middleware (SOCAM) [13] is designed for smart home environment.

Key features of our Context Aware Framework are: - it is an ontology based middleware which can be used for building context aware applications for various domains / applications like health care, intelligent home and smart meeting rooms; It addresses various challenges in developing context aware applications like context data acquisition, context interpretation, context data modeling and reasoning; Context Rules can be dynamically added to the framework. It acts as a communication medium between the sensor layer and the application layer; It is developed using publish-subscribe paradigm and provides event notification service with the actions (turn on/off AC, switch on/off light, send SMS, send E-mail etc.) to be taken based on the context.

7 Conclusion and Future Work

We believe an infrastructure is necessary for building context-aware systems and it should provide adequate support for context modeling and context reasoning. Based on the experience and the feedback on the Context Aware Health Monitoring System developed using the framework, we are improving our reasoning engine with enhanced reasoning capabilities and testing the framework on more metrics like load, reliability, fault tolerance etc. We aim to deploy our framework in buildings to make them smart and energy efficient. Future work includes development of a light weight Java API for Mobile based Context Aware Applications.

References

1. Schilit, B., Adams, N., Want, R.: Context-aware computing applications. In: IEEE Workshop on Mobile Computing Systems and Applications (WMCSA 1994), Santa Cruz, US (1994)
2. Dey, A.K.: Understanding and Using Context. Personal and Ubiquitous Computing Journal (2001)
3. Chen, H., Finin, T., Joshi, A.: A Context Broker for Building Smart Meeting Rooms. In: Proceedings of the Knowledge Representation and Ontology for Autonomous Systems Symposium, AAAI Spring Symposium (March 2004)
4. Son, H., Li, X.: PARMI: A Publish/Subscribe Based Asynchronous RMI Framework for Cluster Computing. In: Perrott, R., Chapman, B.M., Subhlok, J., de Mello, R.F., Yang, L.T. (eds.) HPCC 2007. LNCS, vol. 4782, pp. 19–29. Springer, Heidelberg (2007)
5. Eugster, P.T., Felber, P.A., Guerraioui, R., Kermarrec, A.-M.: The Many Faces of Publish/Subscribe. ACM Computing Surveys 35, 114–131 (2003)
6. Bricon-Souf, N., Newman, C.: Context awareness in health care: A review. International Journal of Medical Informatics 76(1), 2–12 (2007)
7. Sridevi, S., Bhattacharya, S., Pal Amutha, K., Madan Mohan, C., Pitchiah, R.: Context Aware Health Monitoring System. In: Proceedings of International Conference on Medical Biometrics, ICMB 2010, Hong Kong, June 28-30 (2010)
8. Bardram, J.E.: The Java Context Awareness Framework (JCAF) – A Service Infrastructure and Programming Framework for Context-Aware Applications. In: Gellersen, H.-W., Want, R., Schmidt, A. (eds.) Pervasive 2005. LNCS, vol. 3468, pp. 98–115. Springer, Heidelberg (2005)
9. Kovács, L., Mátételki, P., Pataki, B.: Service-oriented Context-aware Framework. In: The 4th European Young Researchers Workshop on Service-Oriented Computing, Pisa, Italy (2009)
10. Cugola, G., et al.: On Context-Aware Publish-Subscribe. In: Proceeding of Second International Conference on Distributed Event Based Systems, DEBS 2008, Rome, Italy (2008)
11. Gehlen, G., Aijaz, F., Muhammad, S., Walke, B.: A Rule Based Publish/Subscribe Context Dissemination Middleware. In: Proceedings of Wireless Communications and Network Conference, WCNC 2007, Hong Kong, China, p. 6 (2007)
12. Salber, D., Dey, A.K., Abowd, G.D.: The context toolkit: aiding the development of context-enabled applications. In: Proceedings of the SIGCHI Conference on Human Factors in Computing Systems: the CHI Is the Limit, Pittsburgh, Pennsylvania, United States, May 15-20 (1999)
13. Gu, T., Pung, H.K., Zhang, D.Q.: A service-oriented middleware for building context-aware services. Journal of Network and Computer Applications 28 (2005)

MOHA: A Novel Target Recognition Scheme for WSNs

Mohammed Al-Naeem and Asad I. Khan

Clayton School of IT, Monash University,
Clayton VIC 3800, Australia
{Mohammed.Al-Naeem,Asad.Khan}@monash.edu

1 Introduction

Macroscopic Object Heuristics Algorithm (MOHA) is a one-shot learning associative memory method for target recognition in wireless sensor networks. This method is able to address pattern displacement and pattern rotation issues. This scheme is also capable of reducing the power and memory consumptions of wireless sensor networks. The experimental results show that the proposed scheme can effectively handle pattern displaced and pattern rotation problems.

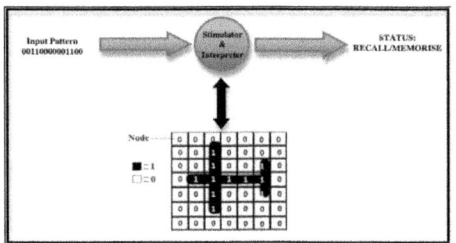

Fig. 1. MOHA architecture

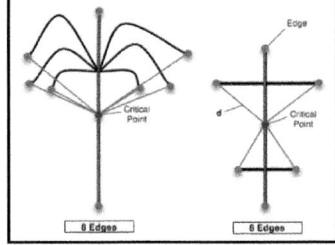

Fig. 2. The edges and the critical points of two targets

2 MOHA Scheme

MOHA builds on the novel one-shot learning associative memory (AM) concept of the Graph Neuron (GN) for template matching in wireless sensor networks. The pattern for the target is stored in a graph-like data structure, which is the same as that of the GN's bias array [1], [2]. The overall architecture of the MOHA algorithm is shown in Figure 1.

The Stimulator and Interpreter (SI module) assigns the target pattern to all nodes in the wireless sensor network, and it consequently obtains the information from the nodes at the edges of the pattern. After that, the SI determines the critical point and calculates the average distance between the critical point and the edges of the pattern. Finally, the SI checks the bias array to determine whether to recall or memorise the pattern (see Figure 2).

3 Evaluation of the Proposed Scheme

In order to determine the MOHA's capability to recognise targets wherever they appear on the wireless sensor network field, we chose 3 different targets with different sizes, and inserted them in a 50×50 bits wireless sensor network field. After that, we tested all possible places where these targets could appear and checked each time whether MOHA could recognise these or not. Table 1 (a) shows that wherever the targets appear on the wireless sensor network field, MOHA could detect and recognise these at the success rate of 100%.

We also tested our scheme ability to recognise targets when these were rotated by 90°, 180°, and 270°. We used the same targets from the first test. The results showed that our scheme was able to recognise these targets after rotation (as seen on Table 1 (b)).

Table 1. a) Results from the first series testing. b) Results from the second series testing.

(a) Images	●	■	▲	(b) Images		●	■	▲
Number of possible places	1122	841	1558	Start position	NE	22	4	13
					AvgD	4.92	7	5.09
Number of Edges (NE)	22	4	13	90° Rotation	NE	22	4	13
					AvgD	4.92	7	5.09
Average Destination (AvgD)	4.92	7	5.09	180° Rotation	NE	22	4	13
					AvgD	4.92	7	5.09
States	Full recalls at all places	Full recalls at all places	Full recalls at all places	270° Rotation	NE	22	4	13
					AvgD	4.92	7	5.09

4 Conclusion

There are three main advantages of using this algorithm. Firstly, only nodes at the edges are required to recognise the target pattern, and the rest of the nodes will be in sleep mode. This will reduce the power consumption. Secondly, there is no need to store any pattern data in the sensor nodes, this will address the nodes memory limitation issue. All pattern data will be stored in the SI bias array, which can be created at the base station or at a more powerful node. Since the pattern data is not stored in the sensors, this algorithm will be able to detect and recognise the target when it appears at different locations or different angles within the wireless sensor network field.

References

1. Khan, A.: A peer-to-peer associative memory network for intelligent information systems. In: The Thirteenth Australasian Conference on Information Systems. AIS Electronic Library, Melbourne (2002)
2. Nasution, B.B., Khan, A.I.: A Hierarchical Graph Neuron Scheme for Real-Time Pattern Recognition. IEEE Transactions on Neural Networks 19, 212–229 (2008)

Policy-Based Personalized Context Dissemination for Location-Aware Services

Yousif Al Ridhawi, Ismaeel Al Ridhawi, Loubet Bruno, and Ahmed Karmouch

School of Information Technology and Engineering (SITE).
University of Ottawa, PO Box 450,
Ottawa, Ontario, K1N 6N5, Canada
{yalri098,ialri083}@uottawa.ca, karmouch@site.uottawa.ca

Abstract. This paper presents a policy-based context-level negotiation protocol and context-aware system architecture to personalize consumer-received context information through negotiations. Location tracking and prediction empower the system to shape contextual information delivered to users according to their expected needs.

Keywords: Context, negotiation, location prediction, context-aware architecture.

1 Introduction

This paper provides an enhanced context-aware system architecture expanding our earlier work [1],[2],[3] to evaluate and enforce CLAs through policies, and enhance *Client Views* generation, modification, and adaptation by tracking and predicting client location. The remainder of this paper is organized as follows. In section 2 we provide our modified an enhanced context-aware system architecture. In section 3 we present our CLA policy management. In section 4 we discuss some related work. Finally in section 5 we discuss our conclusion and future work.

2 Context-Aware System Architecture

Readers are advised to refer to [2] for our complete ontology-based context model used in our system architecture. A tri-layered system architecture was developed by our group and presented in [1]. The three levels were a context acquisition layer, a management layer, and a consumer layer. Since then, the system has been improved, and the updated architecture is provided in Figure 1. The management layer design has been expanded to include a new multi-sourced location tracking and prediction component for initiating negotiations with clients, predicting future movement, and adapting CVs to predicted paths within system's coverage area. The CIC is the *CLA Policy Management* unit responsible for attaining CLAs and generating respective internal policies. The policies are in turn enforced to supply clients with needed contextual knowledge. Further negotiations affecting the original CLA must be reflected within their respective policies; hence the need for a *Policy Update* unit.

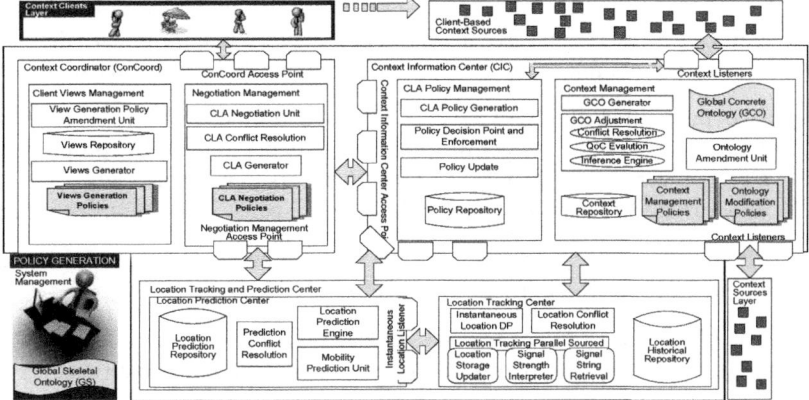

Fig. 1. Three-layered context-aware system architecture with location tracking enhancements

All established CLAs are utilized by the management system to modify existing CVs reflecting provider's changing abilities to supply context information. As CVs are modified, the provider must renegotiate some CLAs to reflect the changes. Modifications affixed to existing CLAs must be promoted into the respective CLA policies generated by the *CLA Policy Management* unit in the CIC. Our context-level negotiation protocol has been described in our earlier works [2], [3].

3 CLA Policy Management

All CLAs are translated into policies for enforcement and conflict resolution. The Policy Decision Point (PDP) examines applicable policies and determines actions to take, and the Policy Enforcement Point (PEP) enforces the outcome of policy decisions. ConCoord's CLA Policy Generator converts established CLAs into a set of policies that are enforced by the PDP in the CIC. In our system, the generator had as input a structure representing a CLA (from ConCoord Access Point) and a single external function call (from Policy Repository) to convert the CLA into policies.

Our choice of policy language must consider that all rules generated from our CLAs have a form of *Event-Condition-Action*. Additionally, users must be able to define their own operations added to existing ones, hence our decision to use Ponder[4]. Ponder policies act on *Managed Objects* representing a part of the system to control. The *Ponder toolkit* with its Java-based APIs allows writing new managed objects and creating operations for these objects. Two types of managed objects exist:

1. *Managed objects shared between all policies in the system.* It has two shared managed objects with one instance per CLA Policy Generator; GCO with context access operations, and CICAP with methods to send context information to clients.
2. *Managed objects created specifically for a CLA.* Two managed objects are created for each CLA. First a CLA object containing information about CLA clients, and methods to activate/deactivate the CLA. Second is a CV object representing a CV.

Every type of event carries information used in the condition and action parts of policies. Additionally, every event type must represent an element that triggers behaviour of the system thus executing a set of policies. Four elements are defined:

1. *Time*: contains information on the current date and time.
2. *GCO_Change*: generated every time a change occurs in the GCO property.
3. *CV_Change*: generated every time a change occurs to a Client View.
4. *CV_Request*: generated every time a client requests its CV from the provider.

The CLA policies we have implemented thus far are aimed at describing all the CLA applicability conditions which involve two steps.

1. Creating elementary policies to activate/deactivate the CLAs.
2. Grouping the policies to obtain complete CLA applicability conditions.

Some policies can take the following form: *at the occurrence of a TIME event, if the time matches START/END date and time, then ACTIVATE/DEACTIVATE the CLA*. Continuous CLAs, as seen in [3], require regular activations and deactivations and can be achieved through *time patterns*. For example, activating a CLA every Monday from the year 2010 at 8:30, the time pattern will be: *"year: 2010; month: -1; day: -1; day of the week: Monday; hour: 8; minute: 30; second: 0"*, -1 marks irrelevant data.

Logical Applicability Conditions; for both WHILE and START/END, the activation and deactivation of the CLA is triggered by logical conditions on the GCO. In both cases the following policy is used: *At any change in the GCO, if the condition C is true, then activate/deactivate the CLA*. Generating policies for the Hybrid CLA Validity Conditions, two policies are needed to activate/deactivate the CLA, and other policies are needed to activate/deactivate these policies. These can be logical conditions (WHILE or Start/END) embedded within a Periodic condition (Continuous or Non_Continuous). Other policies have been established for Continuous Periodic Hybrid Conditions, and Conditional-Periodic Hybrid Validity Conditions.

Context Request policies have also been implemented and divided into 3 types: *Notification_Request, Property_Value Request*, and *Class_Value_Request*. Refer to [3] for further details. Each context request can have an activation condition guarding delivery of its context. These conditions are translated into Activation policies similar to those of CLA Activation Policies and hence will not require further description. Delivery of CVs follows three types of policies; *No_Updates, Periodic_Delivery*, and *On_Change*. Periodic policies are similar to those of periodic CLA activation policies while the On_Change is similar to the START Trigger policies.

Managed Objects and Policies are stored into different domains and sub-domains to avoid mix-ups. For each CLA, a domain *CLA_#XX_Domain* is created, where *XX* is the ID of the CLA. This domain contains the *CLA_#XX* managed object from the CLA type to activate/deactivate the CLA, the CLA policy or policy group, the CV Policy or Policy Group, the Request and activation sub-domains containing all request policies and activation policies respectively. Figure 2a illustrates the CLA Domain.

The policy generator consists of several modules, figure 2b. During system initialization the Coordinator creates event templates and stores them in the Event Repository. At runtime, and for every established CLA, the Coordinator sends a description of the CLA to the Generator that creates domains and managed objects related to the CLA. The Coordinator also sends a description of the policy to the

ECA Policy Generator which creates an empty policy. The Generator then picks the right Event Template and adds it to the policy and sends the information to the Condition Block Generator which returns the condition block back to the Generator. The ECA Condition Block Generator adds the condition and action block to the policy and returns it to the CLA Generator which sends to the Coordinator a set with all the managed objects and policies. The Coordinator interacts with the clients and has the responsibility of executing the necessary policies to enforce established CLAs.

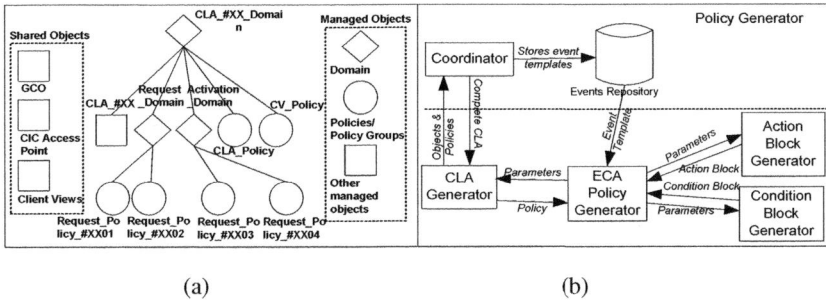

Fig. 2. a) CLA Domain with all sub-domains b) Policy Generator archiecture

We can adapt CVs according to current clients' locations and their predicted motion paths. Responsible components are located within the Location Tracking and Prediction Center (fig. 1) and a complete description of this component was presented in [5]. Accordingly, the steps of adapting CVs are:

1. Create an initial set of CVs based on policies entered by system management.
2. Update a set of existing policies to fit the context access rights presented within the Client's Profile (CPs are received during client-provider negotiations).
3. Track/Predict the user's location/path and update existing CV policies to reveal/hide context according to the client's needs and rights, as well as the availability of contextual sources.

Our system also involves prediction methods for service pre-configuration and enhancement. Multiple position sensing technologies were incorporated to enhance current user location estimation. Location is identified through signal strengths emitted from WLAN access points and RFID readers. The first step consists of discovering surrounding access points and measuring WLAN signal strengths sent to mobile devices. The second step improves accuracy by active RFID tagging, where signal strengths from the RFID reader is measured and transferred to a 2-D map.

WLAN location tracking unit uses NetStumbler to gather information about available access points and emitted signal strengths. The values are compared with those stored in the repository to provide a signal strength map. Once the offline process of information gathering is complete, live user tracking is initiated. Signal strengths from available access points are compared to those in the repository. Once a location is estimated, it is stored in the location history repository.

The RFID location tracking unit uses RFID readers and active tags operating at a frequency of 433.92MHz, giving a range of up to 150 feet. Each reader is responsible for its zone, where it signs the user's RFID tag in and out as the client moves into and out of its coverage area. When a client enters a coverage area, the profile is obtained from the repository and tracking begins. Signal strengths emitted from the RFID reader are gathered and plotted onto an RFID signal strength map. Those values are stored in the repository, where they are compared to find the user's estimated location. The locations are also stored in the location history repository for future use. Location Conflict Resolver Center aims at resolving conflicts between the two sets of location estimates. If the two methods estimate conflicting locations, conflict detection is triggered to choose the candidate closest to previous location.

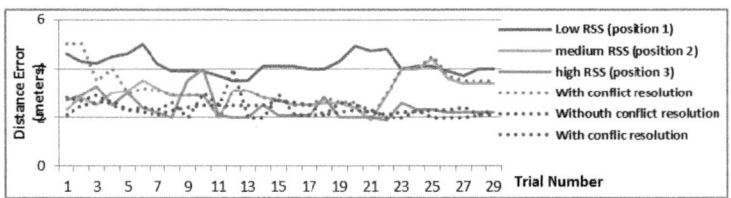

Fig. 3. System results with/without RFID and conflict resolution

4 Experimental Results

A prototype of our system was built and results were presented in [2]. Since then, a testing scenario was conducted at the University of Ottawa to simulate benefits of the dual location-tracking method. The system was implemented using Java built on the same server onto which the provider was activated. This was a 2.00 GB, 3.6 GHz Intel Pentium D station. Clients were installed on 2.0 GHz Intel Pentium Centrino laptops with an RF Code M100 active RFID tag. Information regarding nearby access points and received signal strength values were gathered using NetStumbler by all clients. Once user's location was determined, this information was displayed on a UI containing a map of the client's surrounding area. Since RSS decreases with distance and presence of obstacles, we enhanced our system through RFIDs.

Location tracking experiments were conducted over three different positions with low, medium and high RSS values respectively. Tests in the three locations were done first without utilizing conflict resolution or RFID tags, and then were repeated with the use of our conflict resolution center and RFID tags. As illustrated in figure 3, reductions in the overall distance errors between actual user location and the deduced user location were seen when the dual mode of RSS and RFID were used. Similar results were experienced in tests at positions 2 and 3.

5 Related Work

CASS [6] is a server-based middleware separating reasoning from context-aware applications, and provides event-based supply of context to mobile devices. CASS lacked a description on how context knowledgebase was structured and modeled, and there was no indication to the syntax and semantics of received context notifications.

Gaia's [7] distributed middleware permits development of context-aware applications and acts as a coordinator between software entities and network devices.

Context knowledge within Gaia is presented to applications through a Context File System (CFS) in the form of directories whose path components represent context types and values. Gaia is unsuitable for an environment characterized by mobility. Querying the CFS for context information requires context-aware applications to be aware of the directory structure and path to the needed context.

Hydrogen [8] provides a distributed solution with a firmware located on each device to share context knowledge with other devices. However, Hydrogen lacks two important components: an ontology on which context information is modeled, and a protocol by which context is shared between context servers located on different devices. The missing protocol for sharing context between devices is of particular interest since we provided a negotiation protocol for establishing CLAs. that could greatly improve Hydrogen architecture, by establishing CLAs between servers located on each mobile device giving it ability to acquire context outside its limited abilities.

6 Conclusion

The goal of our work was to develop a context-level negotiation protocol to establish CLAs in context-aware systems. We proposed a tri-layer architecture composed of context sources, providers, and consumers. The design allowed context-level negotiations with Clients and CLA enforcement through policy generation. The system also provided location tracking and prediction components to enhance CLA negotiation. We believe that the protocol can sufficiently meet location tracking and prediction requirements. However, we continue to improve on this system and protocol by adding new options and capabilities as our research progresses.

References

[1] Al Ridhawi, Y., Karmouch, A.: Ontology-Based Context-Level Agreements and Negotiations Protocol. In: International Conference on Intelligent Environments, USA (2008)
[2] Al Ridhawi, Y., Karmouch, A.: Ontology-Based Negotiation Protocol and Context-Level Agreements. In: 4th International Conference on IEs, pp. 1–8 (2008)
[3] Al Ridhawi, Y., Harroud, H., Karmouch, A., Agoulmine, N.: Policy Driven Context-Aware Services in Mobile Environments. In: IIT 2008 International Conference on Innovations in Information Technology, pp. 558–562 (December 2008)
[4] Damianu, N., Dulay, N., Lupu, E., Sloman, M.: The Ponder Policy Specification Language. In: Workshop on Policies for Distributed Systems and Networks, pp. 18–39 (2001)
[5] Al Ridhawi, I., Aloqaily, M., Karmouch, A., Agoulmine, N.: A location-aware user tracking and prediction system. In: Global Information Infrastructure Symposium, GIIS 2009, June 23-26, pp. 1–8 (2009)
[6] Fahy, P., Clarke, S.: CASS-Middleware for Mobile Context-Aware Applications. In: Workshop on Context Awareness, MobiSys 2004 (2004)
[7] Roman, M., Hess, C., Cerqueira, R., Anand, R., Campbell, R.H., Nahrstedt, K.: A Middleware Infrastructure for Active Spaces. IEEE Pervasive Computing, 74–83
[8] Devlic, A., Reichle, R., Wagner, M., Pinheiro, M.K., Vanrompay, Y., Berbers, Y., Valla, M.: Context Inference of User's Social Relationships and Distributed Policy Management. In: Proceedings of IEEE International Conference on Pervasive Computing and Communications, pp. 1–8 (2009)

Automatic Generation of Radio Maps for Localization Systems

Ahmed Eleryan[1], Mohamed Elsabagh[1], and Moustafa Youssef[1,2]

[1] Dept. of Computer Engineering, Alexandria University, Egypt
{ahmed.eleryan,mohamed.elsabagh}@alex.edu.eg
[2] Dept. of Computer and Systems Eng., Alexandria University and E-JUST, Egypt
moustafa.youssef@ejust.edu.eg

Abstract. In this paper, we present the design and evaluation of a system that automatically constructs accurate radio maps for both device-based and device-free WLAN localization systems. The system uses 3D ray tracing enhanced with the uniform theory of diffraction (UTD) to model the electric field behavior and the **human shadowing effect**.

1 Introduction

Both device-based and device-free localization systems usually require the construction of a radio map where collected signal strengths from different streams at different selected locations in the area of interest are tabulated. Using manual calibration, current methods of radio maps construction are therefore tedious and time-consuming, which emphasizes the need for a method to automatically construct the radio maps for an area of interest.

In this paper we present a system which can automatically construct an accurate radio map for a given 3D area of interest. It is unique in supporting automatic radio map generation for **both device-based and device-free** localization systems, which require high accuracy. To our knowledge, this system is the first to consider radio map generation for device-free systems and the first to consider **human body effect** on the generated radio map.

2 System Architecture and Evaluation

The system combines ray tracing with UTD[2] to model both the RF propagation and human shadowing effect. Fig. 1 shows the architecture of the system. Previous work in human modeling has shown a strong correlation between the RF characteristics of the human body and a metallic circular cylinder [1] in indoor radio channels. Therefore, we use a metallic cylinder to model the human body with radius 0.15 m, and height 2 m [1]. For the device-free case, due to the lack of a perfect 3D model, the predicted RSS deviates from the measured ones by a constant offset. To make up for this, the system can optionally be fed with only a *single* sample representing the measured signal in the absence of humans in the environment. This sample is used to compensate for any deviation.

We present only our evaluation of the device-free radio map generation. The experiment was conducted in a typical apartment with an area of $700 ft^2$. 44 locations were chosen and are illustrated in Fig. 2. We collected 60 samples for

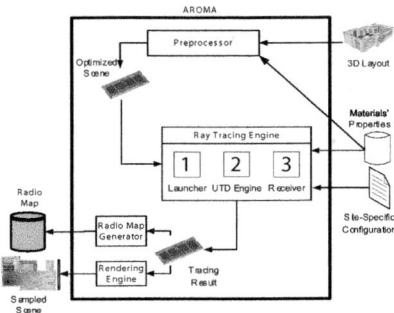 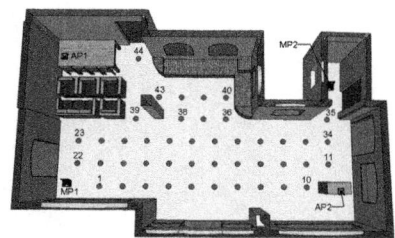

Fig. 1. System architecture

Fig. 2. Device-free experiment layout. The figure highlights the locations of APs, MPs, and radio map locations.

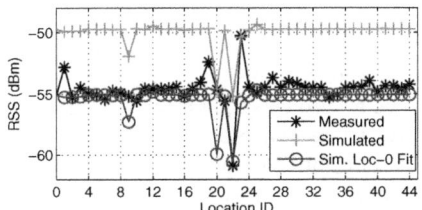 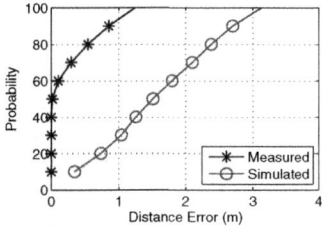

Fig. 3. RSS from AP1 by MP1

Fig. 4. Performance of the MED classifier

each location. A new location is introduced, namely location-0, which represents the environment in the absence of humans. Fig. 3 shows the simulated versus measured RSS for AP1/MP1. We show only one stream due to space limitations. The results show that simulated values are close to the measured values with a maximum average absolute error of 2.77 dBm for all streams. The accuracy of the generated radio map (Fig. 4) is then judged by comparing the average distance error (in meters) of a Mean Euclidean Distance classifier when trained by the manually measured radio map and the automatically generated one. The average distance errors are found to be 1.24 m and 3.13 m, respectively.

3 Conclusion

This paper introduced a system capable of generating site-specific radio maps for device-based and device-free localization systems. Our experience showed that it achieves its goals of: high accuracy, and minimal user overhead.

References

1. Ghaddar, M., Talbi, L., Denid, T., Charbonneau, A.: Modeling Human Body Effects for Indoor Radio Channel using UTD. In: Canadian Conference on Electrical and Computer Engineering. IEEE, Los Alamitos (2004)
2. Mcnamara, D.A., Pistorius, C.W.I., Malherbe, J.A.G.: Introduction to the Uniform Geometrical Theory of Diffraction. Artech House Publishers, Boston (1990)

Monitoring Interactions with RFID Tagged Objects Using RSSI

Siddika Parlak and Ivan Marsic

Department of ECE,
Rutgers University, NJ, USA
parlak@rci.rutgers.edu, marsic@ece.rutgers.edu

Abstract. In this paper, we present SVM and HMM-based methods for monitoring interactions with passive RFID tagged objects. We continuously track the motion status of an object and declare the status as standing still, randomly moving or linearly moving. Inspired by phone transition modeling in speech processing, each interaction type is represented with two sub-states to handle transitions and continuity. Experiments were designed to simulate our target application: monitoring interactions with medical equipment during trauma resuscitation. Our system identified interaction status with 85% accuracy using an HMM. The most useful feature for discrimination was the difference between the average RSSI of two consecutive windows.

Keywords: interaction detection, RFID, ubiquitous computing, trauma resuscitation.

1 Introduction

Inferring activities based on the used objects is an efficient method for high-level activity recognition. However, automatically detecting the set of used objects is challenging due to sensing errors and spurious interactions. In most of the previous work, objects have been identified manually [7], with motion sensors [3] or with near field detectors [2, 3]. Manual labeling requires human effort. Motion sensors are not applicable when small and inexpensive objects need to be tracked. Near field detectors are usually worn or held by humans, which can be obtrusive for chaotic and high-risk environments. Another limitation is the uncertainty in the interaction-to-usage relationship: an interaction does not always signal the item usage. Consider, for instance, an emergency room where a nurse fetches a thermometer to measure the patient's temperature. Since the patient has an oxygen mask on his face, the nurse cannot obtain the temperature at the moment so she leaves the thermometer on the patient bed. After several minutes, she takes the thermometer again and measures the temperature. Because the first interaction in this example is clearly a false alarm, the assumption that any interaction indicates usage is not always correct.

In this work, we develop a minimally intrusive interaction monitoring system using passive Radio Frequency Identification (RFID) tags and fixed readers to identify the type and duration of object interactions. By identifying both the type and the duration, it is possible to filter actual item usage instants. If the detected interaction interval is much shorter than the expected usage interval, we conclude that the item might be interacted accidentally and not actually in-use (Figure 1). In terms of sensing, we prefer a passive UHF RFID system because passive tags are small and inexpensive, enabling small and inexpensive objects to be identified. Fixed readers scan the environment in an unobtrusive way without any human intervention. On the other hand, inferring motion with a passive RFID system is challenging due to the sensitivity to orientation change, environment characteristics and occlusion.

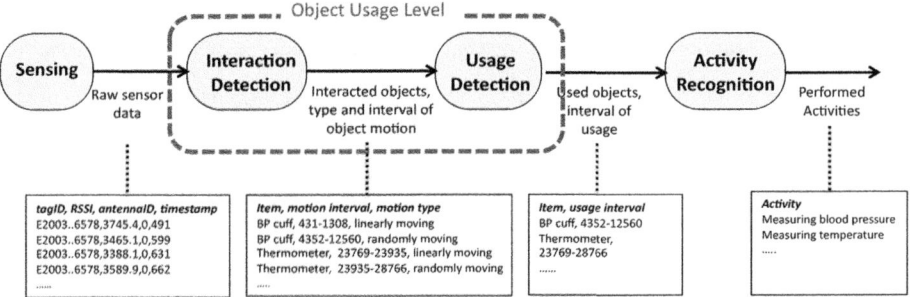

Fig. 1. Block diagram of an overall activity inference system. The interaction detection module is developed in this work.

Our target application is in healthcare domain, specifically Trauma Resuscitation[1]. Because it includes many interactions with medical equipment, trauma resuscitation has a high potential to benefit from object-based activity inference and interaction monitoring [8]. Passive RFID is a suitable technology for trauma resuscitation domain because the medical items for tagging vary in the number and size. Fixed readers are more convenient than wearable readers because urgency of the situation during resuscitation events may make the workers forget or ignore wearing readers.

By visually analyzing the trauma resuscitation videos, we observed that there are three basic types of object motion: *standing still*, *moving linearly* or *moving randomly*. For most items, a relatively long duration of random movement signals item usage. Based on this observation, we define our goal as labeling a sequence of object movements with three labels: *still*, *moving linearly* or *moving randomly*. Unlike the previous work, which used the same RFID technology [1, 4], we continuously monitor the interaction status, rather than declaring the interaction events only. To our knowledge, continuous monitoring has been attempted using technologies such as GPS [5] and Wi-Fi [6], but not with the passive RFID technology. GPS based algorithms are not appropriate for trauma resuscitation application due to high temporal and spatial granularity of GPS-tracked motion. Although Wi-Fi can provide finer granularity, it is not appropriate because of the aforementioned size and cost limitations.

---

[1] Trauma resuscitation consists of a series of tasks performed to identify and immediately treat life-threatening injuries.

2 Methodology

We simulated three motion types—standing still, moving randomly and moving linearly—by interacting with an RFID tagged object (Experimentation details are given in Section 4.1). As seen in Figure 2, received signal strength indication (RSSI) sequence has a distinct pattern depending on the movement type.

For feature extraction, we processed the RSSI sequence with the classical sliding window technique. Features are selected to focus on the "changes" in the signal strength, as required by motion detection applications. The features we used were: trend, standard deviation, median of difference, Δ(mean) and Δ(standard deviation).

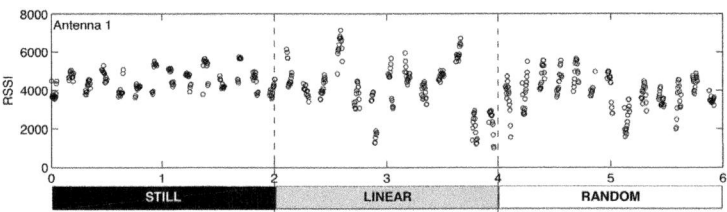

Fig. 2. RSSI values captured by a ceiling mounted antenna for a 60-second interval. Corresponding motion types, determined manually, are shown in the bars at the bottom.

We developed HMM and SVM based methods for assigning a motion label to each feature. We built two different HMMs: in the first one, named as the "uni-state representation", each motion type is represented with a single state. In the second HMM, named as the "bi-state representation", each type of motion is represented with a two-state left-to-right sub-HMM for modeling the motion transitions properly (Figure 3). *Main* is the core sub-state, representing the behavior of the state to which it belongs. *Exit* represents a pre-phase for out-transition. This idea is adopted from the modeling of phone transitions in speech recognition.

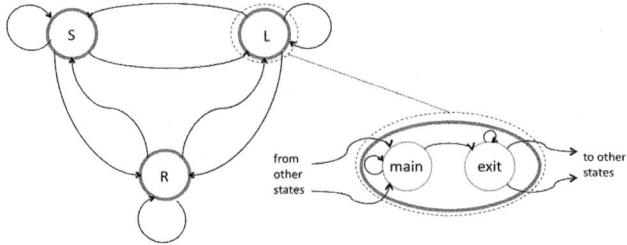

Fig. 3. HMM Topology with *main* and *exit* sub-states (bi-state representation). The dotted circle represents a magnified view of the state *moving linearly* (L).

For both uni-state and bi-state HMMs, self-transition probabilities are estimated from the expected motion durations; the other transition probabilities are distributed among the other allowed transitions equally. (For example in Figure 3 allowed transitions from the *main* state are self-transition and transition to the *exit* state.) Observations are modeled as Gaussian mixtures with parameters estimated from the training data based on the Maximum Likelihood principle. Initial state distribution is uniform among the three *main* sub-states of states L, R and S.

The sliding window technique captures the temporal relations in the sequence of observed movements and allows nontemporal learning algorithms to be used on sequential data. (It does not capture temporal relations in the hidden state sequence.) To classify motions, we used a Support Vector Machine (SVM), which is known as a maximum margin classifier with very low generalization error. Preliminary experiments showed that the label sequence obtained with the SVM included many spurious transitions, which correspond to short-term, frequently changing motions. An HMM-based post processing was applied on the SVM output to make the label sequence smoother.

3 Experimental Results

3.1 Experimental Setup

Our goal is to label a sequence of object movements as *still*, *moving linearly* or *moving randomly*. To record the RSSI dataset, an RFID tagged cartoon box (6"×4.5"×3") was interacted with as follows:

- Not interacting at all (object standing still)
- Holding the box while walking around a table and from sidewalls to the table with an approximate speed of 40 inches/second (object moving linearly)
- Standing at the same position and playing with the box: rotating and occluding it by hand (object moving randomly around the same position).

The three interactions were performed continuously in different order. In total, our training dataset consisted of 60 interaction sequences, each of length 60 seconds.

The experimental equipment consisted of an Alien RFID reader, circularly polarized antennae and passive tags (Squiggle). The environment was designed to match a typical trauma resuscitation room. One of the antennas was ceiling-mounted, facing the center of a plastic table (mimicking a patient bed). Two antennas were mounted on the perpendicular sidewalls. In this way, our aim was to detect movements in all three dimensions. Moreover, considering that the trauma team gathers around the patient during resuscitation and frequently occludes the side antennas, a ceiling mounted antenna is crucial in our design. The laboratory included other objects that caused multipath and other adverse conditions for RF propagation.

Our evaluation metric is the percentage of accurately predicted labels in a sequence ((true positive + true negative)/total number of labels). The performance of each method is reported by averaging the accuracy rate over all sequences.

3.2 Experimental Results

We investigated the motion labeling performance using 10-fold cross-validation. In the testing data fold, the accuracy was calculated for each sequence individually and summarized in the box plot in Figure 4, on the left side.

The HMM-based method achieved an average accuracy of 77.9% and outperformed the SVM-based model (73.66%). The higher accuracy variation in the HMM-based method can be explained as follows. In a nontemporal model, all instances are labeled independently, whereas in an HMM the adjacent label predictions are dependent. An erroneous measurement, as well as a successful one, affects not only the current label, but also the subsequent ones.

The best score (83.01%) was obtained when the SVM output is smoothed with an HMM-based postprocessing. This technique combines the max-margin learning capability of SVMs with the temporal representation of HMMs. An analysis on the confusion matrix showed that most of the misclassifications are among the linear and random movement types. If we were to make a moving/not-moving decision only, the accuracy would be 89.9%. The accuracy of binary motion sensing using WLAN RSSI was reported to be 87% in [6].

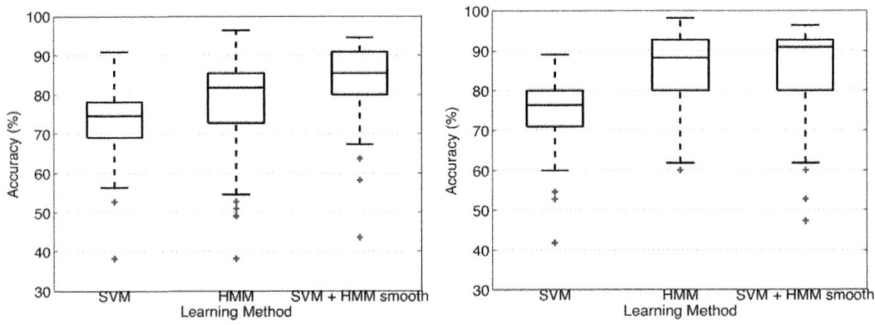

Fig. 4. Comparison of accuracy rates for various methods with uni-state (left) and bi-state (right) representations. (Central mark in the boxes: median; Edges of the boxes: 25th and 75th percentiles; Whiskers: most extreme data points not considered outliers; Plus sign: outliers.)

Next we evaluate the bi-state approach both in HMMs and SVMs. Incorporation of the *exit* state provided improvement for all methods by increasing the resolution of classification (Figure 4 - right). The improvement for the HMM (7.5%) was much higher compared to SVM (0.7%) because the idea of bi-state representation is meaningful when applied to a model that captures temporal correlations. The improvement for the SVM + HMM smoothing (1.4%) was smaller than HMM (7.5%) but higher than SVM (0.7%). This is because HMM fixes transitions on the final predicted sequence but the lost temporal information in the SVM stage cannot be restored. Moreover, as depicted in Figure 4 - left, accuracy variation of the bi-state HMM was lower compared to the uni-state HMM.

Discriminative Power of Individual Features

We also investigated which feature(s) contributed the most to the motion discrimination. In each test, one of the features was ignored and the accuracy was calculated with the bi-state HMM method. We also conducted paired t-tests to evaluate statistical significance.

None of the features contributed negatively to the discrimination. *Delta mean* was the most useful feature, whereas *delta standard deviation* had no effect in any setting and could be discarded. *Trend, standard deviation* and *median of difference* helped only in particular cases, such as increased tag population. Since they do not cause any degradation in the other cases, it is legitimate to include them in the feature vector.

4 Conclusions

This paper describes our work on monitoring of interactions with RFID tagged objects. We developed a continuous and multinomial labeling scheme that provides enhanced information for activity inference. Sensor and model selection, and the design of experiments were all made with respect to our target application: monitoring of the interactions with medical equipment during trauma resuscitation. However our results are generalisable to other high-risk work environments. Best scores were achieved with an HMM, where each movement was represented with a two-state left-to-right sub-HMM to model transitions. We also analyzed the individual contribution of each feature coefficient. Trend, standard deviation, median of difference and delta mean were observed to be useful features for detecting motion using RFID RSSI. Our future goal is to test the system in more challenging conditions, such as high tag population and people movement in the environment.

References

1. Fishkin, K.P., Jiang, B., Philipose, M., Roy, S.: I Sense a Disturbance in the Force: Unobtrusive Detection of Interactions with RFID-tagged Objects. In: Davies, N., Mynatt, E.D., Siio, I. (eds.) UbiComp 2004. LNCS, vol. 3205, pp. 268–282. Springer, Heidelberg (2004)
2. Philipose, M., Fishkin, K.P., Perkowitz, M., Patterson, D.J., Fox, D., Kautz, H., Hahnel, D.: Inferring Activities from Interactions with Objects. IEEE Pervasive Computing, 50–57 (2004)
3. Logan, B., Healey, J., Philipose, M., Tapia, E.M., Intille, S.: A long-term evaluation of sensing modalities for activity recognition. In: Krumm, J., Abowd, G.D., Seneviratne, A., Strang, T. (eds.) UbiComp 2007. LNCS, vol. 4717, pp. 483–500. Springer, Heidelberg (2007)
4. Ravindranath, L., Padmanabhan, V.N., Agrawal, P.: SixthSense: RFID-based Enterprise Intelligence. In: Proceeding of the 6th International Conference on Mobile Systems, Applications, and Services, pp. 253–266. ACM, Breckenridge (2008)
5. Patterson, D., Liao, L., Fox, D., Kautz, H.: Inferring High-Level Behavior from Low-Level Sensors. In: Dey, A.K., Schmidt, A., McCarthy, J.F. (eds.) UbiComp 2003. LNCS, vol. 2864, pp. 73–89. Springer, Heidelberg (2003)

6. Krumm, J., Horvitz, E., Redmond, W.A.: Locadio: Inferring Motion and Location from Wi-Fi Signal Strengths. In: Proceedings of the International Conference on Mobile and Ubiquitous Systems: Networking and Services (2004)
7. Favela, J., Tentori, M., Castro, L., Gonzalez, V., Moran, E., Martínez-García, A.: Activity Recognition for Context-aware Hospital Applications: Issues and Opportunities for the Deployment of Pervasive Networks. Mobile Networks and Applic's, 155–171 (2007)
8. Sarcevic, A., Marsic, I., Lesk, M.E., Burd, R.S.: Transactive memory in trauma resuscitation. In: Proceedings of the 2008 ACM Conference on Computer Supported Cooperative Work, pp. 215–224. ACM Press, San Diego (2008)

AmICA – A Flexible, Compact, Easy-to-Program and Low-Power WSN Platform

Sebastian Wille[1], Norbert Wehn[1], Ivan Martinovic[2], Simon Kunz[3], and Peter Göhner[3]

[1] Microelectronic Systems Design Research Group, Uni. of Kaiserslautern, Germany
{wille,wehn}@eit.uni-kl.de
[2] Distributed Computer Systems, University of Kaiserslautern, Germany
martinovic@informatik.uni-kl.de
[3] Institute of Industrial Automation and Software Eng., Uni. of Stuttgart, Germany
{simon.kunz,peter.goehner}@ias.uni-stuttgart.de

Abstract. In this paper, we present *AmICA*: a flexible, compact, easy-to-program, and low-power WSN platform. Developed from scratch and including a node, a basic communication protocol, and a debugging toolkit, it assists in a user-friendly rapid application development. Our analysis shows that *AmICA* nodes are 67% smaller than BTnodes, have five times more sensors than Mica2Dot and consume 72% less energy than the state-of-the-art TelosB mote in sleep mode.

Keywords: Wireless Sensor Networks, WSN Platform Design, Real-world application, AmICA Node.

The key requirements for a WSN node are flexibility, usability, compactness, high integration of sensors, high transmission range, and power-efficiency. By follwing these requirements as our main design objectives, we developed AmICA nodes. The *AmICA nodes* can be equipped with up to five sensors and two actors (see Fig. 2), have a transmission range up to 280m LOS, are as small as a coin (see Fig. 1) and consume only 1.6µA in sleep mode. More technical details and a comparison to state-of-the-art platforms are depicted in Table 1. Our flexible software stack together with free accessible compilers, and the flexible radio module enable rapid application development and open up a wide usage of the *AmICA platform*. A debugging toolkit (see Fig. 1) allows amongst others to record communication packets, scans for nodes, and re-configure and -program them wireless. Additionally, a new basic communication protocol called *AmICA node protocol* can be used for any single-hop short-distance network.

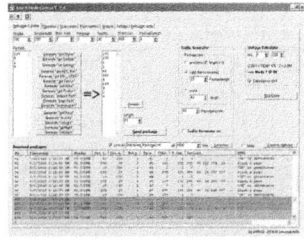

Fig. 1. Debugger toolkit (above), and *AmICA node* and a $0.25 coin (below)

A real-world Ambient Assisted Living (AAL) application running since 18 month evaluates the use of *AmICA* for a low duty-cycle application, where power-efficiency belongs to the most important application constraints. A real-world, high duty-cycle sport application exploits the in-network-processing capabilities, the small footprint, and the hardware robustness of the nodes.

Circuit diagrams, C libraries, software, protocol definitions and the debugger toolkit can be downloaded at www.amica-system.com as well as an elaborate technical report ([4]). Some assembled nodes can be provided on request.

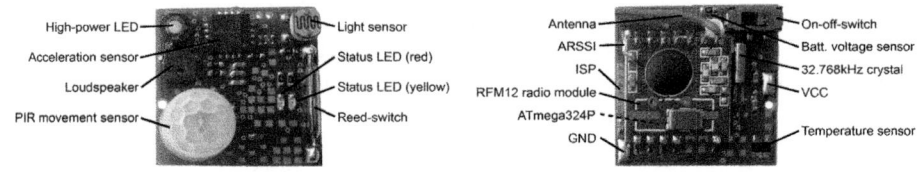

Fig. 2. Fully equipped *AmICA* node; top (left) and bottom (right) view

Table 1. Comparison between Mica2Dot ([1]), TelosB ([3]), BTnode ([2]) and *AmICA* node. [1]v1.1: 0.9-4.4/3.3-5.5V; [2]7 byte header, 4 byte payload per packet @ 3V

	Mica2Dot	TelosB 2420CA	BTnode rev3	AmICA 1.0
Size	⌀25x6mm	65x31x6mm	58x33x7mm	25x25x6mm
Flash/RAM	128KB/4KB	48KB/10KB	128KB/64KB	128KB/16KB
Sensors/actors	1/1 (on-board)	4/3 (on-board)	0/4 (on-board)	5/4 (on-board)
Freq. / Mod.	Sub-GHz FSK	2.4GHz OQPSK	Sub-GHz FSK	Sub-GHz FSK
Data rate	0.6-76.8kbps	250kbps	0.6-76.8kbps	0.6-115.2kbps
Link budget	112dBm (max.)	94dBm (max.)	112dBm (max.)	114dBm (max.)
Input voltage	2.7-3.3V	1.8-3.3V	0.5-4.4/3.6-5.0V	2.6-3.6V[1]
Only RTC	48µW	15.3µW	9000µW	4.2µW
MCU active	24mW	5.4mW	36mW	17.1mW
+ radio tx	81mW@5dBm	58.5mW@0dBm	93mW@5dBm	85.5mW@5dBm
+ radio rx	30mW	74.4mW	75mW	55.5mW
E/bit [µJ]	2.42 @ 38.4k	0.23 @ 250k	2.42 @ 38.4k	0.74 @ 115.2k
⌀ E/30 pck./hr.[2]	62.1µW	18.5µW	9020.2µW	16.2µW
w. 2x2000mAh	up to 45 month	up to 150 month	up to 0.3 month	up to 171 month

References

1. UC Berkeley. Mica2Dot platform (2002),
   http://tinyos.net/scoop/special/hardware
2. Beutel, J., Dyer, M., Hinz, M., Meier, L., Ringwald, M.: Next-generation prototyping of sensor networks. In: SenSys 2004: Proceedings of the 2nd International Conference on Embedded Networked Sensor Systems, pp. 291–292. ACM, New York (2004)
3. Polastre, J., Szewczyk, R., Culler, D.: Telos: enabling ultra-low power wireless research, pp. 364–369 (April 2005)
4. Wille, S., Wehn, N., Martinovic, I., Kunz, S., Goehner, P.: Amica - design and implementation of a exible, compact, easy-to-program and low-power wsn platform. Technical report (October 2010),
   http://ems.eit.uni-kl.de/uploads/tx_uniklwehn/AmICATechReport2010.pdf

The Use of GPS for Handling Lack of Indoor Constraints in Particle Filter-Based Inertial Positioning*

Thomas Toftkjær and Mikkel Baun Kjærgaard

Aarhus University, Denmark
{toughcar,mikkelbk}@cs.au.dk

Abstract. Particle filter-based inertial positioning promises infrastructure-less positioning, but previous research have not provided an understanding of, how the positioning accuracy of such systems depends on the layout of building structures. This poster presents initial result for the impact of the layout of building structures on the positioning accuracy using a particle filter-based inertial positioning system named Pro-Position. We also consider methods for using GPS positioning with particle filter-based inertial positioning to improve accuracy in areas, where positioning is poor because of lack of constraints from the layouts of the building structures.

Keywords: Inertial Positioning, Particle Filter, GPS, Error Sources.

1 Infrastructure-Less Positioning Using Pro-Position

Applying the visions of ubiquitous computing to a variety of domains require indoor positioning independently of local infrastructures. Examples of such domains are fire fighting, search and rescue, health care and police work. A second reason is, that in many indoor areas satellite systems such as GPS, does not provide pervasive coverage and provides a poor accuracy when available [4]. Pedestrian inertial positioning is a technology promising infrastructure-less positioning by depending solely on sensing movement. The primary focus has been devoted to foot-mounted [1,5] and waist-mounted [3] inertial systems. In this work we present and evaluate an inertial positioning system called Pro-Position and we extend the system using GPS positioning to improve accuracy in large open indoor areas.

Pro-Position is a particle filter-based inertial positioning system for both indoor and outdoor positioning. Pro-Position consists of three elements: Firstly a waist-mounted Honeywell DRM 4000 that measures movement and a high sensitivity receiver U-Blox EVK-5H that measures absolute positions, secondly, layouts of building structures in 2.5 dimensions similar to those used in [5] and, thirdly, a particle filter. The Pro-Position particle filter is a Sampling Importance Resampling(SIR) algorithm using a fixed number of particles in each resampling step, which is done to maintain approximately the same computational speed for all iterations of the algorithm. The resampling strategy chosen is the systematic resampling algorithm [2].

---

* The authors acknowledge the financial support granted by the *Danish National Advanced Technology Foundation* under J.nr. 009-2007-2.

2 Accuracy of the Pro-position System and GPS Improvements

The results from our measurement campaign in three different buildings are listed in Table 1, the results for each path are divided into inertial sensor readings and Pro-Position results, for the median error. The results highlight that Pro-Position improves accuracy and that there are variations from building to building but also outliers for each building depending on sensor errors and human behaviors.

Table 1. Comparing inertial sensor readings with Pro-Position results for positioning errors given as the meter distance between the ground truth and the estimated position

		Mall 1	Mall 2	Mall 3	*Mall*	OO 1	OO 2	OO 3	*OO*	RO 1	RO 2	RO 3	*RO*
50%	Sensor	18.34	17.47	25.28	*21.40*	8.03	11.27	10.56	*9.19*	9.12	10.59	8.48	*9.30*
	PP	3.41	4.27	5.12	*4.61*	2.95	8.58	3.87	*5.13*	2.52	3.09	3.87	*3.16*

Pro-Position make use of GPS when available, firstly, for providing the initial position and, secondly, to correct the particle filter positions. The correction is only applied if there are any GPS events with a low estimated error and in the case of multiple events the system uses the one with the lowest GPS receiver estimated horizontal error. Figure 1 shows an example of the improvements for one of the paths in the evaluated mall.

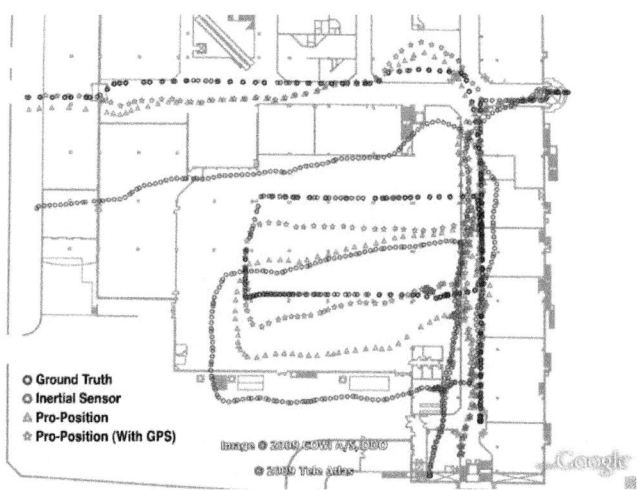

Fig. 1. Results from inertial sensor measurements, Pro-Position with and without GPS together with the ground truth path in a shopping mall

References

1. Foxlin, E.: Pedestrian tracking with shoe-mounted inertial sensors. IEEE Computer Graphics and Applications 25(6), 38–46 (2005)
2. Hol, J.D., Schon, T.B., Gustafsson, F.: On Resampling Algorithms for Particle Filters. In: Proc. of the IEEE Nonlinear Statistical Signal Processing Workshop, pp. 79–82 (2006)

3. Judd, T.: A Personal Dead Reckoning Module. In: Proceedings of the 10th International Technical Meeting of the Satellite Division of the Institute of Navigation, pp. 47–51 (1997)
4. Kjærgaard, M.B., Blunck, H., Toftkjær, T., Christensen, D.L., Grønbæk, K.: Indoor Positioning Using GPS Revisited. In: Proc. of the 8th Int. Conf. on Pervasive Computing (2010)
5. Woodman, O., Harle, R.: Pedestrian localisation for indoor environments. In: Proceedings of the 10th International Conference on Ubiquitous Computing, pp. 114–123 (2008)

F4Plan: An Approach to Build Efficient Adaptation Plans*

Francoise André[2], Erwan Daubert[1], Grégory Nain[1],
Brice Morin[1,**], and Olivier Barais[2]

[1] INRIA, Centre Rennes - Bretagne Atlantique, Campus de Beaulieu, 35042 Rennes, France
{Erwan.Daubert,Gregory.Nain}@inria.fr, Brice.Morin@sintef.no
[2] University of Rennes1, IRISA, Campus de Beaulieu, 35042 Rennes, France
{Francoise.Andre,Olivier.Barais}@irisa.fr

Abstract. Today's society increasingly depends on software systems subject to varying environmental conditions imposing that they continuously adapt. A dynamic adaptation reconfigures a running system from a consistent state into another consistent state. To achieve this goal, a reconfiguration consists in executing a set of actions leading from source to target configuration. The planning of actions has often been neglected in adaptation mechanisms, leading to naive sequential schedules statically predefined. EnTiMid, a ubiquitous software system for assisted living, is one of these adapting systems using basic adaptation plan. This situation may cause problems when considering adaptations involving large set of actions and/or devices, particularly for distributed service-based applications. We propose a framework to ease the integration of different planning algorithms that produce more efficient adaptation plan than an ad-hoc algorithm.

1 Introduction

Large companies, banks, airports, buildings and even houses increasingly depend on software systems. These systems are continuously impacted by changes in their execution environment (infrastructural variation or a modification of the user requirements).

In this domain, EnTiMid is a home-automation software system to assist elderlies in their everyday life. Typically the system has to deal with dozens of devices, different needs per user or frequent changes in the ubiquitous environment. To address the combinatorial explosion of the number of potential configurations, engineers can develop such Dynamically Adaptive Systems (DASs) as Dynamic Software Product Lines (DSPLs) [5] by defining several variation points. Depending on the context, the system dynamically chooses suitable variants to realize those variation points. These variants may provide better quality of service, offer new services that did not make sense in the previous context, or discard some services that are no longer useful.

---

* The research leading to these results has received funding from the European Community's Seventh Framework Program FP7 under grant agreements 215412 (DiVA, http://www.ict-diva.eu/) and 215483 (S-Cube, http://www.s-cube-network.eu/).
** Now at SINTEF ICT, Oslo, Norway.

In previous work [9] using Aspect-Oriented Modeling (AOM) techniques, we can explicitly build a model of the configuration which is suitable for the current context, with no need for the designers to specify the whole set of configurations in extension. Using Model-Driven Engineering (MDE) techniques (model comparison) we were able to infer safe but sub-optimal migration plans to dynamically adapt the system. In particular, the simple heuristic we used tend to maximize the unavailability of services when some components have to be stopped and restarted to achieve a safe adaptation.

In this paper we propose to improve adaptation systems by the use of planning algorithms to build more efficient reconfiguration plans than already existing ones. We illustrate this work on the EnTiMid system.

This paper is structured as follow. Section 2 starts with a presentation of EnTiMid and discusses our previous model-driven approach for designing and executing DAS. Then Section 3 details F4Plan, our proposal for designing and developing an efficient planning phase. Section 4 illustrated our approach on EnTiMid and demonstrates its benefits. Section 5 concludes this article and presents our future works.

2 Background

EnTiMid [10] is an ubiquitous software system built over an OSGi execution platform and developed in an assisted living context. The aim of this system, is to offer a level-sufficient abstraction of the devices in the home, making it possible for highlevel services to interact with physical devices (such as lights, heater or temperature sensors).

To address the problems of heterogeneity and dynamicity encountered in such an ubiquitous system, we proposed an Aspect-Oriented and Model-Based approach to tame the complexity of Dynamically Adaptive Systems (DAS) [8,9]. The overall approach is illustrated in Figure 1.

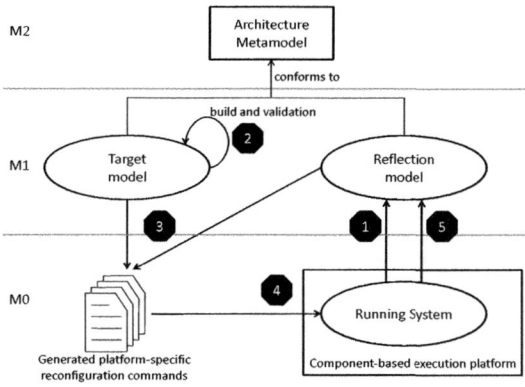

Fig. 1. Overview of our Approach to Dynamic Adaptation

The key idea is to keep an architectural model synchronized with the running system. This reflection model, which conforms to the architecture metamodel, is updated (step 1) when significant changes appears in the running system (addition/removal of components/bindings). It is important to note that the reflection model can only be modified according to runtime events. When a change has been made, a copy of this reflection model is used to build a target architectural model using model transformation or aspect model weaving. When a target model is derived, it is validated (step 2) using classic design-time validation techniques. This new model, if valid, represents the target configuration the running system should reach. Then, the adaptation plan to switch from the current to the target configuration is automatically generated. To do so, we first perform a model comparison between the source configuration (the reflection model) and the target configuration (step 3). This comparison produces an *ordered* set of adaptation actions. This safe sequence of actions is then submitted (step 4) to the running system in order to actually reconfigure it. Finally, the reflection model is automatically updated and becomes equivalent to the target model (step 5).

The sequence of atomic actions is ordered according to the following plan description: (1) components (that should be stopped) are stopped, according to the client/ server dependencies (clients are stopped before the servers), (2) bindings are removed, (3) components are removed, (4) attributes of already present components are updated, (5) components are added and their attributes are set, (6) bindings are added, (7) components are (re-)started, according to the client/server dependencies (servers are started before the clients).

In case of a large number of actions, this simple heuristic tends to maximize the unavailability of components, since the " Stop Component" actions are always executed at the beginning of the adaptation, and the "Start Component" actions are always executed at the end. To allow more efficient scheduling of actions we propose a new methodology based on general purpose planning algorithms. Our objective is to fit different possible needs concerning the planning phase. The next section concentrates on that proposal.

3 A Generic Approach for Planning Adaptation Actions

3.1 Motivations

The MAPE[6] model defines four steps to do dynamic adaptation at runtime: the Monitoring, the Analysis, also called *Decision*, the Planning and the Execution. We define the semantic of these steps as follow. First, the Monitoring is used to detect changes inside the application or in its execution context (step 1 in Figure 1).When significant changes are detected, the Monitoring triggers the Analysis. This phase consists in deciding if adaptation is necessary to maintain the functionalities of the system. If adaptation is needed, the Decision also chooses the adaptation strategy that should be used (step 2). Once done, the Planning phase selects actions to execute and schedule them according to the chosen strategy (step 3). The execution of the selected actions is the last phase of the model (step 4). In this paper, we focus on the Planning phase.

As adaptation is performed at run-time, the time needed to actually perform the adaptation have to be minimized. Therefore, Planning is an important step of the MAPE model. It defines the actions necessary to properly apply the adaptation strategy, and orders the actions to ensure the consistency of the adaptation execution and minimize the time. Indeed some actions may be dependent of some others. For example, it is not possible to start component if its bindings are not already set and its attributes changed.

At the opposite some actions can be independent, leading to a partial order between them (e.g.: two components can be started at the same time).

In our preliminary planning method described in the section 2, a static total order is defined on the different types of actions. This order can only lead to build a sequential schedule. Thus considering a distributed EnTiMid platform, whatever the number of components involved in an adaptation and their location, all the necessary "stop component" commands should be performed before to execute all the "remove binding" commands and so on.

Such an adaptation method consumes more time than needed because for example, the adaptation engine have to wait for all the "component stop" commands to be executed before launching the "component start" commands. Moreover, in a distributed and asynchronous system, synchronization operations between the different platforms should be added to enforce the sequentiality. Also, in this preliminary planning implementation it is not possible to add information or constraints on actions or on sequence of actions to give useful indications for the execution phase(e.g.: non-functional data as the execution time of an action or the among of resources used).

Research works on planning methods such as Artificial Intelligence planning, Motion planning or Control theory, have produced some algorithms that overcome these limitations. In this paper, we propose an architecture for the planning phase to use, according to the needs, one of these algorithms in adaptation system for Service-Oriented Architecture. Most of times a planning phase using one of these algorithms is executed as follow.

It takes the strategy issued from the decision phase as input. This strategy consists of a *source configuration* and a *target configuration* (i.e. the current and the desired state of the system).

An *initial state* and a *goal state*, both given to the planning algorithm, are deduced from the strategy. These states are described in a language that depends on the planning algorithm used. The *domain* of actions, last input needed by the planning algorithm, represents all the possible actions.

In the following we describe our design proposition for the planning phase.

3.2 Our Proposition: F4Plan (Framework for Planning)

As previously said, several planning algorithms exist, each one has its own characteristics. Therefore, to design an adaptation system, a planning algorithm has to be chosen among all existing ones according to the planning objectives. These objectives can be about minimizing the time spend in the planning phase and in that case choose a very simple planning algorithm, even if the resulting schedule is not the best one. In case there exists only one processor to execute the adaptation actions, it is not useful to select an algorithm that may exhibit some parallelism in the schedule. At the opposite it may be preferable to choose an algorithm that will spend some time to obtain the most efficient schedule if the adaptation actions are long and some of them may be executed simultaneously.

Each planning algorithm has its own dedicated language to express the initial/goal states and the domain of actions. In order to keep a coherent chain from the source and target configurations to the initial and goal states (i.e.: from the decision algorithm outputs to the planning algorithm inputs), a language translation is necessary.

We do not want to impose the choice of a specific decision algorithm nor a specific planning algorithm because this choice may depend on changing environmental constraints. For instance sometimes adaptation can concern only few elements, geographically closed, with the objective to quickly adapt. In that case, a simple planning/decision algorithm will be chosen. Sometimes the objective can be to perform proactive adaptation, involving a large set of distributed elements. The preference will then be for a more powerful planning/decision algorithm.

Thus, to ease the work of the final developer, we offer a set of translators from a description language to another. To face the combinatorial explosion of the number of translators, we propose to use a very common and powerful planning language, PDDL [4], as a pivot language.

At the end of the planning phase, the algorithm returns a schedule (a plan) that is used by the last phase of the MAPE model (the execution) to concretely realize the strategy.

An illustration of the benefits of using an efficient planning algorithm and a coherent chain of translation between decision and planning phases is given in section 4 to adapt the Ambient Assisted Living Application based on EnTiMid previously presented in 2).

4 Illustration

This section illustrates the use of F4Plan with a simple example into the EnTiMid system.

In the following, the changes operated on the EnTiMid platform located on the elderly person's house when the night comes will be depicted.

During the day the elderly person, in case of major problem, has to his/her disposal a remote control with a single button (the *SOS* component), that will send an emergency message via SMS using the *SMS* component. In addition, to confirm the person that the SMS has been sent, a *Text-To-Speech* component emits a message. To connect these three components a *dispatcher1* component is involved to trigger in parallel the SMS sending and the vocal message emitting.

At night, when the person signals a problem, the house will be enlighten so that the person can realize what happened and get back its landmarks. As a consequence, lights must be manageable by the system in this configuration and N *light* components are added. These lights are bound with the remote control with the dispatcher1 already available into the platform. A component, here called *central lights command*, is also added to enable the elderly person to switch off the lights when everything get back to normal. A new dispatcher (called *dispatcher2*) is needed to connect lights with the central lights command.

The target configuration to reach is obtained using the mechanism described in section 2. In this scenario, a small subset of actions available to reconfigure the EnTiMid platform is considered. Only, **add component**, **add binding**, **start component** and **stop component** are used.

In this use case F4Plan uses a planning algorithm called GraphPlan [1]. This algorithm uses a PDDL subset called STRIPS [3] as input language.

Therefore, a translation between the decision output language, in our case the ART home made language and STRIPS is needed. Using our implementation, this translation is done with two translators automatically found, communicating through the PDDL pivot language. When the translation is done, the planning algorithm is executed.

At the end GraphPlan returns a partial ordered set of actions (Figure 2).

```
Step 1:
    Add Component "central lights command"
        // ... // Add Component "lightN"
        // Stop Component "dispatcher1"
Step 2:
    Add Binding "dispatcher1Tolight1"
        // ... // Add Binding "dispatcher2TolightN"
Step 3:
    Start Component "dispatcher1"
        // ... // Start Component "lightN"
```

Fig. 2. partial-order set of actions

To sum up, GraphPlan is well adapted for our Ambient Assisted Living use case, showing a substancial gain on the resulting plan reducing the number of steps (from 4N+6 to 3). Meanwhile, some other recent algorithms can provide the same kind of result with better performance.

The translation mechanism, associated with our proposition also implies a cost.

In all cases, the use of such not trivial planning algorithms has a cost. A tradeoff between this cost and benefit during the execution of plans needs to be done. Here we choose more complex planning algorithms to build more efficient plans and reduce their execution.

5 Conclusion

Nowadays, most software developments should consider the issue of their adaptation to the dynamism of the execution environments. However current solutions for adaptation are most often ad hoc and in consequence are not satisfying as long term solutions.

In this paper, we focus on the planning phase which has till now received little attention whereas it is an important phase especially in distributed environments. Most adaptation systems use a very simple ordering of actions even though many general purpose planning algorithms exist.

F4Plan allows the use of already developed planning algorithms. In that way, it is possible to take advantage of research works already done on planning methods such as Artificial Intelligence planning, Motion planning or Control theory. Such general purpose algorithms have already been applied in contexts close to our, in particular in applications for components deployment on computational Grids [7,2]. Contrary to these approaches, our proposal does not impose a specific algorithm for the adaptation system and is self-adaptable, so that a choice between different planning algorithms, for example an algorithm looking for parallelism and a much simpler one, is always possible. So we are able to compute an efficient, coherent and valid plan to apply the decision strategy according to the constraints like time duration or resource consumption. The characteristics of the environment, in particular the distribution of all resources, either software or hardware, can also be taken in consideration to schedule and parallelize the actions.

F4Plan also provides an automatic way to translate the decision output into the planning input, offering several translators around a pivot language.

In our future work, we intend to study the distribution of the planning algorithms. This could be interesting to reduce the computation time of the schedule. Moreover in some cases, part of the schedule can be locally computed when it only involves local actions in a distributed environment.

Another subject of interest is the dynamic discovery of available adaptation actions. Indeed we currently use statically defined types of actions but in a large scale world of services, some actions can only be identified dynamically.

References

1. Blum, A.L., Furst, M.L.: Fast Planning Through Planning Graph Analysis. Artificial Intelligence 90, 1636–1642 (1995)
2. Deelman, E., Singh, G., Su, M.H., Blythe, J., Gil, Y., Kesselman, C., Mehta, G., Vahi, K., Berriman, G.B., Good, J., et al.: Pegasus: A framework for mapping complex scientific workflows onto distributed systems. Scientific Programming 13(3), 219–237 (2005)
3. Fikes, R., Nilsson, N.J.: STRIPS: A new approach to the application of theorem proving to problem solving. Artificial Intelligence 2(3/4), 189–208 (1971)
4. Ghallab, M., Isi, C.K., Penberthy, S., Smith, D.E., Sun, Y., Weld, D.: PDDL - The Planning Domain Definition Language. Technical report, CVC TR-98-003/DCS TR-1165, Yale Center for Computational Vision and Control (1998)
5. Hallsteinsen, S., Hinchey, M., Park, S., Schmid, K.: Dynamic Software Product Lines. IEEE Computer 41(4) (April 2008)
6. Kephart, J.O., Chess, D.M.: The Vision of Autonomic Computing. Computer 36(1), 41–50 (2003)
7. Kichkaylo, T., Ivan, A., Karamcheti, V.: Constrained component deployment in wide-area networks using AI planning techniques. In: Intl. Parallel and Distributed Processing Symposium (2003)
8. Morin, B., Barais, O., Jézéquel, J.-M., Fleurey, F., Solberg, A.: Models@ Run.time to Support Dynamic Adaptation. time to Support Dynamic Adaptation. Computer 42(10), 44–51 (2009)
9. Morin, B., Barais, O., Nain, G., Jézéquel, J.-M.: Taming Dynamically Adaptive Systems with Models and Aspects. In: 31st International Conference on Software Engineering (ICSE 2009), Vancouver, Canada (May 2009)
10. Nain, G., Daubert, E., Barais, O., Jézéquel, J.-M.: Using MDE to Build a Schizofrenic Middleware for Home/Building Automation. In: Mähönen, P., Pohl, K., Priol, T. (eds.) ServiceWave 2008. LNCS, vol. 5377, pp. 49–61. Springer, Heidelberg (2008)

Context Acquisition and Acting in Pervasive Physiological Applications[*]

Andreas Schroeder, Christian Kroiß, and Thomas Mair

Ludwig-Maximilians-Universität, München, Germany
{schroeda,kroiss}@pst.ifi.lmu.de

Abstract. Physiological computing means using physiological sensors in computing. This is a natural and promising continuation of pervasive computing: as smart devices begin to permeate the environment, they can be used to collect information about the user's emotional, cognitive and physical state to improve the context-awareness of applications. Creating pervasive physiological computing applications is hard, however. We propose a software framework that simplifies the creation of these applications by providing a first design as well as support for processing sensor data, distributing analysis results, and decision making under the uncertainty that arise in physiological computing. We illustrate the presented framework with the personalized affective music player, a context-aware physiological application that plays music to guide the mood of a user into a pre-defined direction.

Keywords: Pervasive Adaptation, Pervasive Computing, Physiological Computing, Software Engineering.

1 Introduction

The next step after pervasive computing [8], that is to say, overcoming desktop-oriented IT environments, is to include physiological information becoming available through the entailed proximity of computer devices to the human. A plethora of physiological sensors exist allowing computers to comprehend the user's emotional, cognitive and physical states. Approaches leveraging these source of inputs are known as physiological computing [1,7], or affective computing [12], when restricted to emotional state. Context-aware interactive applications benefit from physiological computing by gaining a better understanding of their users and thus offering improved services.

Creating pervasive physiological applications is no easy task, however. Data from physiological sensors need to be handled appropriately: signal quality and noise, personal variance, and sometimes just the large amount of data makes this a non-trivial issue. Algorithms and techniques for psychophysiological inference, that is to say, extracting a user's emotional, cognitive and physical state from sensor data are still being researched [7]. Therefore, physiological computing applications need to adapt swiftly to new insights in the domain as well as new emerging algorithms and techniques.

In this paper, we describe how a framework approach can support the development of pervasive (i.e. especially context-aware) physiological applications (Sect. 4). We have

---

[*] This work has been partially supported by the EC project REFLECT, IST-2007-215893.

designed and implemented the component-based REFLECT framework that supports the software engineers in designing and creating the analysis part (Sect. 4.1) and the decision making part (Sect. 4.2) of a pervasive physiological application, as well as testing and understanding support offered by the accompanying tooling (Sect. 4.3). We use an implementation of the personalized affective music player [9] (Sect. 3) as a running example to motivate the framework support. Finally, we review related work in Sect. 5 and conclude in Sect. 6. First of all however, we start with a discussion of the challenges in pervasive physiological computing (Sect. 2).

2 From Sensing to Acting - A Software Engineering Perspective

From a bird's eye view, a pervasive physiological system is a software agent that interacts with the physical world and the user through sensors and actuators. The agent translates input that it receives through sensors – often called percepts – to actions on actuators that the agent controls [13]. In a pervasive physiological computing setting, the following challenges arise in addition to correctly interpreting percepts: Firstly, the continuous and noisy character of the data data provided by physiological sensors must be handled properly. Secondly, the results of interpreted data must be distributed to other devices in the pervasive computing environment in order to share and combine knowledge to create a more complete picture of the system's physical context. Thirdly, hidden or unknown parameters and incomplete knowledge in the physiological domain may lead to seemingly inconsistent reactions of the environment and the user to actions performed by the system.

Besides the algorithmic challenges, organizational challenges also arise when building pervasive adaptive applications. Software designs for pervasive adaptive applications must support different activities, namely 1) exploring and experimenting with new algorithms and new techniques, 2) testing and validation of the system and its parts, and 3) rapid creation of software artefacts.

Exploration and experimentation and *testing and validation* must be supported since the algorithms that will actually allow to extract knowledge about the emotional, cognitive, and physical state of a human from basic sensors and features are still a field of active research [7]. A software design or software framework must therefore support algorithm exchange and experimentation at low cost. *Rapid creation of software artefacts* is crucial while still being in the process of understanding the application domain and identifying the hard problems within it. Even later, while creating a product, being able to swiftly create the software artefacts may turn out as crucial for reducing the product's time to market.

3 Affective Music Player Example

The affective music player (AMP) is a pervasive physiological application whose concepts stem from [9]. It functions as a closed loop, repeatedly measuring the current mood of the user and selecting music from the user's own music database in order to influence the user towards a user-defined target mood. In this context, mood is understood as a long lasting (i.e. minutes to days) affective state with no clear cause or origin [16].

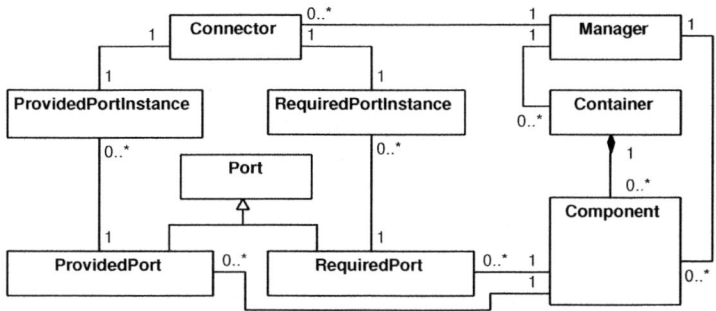

Fig. 1. REFLECT component metamodel

Moods are different from emotions, which are short lasting (i.e. seconds to minutes) affective processes related to an event in the environment or a thought [5].

We built a fully functional prototype of the AMP using the REFLECT framework which is described in the following sections. The AMP serves as a running example to demonstrate the data handling and action selection support the REFLECT framework provides.

4 Framework Support

Handling the challenges described in Sect. 2 without a good design is a daunting task; every step in the process of creating a full system must be carefully thought through from beginning to end with little or no guidance in the design and implementation process. The framework approach followed in the REFLECT framework is one possibility to guide this process.

The basic structure of the REFLECT framework is component-based (see Fig. 1). That is to say, every entity that is created for the REFLECT framework is a component or framed within the context of components: Functionality is encapsulated in *components*, and function groups are arranged within *component containers*. Components communicate to each other over *required* and *provided ports*, which describe the required and provided functionality, respectively. Thus, a component can offer different functionality through different ports; additionally, a component can provide the same functionality multiple times through *multiple instances* of a provided port. Similarly, a required port can collect and bind a set of provided ports. *Connectors* track the binding of ports. Finally, a single *manager* provides access to all components, connectors, and containers. Using components as the basic underlying structure of a system encourages a clear architecture with well-defined interfaces, which is beneficial for rapid development as well as testing.

Fig. 2 shows the basic architecture of applications built on top of the REFLECT framework. Sensors provide transient data that is stored in the context store, that is used by analyser components to extract more abstract features that are stored again in the data context (see Sect. 4.1). The application uses data available in the data context, may make use of persistent data available in the data store (e.g. user preferences), and rater components (see Sect. 4.2) to make decisions on how to control actuators.

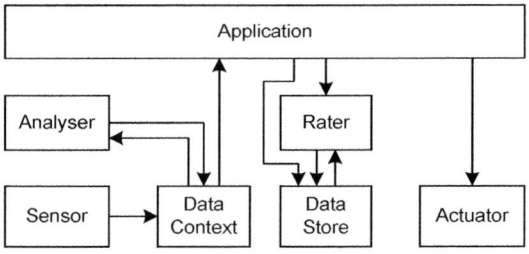

Fig. 2. REFLECT application structure

The REFLECT framework is implemented in Java and built on top of OSGi [2]. It makes use of Java annotations and internal domain specific languages for e.g. container specifications. The implementation itself is focused on ease of use and leverages the Eclipse OSGi tool support [11] wherever possible.

4.1 Handling Sensor Data

Extracting meaningful features from sensor data often requires not only the latest sensor reading, but a set of samples. In the REFLECT framework, this is supported by data windows. These data windows are realized as flexible ring buffer and guarantees to store all recent samples within a fixed time length, and discard old measurements as soon as they exceed the data window time length. Often, a fragment of a data window with specified relative length needs to be created, e.g., to retrieve the last five seconds of data input from a larger data window. The REFLECT framework offers *slices* to create relative fragments of a data window. A slice is defined through start and end points relative to the start and end of the underlying data window. For example, to retrieve the last five seconds of input, a slice is specified as starting at $end - 5s$, and ending at end, where end refers to the end of the underlying data window. Programmatically, start and end are referenced through positive and negative indexes, respectively. Creating a slice with the last five seconds is done programmatically with `window.slice(-5, 0, TimeUnit.SECONDS)`.[1] A slice is live, that is to say, new data added to the underlying data window lets the slice grow and/or shift.

For example, the affective music player infers mood from skin temperature and skin conductance level – two psychophysiological measures known to be related to autonomous nervous system activity [16]. Both skin temperature and skin conductance measurements are stored in data windows in the data context. Both data sources are known to be highly dependent on sensor placement and environment temperature, and in addition, the amplitude of changes in skin temperature and skin conductance have a high interpersonal variation [12]. Therefore, the sensor inputs need to be normalized, which can be easily performed using the data window framework.

---

[1] Interestingly, zero must reference the data window start point when used as start parameter, and also reference the data window end when used as end parameter. Otherwise, it would be impossible to create slices both starting and ending with the underlying data window.

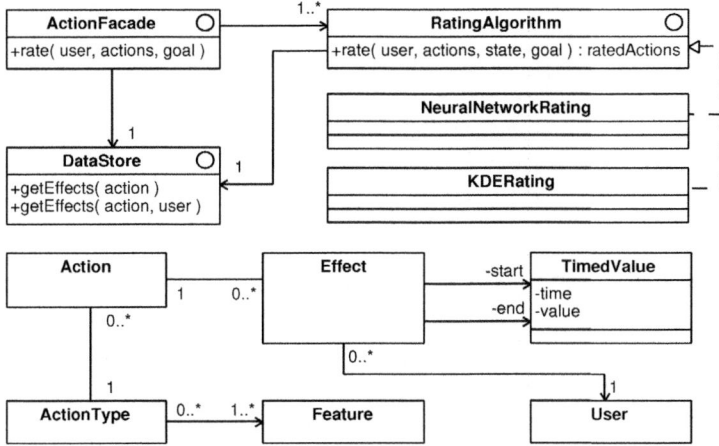

Fig. 3. Rating framework

Including data windows as framework data structure offers several benefits. On the one hand, it avoids the problem of blocking sensor processes when the feature extraction components cannot keep up: As opposed to waiting queues, a data window cannot fill up to its maximum capacity and block the sensor process. Instead, it will always present the most recent data to the feature extraction algorithm using the data window.

In addition to this beneficial process separation, data windows offer a clear concept for distributing data. The process of replicating the content of a data window that is fed by either sensors, feature extraction algorithms or other processes, constitutes an easily understandable and flexible data sharing concept. To put it another way, the process separation offered by data windows allows to distribute processes almost transparently. Within a REFLECT framework instance, a system assembly may specify to replicate the content of a remote data window for local use.

The data context manages all data windows that exist in one framework instance and offers a central point of access. Compared to a connection-oriented design in which a data window may be accessed only by the data producer and its consumers, this design simplifies the access to data windows for new data consumers, data window inspection, and data distribution.

4.2 Selecting Actions

In a pervasive adaptive setting, the problem often arises that the effects of actions are a priori unknown, and need to be learned over time. This is especially true when interacting with humans: reactions may be highly personal, as it is the case with e.g. reactions to music. These highly personal reactions must be caught and learned at run-time. The REFLECT framework offers support for learning action effects and rating actions in two respects: first, it records and stores the effects of executed actions. Secondly, it offers a framework for rating actions based on past effects (see Fig. 3).

In order to record the effects of an action, it must be known a) what part of the environment is affected by the action, i.e. how the action effect can be measured, and b)

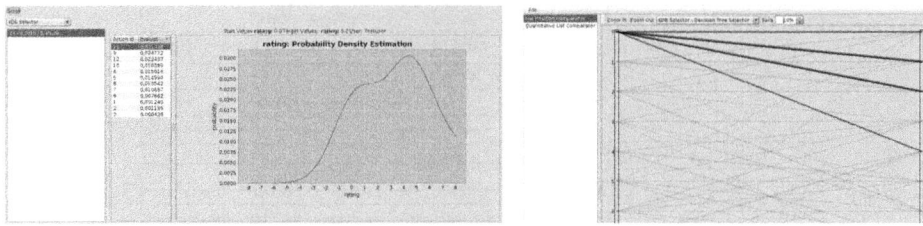

Fig. 4. Tool support for KDE inspection and graphical comparison of rankings

when the action is expected to make an effect. Specifying how action effects are measured is done by grouping actions into action types (e.g. actions consisting of playing a single song are grouped into "a play song action type"), and the affected state features are associated with action types. By specifying only the features that are affected by action types, it is assumed that effects can be tracked uniformly in terms of changes in the affected features. An effect is thereby reduced to pairs of timestamped start and end values. In all applications we created so far, this simplification was feasible. Fig. 3 shows the data model for effects that the framework offers: every effect is associated to a user, as action effects may be highly personal.

Based on the recorded effects in the data store, a rating algorithm can evaluate a set of actions for their fitness to reach a goal state specified by the application. Rating algorithms can be defined using e.g. kernel density estimation techniques (KDE) [14,13] and neural networks [3,13]. The idea in both approaches it to predict the future effect of single actions based on the effects recorded. In the KDE approach, e.g, a Gaussian kernel is fitted on each previously measured effect of an action to create a probability density function for each action. Then, the score of an action is the probability that the action guides the current state into an ϵ-environment around the goal state.

The affective music player uses the action selection framework to select the next song to play once the previous song has ended. It does so on the basis of the past song effects which are recorded in terms of changes in skin conductance level and skin temperature (see Sect. 4.1). In order to select songs, the target mood is brought to desired changes in skin conductance level and skin temperature, and the songs in the user's personalized play list are ranked in terms of their fitness to achieve the requested changes in psychophysiology. The ranking of songs is done using a slight modification of the KDE-based rating algorithm [9].

4.3 Inspecting and Testing

Evaluating the quality of the action selection process can be quite difficult. This is especially true for algorithms like KDE or neural networks, which generally produce results that are hard to interpret. Detailed manual inspection of the internal data and its evolution during simulated runs are often needed. To facilitate this task, the REFLECT framework is accompanied by a tool that provides several visualization options. The tester can use the tool to browse the simulation run and inspect each step, i.e. the updated state of the rating algorithm's internal data and the generated ranking. While the visualization modules for the internal data have to be designed for each rating algorithm (e.g. KDE, see Fig. 4, left), the visualization tool provides a general module for comparing the action rankings generated by two different algorithms (see Fig. 4, right). It shows a graph consisting of two node columns that represent the aggregated rankings. Edges between the two columns show how actions were ranked differently during the simulation runs. The frequency of rank differences is visualized by different line strength of the edges. This way, the user can easily get a qualitative measure of how similar the action selectors behave.

5 Related Work

Several frameworks and middlewares were developed in the context of pervasive computing. The context toolkit [6] proposes an interesting widget approach for the rapid creation of context-aware pervasive applications. Transparent ad-hoc networking and context information distribution through middleware support is proposed in [17]. The Aura architecture [15] offers a design for transparent content migration that allows content to follow its user.

While most approaches provide better support for transparent distribution of data, all approaches define context in the sense that is understood in pure pervasive computing: the user's physical location, time, devices, their interconnections and proximities, and physical features of the environment such as lighting and temperature (e.g. [6] and [17]). Although some definitions of context are so broad that they can include physiological inputs (especially [6]), the specifics physiological computing brings about are not considered. To the best of our knowledge, no framework or middleware exists that specifically supports the design and implementation of pervasive physiological computing applications.

Rainbow [4] and KX [10] are component-based approaches that use probes and gauges to generate and aggregate streams of measurements for a running application. Then, reconfiguration decisions are made by a controller. The probes, however, are software entities used to provide measurement from legacy systems ([10]) or generic components ([4]) instead of physical sensors.

6 Conclusion

Physiological computing [1,7] is a logical consequence of pervasive computing. Leveraging the physical proximity to the human body offered by smart devices pervading our environment allows to create applications taking into account the user's emotional, cognitive and physical state.

We presented a component-based framework that aims at providing a clear design blueprint, and offering support for processing of physiological data, distributing analysis results, and making decisions in the domain of physiological computing. By this, we aim to simplify the creation of pervasive physiological computing applications.

The REFLECT framework has several limitations, however. First, it does not offer transparent distribution of context information. Instead, each framework instance must specifically declare its interest in the data provided by a known remote instance. Furthermore, the action rating support currently operates solely on continuous values, and action effects are considered to be measurable by changes in a defined set of features. Validation of the applicability of the framework in other applications than the personalized affective music player is needed. Finally, tool support for automated testing of decision making algorithms is needed.

Looking further ahead, support pervasive physiological computing applications spanning groups of users, from small to large, may also prove a worthwhile extension of the REFLECT framework. Currently, our framework is focussed on supporting single user (or multiple, but sequential users) applications.

References

1. Allanson, J.: Electrophysiologically Interactive Computer Systems. IEEE Computer 35, 60–65 (2002)
2. OSGi Alliance. OSGi Service Platform Release 4.2 (2009)
3. Bishop, C.M.: Neural Networks for Pattern Recognition. Oxford University Press, Oxford (1995)
4. Cheng, S.-w., Huang, A.-c., Garlan, D., Schmerl, B., Steenkiste, P.: Rainbow: Architecture-based self-adaptation with reusable infrastructure. IEEE Computer 37, 46–54 (2004)
5. Damasio, A.D.: Descartes' error: Emotions, reason, and the human brain. Putman, New York (1994)
6. Dey, A.K., Abowd, G.D., Salber, D.: A conceptual framework and a toolkit for supporting the rapid prototyping of context-aware applications. Human-Computer Interaction 16(2), 97–166 (2001)
7. Fairclough, S.H.: Fundamentals of physiological computing. Interacting with Computers 21(1-2), 133–145 (2009)
8. Hansmann, U.: Pervasive computing: the mobile world. Springer, New York (2003)
9. Janssen, J.H., van den Broek, E.L., Westerink, J.H.D.M.: Personalized affective music player. In: 3rd Int. Conf. Affective Computing and Intelligent Interaction. IEEE, New York (2009)
10. Kaiser, G., Gross, P., Kc, G., Parekh, J.: An Approach to Autonomizing Legacy Systems. In: Wshp. Self-Healing, Adaptive and Self-Managed Systems (2002)
11. McAffer, J., VanderLei, P., Archer, S.: OSGi and Equinox: Creating Highly Modular Java Systems. Addison-Wesley Professional, Upper Saddle River (2010)
12. Picard, R.W.: Affective Computing. MIT Press, Cambridge (1997)
13. Russell, S.J., Norvig, P.: Artificial Intelligence: A Modern Approach. Pearson Education, New Jersey (2003)
14. Silverman, B.W.: Density Estimation for Statistics and Data Analysis. Chapman and Hall, London (1986)
15. Sousa, J.P., Garlan, D.: Aura: an Architectural Framework for User Mobility in Ubiquitous Computing Environments. In: 17th World Computer Congress - TC2 Stream / 3rd Conf. Software Architecture, Deventer, Netherlands, pp. 29–43. Kluwer, B.V., Dordrecht (2002)
16. Thayer, R.E.: The biopsychology of mood and arousal. Oxford University Press, New York (1989)
17. Yau, S.S., Karim, F., Wang, Y., Wang, B., Gupta, S.K.S.: Reconfigurable Context-Sensitive Middleware for Pervasive Computing. IEEE Pervasive Computing 1, 33–40 (2002)

Linking between Personal Smart Spaces

Sarah Gallacher, Elizabeth Papadopoulou, Nick K. Taylor,
M. Howard Williams, and Fraser R. Blackmun

School of Maths and Computer Sciences, Heriot-Watt University,
Riccarton, Edinburgh, EH14 4AS, UK
{S.Gallacher,E.Papadopoulou,N.K.Taylor,
M.H.Williams,F.R.Blackmun}@hw.ac.uk

Abstract. One approach to the development of pervasive systems is based on the notion of Personal Smart Spaces (PSSs). A PSS is implemented as an ad hoc network and may be either fixed or mobile. When one PSS encounters another, communication is established between them. This may be used to alert one user to the presence of another, or a fixed smart space to the presence of a mobile user. There is considerable potential for applications using this type of functionality and it could become an important component of pervasive systems in the future. However, one problem with this is that it is difficult to detect when one PSS is close enough to another to be relevant. The Persist project has built a pervasive system based on PSSs and investigated the problems of their interaction in order to demonstrate this functionality. This paper discusses the problem of proximity and attempts to address it.

Keywords: Pervasive systems, smart spaces, ubiquitous systems, inter-PSS communication.

1 Introduction

One of the key objectives of a pervasive system is to provide support to enable the user to interact easily with and control the myriad of devices in the environment surrounding him/her. Many different researchers have been engaged in finding solutions to different aspects of this problem, and from this research two main approaches have emerged. The first is focused on fixed smart spaces which support users while they are present in the space; typical of this type of system is the Smart Home (e.g. Adaptive House [1], MavHome [2], GAIA [3], Synapse, Ubisec [4], etc.). The second approach is focused on the mobile user and provides support wherever he/she may go. Examples of this type of system include Daidalos [5], Mobilife, Spice, etc.

A novel approach to the development of pervasive systems is based on the idea of a Personal Smart Space (PSS) - this was developed in part to bridge the gap between fixed and mobile pervasive systems. It is based on the use of ad hoc networks as a means of communication within a PSS and for interactions between different PSSs.

The PSS approach has a number of advantages. One of these is the possibility of using the detection of another PSS to trigger particular actions. For example, when one mobile PSS encounters another which it recognises, it could alert the user to the presence of the other user (by name and/or other details). Similarly, when a mobile PSS comes within range of a fixed PSS, the fixed PSS may respond appropriately.

The PSS approach and some of the challenges and new functionality that it offers, have been investigated in the Persist project, a European research project funded under Framework Programme 7. Within Persist a pervasive system has been developed based on PSSs. This system has been demonstrated at several conferences and the open source code of the platform is available to download from the Sourceforge website [6]. In the course of developing this pervasive system one of the problems that was encountered, was the difficulty in detecting when one PSS is close enough to another to be considered relevant and the appropriate action triggered. This paper is concerned with this problem and attempts made to address it.

The next section provides a brief overview of the notion of Personal Smart Spaces while section 3 describes two scenarios used to demonstrate the functionality discussed. Section 4 discusses the issue of how close a PSS needs to be, to be relevant. Section 5 considers the problem of determining proximity and attempts to address this. Section 6 concludes.

2 Personal Smart Spaces

Fixed smart spaces suffer from the problem that while they provide support for the users who occupy them, when a user leaves such a space, the support provided disappears. The net result is that one will end up with "islands of pervasiveness" within which fully functional pervasive services are provided, but outside of which there is limited (if any) support for pervasive features. Although research has also been carried out on systems to support mobile users, the problems here are somewhat different and these two types of system are generally quite independent of each other.

The notion of a Personal Smart Space [7] was introduced to integrate fixed smart spaces and mobile systems in a clean and consistent fashion. The result is that the user will have a degree of pervasive support at all times, which is enhanced by additional functionalities provided by other PSSs whenever these are sufficiently close to the user.

A Personal Smart Space may be defined as "the set of services that are running or available within a dynamic space of connectable devices where the set of services and devices are owned, controlled, or administered by a single user or organisation". More specifically, a PSS must have the following three essential properties:

(1) A PSS must have an "owner". From the above a PSS consists of a collection of devices and services that are owned, controlled or administered by a single user or organisation and that work together on behalf of the person or legal entity that owns it. As such it forms a pervasive subsystem on behalf of its owner. From the point of view of personalisation it maintains a set of preferences of the owner that are used to personalise the behaviour of the PSS and its services, and, by extension, services from another visited PSS, subject to group conflict resolution on those preferences (where two different PSSs have different requirements for the same service at the same time), both proactively and by reacting to changes in the environment.

(2) A PSS may be mobile or stationary. If a PSS is owned by a person, its physical boundary will move around with the user whereas a PSS associated with a building (e.g. railway station, airport, hospital) will be stationary. Both types of PSS have exactly the same architecture but differ in the devices and third party services that they offer. In the case of the mobile user the preferences are those of the user who owns it; in the case of a stationary PSS (e.g. office, railway station, etc.) they are those of the organisation or user (in the case of a smart home) that owns it.

(3) A PSS must be able to identify and interact with other PSSs. By using an ad hoc network, a mobile PSS can interact with other mobile PSSs or with a stationary PSS to exchange information or access services. Access to information and services is governed by a set of rules defining admissibility to the PSS. A simple example of mobile-stationary interaction is when a mobile PSS enters the railway station and asks the station PSS for platform information.

In addition to these three characteristics, a PSS must also possess the general properties that are normally associated with a pervasive system. These include:

(4) A PSS must be context-aware, personalisable and adaptable.

(5) A PSS must be capable of pro-active behaviour.

(6) A PSS must be capable of self-improvement by learning from the user's interactions.

(7) A PSS must provide appropriate measures to protect the privacy of the user.

Thus a PSS can be realised as an ad hoc network which may interact with the networks of other PSSs when these are encountered. This has the advantage of not requiring any fixed infrastructure to be provided by Internet Service Providers or Mobile Network Operators, although it is able to take advantage of infrastructure when it is available. Thus users can deploy their own personal smart spaces, populating them with their mobile and fixed devices.

3 Two Scenarios

The prototype platform developed within Persist has been used to demonstrate the usefulness of this approach in a variety of scenarios. For this paper, two of these scenarios have been selected to illustrate the problem that is being addressed here.

3.1 Memory Support

The first scenario is a hypothetical case in which a PSS may be used in the future to provide memory support to users. The scenario is as follows:

"Arthur sometimes has difficulty remembering people's names. When he is in town, Arthur meets his old friend Bill. Arthur's PSS identifies Bill and tells Arthur Bill's name through his earpiece while displaying relevant information about Bill to help trigger his memory."

For this demonstration two PSSs are used. In Arthur's PSS text to speech generation is used to provide the information discreetly to him (through an earpiece).

We have also experimented with special spectacles to display information to the user although the technology available at this stage is not yet suitable for this.

3.2 Personalised Advertisement

The idea of personalised advertisements targeted at the user as he/she approaches a particular shop has been used by various researchers, e.g. [8, 9]. For our purposes the following scenario was used:

"As Jack approaches the Shopping Mall, the Shopping Mall PSS detects Jack's PSS, and from what is known about Jack's interests, it displays a personalised message to Jack on one of its large screens, alerting him to a special offer that is relevant to his interests."

Again two PSSs are used – in this case a mobile PSS owned by Jack and a fixed PSS in the shopping mall, hosting an advertising service that uses a large screen.

4 Problem of Detecting Proximity

Both of the scenarios in the previous section depend on one of the fundamental ideas underpinning the notion of the PSS, namely that when one PSS approaches another, the two establish communication between themselves. This is essential both for basic operation of the PSS as well as for the additional functionality they provide.

The way in which this is realized is through ad hoc networks. Thus when one PSS comes within range of another, an ad hoc network connection is established between them. The two can then exchange information. This includes an identifier that identifies the PSS – Digital Personal Identifier (DPI) – as well as advertisements, advertising the services that each is prepared to share with the other.

In the case of the memory support scenario, when Arthur's and Bill's PSSs come within range of each other, they will send out their DPIs on the ad hoc network. A third party service for memory support in Arthur's PSS checks whether it recognizes the new DPI and, if it does so, it relays the appropriate information to Arthur via a text to speech conversion package or interface to the special glasses.

A similar situation applies to the advertisement scenario. Again both PSSs identify each other. A third party service running in the shopping mall PSS recognizes Jack and directs an appropriate personalised advertisement to the large screen which it controls.

However, one problem with these two scenarios and with other similar ones in which one PSS is alerted to the presence of another is that of determining when a PSS is close enough to be "relevant". Different factors affect this decision and, in general, the problem is non-trivial. In our investigation we have focused on the distance between the PSSs although one may argue that this will also depend on other contextual factors such as the density of other PSSs around them and their directions of travel relative to each other as well as personal factors such as the relationship between the owners of the two PSSs and the need for them to communicate. However, these additional factors will not be addressed here.

Consider some typical cases:

(1) In the case of the memory support scenario Arthur and Bill need to be close enough to be within "greeting" distance of each other – say 10 metres.

(2) However, suppose that Arthur had agreed to meet Bill and was looking out for him. In this case one would want the third party service to alert Arthur when Bill was within "approaching" distance – say, 50 metres.

(3) In the case of the advertisement scenario Jack needs to be close enough to the screen to have his attention drawn to it. This "viewing" distance may be between, say, 5 and 20 metres depending on the screen size.

(4) If, instead of a shopping mall, Jack was hurrying to the station to catch a train, he might want to establish communication with the station PSS as early as possible to determine whether the train is on time and what platform it is leaving from. This "early contact" distance will be the limit at which the two PSSs can join in an ad hoc network.

Besides these four cases there are obviously others that need to be considered in the future. However, this research did not take this further but instead focused on the problem of separating distance as described in the next section.

5 Problem of Determining Proximity

Besides the complex problem of determining when one should alert one PSS to the presence of another, there is another problem – that of determining when one PSS is close to another. Within the Persist project the issue of proximity has been investigated in depth, utilising techniques such as indoor locationing (Ubisense) and outdoor locationing (GPS) to identify PSS locations and determine distances between them [10],[11]. However, this work is reliant on the availability of locationing infrastructure and hence cannot provide proximity information when such infrastructure is not available. In contrast, the PSS must be capable of providing pervasive support even outside areas of infrastructure support to fulfill the aim of the PSS acting as a bridge between islands of pervasiveness.

To address this problem, two solutions were considered involving ad-hoc network connections with no dependency on external infrastructure beyond the PSS device itself. The solutions, testing and results are described below.

5.1 Using Wireless Network Alone

Initially a simple ad-hoc wireless connection was used to determine proximity. The act of one PSS connecting to an ad-hoc network alerted the other PSSs already on the ad-hoc network that another PSS was close by. However, since different wireless technologies are available, with different ranges, a quantitative analysis was performed to investigate the PSS range over different wireless technologies. Specifically the 802.11 protocols *b* and *g* were utilised for inter PSS communications to investigate their utility for identification of PSSs in close proximity.

Two laptops were used, each acting as a mobile PSS. Each device ran the latest release of the PSS platform under Windows XP. An ad-hoc wireless network was created between the two devices and they were configured with static IP addresses.

For the first set of tests, both devices were configured to use the 802.11b protocol. The PSS platform was launched on each device and one was left stationary while the other was moved away until communication between the two was lost. The distance between the two was then recorded. This process was repeated 10 times with the mobile PSS moving in different directions through the test building each time. For the second set of tests the same process was followed with the devices configured to use the 802.11g protocol.

For the first 10 tests with 802.11b the PSS range lay between 13m and 39m. This variance in range was due to the interference (from walls, floors, etc.) experienced at different locations in the test building. The average PSS range across the 10 tests using the 802.11b protocol was 29m.

For the second 10 tests with 802.11g the PSS range lay between 15m and 39m. Again the variance was due to interference but the average PSS range across the 10 tests was slightly higher than the first set of tests, at 33m. The test results are shown graphically below.

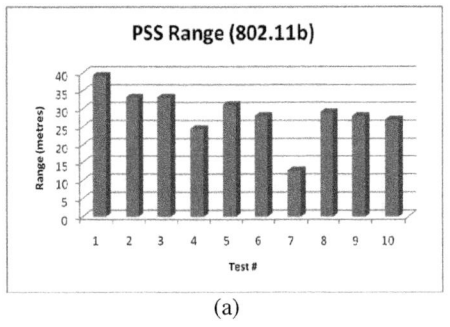
(a)
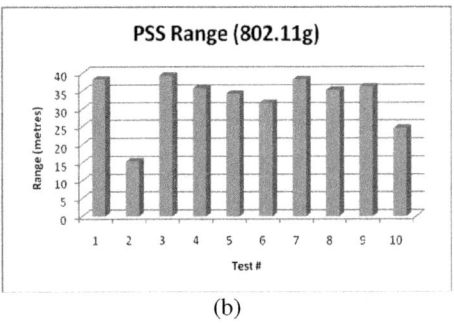
(b)

Fig. 1. Graphs showing the PSS range (in metres) using (a) the 802.11b and (b) the 802.11g wireless LAN protocols

The results show that there is a small difference between the PSS ranges using the wireless b and g protocols. On average the PSS range was wider when using the 802.11g protocol. However, the PSS ranges for both sets of tests are comparable.

Thus, in general, the distance at which one PSS detected another was quite large, and this was suitable for recognition at the limit. However, there was little difference between the two protocols.

5.2 Using Bluetooth in Addition to Wireless

The alternative approach that we investigated, was to use a combination of Bluetooth and ad hoc wireless connection. The aim here was to use the wireless network to establish a connection between two PSSs and hence to determine when they are at the limit with respect to one another. Then the Bluetooth connection could be used to determine when the two PSSs are close to one another.

Although much work has been done on the problem of using Bluetooth with mobile phones for proximity detection (e.g. [12], [13]), its use in more general situations is less straightforward. In particular, two problems were encountered:

(1) Suppose that a Bluetooth connection is established between two PSSs and then one moves out of range of the other with the result that connection is lost. If subsequently the two PSSs come back within range of each other, some operating systems do not provide the functionality to re-establish the connection automatically. Initial testing with Windows highlighted that, although a Bluetooth enabled peripheral device such as a keyboard or mouse can automatically reconnect without issue, there was no option provided for automatic re-establishment of other Bluetooth connections due to pre-configuration of their Bluetooth drivers for security reasons. The Linux OS does allow for Bluetooth connections to automatically reconnect; however, after much effort it became apparent that the task is non-trivial. After initially establishing a TCP IP persisted connection over Bluetooth, network interfaces on the laptop devices would fail to re-establish themselves after losing connection with each other, resulting in non-deterministic reconnection behaviour.

(2) Suppose that a PSS has established an ad hoc wireless connection with more than one other PSS, then if it now finds a Bluetooth connection with one of these, there is no simple way of determining to which PSS the Bluetooth link belongs. Ideally the Bluetooth advertisement should contain some means of identifying the PSS – the most obvious being the related DPI. This would allow other PSSs to perform mappings between DPIs and Bluetooth Ids.

6 Summary and Conclusion

The notion of a Personal Smart Space (PSS) provides a useful approach for building pervasive systems which combines the fixed smart space and the mobile one. One consequence of this approach is that it also provides a basis for additional functionality to identify other PSSs in the near environment and raises the possibility of a range of new types of services based on the interactions between PSSs. This may be one mobile PSS encountering another, or a mobile PSS encountering a fixed one. This paper is concerned with some of the issues around providing this functionality.

In the Persist project a pervasive system has been developed based on PSSs. This system has been used to demonstrate a number of scenarios, in particular in relation to new services. This paper describes two of these. The first is concerned with memory support for recognising people, but could be used in a range of different types of situations, including support for people in the early stages of dementia, partially sighted or blind users, users at large meetings, conferences, conventions, etc. The second relates to recognising passers-by to target personalised advertisement towards them. This too could be used in a variety of other applications to support users in smart buildings.

For these applications it is necessary to determine when two PSSs are in close proximity. This paper is focused on one aspect of this, namely that of the separating distance between two PSSs. Other factors that should also be considered include contextual factors such as the density of other PSSs around them and their directions of travel relative to each other as well as personal factors such as the relationship between the owners of the two PSSs and the need for them to communicate.

Section 5 discusses the problem of determining the proximity of another PSS without relying on additional infrastructure. Two approaches were tried but, although the research is still only at a very preliminary stage, only one was found useful. The problem of determining when a user is nearby (say, within the same room or within 10 metres) remains an issue. It is hoped that this will be further studied in the Societies project.

Acknowledgments. This work is supported by the European Union under the FP7 programme (Persist and Societies projects) which the authors gratefully acknowledge. The authors wish to thank all colleagues in the Persist project developing Personal Self-Improving Smart Spaces. However, it should be noted that this paper expresses the authors' personal views, which are not necessarily those of the Persist consortium. Apart from funding these two projects, the European Commission has no responsibility for the content of this paper.

References

1. Mozer, M.C.: Lessons from an Adaptive House. In: Cook, D., Das, R. (eds.) Smart Environments: Technologies, Protocols and Applications, pp. 273–294 (2004)
2. Youngblood, M.G., Holder, L.B., Cook, D.J.: Managing Adaptive Versatile Environments. In: 3rd IEEE Int. Conf. on Pervasive Computing and Communications (PerCom 2005), pp. 351–360 (2005)
3. Román, M., Hess, C.K., Cerqueira, R., Ranganathan, A., Campbell, R.H., Nahrstedt, K.: Gaia: A middleware infrastructure to enable active spaces. IEEE Pervasive Computing 1, 74–83 (2002)
4. Groppe, J., Mueller, W.: Profile Management Technology for Smart Customizations in Private Home Applications. In: Andersen, K.V., Debenham, J., Wagner, R. (eds.) DEXA 2005. LNCS, vol. 3588, pp. 226–230. Springer, Heidelberg (2005)
5. Williams, M.H., Taylor, N.K., Roussaki, I., Robertson, P., Farshchian, B., Doolin, K.: Developing a Pervasive System for a Mobile Environment. In: eChallenges 2006 – Exploiting the Knowledge Economy, pp. 1695–1702. IOS Press, Amsterdam (2006)
6. PSS platform open source, http://sourceforge.net/projects/psmartspace
7. Crotty, M., Taylor, N., Williams, H., Frank, K., Roussaki, I., Roddy, M.: A Pervasive Environment Based on Personal Self-improving Smart Spaces. In: Gerhäuser, H., Hupp, J., Efstratiou, C., Heppner, J. (eds.) AmI 2008. CCIS, vol. 32, pp. 58–62. Springer, Heidelberg (2010)
8. Apple's iAd, http://www.independent.co.uk/life-style/gadgets-and-tech/news/how-apples-new-iphone-brings-minority-report-a-step-closer-to-reality-1940723.html
9. Digital Signage Promotion Project, http://www.google.com/hostednews/afp/article/ALeqM5iDd1xzYx7CaahlxkLnvo4Xtcksug
10. Frank, K., Robertson, P., McBurney, S., Kalatzis, N., Roussaki, I., Marengo, M.: A Hybrid Preference Learning and Context Refinement Architecture. In: PERSIST Workshop on Intelligent Pervasive Environments, AISB 2009, Edinburgh, pp. 9–15 (2009)
11. Khider, M., Kaiser, S., Robertson, P., Angermann, M.: A Novel Movement Model for Pedestrians Suitable for Personal Navigation. In: ION NTM 2008, San Diego (January 2008)
12. Kostakos, V., O'Neill, E., Penn, A., Roussos, G., Papadongonas, D.: Making sense of urban mobility and encounter data. ACM Trans. CHI (2010) (to appear)
13. Yoneki, E., Hui, P., Crowcroft, J.: Visualizing community detection in opportunistic networks. In: 2nd Workshop on Challenged networks (CHANTS), Montreal, Canada (2007)

A Smart-Phone-Based Health Management System Using a Wearable Ring-Type Pulse Sensor

Yu-Chi Wu[1], Wei-Hong Hsu[1], Chao-Shu Chang[1], Wen-Ching Yu[1], Wen-Liang Huang[2], and Meng-Jen Chen[2]

[1] National United University, Miao-Li, Taiwan 360
[2] National Kaohsiung University of Applied Science, Kaohsiung, Taiwan 807
{ycwu,cschang,wcyu}@nuu.edu.tw,
mengjen@mail.ee.kuas.edu.tw, bsa1998@ms34.hinet.net

Abstract. In this paper, a mobile e-health-management system is presented which extends authors' previous works on mobile physiological signal monitoring. This system integrates a wearable ring-type pulse monitoring sensor with a smart phone and provides a mobile "exercise-333" health management mechanism. All physiological measurements are transmitted to the smart phone through Bluetooth. The user can monitor his/her own pulse and temperature from the smart phone where the health management mechanism helps him/her to develop a healthy life style: taking exercise 3 times a week and at least lasting for 30 minutes with heart rate over 130 each time. With the popularity and mobility of smart phones, this system effectively provides the needs for mobile health management.

Keywords: health management system, smart phone, Bluetooth, RFID, GPS.

1 Introduction

"Prevention is better than cure." The system proposed in this paper aims to achieve this. At the end of 2008 the elder population was 2.4 million in Taiwan, according to the bulletin report of Taiwan Ministry of Interior, and it was about 10.4% of the total Taiwan population. This percentage has already exceeded the standard for aging society set by the World Health Organization (WHO). It is estimated that in 2025 the elder population in Taiwan will reach more than 20% of the total population and the market revenue of home health care for these elders would reach 300 million dollars in 2010. The "long-distance home health care service" has thus become one of the key emerging businesses in Taiwan.

In recent years, several studies integrating communication and sensor technologies for home health monitoring system have been discussed [1-18], such as monitoring long-term health data to find out the abnormal signs and monitoring the medical record regularly for chronic patients to cut down their treatment frequency, to save doctor's treatment time, and to reduce medical expenses. Based on the sensor and communication technologies used, these systems can be categorized into two systems: immobile and mobile long-distance health monitoring systems. Our previous works [13]-[18] all focused on mobile long-distance physiological signal measuring based on either a single-chip-microprocessor [13]-[15] or a smart phone [16]-[18]. The physiological sensor used was a RFID ring-type pulse/temperature sensor. The measured data can be transmitted via different communication protocols, such as Bluetooth, ZigBee, HSDPA, GPRS, and TCP/IP. In order to meet the requirement for mobile health monitoring system (MHMS), the system design needs to adopt light modular sensors for data collection and wireless communication technology for mobility. The popular smart phones used in people's daily life are the best devices for MHMS.

In this paper, a mobile e-health-management system is presented which extends authors' previous work [13]-[18] on mobile physiological signal monitoring to practice the idea of "Prevention is better than cure." This system integrates a wearable ring-type pulse monitoring sensor with a smart phone and provides a mobile "exercise-333" health management mechanism. The user can monitor his/her own pulse and temperature from the smart phone where the "exercise-333" health management mechanism helps him/her to develop a healthy life style: taking exercise 3 times a week and at least lasting for 30 minutes with heart rate over 130 each time. With the popularity and mobility of smart phones, this system effectively provides the needs for mobile health management.

2 System Architecture

For the descriptive purpose, the monitoring system proposed in [16]-[18] is presented here again. Fig. 1 is the system architecture where the ring-type sensor measures pulse/temperature and transmits the physiological data to the reader using wireless RF, the Bluetooth connected to the reader then passes the data to the smart phone, and the smart phone collects/displays the physiological data and also transmits data to the remote medical station using GPRS, HSDPA (3.5G), WiFi, or WiMax. The GPS built in the smart phone can provide the position information of the monitored person so that the medical personnel can be dispatched to the right location more promptly in an emergency situation. Figs. 2-5 show the photos of RFID pulse/temperature sensor tag (Ring), RFID reader, Bluetooth RS232 adaptor, and the integration of Bluetooth adaptor, RFID tag (Ring), and RFID reader, respectively. This ring sensor developed by Sinopulsar Technology Inc. is non-invasive, portable, and mobile. It can measure pulse and temperature signals which are processed by a built-in microcontroller. It uses optical sensor to detect heart rate and has anti data collision mechanism. Physiological data are then transmitted by RF wireless transmission with FSK modulation using UHF ISM band (up to 50 meters) to its RFID reader with a RS232 port. The data communication between RFID reader and the smart phone is through Bluetooth. HL-MD08A (Bluetooth RS232 Adaptor manufactured by Hotlife Technology) is used in this system. It supports a wide range of Baud rates from 1.2K to 921.6K bps. Fig. 6 shows the photo of the smart phone (ASUS P552W) used in this system. Its operating system is Windows Mobile 6.1 operating system with built-in GPS, and it supports HSDPA 3.6Mbps, EDGE, GPRS, and GSM 900/1800/1900. The screen in Fig. 6 depicts the physiological data monitoring GUI where two temperature values measured by RFID reader and Tag and pulse data are shown.

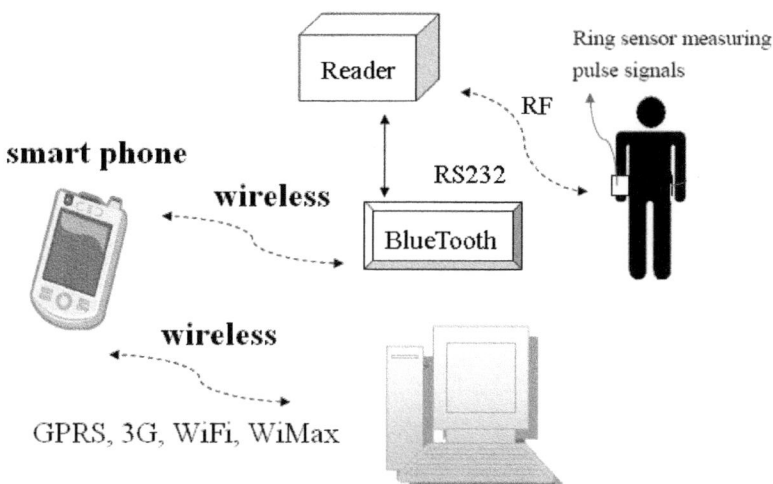

Fig. 1. MHMS Architecture

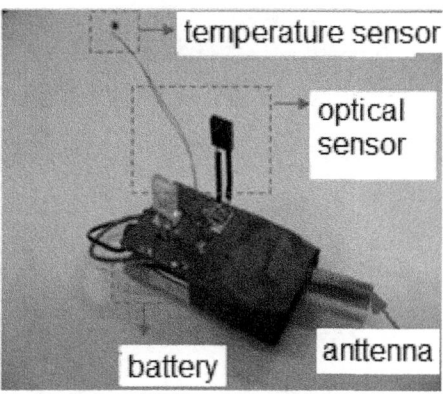

Fig. 2. RFID tag (Ring)

Fig. 3. RFID reader

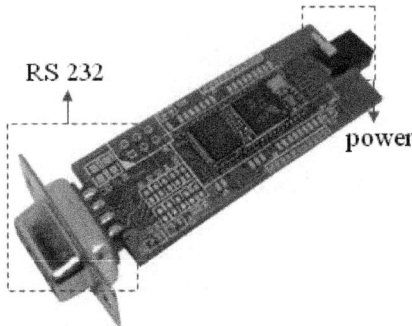

Fig. 4. Bluetooth RS232 Adaptor

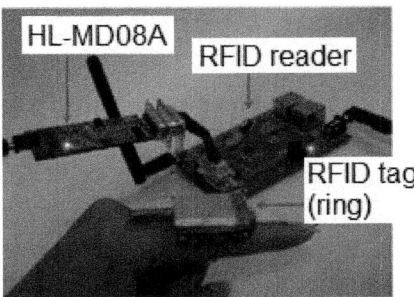

Fig. 5. Bluetooth adaptor, RFID tag (Ring) & RFID reader

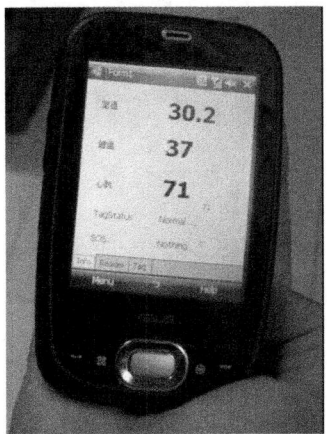

Fig. 6. ASUS P552W smart phone

The GUI programs developed on the smart phone and on the remote medical station were coded in Visual C#. Microsoft .Net compact framework 3.5 was installed on the smart phone for running the client APs, and Windows Mobile 6 SDK, smart phone emulator, and Cellular Emulator were installed on the PC for developing the client APs.

Fig. 7 depicts the physiological data monitoring GUI on the smart phone emulator (on PC side), which is the same GUI shown on the smart phone screen in Fig. 6. The SOS message box indicates the status of the SOS button on the ring Tag. If this button is pressed, the smart phone will dial the pre-set emergency phone number(s) to send out SMS with the GPS position information to other people for help. Fig. 8 presents the 3G communication of smart phone and the webpage of remote medical station. The 3G connection setup page of the smart phone is shown on the left side and the webpage of the server is on the right side where temperature values, pulse, and Google map are displayed. Based on the GPS position information sent from the smart phone, the Google map helps the medical staff to locate promptly the monitored user who needs further assistance.

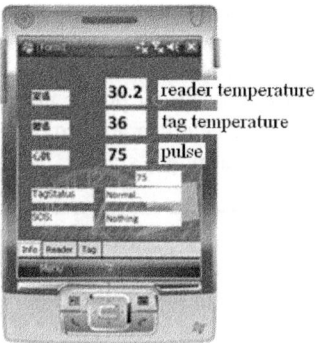

Fig. 7. Physiological data monitoring GUI

Fig. 8. 3G communication setup and the webpage of remote medical station

3   333 Health Management Mechanism

Compared to the traditional medical care that people receive medical treatment after they feel ill, an effective health management program can provide prevention of illness in a more aggressive manner—prevention is better than cure. Especially for people working in present modern high-tech society under lots of pressures and lacking exercise, such a preventive health management becomes essential. According to the report conducted by a hospital in Taiwan [19], exercise-333 can effectively prevent cardiovascular diseases. The concept of "exercise-333" is quite simple; i.e., taking exercise 3 times a week and lasting for more than 30 minutes with heart rate over 130 each time. The health management mechanism developed in this paper, based on the system architecture discussed in Section 2, can help to remind the user to develop such a healthy life style.

Fig. 9 depicts the exercise-333 health management GUIs on the smart phone. In the first GUI, the user can set up 3 weekdays as checking points and at each checking point it will show the progress status to remind the user. In the second GUI, at the end of each day it will show the user whether his/her heart rate has ever been over 130 for more than 30 minutes. And at the end of each week, this GUI also shows the condition whether the user has accomplished exercise-333. With these two smart-phone GUIs, the user can constantly receive reminders and check his/her exercise status. Based on the fact that in Taiwan every 100 people have 108 cellular phone numbers [20], the high popularity of smart phone makes the presented health management system effective and convenient to help people on developing a healthy life style.

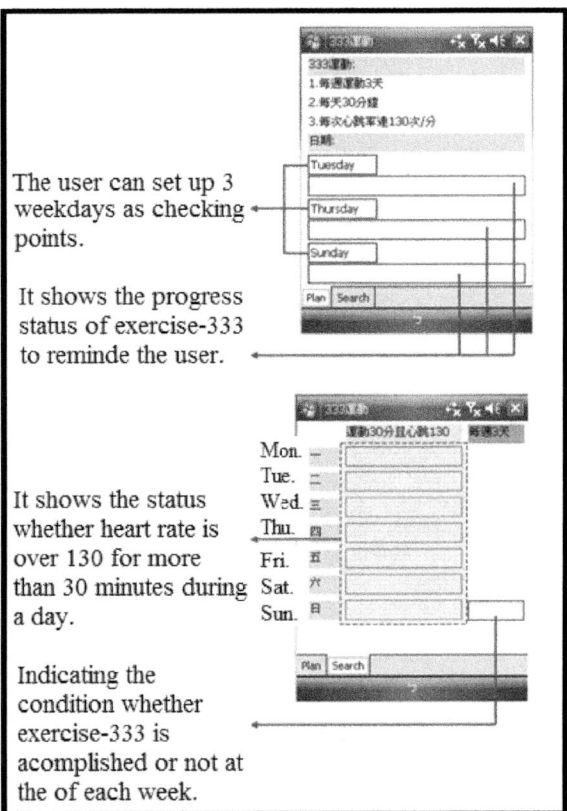

Fig. 9. Exercise-333 health management GUIs

4 Conclusions

In this paper, a mobile e-health-management system has been presented. This system is an extension of our previous works on mobile physiological signal monitoring and practices the idea of "Prevention is better than cure." The presented system consists of a wearable ring-type pulse sensor and a smart phone to provide a mobile "exercise-333" health management mechanism. The user can monitor his/her own pulse and temperature from the smart phone where the "exercise-333" health management mechanism helps him/her to develop a healthy life style: taking exercise 3 times a week and lasting at least for 30 minutes with heart rate over 130 each time. With the popularity and mobility of smart phones, the presented system effectively provides the needs for mobile health management.

Acknowledgments. The authors would like to express their gratitude to Sinopulsar Technology Inc. for free use of their RFID ring-type pulse sensor in this project. Financial support from the National Science Council (project # NSC 99-2622-E-239-004-CC3) is acknowledged as well.

References

1. Yu, S.A., Lu, S.S., Lin, C.W., Wang, Y.H.: Personal Electronic Nurse. Scientific Development 393 (2005) (in Chinese)
2. Chang, K.S.: Embedded Electrocardiogram Measurement System Design and Its Application to Personal Remote Health Care, pp. 32–37 (2004) (in Chinese, Master thesis, National Cheng Kung University, Taiwan)
3. Lin, J.L.: Development of Wireless Sensor Network for Home Health Care (2005) (in Chinese, Master thesis, National Chiao Tung University, Taiwan)
4. Wu, J.L.: Implementation of a Portable Wireless Physiological Signal Measurement System (2004) (in Chinese, Master thesis, Southern Taiwan University, Taiwan)
5. Lin, T.H.: A Mechanism Integrating Electrocardiogram Compression and Error Protection and Its Application to the Bluetooth Transmission in the Home Care System (2004) (in Chinese, Mater thesis, Chung Yuan Christian University, Taiwan)
6. Shu, Y.L.: Development of Intelligent Maintenance System for Establishing the Quality of the Elder's Life, Engineering. Science and Technology Communication 84 (2005) (in Chinese)
7. Ye, C.F.: A PDA-based Home Care System (2006) (in Chinese, Master thesis, National Chiao Tung University, Taiwan)
8. Lee, R.G.: A Mobile Care System with Alert Mechanism. IEEE Transactions on Information Technology in Biomedicine 11(5), 507–517 (2007)
9. Lee, R.G., et al.: A Mobile-care System Integrated with Bluetooth Blood Pressure and Pulse Monitor, and Cellular phone. IEICE Transactions on Information and Systems E89-D(5), 1702–1711 (2006)
10. Lee, R.G., et al.: A Mobile-care System over Wireless Sensor Network for Home Healthcare Applications. Biomedical Engineering-Applications, Basis and Communications 19(2), 85–90 (2007)
11. Lee, R.G.: Design and Implementation of a Mobile-care System over Wireless Sensor Network for Home Healthcare Applications. In: Proceedings of Annual International Conference of the IEEE Engineering in Medicine and Biology, pp. 6004–6007 (2006)

12. Chen, C.M., et al.: Web-based Remote Human Pulse Monitoring System with Intelligent Data Analysis for Home Healthcare. In: IEEE International Conference on Cybernetics and Intelligent Systems (2008)
13. Wu, Y.-C., et al.: Physiological Signal Measuring System via Multiple Communication Protocols. Journal of United University 7(1), 251–265 (2010) (in Chinese)
14. Wu, Y.-C., Chen, P.-F.: Multi-channel Data-acquisition and Controller for Mobile Health Monitoring System with HSDPA and GPS. In: The Second International Conference on the Applications of Digital Information and Web Technologies, ICADIWT 2009 (2009)
15. Wu, Y.-C., Chen, P.-F.: Multi-channel Data Acquisition and Controller for Mobile Health Monitoring System with GPRS and GPS. In: 2009 International Conference on e-Technology (2009)
16. Wu, Y.-C., et al.: A Mobile e-Health Management System. Journal of Technology (in Chinese) (revised)
17. Wu, Y.-C., et al.: Mobile Physiological Measurements and Health Management System. In: 2010 International Conference on e-Commerce, e-Administration, e-Society, e-Eduction, and e-Technology (2010 e-Case & e-Tech) (2010)
18. Wu, Y.-C., et al.: A Mobile Health Monitoring System Using RFID Ring-type Pulse Sensor. In: The 8th International Conference on Pervasive Intelligence and Computing (PICom 2009), pp. 317–322 (2009)
19. Cheng-Ching Hospital Medical Care Center: Exercise-333 and Vegie-579 Can Prevent Cardiovascular Diseases (in Chinese),
    http://www.uho.com.tw/hotnews.asp?aid=5628
20. Lin, Y.-S.: High Ownership Rate of Mobile Phones Brings New Media Era. Electronic Commerce Times (2008),
    http://www.ectimes.org.tw/shownews.aspx?id=080622225140

An Adaptive Driver Alert System Making Use of Implicit Sensing and Notification Techniques[*]

Gilbert Beyer[1], Gian Mario Bertolotti[2], Andrea Cristiani[2], and Shadi Al Dehni[1]

[1] Ludwig-Maximilians-University Munich,
Programming and Software Engineering,
Oettingenstr. 67, 80538 Munich, Germany
{gilbert.beyer,shadi.aldehni}@ifi.lmu.de
[2] Università di Pavia,
Dip. Informatica e Sistemistica,
Via Ferrata 1, 27100 Pavia, Italy
{andrea.cristiani,gianmario.bertolotti}@unipv.it

Abstract. In this paper we present an adaptive driver alert system that uses passive techniques for extracting psycho-physiological features from the user, and a head-up display actuator that hands preprocessed information about the driving behavior back to the user. The paper starts with background information on driver inattentiveness. That followed we present our conception of an adaptive loop with a view to improve the driver's attention. We give an overview on current research on sensor and actuator techniques that support our adaptation strategy attaining data about user drowsiness and distraction. Then we describe a suitable hardware and software solution for the proposed system, using a head-up display and vision-based sensor techniques. We close describing first tests of our system in the lab and road tests with a Ferrari car.

Keywords: driver assistance systems, sensor-actuator supported interaction, psycho-physiological sensing, computer vision, adaptive control, implicit interaction, adaptive user interfaces, head-up displays.

1 Introduction

Driver alert systems have the purpose to shift the driver's attention back to the driving task after being shifted away from it by an event, activity, object or person inside or outside the vehicle [1]. Driver distraction is one form of driver inattention. It can be either caused by the driver himself, by the vehicle condition, or from environmental factors like the traffic situation [2]. For user-centric adaptive sensor-actuator systems, distractions are of special interest, that whether derive from the driver's psycho-physiological state (such as fatigue or drowsiness) or manifest themselves in the long-term behavior and repeated reactions to distractions caused by secondary tasks or social situations (such as continuously looking away from the driving direction when conversing with the co-driver).

---

[*] This work has been partially sponsored by the EC project REFLECT, IST-2007-215893.

According to official statistics, car drivers' fatigue is one of the causes of roughly 1% serious accidents. Anyway, it is presumable that the real number is even higher, since in this kind of analysis it is often impossible to determine the effective cause of the crash. Some European studies say that 24-33% of fatal crashes are due to driver's fatigue [3,4]. In the United States about 100000 accidents are caused by drivers falling asleep [5]. Distractions caused by the driver's behavior or by persistent external influences are another important cause of crashes. Examples for such distractions are searching an object under the seat, adjusting the car stereo or all kinds of passenger-related distractions [6,7]. The NHTSA reports that driver distraction is the cause of 40% of accidents in which involved cars go off road.

The given examples show that, in order to increase the safety of the driver and the passengers, systems which also monitor such long-term distractions would be useful. We here propose an adaptive driver alert system dealing with such problems making use of implicit sensing and notification techniques.

2 Design of an Adaptive Driver Alert System

The system we propose monitors persistent or long-term distractions, to be able to react in an anticipatory way and notify the driver gradually and in a gentle way before the driving situation even gets critical. In the following we first describe the overall adaptation idea of our system, and then deduce the requirements of sensors and actuators that can realize such an adaptive driver alert system.

2.1 Adaptation Strategy

Our basic idea is a feedback loop where the level of driver's distraction is attained by measuring indicators such as drowsiness and eyes-off-the-road time with the help of vision sensors, and is regulated by implicit visual actuators in the viewing direction. Implicit feedback means in this regard, that the feedback is given long-term: If the driver has continuously shown signs of drowsiness or distraction over a certain time period, he is given a visual notification of his long-term behavior. According to [2], every modality (vision, hearing, touch) has its own temporal behavior, and we propose that for long-term notification unobtrusive visual information in the viewing direction is most suitable.

On the sensing side, with regard to the long-term feedback visualization that we propose, the analysis of properties is of special interest that are not indicating immediate but gradually increasing driver distraction. In this work we focus on drowsiness (eyes-closed time and frequency) and distractions caused by secondary tasks or social situations in the car, that express themselves in repeatedly looking away from the driving direction (eyes-off-the-road time and frequency). The eyes-closed time is attained via an accordant algorithm, and for the eyes-off-the-road time we currently use the head position of the driver.

In a further step we combine this data on the user state with information on the car state. The purpose of such a combination shall be explained by the information value of the head position recognition: Looking right to the co-driver while driving fast on a curvy country-road can be classified as risky behavior, while looking right while waiting at a crossing is allowed (or may even be required, e.g. before turning into a road). To determine the car state, telemetry data such as velocity, lateral and longitudinal acceleration are taken from the CAN-Bus. An accordant analysis component combines this telemetry data and the psycho-physiological data of the user state, producing new decisions about the driver's behavior. The output of the combined states is visualized by an appropriate visualization in the driving direction.

2.2 Sensor-Actuator Support of the System

For the long-term visual actuator we propose to use a head-up display. The term head-up display (HUD) usually describes a transparent display that presents information without requiring the user to look away from his usual viewpoint. This viewing situation is innate to activities where an essential primary task requires the full attention of the viewer, like the driving task where user is required to look most of the time on the road, to not miss critical events in the driving environment.

The basic idea behind our adaptive head-up display actuator is a transparent windshield visualization that adapts to the combined user-car state. The feedback on the driver is then carried out by presenting implicit information in the viewing direction. In contrast to explicit, attention-grabbing warnings this information should be unobtrusive, consisting of rather ambient, continuously changing visualizations at the bottom of the windshield. Once the driver has understood the meanings of these ambient visualizations, they should influence his situational awareness and bring him to adapt his driving style accordingly.

On the sensor side, we decided to use cameras, as vision-based techniques operate non-invasively and support unobtrusive, implicit interaction. In last years many works have dealt with the development of car safety systems that employ sensor technologies that are "intrusive". For example to detect the vigilance level of drivers, mostly physical sensors or electrodes are used that must be positioned on the body of the subjects to attain the needed information (e.g. electromyogram, respiration, skin conductance monitoring, electro-oculography). Electro-oculography (EOG) has been used to study spontaneous eyelid closure, providing a series of interesting parameters that can be taken as indicators in fatigue diagnostics and in drowsiness detection [8]. Although techniques like EOG provide very detailed information about one's vigilance level, they could hardly be used in real systems, since they are intrusive, and not easily performed in a non-laboratory environment. Recent studies have shown that also non-intrusive techniques can be used to extract information related to drivers' vigilance level. Video-cameras and image elaboration algorithms, for example, are now broadly used to detect the face and the eyes of a subject, in order to detect driver's inattention [9]. The analysis of drivers' behavior, through the elaboration of car telemetry data, is another technique which can give important clues about pilot's vigilance level. Experiments conducted in simulated environments showed, for example, the existence of a relationship between drowsiness and driving "errors".

3 Related Work

In literature, a number of papers describe camera-based systems for detecting driver's inattention. Most of reported results show very good performances while tests are performed in simulated conditions or in laboratory environments (see [10] as an example). Few of presented systems have been tested in real vehicles.

While dealing with the study of driver vigilance level, it is important to mention the AWAKE project (System for Effective Assessment of Driver Vigilance and Warning According to Traffic Risk Estimation) [4], since it is probably the main reference for researches in this field. They proposed a multi-sensor approach to the problem, since the system monitors eyelid closures through a camera, and it also acquires information from a steering grip sensor, a lane tracking sensor. Furthermore gas/brake and steering wheel signals are used as indicators of the driver behavior. To close the loop, a "Driver Warning System" is used to warn the driver through visual and haptic means according to the type of AWAKE warning. In this case, visual elements are located at an external box on top of the dashboard or at the rear view mirror. Other works suggest that visual information or alarms are provided to the driver by means of HUDs, allowing him to keep his eyes on the road [11]. There already have been proposals for adaptive HUDs in automobiles that visualize the state of the car or that adapt the amount of items to be displayed by only showing the most useful information at any time [12,13]. Yet these works did neither deliver detailed answers on how such feedback information should be presented, nor what kind of data should be processed for that purpose.

4 Hardware Prototype

The proposed hardware solution consists of a web camera as sensor device and a head-up display as implicit information device. The camera is installed near the viewing direction of the driver to get optimal image details of the required facial features. Several positions were evaluated, and the position behind the steering wheel was figured out as the most applicable as it doesn't constrain the driver's view on the road (see Fig. 1a). For the head-up display a hardware solution was designed that fits the demands of the cockpit of our test car, a Ferrari California. To physically integrate the adaptive HUD, an additional device had to be designed. Display technologies that can be used in HUD systems include different kinds of emissive displays (e.g. LED) or backlit LCDs, that reflect the screen image towards the windshield or special transparent reflective surfaces attached to the windshield. The first idea was to use such flat panel screens, but the available products showed hardly to fit into the narrow space between the windshield and the dashboard of the Ferrari. Other problems were low brightness and the presence of backlight glow in the reflected image. Another

Fig. 1. a, Camera sensor position behind the steering wheel. b, Head-up display projection hardware below the windshield. c, Icon for visualizing the eyes-off-the-road state.

idea was to use a micro projector and let it beam directly onto a reflective surface attached to the windshield, but here the driver could be blinded by the direct light of the beamer lens. The best solution showed to be a setup where the projector beams the light through a channel (to prevent backlight reflections) towards a highly reflective surface that in turn reflects the image towards the windshield (see Fig. 1b and 1c).

5 Implicit Sensing Software

This section describes the algorithms that have been developed in order to detect potentially dangerous situations related to drowsiness and distraction while driving. They use image elaboration techniques to detect and measure the duration of driver's eye blinks and to monitor his head position.

5.1 Drowsiness Detection

We developed an algorithm which combines the Viola-Jones [14] and the template-matching techniques to track one eye of the subject. At the beginning, the first technique is used in order to identify the driver's face within the acquired frames, defining a rectangular area which contains it. That area becomes the region of interest inside of which the eyes will be searched for. To reduce the computational load, only the right eye is tracked. Once again, the Viola-Jones algorithm is used to identify a rectangular region enclosing the eye. Then, an area A centered on the eye position previously found and double-size with respect to the eye rectangle is defined (Fig. 2).

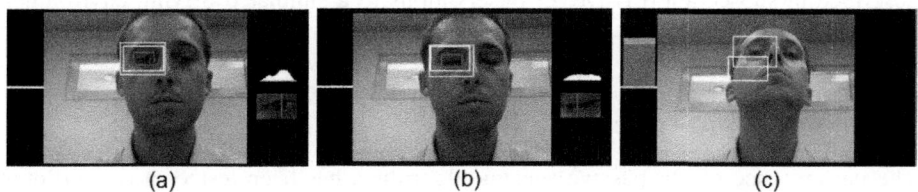

Fig. 2. The A area (delimited by the largest rectangle) is the region of interest in which a combination of Viola-Jones and template-matching techniques is used to detect driver's eye. In (a) subject's right eye is found (smallest rectangle); in (b) an eye blink is correctly identified; in (c) the algorithm detects that the subject has moved his head up with respect to previous frame.

The Viola-Jones algorithm is applied to A: if the eye is found, then a copy of it is saved and used as a template; if not, a template-matching technique is used, basing on last saved template. If a certain number of wrong matchings occur, then a re-initialization procedure is performed. Subsequently, the eye image is converted to grayscale and thresholded, obtaining a binary picture. The latter is further processed to eliminate the gaps generated by pupil reflections and to reduce the noise due to shadows or portions of eyebrows. The vertical projection histogram of the resulting image is then elaborated to determine the pupil's position and the eye openness level. An eye which remains closed for more than a pre-defined amount of time (which is about 500 ms) is considered as an indicator of possible drowsiness (the driver could be falling asleep). If this event occurs frequently, the system gives the driver a visual notification in form of an icon on the head-up display. Only if the driver is closing his eyes for a critical period (e.g. 1s) the system will react by giving a proper audio alarm.

5.2 Head-Position Detection

The eye test described in the previous subsection is made only when the face of the driver is about still and directed forward. The identification of this "rest" posture is performed by averaging the eye location over last n frames. Then, the eye status test is performed on a region which is centered in this location. Vertical and lateral positions of the eye within this region give an indication of how much the pilot's head is tilted and turned. This information is used to assess the driver's level of attention: if his head is up/down or left/right turned, the system assumes that he is not paying attention to the street (e.g. when being distracted by a passenger). If this event occurs frequently, the system gives the driver a visual notification in form of an icon on the head-up display that gradually inclines in brightness (see Fig. 1c). Only if the driver looks away for a critical amount of time, i.e. his head is up/down or left/right turned for more than a pre-defined time-out, the gentle visual information will be intensified by an immediate audio warning by the car speakers.

6 Experiments and User Analysis

Three kinds of experiments have been carried out in order to test the adaptive driver alert system. Tests to evaluate performances of implicit sensing algorithms are described in subsection 6.1. Driver observations in a simulated driving environment and testing sessions of the adaptive head-up display on a Ferrari California are reported in subsection 6.2.

6.1 Passive Sensing Performance Tests

The performance of our passive sensing algorithms has been tested in two different environments: a laboratory room, and the Ferrari California car. During the experiments, performed in daylight conditions except for the lab environment (which was lighted only by a fluorescent light), the tester, a 25 years old male, was asked to "behave normally" while driving the car or while sitting at a desk, in front of a PC screen. A Hercules Classic Silver USB web-camera (maximum 30 fps, resolution 640 x 480) was connected to a Dell Vostro 1310 notebook (Intel® Core 2 Duo T8100 (2.1GHz), 2GB RAM, NVIDIA GeForce 8400M GS, Windows XP™). Our preliminary tests focused on the recognition of slow/fast blinks and the accuracy of tracking of slow/fast head movements. Slow blinks were correctly identified in 99.1% (room) and 98.3% of cases (in the Ferrari California car), while fast blinks were clearly detected in 98.6% (room) and in 97.3% of cases (in the Ferrari California car). Slow head movements were correctly detected and tracked in 99.7% (room) and in 99.0% of cases (in the Ferrari California car), while fast head movements were correctly identified and tracked in 96.1% (room) and in 96.0% of cases (in the Ferrari California car).

Results showed different kinds of errors: the tracking is not performed correctly in successive frames; shadows or occlusions hamper the eye/face detection; blinks are not recognized; particular head positions cause erroneously detected blinks. Performed tests are just preliminary; nevertheless, obtained results are promising and further experiments (with more testers, lighting conditions) are already planned.

6.2 Driver Observations in the Lab and Driving Tests at Ferrari

To develop our adaptive sensor-actuator solution and to be able to perform preliminary user tests in the lab, a driving simulator was build. This driving simulator consists of two seats, a steering wheel mounted to a console, a plasma display to visualize a driving simulation, and a web camera connected through USB to obtain psycho-physiological data from test persons (see Fig. 3a). To obtain telemetry data that we later get in road tests from the CAN-Bus of the car, we used and adapted a suitable driving simulation [15]. With the driving simulator we performed first user tests. Therefore we invited employees and students to the lab and let them perform test runs that confronted them with route sections of different degrees of difficulty and demands for attentiveness. These tests showed that test persons in most cases adapted their driving behavior according to the adaptive visualizations.

The adaptive head-up display could be tested and refined during an onboard installation and testing session on a Ferrari California. The in-car version of the HUD was installed at an applicable position below the windshield. Several test runs were conducted with a professional test driver, and the driver was surveyed during and after the test runs on the quality of perception of the icons during different driving and lighting conditions. These tests produced significant findings that poured in the refinement of the hardware, the improvement of the HUD position and the visual representation. For practical and security reasons however, we were not able to conduct realistic user tests on drowsiness in the lab and during test drives.

Fig. 3. a, Driving simulator with camera sensor. b, User tests in the lab. c, Testing of the adaptive head-up display in the Ferrari car.

7 Prospects and Future Work

In this paper we presented an adaptive driver alert system making use of passive, camera-based sensor techniques and implicit windshield visualizations. We presented the overall concept of our adaptive loop, hardware solutions to realize an unobtrusive loop on the sensor and the actuator side, and software algorithms that can measure the required features of user inattentiveness. We deployed a prototype system and observed users in a lab environment as well as during tests on the road.

These first tests with our prototype show that the overall concept of our adaptive driver alert system works quite well, properly detecting potentially dangerous driving behaviors and making the driver aware of them through gentle alert signals.

A field study of the prototype will be the next step in investigating the properties of the system. Moreover, we'll test our implicit sensing software with a nightvision webcam, in order to make our adaptive driver alert system efficient also during nighttime.

References

1. Treat, J.R.: A study of precrash factors involv. in traffic acc. HSRI, Research Rev.10 (1980)
2. Riener, A.: Sensor-Actuator Supported Implicit Interaction in Driver Assistance Systems. Vieweg+Teubner Verlag, Wiesbaden (2010)
3. Streff, F., Spradlin, H.: Driver distraction, aggression, and fatigue: a synthesis of the literature and guidelines for Michigan planning. Ann Arbor 1001, 48109 (2000)
4. A.C. (IST 2000-28062): Awake-system for effective assessment of driver vigilance and warning according to traffic risk estimation (September 2001-2004)
5. Bergasa, L., Nuevo, J., Sotelo, M., Barea, R., Lopez, M.: Real-time system for monitoring driver vigilance. IEEE Trans. on Intell. Transp. Systems 7(1), 63–77 (2006)
6. Stutts, J., Feaganes, J., Reinfurt, D., Rodgman, E., Hamlett, C., Gish, K., Staplin, L.: Driver's exposure to distractions in their natural driving environment. Accident Analysis and Prevention 37(6), 1093–1101 (2005)
7. Sheridan, T.: Driver distraction from a control theory perspective. Human Factors 46(4), 587 (2004)
8. Ingre, M., Åkerstedt, T., Peters, B., Anund, A., Kecklund, G.: Subjective sleepiness, simulated driving performance and blink duration: examining individual differences. Journal of Sleep Research 15(1), 47–53 (2006)
9. Bergasa, L., Buenaposada, J., Nuevo, J., Jimenez, P., Baumela, L.: Analysing Driver's Attention Level using Computer Vision. In: ITSC 2008, pp. 1149–1154 (2008)
10. Ji, Q., Yang, X.: Real-time eye, gaze, and face pose tracking for monitoring driver vigilance. Real-Time Imaging 8(5), 357–377 (2002)
11. Karvonen, H., Kujala, T., Saariluoma, P.: In-Car Ubiquitous Computing: Driver Tutoring Messages Presented on a Head-Up Display. In: Ikeda, M., Ashley, K.D., Chan, T.-W. (eds.) ITS 2006. LNCS, vol. 4053, pp. 560–565. Springer, Heidelberg (2006)
12. Dienelt, M.: Adaptive Informationsdarstellung für ein Head-Up Display. In: Stary, C. (ed.) Mensch & Computer 2005, pp. 1148–1158. Oldenbourg Verlag, München (2005)
13. Blanco, V., Goldthwaite, F., Baumert, D.: Adaptive heads-up user interface for automobile (2007),
    http://assignments.uspto.gov/assignments/q?db=pat&pub=200701 94902
14. Viola, P., Jones, M.J.: Rapid Object Detection using a Boosted Cascade of Simple Features. In: IEEE Computer Society Conf. on Computer Vision and Pattern Recog., vol. 1 (2001)
15. Wymann, B.: TORCS, The open racing car simulator (2010),
    http://torcs.sourceforge.net/

Defining the Criteria for Supporting Pervasiveness in Complex Adaptive Systems

Shiva Mir

University of Technology Sydney, Sydney, Australia
shiva.mir@student.uts.edu.au

Abstract. This paper, will explore potential contributions from the study of complexity theory to the area of pervasiveness of information systems. It offers insights for enriching traditional approaches of information system design that could lead to better framework for supporting pervasiveness in complex environment. Complex environment makes new problems for pervasiveness because it is hard for users to be aware of change. So to be aware of change is what this paper tries to introduce as a new criterion. First it describes complexity theory as a metaphorical devise which gives new insights of the environment. Second a short review of implication of these concepts to complex systems modeling especially on ways to define requirements that should be provided by infrastructure, which tend to be initially incomplete. Then it suggest how complexity theory can provide the guidelines and concepts needed to be aware of change and defines new research directions for the future that uses these guidelines for building an infrastructure of pervasive adaptive systems.

Keywords: Pervasive adaptive system, complexity theory, system design.

1 Introduction

This paper, offers insights for enriching traditional approaches of information system design that could lead to better framework for supporting pervasiveness in complex environments. It describes important issues in system and identifies concepts from complexity theory which can be applied to information systems to support pervasiveness and user awareness of change. It is organized as follows: it starts with describing the trends in systems and their correspondence to pervasive adaptive systems. This is followed by an overview of ways of applying complexity theory in computing systems and then introducing concepts driven from complexity science which can be used in information systems. The last section describes future work to define the kind of concepts needed to support pervasiveness and user awareness in complex adaptive environments.

Regarding some challenges, moving from desktop computers to designing pervasive adaptive systems, the research scope of system architecture has to become more interdisciplinary. The issues which designers usually face are due to increases in computing speed, decreasing computing cost and changes in technological, task, and organizational environments where more user-centric pervasive systems are either produced or deployed. They need to know how to build an infrastructure to support people remain aware in complex environment. They also need to attend more to the dynamic composition of the system and environment.

According to the literature, it seems that information system scholars are ready to take a paradigm shift for moving toward better theoretical and methodological platform suited for pervasiveness, user awareness of change in information system design and developments. To achieve this goal, complexity theory can be outlined and influence information system professionals in order to meet these goals. So the questions regarding the application of complexity theory to pervasive adaptive systems are: what is the impact of complexity theory on designing infrastructures? How it would be possible to bring complexity theory concepts in to the complex system architectures? How to ensure that they can be modelled? How to research information systems as complex adaptive systems, in the concept of user-centric pervasiveness and system adaptability?

2 Complexity Theory

A complex system is comprised of large number of different systems which obey some rules that connect them interactively with other parts. The systems can cause the system as a whole to display emergent patterns at the global level. The emergence cannot be predicted from properties of each part. There are two groups of complex systems. One is the complex system which consists of parts which are not complex and are called agents. These agents manage by unchanging rules. The other group of complex systems consist of other complex systems capable of learning and simply means that they are governed by rules that they are involved[12].

Bill McKelvey[1] claims that complexity theory is relatively a new way of thinking about the systems of interacting agents such as companies. Such systems are not centrally controlled, they rest on the idea that order emerge through the interactions of companies and agents[1]. So the complexity theory seeks to explain the processes of interaction leading to emergent structure rather than the effects of enteric forces of objects. In complexity science we study systems as a whole without the simplifications, and observing the interactions between the elements more than the elements itself[4]. A complex system is not categorized a new class of systems. Complex systems are comprised of group of interacting entities which emerge through these interactions and the behavior of the system is not predefined. Different scholars have defined different characteristics for complex systems. For example Lucas[5] defined fourteen characteristics for complex systems. These characteristics are: autonomous agents, nonstandard, co evolution, self modification, downward causation, self reproduction, undefined values, fitness, non uniform, nonlinear, emergence, attractors, phase changes and unpredictability[5]. He categorized these characteristics to three groups, such as autonomous agents, undefined values and non linearity. Another definition is that, complex system consist of a large number of self organizing agents that interact in a dynamic and nun-linear fashion that share a path dependent history. Some information systems can be considered as complex systems, which contain large number of independent systems and agents which interact locally and randomly to produce a goal oriented behaviour[13][1]. It is manifested by the continuous changes in complex environments due to changing organizational goals driven by the external competitive environment [2]. So because of its adaptability to change, complexity theory can be a good foundation for systems design architecture and infrastructures. There are many concepts in complexity theory like: self organisations, learning, knowledge, co-evolution, adaptation, emergent, responding to change, chaos and lack of central control that can be translated to the information system architecture and infrastructure context.

3 Complex Adaptive System Theory

The meaning of complex adaptive information system comes from such systems in the nature. These systems are composed of huge number of components that interact locally to produce global behavior. The random and local interactions of components lead to the emergence and goal oriented behavior of the whole system. And the adaptation means the ability to accommodate influences from changing environments without disintegrating[13]. These systems have no central control and cannot be equilibrium. Kovacs[13], defines complex information system as a large number of memes, micro-strategies and users. All of these parts interact with each other locally without global control. The bits of information that can be stored and be processed are called memes. Information system is composed of micro-strategies which take memes from outside and, processes, stores and retrieve them later. The complex adaptive information system cannot be built, it needs to grow and evolve. And have highly decentralized interactions with its components. By understanding information system architecture and infrastructural design as a complex activity, it would be possible for designers to improve their innovative solutions. Complexity theory is a promising framework that accounts for the dynamic evolution of information system and the complex interactions among industry actors. By conceptualizing information systems as chaotic systems, a number of design implications can be developed. Long term forecasting is almost impossible for chaotic systems, and dramatic change can occur unexpectedly.

4 Conceptual Foundation

We utilize the concepts which arise from complexity theory in user-centric and pervasive information system architecture. These concepts can give a metaphorical device for users to be aware of change. They can be described as follows:

- **Chaos:** A system which is deterministic, but it is hard to be predicted technically is called a chaotic system. The behavior of a chaotic system in future times depends on its sensibility to initial conditions. It means that you cannot define patterns to model its behavior. Models of chaotic system describe the dynamics of a few variables which reveals some special characteristics[7].
- **Self-organizing:** Self-organizing is the ability of a complex system to generate new behavior and structure is called. Self-organizing makes the information system as an open system with continuous flow of energy and

resources pass through it. In a self-organized process, the components reorient and restructure their relationships with other components[7]. This action introduces the relationship packages. Designing modules and defining the interaction between them are important issues in information system design concepts.

- **Learning:** Complexity theory is the key to understand how knowledge naturally unfolds in living systems. This process offers a solid foundation on which we can build tools and techniques for use in the real world. It illustrate the nature and role of cognition in living systems, and shows: (i) how knowledge take place in human organizations and (ii) how learning happens in organization[6]. Learning is very important in organizations and allows evolution to higher forms and behaviors. This requires a lot of shared information in form which is easily accessible to everyone.
- **Agent:** A living part of a complex system is called an agent. Agents interact and affect each other, and can have high degree of creativity which cannot be precisely predicted. The perspectives which arise from this concepts are:
  - Representation of agents communicating with other agents to complete the task.
  - The distribution of agents in the organizations and changing as agents communicate.
  - The ability of the agent to make decisions in response to the environment and to the action of other entities.
  - The active agent rules, the reason why agents follow some rules and not others and When do agents' rules change
  - The kinds of emergent social phenomena arise from interacting and learning agents
  - The role which do contextual energy differentials (adaptive tension) play in motivating agent behaviors
  - Managing agents and get them to produce more economically viable teams, new product developments, entrepreneurial ventures, and generally, more effective socioeconomic and/or organizational (complex adaptive) systems

Evolution and Co-evolution: Co-evolution is a multi-level phenomena within an organization as well as between organizations. It involves interactions between different species and adaptive moves by the members of one species will deform the fitness landscape of the other species with which it is co-evolving [11]. There are two kinds of evolutions: i) the gradual development of a system, (ii) the development over generation. This means that adaptation can happen during the development of a system or while components of system develop [13]. Merali and McKelvey[8] look at strategic alignment as a co evolutionary process and tried to identify the need for dynamic approach to strategy development between information system and business. They considered various levels involved in strategic alignment. These approaches are individual, operational and strategic.

Thompson[10] classified different types of co evolutionary relationships. He also tried to identify the primary forms of selection influencing each type. Future information system research applying a co evolutionary framework can be enhanced with incorporation of theories from the social sciences and deep consideration of the nature of the interactions.

- **Adaptation:** New users with different ideas and functionality, bring new requirements that the designed system cannot accommodate. These new ideas change the perspective of the organization regarding the requirements of the information system [1]. In order to manage the complexity, organizations have started to learn the benefit of being adaptive in their behavior. For instance Sheffi and Rice[9] tried to represent adaptive firm behavior in a cellular telephone supply network. They modified designs of the handsets where possible and secured worldwide manufacturing capacity from Philips to ensure a steady supply of the specific product. Meanwhile the direct interaction between top management of Nokia and Philips further enhanced the ability of Nokia to adapt in the future[9].
- **Emergent:** Emergence describe that the macroscopic properties of a system arise from microscopic properties. These microscopic properties are interactions, relationships, structures and behaviors[7]. So defining emergence as an overall system that comes out of the interaction of many participants behavior cannot be predicted from the knowledge of what each component of the system does in isolation.
- **Non-standard:** Complex systems are non-standard as well. They contain structures in space and time. Their part freedoms will allow varying associations or movement, permitting clumping and changes over time. Thus initially homogenous systems will develop .self-organizing structures dynamically; order – and thereby value – increases over time rather than decreasing as expected in conventional thought[3].
- **Downward causation:** This means that the existence and properties of the parts themselves are affected by the emergent properties - or higher level systemic features – of the whole, which form constraints or boundary conditions on the freedom of the constituents[3][2].

5 Summary and Future Work

This paper has explored current ideas in the field of complexity theory focusing in particular on information systems. It tries to explore potential contributions from the study of complexity theory to the area of pervasiveness of information systems. It offers insights for enriching traditional approaches of information system design that could lead to better framework for supporting pervasiveness in complex environment. Complex environment makes new problems for pervasiveness because it is hard for users to be aware of change. So to be aware of change is what this paper tries to introduce as a new criterion. It is considered as a source of concepts for user-centric and pervasive information systems in complex environments. Complexity theory offers a powerful set of methods for explaining non-linear, emergent behavior in complex environments. It offers us concepts and tools for building multi-level representations of different environments. It helps us to make sense the dynamics and emergence behavior of the complex environments and offer insights regarding building effective infrastructure for pervasive adaptive systems and increase user awareness of environments.

The literature indicates that there is much work to be done to deepen the theoretical framework and to re-conceptualize adaptive system architecture. In order to deepen our understanding of sources of complexity in information system, we need to come up with a concrete definition of complexity theory. Current definitions vary according to the research fields, so future work could be: defining complex system characteristics in the field of information system infrastructure, drawing more concepts from complexity theory, developing practical implication for user-centric pervasive complex adaptive system.

We need to be more emphasis on providing the infrastructure that allows systems to evolve in a self-organizing manner. So regarding to this concept, working on the following concepts should be taken in to considerations:

- Developing a better definition of complexity and how it can be operationalized and be considered in architectural and infrastructure aspects.
- Ways to specify the user-centric concepts in IS and integrate them into functional specifications.
- The kinds of infrastructure needed and how to include them in specifications.

References

1. Benbya, H., McKelvey, B.: Toward a complexity theory of information systems development. Information Technology & People 19(1), 12–34 (2006)
2. Benbya, H.: McKelvey: Using coevolutionary and complexity theories to improve IS alignment: a multi-level approach. Journal of Information Technology 21(4), 284–298 (2006)
3. Bertelsen, S.: Construction as a complex system, pp. 11–23 (2003)
4. Kauffman, S.: Antichaos and adaptation. Scientific American 265(2), 78–84 (1991)
5. Lucas, C.: The Philosophy of Complexity (2000), retrieved from http://www.calresco.org/lucas/philos.htm
6. McElroy, M.W.: Integrating complexity theory, knowledge management and organizational learning. Journal of Knowledge Management 4(3), 195–203 (2000)
7. Merali, Y.: Complexity and Information Systems: the emergent domain. Journal of Information Technology 21(4), 216–228 (2006)
8. Merali, Y., McKelvey, B.: Using Complexity Science to effect a paradigm shift in Information Systems for the 21st century. Journal of Information Technology 21(4), 211–215 (2006)
9. Sheffi, Y., Rice, J.: A supply chain view of the resilient enterprise. MIT Sloan Management Review 47(1), 41 (2005)
10. Thompson, J.: Specific hypotheses on the geographic mosaic of coevolution. Am. Nat. 153, S1–S14 (1999)
11. Vidgen, R., Wang, X.: From business process management to business process ecosystem. Journal of Information Technology 21(4), 262–271 (2006)
12. Clegg, S.: The Sage handbook of organization studies. Sage Publications Ltd., Thousand Oaks (2006)
13. Kovács, A.I., Ueno, H.: Towards complex adaptive information systems. In: Proceedings of the 2nd International Conference on Information Technology for Application, Citeseer (2004), http://www.alexander-kovacs.de/kovacs04icita.pdf

User Centric Systems: Ethical Consideration

Nikola Serbedzija

Fraunhofer FIRST Berlin
nikola@first.fraunhofer.de

Abstract. User-centric systems call for novel methods and tools for a tight and implicit man machine interaction. Often highly personalized system are sensitive and reactive to user psychological, social and physical situation. However providing such a support may cross the barriers of our privacy or may have impacts on us that we do not necessarily agree upon.

Keywords: pervasive adaptation, ambient assistance, ethical issues, privacy protection.

1 Introduction

Nowadays, computers have become an integral part of our everyday surroundings making it necessary to re-thing and re-design ways in which we cooperate with computers. To make a control system a genuine companion in everyday life, it should be enriched with some adaptation capabilities to adjust its functioning to the users' needs. Otherwise, a modern homo *technicus*, as intrinsically adaptive species may, due to his permanent and enthusiastic exposure to ridged artificial systems, experience unpleasant retrogressive changes.

A perfect man-machine ccollaboration should minimise explicit interaction and maximise the functionality of the system. The goal is to avoid giving commands to a control system, but rather enabling the system to understand what is needed in given circumstances. In a similar way that text editing has been revolutionized by the „what you see is what you get" principle, the motto "what you need is what you get"[1] is radically changing the landscape of man-machine interface.

As an example of user-centric approach to man machine interaction, the reflect system is described, followed by some examples of its use in practice. Afterwards, impacts of new technology, such as privacy and trust issues, are discussed and ideas for further work are given.

2 Reflective Approach

The approach described here is called reflective as it observes people in their activities and reflects their need by adjusting the control system accordingly. Observation is done through numerous sensor devices that collect information. Based on the collected information, an analysis is performed that results in emotional, cognitive and physical diagnosis. Taking into account the detected user state and the system goals, the functioning is adapted to the users' needs, making the control system naturally embedded into wider surroundings.

As an illustration one application scenario is given, followed by some requirement analyses, psychological consideration and implementation issues.

2.1 User-Centric Applications

An example of user-centric systems is the driving assistance. Especially in the situation of a longer lonely drive or in a tense traffic situation, the driving assistance may substitute a missing co-driver and improve both safety and pleasure in driving. The reflective vehicle [2] implements such scenario and is equipped with numerous sensors for detecting facial expressions, postures and psychophysiological measures (electrocardiogram, heart rate and heart rate variability, respiration, etc). Further sensors is a vehicular data bus system (e.g. CAN bus) that offers numerous real-time date about the driving and engine condition (pedal pressure, wheel corrections, speed, etc). The sensors are used to monitor and diagnose the driver's state in a specific driving situation. The vehicle is also set with actuator devices like "reflective console" that give warnings, reflective media player and reflective seat (that re-shapes to match the driver's comfort) [2]. The reflective vehicle actually plays a role of a friendly co-driver that observes the driver, assists in driving and makes the trip safer and more comfortable.

The reflective control system that implements the above mentioned scenario, has to be capable of adaptation, featuring dynamic adjustment of its functioning according to the real-time situation. Different types of adaptation are considered:

- Immediate adaptation – a capability of reacting in a moment when a sudden change occurs (e.g. mastering difficult driving situation)
- Short term adaptation – a capability to adjust to a more complex situation which requires some time to be comprehended (e.g. improving overall emotional state and comfort of the driver)
- Long term adaptation – a capability to learn over a longer period of time and adjust to a personal needs of individual users (make the above adaptation personalized and relative to the individual driver)
- Pervasive adaptation – a capability to exchange knowledge with other systems and to act adaptively in different settings and in any situation.

Reflective systems aim at all above mentioned varieties of adaptation which constitute the major system requirements.

2.2 Psychological Background

One of the most striking features of nature that ensures evolution and progress is adaptation: a capability to self adjusts according to changing conditions. In effort to mimic the adaptation in artificial system and deploy it naturally in man-machine interface, this approach takes the biocybernetic loop as a starting point. Originally, the loop [3] describes how psychophysiological data regarding the status of the user are captured, analyzed and converted to a computer control input in real-time. The function of the loop is to monitor changes in user state in order to initiate an appropriate adaptive response.

Reflective approach extends the original concept of biocybernetic loop [4] to a wider set of input information allowing for a composite analyses and decision making. It takes results of affective computing and combines it with higher level understanding of social and goal–oriented situations. The approach is multi modal as it takes into account different kinds of information, processes them in multiple loops and at different time scales. There are three major phases of a single loop: sense, analyses and activate. These phases are repeated endlessly, where each consecutive cycle takes into account the effects of the previous one performing constant self-tuning and optimization.

Based on a better understanding of both personal involvement and social and behavioural situation of the user, a reflective system may offer adaptive control of different types and different time scales: immediate, short-term adaptation, long-term adaptation. It also supports pervasiveness, as an orthogonal dimension to previous type of adaptation, providing interfaces to other similar systems, anywhere and at any time.

2.2.1 Immediate Adaptation

Immediate adaptation is a prompt system, reaction to any phenomenon taht system becomes aware of. Awareness implies knowledge gained through one's own perceptions or by means of information. At a fundamental level, awareness describes the 'field of view' of the system, i.e. the range of events that the adaptive system is capable of responding to.

In a driving situation, any driver may experience difficulties in navigating through narrow lanes in a stressful traffic situation (e.g. reduced visibility or ice). The system becomes aware of this change in driver state via increased heart rate or reduced heart rate variability as well as frequent corrections of the steering wheel detected via vehicular CAN bus. In this case, driving assistance disables incoming phone calls and reduces the volume of the car entertainment system.

2.2.2 Short Term Adaptation

Adjustment is a typical example for the adaptation seen as an act of change, "so as to become suitable to a new or special situation". A system adjustment at a moderate time scale is implemented through the biocybernetic loop that monitors behaviour with respect to a certain goal or in achieving a planned state. In the application domain it may be best illustrated by achieving comfort (positive physical state) in driving process.

The reflective vehicle may be capable of influencing disposition of the driver, to prevent negative moods such as boredom or anger and to promote positive moods of high activation (alertness) and positive affect (happiness). The driver's mood is accessed via cameras for facial recognition combined with psychophysiology. However, changes in mood take place over several minutes and the system requires a moderate time scale in order to detect a negative mood such as anger. Once diagnosed the driver's emotional state is influenced by appropriate music and lighting.

2.2.3 Long Term Adaptation

Long term adaptation or evolution can be defined as a gradual process in which something changes into a different and improved form - a modification that happens over a longer period of time. An aspect of evolution is a dynamic form of personalization where both system and user respond to repeated exposure to one another.

In the vehicular emotional adaptation, the system monitors the user state in order to formulate an appropriate adaptation which also includes keeping record (in database) on driver's personal reaction to the assistance offered by the system. This personalized database may be subsequently used to tailor system behaviour to the individual over hours, days, weeks, and months. Therefore, the system must continue to evolve and to experiment with different adaptive responses in order to "refresh" the interface with the user at a longer term.

2.2.4 Pervasiveness

Pervasiveness denotes the quality of filling or spreading throughout. Pervasiveness of a reflective system is an orthogonal concept to adaptation that introduces collaboration and data exchange across different reflective systems. In addition, pervasive technology brings the potential to enrich information about the current state of user by pooling information across systems.

In vehicle, reflective assistance have capability to connect to other control system from traffic or city infrastructure and with data exchange (taking into account knowledge about user, her/his preferences and goals) provide better services.

2.3 Reflective Approach

To accomplish the requirements of pervasive adaptive systems, a service- and component-oriented middleware architecture, has been designed and developed [5,6,7]. It insures a dynamic and re-active behaviour featuring different biocybernetic loops. The reflective software is grouped into three layers: (1) Tangible layer - a low-level layer that controls sensor and actuator devices; (2) Reflective layer - a central layer - that combines input from lower layer to evaluate user states and trigger system (re-)action, according to the application goals. (3) Application layer - a high level layer that defines application scenario and system goals. By combining low and high level components from other layers, application layer runs and controls the system.

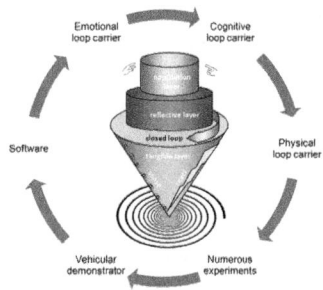

Fig. 1.

Fig. 1 Reflective framework illustrating the reflect layered architecture and the process of the reflect application development and deployment In the middle of the picture a layered architecture is metaphorically pictured as a spinning top, as it operates in endless loop, performing biocybernetic processing. In the outer circle, a development and deployment process is indicated featuring emotional cognitive and physical experience (loop carriers). Pervasiveness is exercised through numerous experiments and case studies and vehicular demonstrator brings all together into a pervasive and adaptive application [5].

2.4 Reflective Vehicle

The reflective vehicle system is a combination of all three major experiences modules configured into a single pervasive reflective application. The resulting system helps the driver throughout the ride, observing emotional, cognitive and physical condition and actively trying to assist in the process of driving.

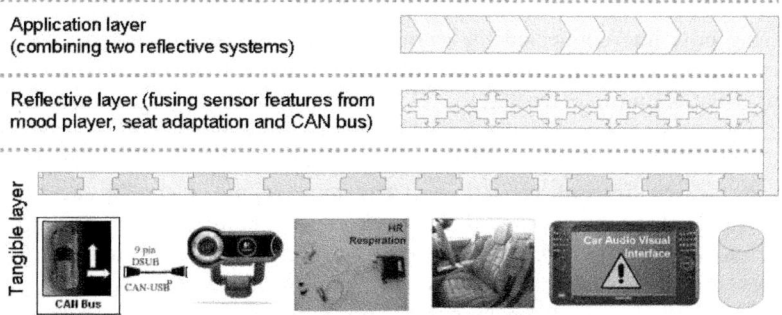

Fig. 2. Reflective vehicle configuration

Figure shows the reflect middleware configured for vehicular assistance. The sensors used are: CAN bus, cameras, physiological measurement devices, equipped with reflective interface. The actuators are: adaptive seat and the 3 board computers (playing the role of the system monitor, mobile phone and media player).

Fig. 3. Reflective cockpit

Figure shows the cockpit of the reflective vehicle with three on board computers, adaptive seat and cameras installed in front of driver. System has been tested in practice and effectively justifies the pragmatic orientation of reflective approach [5].

3 Impact - The Use and Consequences

The ultimate goal of reflective approach is to make future control systems genuinely friendly and personalized to suit the needs of individual users. To achieve this goal it is not enough to create interfaces that suit an average users, but to make it really serves the personal needs, taking into account current necessities as well psycho-physiological state of the persons being involved in real-time. At one hand, emerging systems that are reactive to human senses can be interpreted as our technical "sixth sense", at another we should be aware that our personal identity is being exposed to the interconnected world whose meta-goals may be different from our expectations. This raises ethical concerns: Can we trust the systems? Do they respect our privacy? What is a long term impact of such systems?

Knowing how interconnected the digital world is[8], recording what we buy and eat, where we travel, sleep, rent-a car (e.g. credit card institution), what we read (e.g. Amazon), what we write or visit at the Web (our internet provider), how much fun we had (e.g. Flickr), how we socialise (e.g. Facebook), how our biometric signature look like (e.g. custom control), our medical record (health insurance smart cards) we can fully relax and let "Infosphere" [9] assist in most of our every day activities. The digital divide is sharpening, "digitalize or perish" seems to be the motto – those who are left behind the digital curtain belong to third (digital) community. With the technology described here as well as with neuro-science that is making a huge progress in the domain of brain-computer interface we are about to submit the last fence of our privacy to the "digital consideration". Namely our feelings, mental constitution, even our thoughts.

Certainly, the use of modern technology facilitates the life, provides efficiency, comfort and smooth communication in a way it could not be even perceived a few decades ago. Especially the applications in medicine and ambient assisted living domain represent clear justification of the recent achievements. Nevertheless, a potential for misuses is wide and what is even more disturbing, neither possibilities nor risks are properly understood.

In that context the observation given here can be described as "nano-Ethics" focusing only at a small segment of the problem – where technology may help. For example, making reflective systems closed by strict separation of psycho-physiological and administrative data, and making it technically impossible to exchange information with other systems – would be a step in right direction in protecting the privacy.

The problem of the impact of the new technology to our life and society as a whole is neither new nor unsolvable [10]. Due to rapid developments it is sharpening though, and needs to be addressed from different viewpoints involving wide and cross-disciplinary discussion. The roles are traditionally divided among:

- artists who picture the universe in a free and imaginative way [11] with a free mind calling for re-thinking, re-consideration, re-involvement, esthetic, *l'art-pour-l'artism*, ...
- scientists and technology providers who are not only responsible for the development of novel ideas but also for hints on how to technically deal with possible ethical concerns [12].
- philosophers and sociologists as leaders in considering the impacts of new technology to the society as a whole [9,11].
- low makers and politicians to insure efficient legal background and deployment [13].
- practitioners and industry to respect regulation and ethical norms while making commercial and other use of modern technology.

In one of his articles, Floridi [9] introduces a neologism *Infosphere* as a collection of informational entities that inevitably constitutes the environment that supports our life. To further understand the impact of the *Infosphere* to us (as *Infogs* – informational organisms), the author introduces "*re-ontologization*", referring to fundamental transformation of our environment as a consequence of digitalization and *Infosphere*. Such a complex approach allows for thorough reasoning about the meaning, impact and ethics of *Infosphere* as a non-avoidable part of our ecological system.

4 Conclusion

The paper presents a novel approach in building smart systems illustrating both its technical background and its applicability. Reflective approach, being highly user-centric, makes a person's behaviour a part of processing loop and body area network an integral part of a wider system. However, a fast and uncontrolled deployment of smart technology may bring dangers and have negative impact on us. The paper further discussed how to prevent that the "digital victory" does not turn into a *Pyrrhic* one, as the use of massively interconnected digital devices may endanger our privacy exposing all aspects of our behaviour from the everyday activities, work competences, habits, feelings, intentions up to our inner thinking. Negative consequences may be prevented, if the design of new technology takes into account privacy concernes and integrated protection mechanism into the systems in all phases, namely design, development and deployment.

It seems that the one of the main controversies a modern society is confronted with is not the rapid technology development which sometimes goes even beyond science-fiction, but rather the slow pace at which humanities answer, or are allowed to answer to the new moral challenges. Their main task remains to be to warn about possible nightmare scenarios. Orwell's greatness (Orwell, 1949) cannot be measured by the fact that he envisaged many abuses already present nowadays, but rather by his warnings which helped that his worst predictions haven't (yet) became true. Therefore, an interdisciplinary approach is a *conditio sine qua* non in all future research programmes.

The spectrum of further challenges that still needs to be researched crosses various disciplines. Psychology needs to offer more expertise on diagnosing different emotional and cognitive states. Better methods and tools are needed to analyze and filter-out raw sensor data in order to get clear features (that may lead to more precise determination of different psychological states). Social sciences and philosophy should investigate the impact that such user-centric and personalized approach may have on us, as it is clear that we also do change and evolve through constant exposure to technical systems that surround us. Finally, technology providers need to take into account not only technical benefits that envisaged systems should bring, but to try to find technical solutions that support privacy, trust and insure positive impact to us as biological and social beings.

Acknowledgement. Most of the work presented here has been has been done under the REFLECT project (project number FP7-215893) funded by the European Commission within the 7th Framework Programme, pervasive adaptation initiative.

References

1. Serbedzija, N., Bertolloti, G.-M.: Adaptive and Personalized Body Networking. In: Proc. Bodynets 2010, Corfu, Greece (September 2010)
2. Serbedzija, N., et al.: Vehicle as a Co-Driver. In: Proceedings First Annual International Symposium on Vehicular Computing Systems, ISVCS 2008, Dublin, Ireland, July 22-24 (2008)
3. Pope, A.T., Bogart, E.H., Bartolome, D.S.: Biocybernetic system evaluates indices of operator engagement in automated task. Biological Psychology 40, 187–195 (1995)
4. Fairclough, S.H.: Fundamentals of physiological computing. Interacting with Computers 21, 133–145 (2009)
5. REFLECT, REFLECT project - Responsive Flexible Collaborating Ambient (2010), http://reflect.first.fraunhofer.de
6. Schroeder, A., Zwaag, M., Hammer, M.: A Middleware Architecture for Human-Centred Pervasive Adaptive Applications. In: 1st PerAda Workshop at SASO 2008, Venice, Italy, October 21 (2008)
7. Beyer, G., Hammer, M., Kroiss, C., Schroeder, A.: Component-Based Approach for Realizing Pervasive Adaptive Systems. In: Workshop on User-Centric Pervasive Adaptation, UCPA 2009, Berlin, Germany, April 27 (2009)
8. Norman, D.A.: The Design of Future Things. Basic Books, New York (2007)
9. Floridi, L.: A look into the future impact of ICT on our lives. Information Society 23(1), 59–64 (2007)
10. Orwell, G.: Nineteen Eighty-Four._A novel. Secker & Warburg, London (1949)
11. Hettinger, E.: Justifying intellectual property. Philosophy and Public Affairs 18, 31–52 (1989)
12. Rulon, M.: New Technology Raises Privacy Concerns, USA Today, http://www.usatoday.com/tech/news/surveillance/2006-08-31-rfid-privacy_x.htm (accessed April 30, 2009)
13. EC. The European Group on Ethics in Science and New Technologies to the European Commission. Ethically Speaking (9) (2009)

Author Index

Abawajy, Jemal   308, 310
Abreu, Renato   51
Agarwal, Vikas   13
Agüero, Jorge   78
Ahammed, Farhan   149
Åhlund, Christer   330
Aihara, Toru   198
Al Dehni, Shadi   417
Allard, Fabien   305
Almuhaideb, Abdullah   224
Al-Naeem, Mohammed   364
Al Ridhawi, Ismaeel   366
Al Ridhawi, Yousif   366
André, Francoise   386

Bagrodia, Rajive   261
Banâtre, Michel   305
Barais, Olivier   386
Berbers, Yolande   39
Bertolotti, Gian Mario   417
Beyer, Gilbert   417
Bhattacharya, Sayantani   358
Blackmun, Fraser R.   401
Borcea, Cristian   210
Boyali, Ali   137
Bruneau, Julien   347
Bruno, Loubet   366

Cao, Guohong   26
Cardoso, Kleber V.   237
Carrascosa, Carlos   78
Chang, Chao-Shu   409
Chehri, Abdellah   114
Chen, Hanhua   102
Chen, Meng-Jen   409
Chen, Penghe   340
Chevrier, Vincent   273
Chi, Caixia   26
Ciarletta, Laurent   273
Collins, Justin   261
Consel, Charles   347
Couderc, Paul   305
Cristiani, Andrea   417
Curtmola, Reza   210

Dantas, Priscilla   350
da Silva, Marcel W.R.   237
Daubert, Erwan   386
Delicato, Flávia C.   350, 353
de Oliveira, Tibério M.   237
de Rezende, José Ferreira   237
Dick, Robert P.   342

Eleryan, Ahmed   372
Elsabagh, Mohamed   372

Fallah, Yaser   125
Farjow, Wisam   114
Fernando, Harinda   310
Festor, Olivier   273
Fleisch, Elgar   63
Fry, Michael   1

Gallacher, Sarah   401
Gerber, Simon   1
Göhner, Peter   381
Gonçalves, Glauco   51
Goyal, Sunil   13
Gu, Tao   102
Gunasekera, Kutila   90
Gupta, Akhilesh   297

Hadj Sadok, Djamel   51
Hamvas, Áron   285
Hannourah, Wail Masry   347
Hossain, A.K.M. Mahtab   174
Hsu, Wei-Hong   409
Huang, Ge   350, 353
Huang, Wen-Liang   409

Imran, Nomica   303
Inoue, Masugi   344
Ito, Minoru   324

Joshi, Anupam   297
Julián, Vicente   78

Kamijo, Koichi   198
Kanchanasut, Kanchana   174
Karmouch, Ahmed   366
Kavakli, Manolya   137
Kay, Judy   1

Author Index

Kelner, Judith 51
Khan, Asad I. 303, 364
Kishi, Yoji 344
Kjærgaard, Mikkel Baun 162, 383
Krishnaswamy, Shonali 90
Kroiß, Christian 393
Kummerfeld, Bob 1
Kunz, Simon 381

Lacan, Jérôme 249
Le, Phu Dung 224
Leclerc, Tom 273
Liu, Guimei 102
Lochin, Emmanuel 249
Loke, Seng Wai 90, 332
Loock, Claire-Michelle 63
Lu, Jian 102

Mair, Thomas 393
Marsic, Ivan 374
Martinovic, Ivan 381
Mat Deris, Mustafa 308
Matsunaka, Takashi 344
Mattern, Friedemann 63
Md Fudzee, Mohd Farhan 308
Mekbungwan, Preechai 174
Miceli, Claudio 350
Micskei, Zoltán 285
Mir, Shiva 425
Mitra, Karan 330
Mittal, Sumit 13
Morin, Brice 386
Mouftah, Hussein 114
Mukherjea, Sougata 13
Murase, Masana 198

Nain, Grégory 386
Nakauchi, Kiyohide 344
Nassar, Rachad 312
Nitu, Irina 285

Oliveira, Luciana 51
O'Malley, Marcia 347
Orgun, Mehmet A. 186
Ott, Max 149

Papadopoulou, Elizabeth 401
Parlak, Siddika 374
Petander, Henrik 249
Pingali, Gopal 297

Pink, Glen 1
Pires, Paulo F. 350, 353
Pirmez, Luci 350
Ponzo, John 13
Preuveneers, Davy 39
Pung, Hung Keng 340

R., Pitchiah 358
Rebollo, Miguel 78
Rivière, Nicolas 285
Rodrigues, Taniro 350

S., Sridevi 358
Sattar, Abdul 186
Scheuermann, Peter 342
Schroeder, Andreas 393
Sen, Shubhabrata 340
Sengupta, Raja 125
Serbedzija, Nikola 431
Shah, Fenil 13
Shankaran, Rajan 186
Siebert, Julien 273
Simoni, Noëmie 312
Srinivasan, Bala 224
Staake, Thorsten 63
Sun, Weihua 324

Taha, Walid 347
Taheri, Javid 149
Takamatsu, Yu 324
Talasila, Manoop 210
Tao, Xianping 102
Taylor, Nick K. 401
Toftkjær, Thomas 383
Torabi, Torab 332
Tournoux, Pierre-Ugo 249
Trajcevski, Goce 342
Twamley, Jason 137

Umezawa, Takeshi 344

Varadharajan, Vijay 186
Vo, Chuong C. 332

Waeselynck, Hélène 285
Wang, Liang 102
Warabino, Takayuki 344
Wasinger, Rainer 1
Wehn, Norbert 381
Weiss, Markus 63
Wille, Sebastian 381

Williams, M. Howard 401
Wong, Wai Choong 340
Wu, Yu-Chi 409

Xu, Zhi 26
Xuan, Yiguang 125
Xue, Wenwei 340

Yamauchi, Yukiko 324
Yasumoto, Keiichi 324

Yazji, Sausan 342
Youssef, Moustafa 372
Yu, Wen-Ching 409

Zaslavsky, Arkady 90, 330
Zhang, Junqi 186
Zhao, Bo 26
Zhu, Sencun 26
Zomaya, Albert Y. 149, 350, 353

GPSR Compliance

The European Union's (EU) General Product Safety Regulation (GPSR) is a set of rules that requires consumer products to be safe and our obligations to ensure this.

If you have any concerns about our products, you can contact us on ProductSafety@springernature.com

In case Publisher is established outside the EU, the EU authorized representative is:

Springer Nature Customer Service Center GmbH
Europaplatz 3
69115 Heidelberg, Germany

Batch number: 09478804

Printed by Printforce, the Netherlands